Reinforced and Prestressed Concrete

3rd edition

F K Kong

MA, MSc, PhD, CEng, MICE, FIStructE
Professor of Structural Engineering
University of Newcastle upon Tyne

R H Evans

CBE, DSc, D ès Sc, DTech, PhD, CEng, FICE, FIMᴇᴄʜE, FIStructE
Emeritus Professor of Civil Engineering
University of Leeds

 Van Nostrand Reinhold (UK) Co. Ltd

© 1975, 1980, 1987 F. K. Kong and R. H. Evans

All rights reserved. No part of this work covered
by the copyright hereon may be reproduced or used
in any form or by any means — graphic, electronic,
or mechanical, including photocopying, recording,
taping, or information storage or retrieval systems —
without the written permission of the publishers

First published in 1975 by
Van Nostrand Reinhold (UK) Co. Ltd.
Molly Millars Lane, Wokingham, Berkshire, England

Reprinted 3 times
Second edition 1980
Reprinted 1981, 1983 (twice), 1985, 1986
Third edition 1987
Reprinted 1987

Typeset in Linotron 202 Times 10/11 pt
by Best-set Typesetter Limited

Printed and bound in Hong Kong

British Library Cataloguing in Publication Data
Kong, F. K.
  Reinforced and prestressed concrete.
  — 3rd ed.
  1. Prestressed concrete construction
  2. Reinforced concrete construction
  I. Title   II. Evans, R. H. (Rhydwyn
  Harding)
  620.1'37   TA439

ISBN 0-278-00016-9

Contents

3 Axially loaded reinforced concrete columns

4 Reinforced concrete beams—the ultimate limit state

5 Reinforced concrete beams—the serviceability limit states

6 Shear, bond and torsion

7 Eccentrically loaded columns and slender columns

8 Reinforced concrete slabs and yield-line analysis

9 Prestressed concrete simple beams

Preface to the Third Edition

The third edition conforms to BS 8110 and includes a new Chapter 12 on microcomputer programs. Like the earlier editions, it is intended as an easy-to-read main text for university and college courses in civil and structural engineering. The threefold aim of the book remains as before, namely:

(a) To explain in simple terms the basic theories and the fundamental behaviour of structural concrete members.
(b) To show with worked examples how to design such members to satisfy the requirements of BS 8110.
(c) To explain simply the technical background to the BS 8110 requirements, relating these where appropriate to more recent research.

Students will find the new edition helpful in their attempts to get to grips with the **why** as well as the **what** and the **how** of the subject.

For the convenience of those readers who are interested mainly in structural design to BS 8110, most of the chapters begin with a **Preliminary note** which lists those parts of the chapter that are directly concerned with BS 8110. However, structural design is not just BS 8110; hence the university or college student should pay attention also to the rest of the book, which has been written with the firm belief that the emphasis of an engineering degree course must be on a sound understanding of the fundamentals and an ability to apply the relevant scientific principles to the solution of practical problems. The authors wish to quote from a letter by Mr G. J. Zunz, co-Chairman of Ove Arup and Partners:

> You will see that generally my comments tend to place emphasis on getting the fundamentals straight. As my experience and that of my colleagues develops, I find more and more that it is the fundamentals that matter and those without a sound training in them suffer for the rest of their careers.

Acknowledgements

Sincere thanks are due to Dr B. Mayfield of the University of Nottingham for indispensable help with Chapter 11; to Dr H. H. A. Wong, formerly

Croucher Foundation Scholar at the University of Newcastle upon Tyne, for much valued collaboration on the new Chapter 12; to BS 8110 Committee members Dr A. W. Beeby of the Cement and Concrete Association (C & CA), Mr H. B. Gould of the Property Services Agency, Dr H. P. J. Taylor of Dow Mac Concrete Ltd and Mr R. T. Whittle of Ove Arup and Partners, for advice on the proper interpretation of BS 8110 clauses; to Mr R. S. Narayanan of S. B. Tietz and Partners for advice on the use of the I.Struct.E. Manual; to Mr B. R. Rogers of the C & CA for advice on structural design and detailing; to past and present students at the Universities of Cambridge and Newcastle upon Tyne for helpful comments and valuable assistance with the worked examples: Mr R. B. Barrett, Mr M. Chemrouk, Mr A. E. Collins, Mr J. Cordrey, Mr P. S. Dhillon, Mr J. P. J. Garner, Mr B. K. Goh, Mr K. H. Ho, Mr A. P. Hobbs, Mr D. A. Ireland, Mr H. P. Low, Mr S. F. Ng, Mr E. H. Osicki, Mr A. R. I. Swai, and Dr C. W. J. Tang.

The authors are grateful to Professor P. G. Lowe of the University of Auckland, Dr E. A. W. Maunder of Exeter University and Mr J. P. Withers of Trent Polytechnic for enlightening comments on parts of the earlier editions. They wish also to record, once again, their gratitude to Dr C. T. Morley of Cambridge University, Mr A. J. Threlfall of the C & CA, Dr C. D. Goode of Manchester University and Dr M. S. Gregory of Tasmania University for their valuable comments on the previous editions, on which the present edition has been built.

Extracts from the DoE's *Design of Normal Concrete Mixes* are included by courtesy of the Director, Building Research Establishment; Crown copyright Controller HMSO. Extracts from BS 8110 are included by kind permission of the British Standards Institution, Linford Wood, Milton Keynes, MK14 6LE, from which complete copies can be obtained. Extracts from the *Manual for the Design of Reinforced Concrete Building Structures* are included by kind permission of the Institution of Structural Engineers, 11 Upper Belgrave Street, London, SW1X 8BH, from which complete copies can be obtained.

The authors wish to thank Mrs Diane Baty for her excellent typing and Mr George Holland for the skilfully prepared drawings for the new edition. Finally, they wish to thank the publisher's editor Mr Mark Corbett and former editors Dr Dominic Recaldin and Mr David Carpenter; the book owes much of its success to their efforts, devotion and foresight.

F.K.K.
R.H.E.

Notation

The symbols are essentially those used in current British design practice; they are based on the principles agreed by the BSI, ACI, CEB and others.

A	= cross-sectional area of member
A_c	= area of concrete
A_{ps}	= area of prestressing tendons
A_s	= area of tension reinforcement; in eqns (6.9–1) and (6.11–6), A_s = area of longitudinal torsion reinforcement
A_s'	= area of compression reinforcement
A_{sc}	= area of longitudinal reinforcement in column; in Chapter 7, $A_{sc} = A_{s1}' + A_{s2}$
A_{sv}	= area of both legs of a link
A_{s1}'	= area of reinforcement near the more highly compressed face of a column section
A_{s2}	= area of reinforcement in the less compressed face of a column section
a	= deflection; moment arm
a_b	= clear distance between bars (Fig. 5.4–1)
a_c	= corner distance (Fig. 5.4–1)
a_u	= additional eccentricity of slender column (eqn 7.4–5)
a_v	= shear span
b	= width of beam or column; effective flange width; width of slab considered
b_v	= width of beam (see eqns 6.2–1 and 6.4–1), to be taken as b for a rectangular beam and as b_w for a flanged beam
b_w	= width of rib or web of beam
d	= effective depth; in Chapter 7, $d = h - d' = h - d_2$ for symmetrically reinforced columns
d'	= depth from compression face to centroid of compression steel; in Chapter 7, d' = concrete cover to centroid of A_{s1}'
d_c	= depth of concrete stress block
d_2	= concrete cover to centroid of A_{s2}
E	= modulus of elasticity
E_c	= modulus of elasticity of concrete

E_s	= modulus of elasticity of steel
e	= eccentricity
e_{add}	= additional eccentricity due to slender column effect
e_{min}	= design minimum eccentricity ($= 0.05h \leq 20$ mm in BS 8110)
e_p	= eccentricity of line of pressure from centroidal axis of beam (sign convention: downwards is positive)
e_s	= eccentricity of tendon profile from centroidal axis of beam (sign convention: downwards is positive)
e_t	= eccentricity of transformation profile from centroidal axis of beam
F	= design load
f	= stress; strength; frequency
f_{amax} (f_{amin})	= maximum (minimum) allowable concrete stress under service conditions, compressive stress being positive
f_{amaxt} (f_{amint})	= maximum (minimum) allowable concrete stress at transfer, compressive stress being positive
f_b	= anchorage bond stress
f_c	= concrete compressive stress at compression face of beam; compressive stress in concrete
f_c'	= concrete cylinder compressive strength
f_{cu}	= characteristic cube strength of concrete
f_k	= characteristic strength (eqn 1.4–1)
f_m	= mean strength (eqn 1.4–1)
f_{pb}	= tensile stress in prestressing tendons at beam failure
f_{pe}	= effective tensile prestress in tendon
f_{pu}	= characteristic strength of prestressing tendon
f_s	= tensile stress in tension reinforcement; steel tensile stress in service
f_s'	= compressive stress in compression reinforcement
f_{s1}'	= compressive stress in column reinforcement A_{s1}'
f_{s2}	= compressive stress in column reinforcement A_{s2}
f_t	= cylinder splitting tensile strength of concrete; principal tensile stress
f_y	= characteristic strength of reinforcement; in eqns (6.9–1) and (6.11–6), f_y = characteristic strength of longitudinal torsion reinforcement
f_{yv}	= characteristic strength of links
f_1 (f_2)	= concrete compressive prestress at bottom (top)face of beam section in service
f_{1t} (f_{2t})	= concrete compressive prestress at bottom (top) of beam section at transfer
G	= shear modulus
G_k	= characteristic dead load
g_k	= characteristic dead load (distributed)
h	= overall depth of beam or column section; overall thickness of slab; in Sections 7.4 and 7.5, h = overall depth of column section in the plane of bending
h_f	= overall! thickness of flange

h_{max} (h_{min})	= larger (smaller) overall dimension of rectangular section
I	= second moment of area
I_c	= second moment of area of cracked section
I_u	= second moment of area of uncracked section
K	= $M/f_{cu}bd^2$ (see eqn 4.6–4 and Tables 4.6–1 and 4.7–2); torsion constant (see eqn 6.8–3 and Table 6.8–1); optional reduction factor in slender column design (see eqns 7.4–6 and 7.5–5)
K'	= $M_u/f_{cu}bd^2$ (see eqns 4.6–5 and 4.7–5)
k_1 (k_2)	= characteristic ratios of stress block (see Figs 4.2–1, 4.4–1, 4.4–4 and 4.4–5)
l	= span length; anchorage bond length; (eqn 6.6–3a) column height; length of yield line
l_e	= effective column height (Table 7.2–1)
l_R	= $l_1 + l_2 + l_3 + \ldots$ where l_1, l_2, etc. are the vectors representing the yield lines that form the boundary to a rigid region
l_u	= ultimate anchorage bond length (Table 4.10–2 and eqn 6.6–3b)
M	= bending moment (sign convention if required: sagging moments are positive)
M_{add}	= additional moment due to lateral deflection of a slender column
M_d	= sagging moment due to dead load in prestressed beam
M_e	= bending moment computed from elastic analysis
M_i	= initial bending moment in column; sagging moment due to imposed load in prestressed beam
M_{imax} (M_{imin})	= maximum (minimum) sagging moment at section considered, due to imposed load
M_0	= ultimate strength in pure bending
M_p	= bending moment due to permanent load; plastic moment of resistance
M_r	= $M_{imax} - M_{imin}$
M_t	= bending moment due to total load; total bending moment including additional moment due to slender column effect
M_u	= capacity of singly reinforced beam (see eqn 4.6–5); ultimate moment of resistance
M_1	= primary moment (sagging) in prestressed beam
M_2	= secondary moment (sagging) in prestressed beam
M_3	= resulting moment (sagging) in prestressed beam: $M_3 = M_1 + M_2$
m	= yield moment per unit width of slab
m_1 (m_2)	= yield moment per unit width of slab due to reinforcement band number 1 (number 2) alone
m_n	= normal moment per unit length along yield line
m_{ns}	= twisting moment per unit length along yield line
N	= compressive axial load
N_{bal}	= compressive axial load corresponding to the balanced

	condition (eqn 7.5–8)
N_{uz}	= capacity of column section under pure axial compression (eqn 7.5–6)
P	= prestressing force at transfer
P_e	= effective prestressing force
P_{emax} (P_{emin})	= maximum permissible (minimum required) effective prestressing force
Q	= point load
Q_k	= characteristic imposed load
q	= distributed load
q_k	= characteristic imposed load (distributed)
r	= radius of curvature; internal radius of hook or bend (see Fig. A–21)
$\dfrac{1}{r}$	= curvature
$\dfrac{1}{r_{cs}}$	= shrinkage curvature
$\dfrac{1}{r_{ip}}$	= instantaneous curvature due to permanent load
$\dfrac{1}{r_{it}}$	= instantaneous curvature due to total load
$\dfrac{1}{r_{lp}}$	= long-term curvature due to permanent load
$\dfrac{1}{r_m}$	= maximum curvature; curvature at critical section
s	= reinforcement spacing
s_v	= longitudinal spacing of links or shear reinforcement
T	= torsional moment
T_i	= torsional moment resisted by a typical component rectangle
T_0	= ultimate strength in pure torsion
V	= shear force (see Fig. 9.2–5 for sign convention where such is required)
V_a	= shear force resisted by aggregate interlock
V_b	= shear resistance of bent-up bars (eqn 6.4–4)
V_c	= shear force resisted by concrete; (in Section 9.6) ultimate shear resistance of concrete section
V_{c0} (V_{cr})	= ultimate shear resistance of concrete section which is uncracked (cracked in flexure)
V_{cz}	= shear force resisted by concrete compression zone
V_d	= shear force resisted by dowel action; dead load shear force
V_p	= shear force due to prestressing (sign convention as in Fig. 9.2–5)
V_s	= shear force resisted by the web steel
v	= design shear stress ($V/b_v d$)
v_c	= design shear stress for concrete only ($= V_c/b_v d$)
v_t	= torsional shear stress

v_{tmin}	= permissible torsional shear stress for concrete only
v_{tu}	= maximum permissible torsional shear stress for reinforced section
v_u	= maximum permissible shear stress for reinforced section
W_k	= characteristic wind load
w_k	= characteristic wind load (distributed)
x	= neutral axis depth
x_1	= smaller centre-to-centre dimension of a link
y_1	= larger centre-to-centre dimension of a link
Z	= elastic sectional modulus
$Z_1 (Z_2)$	= elastic sectional modulus referred to bottom (top) face of section
z	= lever-arm distance
α	= $N/f_{cu}bh$; a ratio; an angle; prestress loss ratio
α_{conc}	= $N(\text{concrete})/f_{cu}bh$
α_e	=modular ratio E_s/E_c
α_{s1}	=$N(A'_{s1})/f_{cu}bh$
α_{s2}	= $N(A_{s2})/f_{cu}bh$
β	= $M/f_{cu}bh^2$; biaxial bending coefficient (Table 7.3–1); bond coefficient (Table 6.6–1); a ratio; an angle; inclination of shear reinforcement or prestressing tendon
β_a	= slender column coefficient (eqn 7.4–5 and Table 7.5–1)
β_b	= moment redistribution ratio (eqns 4.7–1 and 4.7–2)
β_{conc}	= $M(\text{concrete})/f_{cu}bh^2$
β_{s1}	= $M(A'_{s1})/f_{cu}bh^2$
β_{s2}	= $M(A_{s2})/f_{cu}bh^2$
γ	= a ratio; an angle; a partial safety factor
γ_f	= partial safety factor for loads
γ_m	= partial safety factor for materials
ε	= strain
ε_c	= concrete compressive strain at compression face of section
ε_{cc}	= concrete creep strain
ε_{cs}	= concrete shrinkage; shrinkage strain
ε_{cu}	= ultimate concrete strain in compression ($= 0.0035$ for BS 8110)
ε_0	= concrete strain when peak stress is reached
ε_s	= tensile strain in tension reinforcement
ε'_s	= compressive strain in compression reinforcement
ε'_{s1}	= compressive strain in column reinforcement A'_{s1}
ε_{s2}	= compressive strain in column reinforcement A_{s2}
θ	= angle of torsional rotation per unit length
θ_A	= vector representing rotation of rigid region A (sign convention: left-hand screw rule)
ν	= Poisson's ratio
ϱ	= tension steel ratio (A_s/bd)
ϱ'	= compression steel ratio (A'_s/bd)
ϱ_v	= web steel ratio (A_{sv}/bd)

σ = standard deviation

ϕ = bar size; an angle; creep coefficient

ϕ_1 = torsion function (eqns 6.8–1 to 6.8–3); acute angle measured anticlockwise from yield line to moment axis

Chapter 1
Limit state design concepts

Preliminary note: Readers interested only in structural design to BS 8110 may concentrate on the following sections:

(a) *Section 1.2: Limit state design philosophy.*
(b) *Section 1.4: Characteristic strengths and loads.*
(c) *Section 1.5: Partial safety factors.*

1.1 The aims of structural design

This book is concerned with reinforced and prestressed concrete, and since structural engineering is dominated by design, it is appropriate to begin by stating the aims of structural design and briefly describing the processes by which the structural engineer seeks to achieve them [1].

There are three main aims in structural design. First, the structure must be safe, for society demands security in the structures it inhabits. Second, the structure must fulfil its intended purpose during its intended life span. Third, the structure must be economical with regard to first cost and to maintenance costs; indeed, most design decisions are, implicitly or explicitly, economic decisions.

A structural project is initiated by the client, who states his requirements of the structure. His requirements are usually vague, because he is not aware of the possibilities and limitations of structural engineering. In fact, his most important requirements are often not explicitly stated. For example, he will assume that the structure will be safe and that it will remain serviceable during its intended life. The process of structural design begins with the engineer's appreciation of the client's requirements. After collecting and assimilating relevant facts, he develops concepts of general structural schemes, appraises them, and then, having considered the use of materials and the erection methods, he makes the important decision of choosing the final structural scheme—after consultations with the client if necessary. This is followed by a full structural analysis and detailed design, which are often collectively referred to as structural design and which form the subject matter of this book. Having checked, through such analysis and design, that the final structure is adequate under service conditions and during erection, the engineer then issues the specifications and detail drawings to the contractor. These documents are the engineer's instructions to the contractor, who will erect the structure under the engineer's supervision.

In the detailed analysis and design, and indeed in the appraisal of the overall structural scheme, the engineer is guided by codes of practice, which are compendia of good practice drawn up by experienced engineers. Codes of practice are intended as guides to the engineer and should be used as such; they should never be allowed to replace his conscience and competence. Finally, while the engineer should strive to achieve good design and be creative, he must appreciate the dangers inherent in revolutionary concepts; ample experience in the past and in recent times has shown that uncommon designs or unfamiliar constructional methods do increase the risk of failures.

1.2   Limit state design philosophy

The philosophy of **limit state design** was developed mainly by the Comité Européen du Béton (CEB) and the Fédération Internationale de la Précontrainte (FIP), and is gaining international acceptance [2–7]. As stated in Section 1.1, a structure must be designed to sustain safely the loads and deformations which may occur during construction and in use, and should have adequate durability during the life of the structure. The design method aims at guaranteeing adequate safety against the structure being rendered unfit for use. A structure, or part of a structure, is rendered unfit for use when it reaches a **limit state**, defined as a particular state in which it ceases to fulfil the function or to satisfy the condition for which it was designed. There are two categories of limit states:

(a)   An **ultimate limit state** is reached when the structure (or part of it) collapses. Collapse may arise from the rupture of one or more critical sections, from the transformation of the structure into a mechanism, from elastic or inelastic instability, or from loss of equilibrium as a rigid body, and so on.

(b)   The **serviceability limit states** are those of excessive deflection, cracking, vibration and so on.

Normally, three limit states only are considered in design: the ultimate limit state and the serviceability limit states of excessive deflection and cracking under service loads. The structure is usually designed for the ultimate limit state and checked for the serviceability limit states. Structural collapses often have serious consequences; therefore in design the probability of reaching the ultimate limit state is made very low, say, 10^{-6}. Since the loss resulting from unserviceability is generally much less than that from collapse, a probability, much higher than 10^{-6}, of reaching a serviceability limit state may still be acceptable. In limit state design, the engineer's aim is that the probability of each limit state being reached is about the same for all the members in a structure and is appropriate to that limit state.

Limit state design philosophy uses the concept of probability and is based on the application of the methods of statistics to the variations that occur in practice in the loads acting on the structure or in the strength of the materials. Before further discussion of limit state design, it is desirable, therefore, to review some relevant concepts in statistics.

1.3 Statistical concepts

In this section we shall briefly discuss those concepts of statistics which are helpful to our study of limit state design philosophy. For more detailed descriptions of statistical methods and of the theory underlying them, the reader is referred to other specialist texts [8, 9].

Probability
Suppose there is a large number n of occasions on which a certain event is equally likely to happen, and that the event happens on a number m of the n occasions. We then say that the probability of the event happening on any one of the n occasions is m/n, and the probability of its not happening on any one of the n occasions is $1 - m/n$. Probability is thus expressed as a number not greater than 1. A value of unity denotes a certainty of the event happening; a value of zero means an impossibility of the event happening. (Note: We have adopted the above concept of probability because it serves our purpose and its meaning is intuitively clear, at least to structural engineers. However, it should be pointed out that some mathematicians [10] regard this concept as invalid and meaningless.)

Frequency distribution
Table 1.3–1 gives the results of cylinder splitting tensile tests on 100 concrete specimens. The numbers in the table are called the characteristic values of the **variate**; in this case the variate is the tensile strength. The characteristic values can be studied more conveniently if they are rearranged in ascending order of magnitude. Since the numbers are correct to one decimal place, a value of 2.1, for example, may represent any value from 2.05 to 2.14. In Table 1.3–2, the characteristic values are divided into class intervals of 1.45–1.54, 1.55–1.64 ... and 2.65–2.74, and the number of values falling into each interval, known as the frequency in the interval, is also shown. Table 1.3–2, therefore, shows the **frequency distribution** of the tensile strengths, since it shows with what frequencies tensile strengths

Table 1.3–1   Tensile strengths of concrete (N/mm²)

2.1	1.9	2.2	2.5	2.0	1.8	1.9	2.0	2.2	2.0
2.2	1.7	2.0	2.4	1.8	1.9	2.0	1.5	2.4	2.1
2.6	2.0	2.3	2.0	1.7	2.0	2.2	1.5	2.4	2.0
1.8	1.6	2.3	2.0	2.2	2.0	2.2	1.8	2.1	2.2
2.3	1.9	1.8	2.2	1.8	1.7	2.2	1.6	2.7	2.3
1.6	1.8	1.9	2.5	1.9	1.9	2.0	1.7	1.7	2.0
1.8	1.7	2.2	1.7	2.1	2.4	1.9	1.9	2.0	2.0
2.0	2.2	1.9	2.1	2.0	2.4	2.0	2.0	1.8	2.1
1.8	1.7	2.3	1.8	2.0	2.4	1.8	2.0	1.8	2.2
1.9	1.8	2.0	1.6	1.8	2.3	2.5	1.7	2.3	2.0

Table 1.3–2   Frequency distribution

Class interval	Frequency f_i
1.45–1.54	2
1.55–1.64	4
1.65–1.74	9
1.75–1.84	15
1.85–1.94	11
1.95–2.04	23
2.05–2.14	6
2.15–2.24	12
2.25–2.34	7
2.35–2.44	6
2.45–2.54	3
2.55–2.64	1
2.65–2.74	1

of different magnitudes are distributed over the range of the observed values.

In Fig. 1.3–1 the frequency distribution is represented graphically as a **histogram**. *The principle of the histogram is that the area (not the ordinate) of each rectangle represents the proportion of observations falling in that interval.* For example, the proportion of tensile strengths in the interval 1.45–1.54 N/mm² is 2/100 = 0.02, where 100 is the total number of test results; this is represented in Fig. 1.3–1 by a rectangle of the same area, namely, a rectangle of width 1 N/mm² and height 0.02 per N/mm², erected on the base 1.45–1.54 N/mm². If the number of observations (i.e. the test

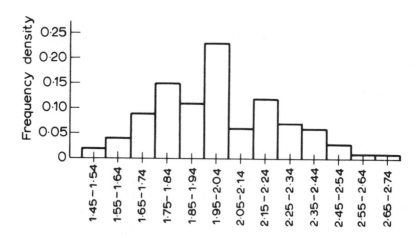

Fig. 1.3–1   Histogram

results in this case) is increased and the class interval reduced, it can be seen that the histogram will resemble a smooth curve; in the limit when the number of observations becomes infinite, the histogram becomes a smooth curve, known as the probability density curve or the **probability distribution curve** (Fig. 1.3–2). The area under the distribution curve between any two points x_1 and x_2 (i.e. the integral of the probability density function $f(x)$ between them) represents the probability that the tensile strength of the concrete will lie between these two values. For a very small increment dx, the function $f(x)$ may be considered to be constant from x to $x + dx$, and the probability that the variate will have a value lying in this small interval is very nearly $f(x)\,dx$, which is the area of the rectangle with height $f(x)$ and width dx. $f(x)$ may therefore be thought of as representing the probability density at x. Note that the probability of the variate having exactly a particular value x is zero; we can only consider the probability of its value lying in the interval x to $x + dx$. Also, the total area under the entire curve of the probability function is unity.

Characteristics of distributions
The sum of a set of n numbers $x_1, x_2, \ldots x_n$ divided by n is called the arithmetic mean, or simply the **mean** of the set of numbers. If $\bar{x}$ denotes the mean, then

$$\bar{x} = \frac{x_1 + x_2 + x_3 + \ldots x_n}{n} = \frac{\sum x}{n} \tag{1.3–1}$$

If the numbers $x_1, x_2, \ldots x_j$ have frequencies $f_1, f_2, \ldots f_j$ respectively, the mean may equally be calculated from

$$\bar{x} = \frac{f_1 x_1 + f_2 x_2 + \ldots f_j x_j}{f_1 + f_2 + \ldots f_j} = \frac{\sum fx}{n} \tag{1.3–2}$$

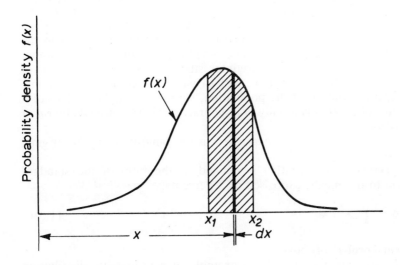

Fig. 1.3–2   Probability distribution curve

It follows that

$$\sum f(x - \bar{x}) = 0 \qquad (1.3\text{-}3)$$

that is, the sum of the deviations from the mean is zero.

We should distinguish between the **sample mean** and the **population mean**. In collecting data concerning the characteristics of a group of objects, it is often impossible or impracticable to observe the entire group. For example, suppose we have a number of precast concrete columns, some of which are tested to destruction to determine the ultimate strength (see Chapter 3). The tested specimens are referred to collectively as the **sample** and the entire group of precast columns is called the **population**. If the sample is representative of the population, useful conclusions about the population can often be inferred from analysis of the sample. But it should be remembered that since we actually test only a limited number of specimens out of the whole population, the mean strength we measure is the sample mean and not the population mean.

The **standard deviation** of a set of n numbers $x_1, x_2, x_3 \ldots x_n$ is denoted by σ and is defined by

$$\sigma = \sqrt{\frac{\sum (x - \bar{x})^2}{n}} \qquad (1.3\text{-}4)$$

where $(x - \bar{x})$ is the deviation of a number from the mean. The standard deviation is therefore the root mean square of the deviation from the mean. If $x_1, x_2, \ldots x_j$ occur with frequencies $f_1, f_2, \ldots f_j$, it is clear that the standard deviation may equally be calculated from

$$\sigma = \sqrt{\frac{\sum f_i(x_i - \bar{x})^2}{n}} \qquad (1.3\text{-}5)$$

When estimating the standard deviation of a population from a sample of it, an error is introduced by assuming the sample mean to be the population mean, the latter being not known in this case. It can be proved [8] that the error so introduced may be corrected by replacing n in the denominators of eqns (1.3–4) and (1.3–5) by $n - 1$. In making a number of observations, we use n if we are interested in the standard deviation of the observations themselves; if we wish to use the observations to estimate the standard deviation of the population, we use $n - 1$. For values of n exceeding about 30, it is usually of little practical significance whether n or $n - 1$ is used.

The standard deviation is expressed in the same units as the original variate x.

The **coefficient of variation** is defined as the ratio of the standard deviation to the mean, expressed as a percentage, viz.

$$\delta = \frac{\sigma}{\bar{x}} \times 100 \qquad (1.3\text{-}6)$$

The normal probability curve

We have seen, in Fig. 1.3–2, an example of a probability distribution curve. In limit state design, and indeed in many branches of engineering, an important probability distribution is the **normal probability distribution**, defined by the equation

$$y = \frac{1}{\sigma\sqrt{(2\pi)}} \exp\{-\tfrac{1}{2}(x - \bar{x})^2/\sigma^2\} \qquad (1.3\text{-}7)$$

where σ is the standard deviation, the exponential $e(= 2.71828)$ is the base of the natural logarithm, and $\bar{x}$ is the mean of the variable x.

Suppose in eqn (1.3–7) the variable x is expressed in terms of another variable z defined by

$$z = \frac{x - \bar{x}}{\sigma} \qquad (1.3\text{-}8)$$

that is to say, z is the deviation from the mean expressed in multiples of the standard deviation. Equation (1.3–7) is then replaced by the following so-called standard form [8, 9]:

$$y = \frac{1}{\sqrt{(2\pi)}} \exp(-\tfrac{1}{2}z^2) \qquad (1.3\text{-}9)$$

(See also Example 1.3–5.) Figure 1.3–3 shows the graph of the standardized eqn (1.3–9), where the range of z is from $-\infty$ to $+\infty$. Of course, the total area bounded by this curve and the x-axis is equal to unity; the area under the curve between $z = z_1$ and $z = z_2$ represents the probability that z lies between z_1 and z_2. For example, 68.26% of the total area is included between $z = -1$ and $+1$; 95.44% between $z = -2$ and $+2$;

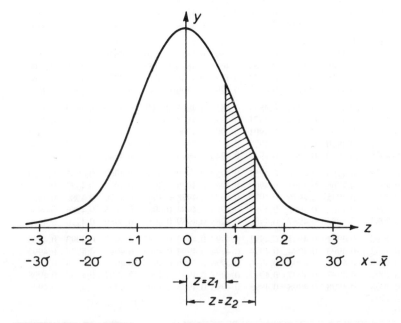

Fig. 1.3–3   Areas under the normal probability distribution curve

and 99.74% between $z = -3$ and $+3$. In particular, very nearly 95% of the area lies between $z = -1.64$ and $+\infty$ (Example 1.3–2); therefore the probability of $x \geq \bar{x} - 1.64\sigma$ is (very nearly) 95%. In other words, **there is only a 5% probability that the value of x would fall below the mean value $\bar{x}$ by 1.64 times the standard deviation**. This last statement is of special interest in limit state design, as we shall see in Section 1.4.

Table 1.3–3 gives the areas under the normal probability distribution curve (Fig. 1.3–3) bounded by the ordinates at $z = 0$ and any positive value of z. The following examples illustrate the use of this table.

Table 1.3–3   Area under the normal probability curve from $z = 0$ to $z = z$

z	0	1	2	3	4	5	6	7	8	9
0.0	0.0000	0.0040	0.0080	0.0120	0.0160	0.0199	0.0239	0.0279	0.0319	0.0359
0.1	0.0398	0.0438	0.0478	0.0517	0.0557	0.0596	0.0636	0.0675	0.0714	0.0754
0.2	0.0793	0.0832	0.0871	0.0910	0.0948	0.0987	0.1026	0.1064	0.1103	0.1141
0.3	0.1179	0.1217	0.1255	0.1293	0.1331	0.1368	0.1406	0.1443	0.1480	0.1517
0.4	0.1554	0.1591	0.1628	0.1664	0.1700	0.1736	0.1772	0.1808	0.1844	0.1879
0.5	0.1915	0.1950	0.1985	0.2019	0.2054	0.2088	0.2123	0.2157	0.2190	0.2224
0.6	0.2258	0.2291	0.2324	0.2357	0.2389	0.2422	0.2454	0.2486	0.2518	0.2549
0.7	0.2580	0.2612	0.2642	0.2673	0.2704	0.2734	0.2764	0.2794	0.2823	0.2852
0.8	0.2881	0.2910	0.2939	0.2967	0.2996	0.3023	0.3051	0.3078	0.3106	0.3133
0.9	0.3159	0.3186	0.3213	0.3238	0.3264	0.3289	0.3315	0.3340	0.3365	0.3389
1.0	0.3413	0.3438	0.3461	0.3485	0.3508	0.3531	0.3554	0.3577	0.3599	0.3621
1.1	0.3643	0.3665	0.3686	0.3708	0.3729	0.3749	0.3770	0.3790	0.3810	0.3830
1.2	0.3849	0.3869	0.3888	0.3907	0.3925	0.3944	0.3962	0.3980	0.3997	0.4015
1.3	0.4032	0.4049	0.4066	0.4082	0.4099	0.4115	0.4131	0.4147	0.4162	0.4177
1.4	0.4192	0.4207	0.4222	0.4236	0.4251	0.4265	0.4279	0.4292	0.4306	0.4319
1.5	0.4332	0.4345	0.4357	0.4370	0.4382	0.4394	0.4406	0.4418	0.4428	0.4441
1.6	0.4452	0.4463	0.4474	0.4484	**0.4495**	0.4505	0.4515	0.4525	0.4535	0.4545
1.7	0.4554	0.4564	0.4573	0.4582	0.4591	0.4599	0.4608	0.4616	0.4625	0.4633
1.8	0.4641	0.4649	0.4656	0.4664	0.4671	0.4678	0.4686	0.4693	0.4699	0.4706
1.9	0.4713	0.4719	0.4726	0.4732	0.4738	0.4744	0.4750	0.4756	0.4761	0.4767
2.0	0.4772	0.4778	0.4783	0.4788	0.4793	0.4798	0.4803	0.4808	0.4812	0.4817
2.1	0.4821	0.4826	0.4830	0.4834	0.4838	0.4842	0.4846	0.4850	0.4854	0.4857
2.2	0.4861	0.4864	0.4868	0.4871	0.4875	0.4878	0.4881	0.4884	0.4887	0.4890
2.3	0.4893	0.4896	0.4898	0.4901	0.4904	0.4906	0.4909	0.4911	0.4913	0.4916
2.4	0.4918	0.4920	0.4922	0.4925	0.4927	0.4929	0.4931	0.4932	0.4934	0.4936
2.5	0.4938	0.4940	0.4941	0.4943	0.4945	0.4946	0.4948	0.4949	0.4951	0.4952
2.6	0.4953	0.4955	0.4956	0.4957	0.4959	0.4960	0.4961	0.4962	0.4963	0.4964
2.7	0.4965	0.4966	0.4967	0.4968	0.4969	0.4970	0.4971	0.4972	0.4973	0.4974
2.8	0.4974	0.4975	0.4976	0.4977	0.4977	0.4978	0.4979	0.4979	0.4980	0.4981
2.9	0.4981	0.4982	0.4982	0.4983	0.4984	0.4984	0.4985	0.4985	0.4986	0.4986
3.0	0.4987	0.4987	0.4987	0.4988	0.4988	0.4989	0.4989	0.4989	0.4990	0.4990
3.1	0.4990	0.4991	0.4991	0.4991	0.4992	0.4992	0.4992	0.4992	0.4993	0.4993
3.2	0.4993	0.4993	0.4994	0.4994	0.4994	0.4994	0.4994	0.4995	0.4995	0.4995
3.3	0.4995	0.4995	0.4995	0.4996	0.4996	0.4996	0.4996	0.4996	0.4996	0.4997
3.4	0.4997	0.4997	0.4997	0.4997	0.4997	0.4997	0.4997	0.4997	0.4997	0.4998
3.5	0.4998	0.4998	0.4998	0.4998	0.4998	0.4998	0.4998	0.4998	0.4998	0.4998
4.0	0.49997									
5.0	0.49999									

Example 1.3–1
Determine the area under the normal probability curve between $z = 1.2$ and $z = 1.94$.

SOLUTION

$$\begin{aligned} \text{Required area} &= (\text{area between } z = 0 \text{ and } z = 1.94) - \\ &\quad (\text{area between } z = 0 \text{ and } z = 1.2) \\ &= 0.4738 \text{ (from Table 1.3–3)} - 0.3849 \\ &= \underline{0.0889} \end{aligned}$$

Example 1.3–2
Determine the probability of a set of observations, believed to be normally distributed, having values that fall below their mean by 1.64 times their standard deviation.

SOLUTION
We have to determine the area under the normal probability curve between the ordinates $z = -\infty$ and $z = -1.64$. The area between $z = -1.64$ and $z = 0$ is, by symmetry, equal to the area between $z = 0$ and $z = +1.64$ and is 0.4495 from Table 1.3–3.

The area between $z = 0$ and $z = \infty$ is one half the total area between $z = -\infty$ and $z = +\infty$ and is therefore equal to 0.5.

Therefore

$$\begin{aligned} \text{area between } z = -1.64 \text{ and } z = \infty &= 0.4495 + 0.5 \\ &= 0.9495 \end{aligned}$$

The area between $z = -\infty$ and $z = -1.64$ is given by

$$\begin{aligned} &(\text{area between } z = -\infty \text{ and } z = +\infty) \\ &- (\text{area between } z = -1.64 \text{ and } z = +\infty) \\ &= 1 - 0.9495 \\ &= 0.0505 \simeq \underline{0.05} \end{aligned}$$

Ans. There is a 5% probability that the value of an observation would fall below the mean value of all the observations by more than 1.64 their standard deviation.

Example 1.3–3
A set of concrete cube strengths are normally distributed with a mean of 45 N/mm^2 and a standard deviation of 5 N/mm^2.

(a) Determine the probability of a random cube having a strength between 50 and 60 N/mm^2.
(b) Determine the range in which we would expect these strengths to fall, with a probability of 99.9%.

SOLUTION
(a) $\bar{x}$, the mean $= 45 \ N/mm^2$
σ, the standard deviation $= 5 \ N/mm^2$

for $x = 50$ N/mm^2, $z = \dfrac{50 - 45}{5} = 1$

for $x = 60$ N/mm^2, $z = \dfrac{60 - 45}{5} = 3$

From Table 1.3–3,

> area between $z = 0$ and $z = 3$ is 0.4987
> area between $z = 0$ and $z = 1$ is 0.3413

Therefore

> area between $z = 1$ and $z = 3$ is $0.4987 - 0.3413 = \underline{0.1574}$

(b)   We want the limits of z between which the area under the normal probability curve is 0.999. Because of symmetry, the area between $z = 0$ and $z = z$ is half of 0.999, i.e. 0.4995. From Table 1.3–3, $z \approx 3.3$. Therefore

$$x = \bar{x} \pm 3.3\sigma = 45 \text{ N/mm}^2 \pm 3.3 \times 5 \text{ N/mm}^2$$
$$= \underline{28.5 \text{ N/mm}^2 \text{ or } 61.5 \text{ N/mm}^2}$$

Ans.   (a)   There is a probability of 15.74% that a random cube would have a strength between 50 and 60 N/mm^2.
    (b)   There is a 99.9% probability that the whole range of cube strengths is from 28.5 to 61.5 N/mm^2.

Level of significance and confidence level
If a set of values are normally distributed, then the probability of any single value falling between the limits $\bar{x} \pm z\sigma$ is the area of the normal probability curve (Fig. 1.3–3 and Table 1.3–3) between these limits. This probability, expressed as a percentage, is called the **confidence level**, or confidence coefficient. The limits $(\bar{x} - z\sigma)$ and $(\bar{x} + z\sigma)$ are called the **confidence limits** and the interval $(\bar{x} - z\sigma)$ to $(\bar{x} + z\sigma)$ is called the confidence interval. The probability of any single value falling outside the limits $\bar{x} \pm z\sigma$ is given by the areas under the two tails of the normal probability curve outside these limits; this probability, expressed as a percentage, is called the **level of significance**.

Example 1.3–4
It is known that a set of test results (which are normally distributed) has a mean strength of 82.4 N/mm^2 and a standard deviation of 4.2 N/mm^2. Determine:

(a)   the 95% confidence limits;
(b)   the strength limits at the 5% level of significance; and
(c)   the strength below which 5% of the test results may be expected to fall.

SOLUTION
(a)   With reference to Fig. 1.3–3 and Table 1.3–3, we wish to determine the limits of z such that the area below the normal curve between

these limits is 0.95; or, half this area is 0.475. From Table 1.3–3, $z =$ 1.96. Therefore the 95% confidence limits are $\bar{x} \pm 1.96\sigma$, i.e.

$$82.4 \text{ N/mm}^2 \pm 1.96 \times 4.2 \text{ N/mm}^2$$

or

$$74.2 \text{ and } 90.6 \text{ N/mm}^2$$

(b) The strength limits at the 5% level of significance are the limits outside which 5% of the results can be expected to fall; in other words, they are the limits *within* which (100% − 5%) = 95% of the results can be expected to fall. Hence these limits are 74.2 and 90.6 N/mm², as in (a).

(c) Here we are interested ony in one strength limit, that below which 5% of the test results can be expected to fall. With reference to Fig. 1.3–4, we want to find a value of z such that the area of the tail to the left of z is 0.05. In Fig. 1.3–4, the shaded area is 0.05 and the blank area is 0.95; the area OABC is therefore 0.95 − 0.5 = 0.45, and we wish to find the value of z corresponding to an area of 0.45.

From Table 1.3–3, an area of 0.4495 corresponds to $z = +1.64$. This value of z is the mirror image of the negative value we want, since Fig. 1.3–4 makes it clear that our z must be negative. Therefore

$$z \text{ required} = -1.64$$

Therefore the required strength limit is $\bar{x} - 1.64\sigma$ or

$$(82.4 - 1.64 \times 4.2) \text{ N/mm}^2 = \underline{75.5 \text{ N/mm}^2}$$

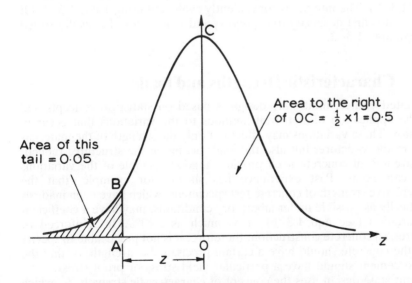

Fig. 1.3–4 Value of z for one tail of the diagram having a prescribed area

This lower strength limit, below which not more than 5% of the test results can be expected to fall, is referred to as the **characteristic strength** in Section 1.4, and is equal to the mean strength less 1.64 times the standard deviation.

Example 1.3–5

A random variable x is known to be normally distributed. Explain how the probability of x assuming any value between the limits x_1 and x_2 can be determined.

SOLUTION

Let $P(x_1 \leq x \leq x_2)$ denote the probability of x assuming any value between the limits x_1 and x_2.

Using the normal probability distribution as defined by eqn (1.3–7),

$$P(x_1 \leq x \leq x_2) = \int_{x_1}^{x_2} \frac{1}{\sigma\sqrt{(2\pi)}} \exp\{-\tfrac{1}{2}(x - \bar{x})^2/\sigma^2\}\, dx$$

The integral on the right-hand side of the above equation cannot be evaluated by elementary means. However, if we set $z = (x - \bar{x})/\sigma$, then $x = \sigma z + \bar{x}$ so that $dx = \sigma\, dz$; also $z_1 = (x_1 - \bar{x})/\sigma$ and $z_2 = (x_2 - \bar{x})/\sigma$. Then the above equation becomes

$$P(x_1 \leq x \leq x_2) = \int_{z_1}^{z_2} \frac{1}{\sigma\sqrt{(2\pi)}} \exp(-\tfrac{1}{2}z^2)\sigma\, dz$$

$$= \int_{z_1}^{z_2} \frac{1}{\sqrt{(2\pi)}} \exp(-\tfrac{1}{2}z^2)\, dz$$

In other words, the required answer is obtained by integrating the area under the standardized normal probability distribution curve, as defined by eqn (1.3–9). The integral is conveniently evaluated using Table 1.3–3. Of course, it is first necessary to express x_1 and x_2 in terms of z, as illustrated by Example 1.3–3.

1.4   Characteristic strengths and loads

As stated earlier, limit state design is based on statistical concepts and on the application of statistical methods to the variations that occur in practice. These variations may affect not only the strength of the materials used in the structure, but also the loads acting on the structure. Indeed, the strength of concrete itself provides a good example of the variations that can occur. Past experience has shown, for example, that the compressive strength of concrete test specimens, which have been made as identically as possible under laboratory conditions, may have a coefficient of variation (see eqn 1.3–6) of as much as $\pm 10\%$. In reinforced or prestressed concrete construction, therefore, it is not practicable to specify that the concrete should have a certain precise cube strength, or that the reinforcement should have a particular yield stress or proof stress. Limit state design uses the concept of **characteristic strength**, f_k, which means that value of the compressive strength of concrete, the yield or proof stress of reinforcement, or the ultimate load of a prestressing tendon, below which not more than a prescribed percentage of the test results should fall. Specifically, BS 8110 defines the **characteristic strength of concrete** as that value of the cube strength below which not more than 5% of the test results may be expected to fall. Similarly, BS 4449 and 4461 define the **characteristic strength of steel reinforcement** as that value of the yield stress below which not more than 5% of the test material may be expected to fall. Similarly, the **characteristic strength of a prestressing tendon** is that value of the ultimate strength below which not more than 5% of the test results may be expected to fall.

Current limit state design philosophy assumes that the strengths of concrete and steel are normally distributed. Hence, from Examples 1.3–2 and 1.3–4(c), it is clear that in British practice the characteristic strength f_k is the mean strength f_m less 1.64 times the standard deviation σ:

$$f_k = f_m - 1.64\sigma \qquad \qquad *(1.4-1)$$

Therefore, for the same specified value of the characteristic strength, the higher the value of the standard deviation, the higher will be the necessary value of the mean strength. In the production of concrete and steel, producton costs are related to the mean strength; a higher value of the mean strength will necessitate the use of a more expensive material. On the other hand, to reduce the standard deviation requires a higher degree of quality control and hence a higher cost. In practice, a compromise is struck between these conflicting demands, in order to achieve overall economy.

In the general context of limit state design, the **characteristic load** is that value of the load which has an accepted probability of not being exceeded during the life span of the structure. Ideally, such a value should be determined from the mean load and its standard deviation. However, because of a lack of statistical data, it is not yet possible to express loads in statistical terms, and in current practice the so-called characteristic loads are simply loads which have been arrived at by a consensus that makes them characteristic loads. For example, in Great Britain the load values quoted in BS 6399 : Part 1 and CP3 : Chapter V : Part 2 are accepted by BS 8110 as being the characteristic loads.

1.5 Partial safety factors

In limit state design, the load actually used for each limit state is called the **design load** for that limit state and is the product of the characteristic load and the relevant **partial safety factor for loads** γ_f:

$$\text{design load} = \gamma_f \times \text{characteristic load} \qquad (1.5-1)$$

The partial safety factor γ_f is intended to cover those variations in loading, in design or in construction which are likely to occur after the designer and the constructor have each used carefully their skill and knowledge. It also takes into account the nature of the limit state in question; in this respect

* Readers should pay special attention to equations with **bold** numbers.

there is an element of judgement and experience related to the relative values a community places on human life, permanent injury and property damage as compared with a possible increase in initial investment.

Similarly, in the design calculations the **design strength** for a given material and limit state is obtained by dividing the characteristic strength by the **partial safety factor for strength** γ_m appropriate to that material and that limit state:

$$\text{design strength} = \frac{1}{\gamma_m} \times \text{characteristic strength} \qquad (1.5\text{--}2)$$

Although it has little physical meaning, the **global factor of safety**, or **overall factor of safety**, has been defined as the product $\gamma_m\gamma_f$. The value assigned to this factor (indirectly, through the values assigned to γ_m and, particularly, γ_f) depends on the social and economic consequences of the limit state being reached. For example, the ultimate limit state of collapse would require a higher factor than a serviceability limit state such as excessive deflection. Table 1.5–1 shows the γ_f factors specified by BS 8110 for the ultimate limit state. *These γ_f values have been so chosen as to ensure also that the serviceability requirements can usually be met by simple rules* (see Chapter 5). In assessing the effects of loads on a structure, the choice of the γ_f values should be such as to cause the most severe stresses. For example, in calculating the maximum midspan moment for the ultimate limit state of a simple beam under load combination (I), γ_f would be 1.4 and 1.6 for dead and live load, respectively. However, in calculating the maximum midspan moment in the centre span of a three-span continuous beam, the loading would be $1.4G_k + 1.6Q_k$ on the centre span plus the **minimum design dead load** of $1.0G_k$ on the exterior spans.

Table 1.5–2 shows the γ_m values specified by BS 8110. These values take account of (1) the importance of the limit state being considered and (2) the differences between the strengths of the materials as tested and those of the materials in the structure.

It is worth noting that the selection of γ_f values in BS 8110 is largely empirical; the γ_f values have been so chosen that, in the design of common structures, much the same degree of safety is achieved as for similar structures designed in accordance with those earlier codes of practice (CP 114, CP 115 and CP 116). The authors wish to quote Heyman [11]:

Table 1.5–1   Partial safety factors for loads γ_f: ultimate limit state (BS 8110:Clause 2.4.3.1)

| Combination of loads | Dead | | Imposed | | |
	Adverse	Beneficial	Adverse	Beneficial	Wind
		when effect of load is			
(I) Dead + imposed	1.4	1.0	1.6	0	—
(II) Dead + wind	1.4	1.0	—	—	1.4
(III) Dead + wind + imposed	1.2	1.2	1.2	1.2	1.2

Table 1.5–2   Partial safety factors for strength γ_m: ultimate limit state
(BS 8110:Clause 2.4.4.1)

Reinforcement	1.15
Concrete	
Flexure or axial load	1.50
Shear strength without shear reinforcement	1.25
Bond strength	1.40
Others (e.g. bearing stress)	≥ 1.50

'The empirical assignment of values to partial load factors (γ_f) in this way seems sensible and acceptable; in the absence of precise information it is right to make use of experience. But it is wrong to forget that the numerical work *has* been arranged empirically, and to come to the belief that the values of partial load factor found to give good practical results actually correspond to a real state of loading'.

1.6   Limit state design and the classical reliability theory

In limit state design, probabilistic concepts are for the first time accepted explicitly. It is now formally accepted that no structure can be absolutely safe. The difference between a 'safe' and an 'unsafe' design is in the degree of risk considered acceptable, not in the delusion that such risk can be completely eliminated [12–15]. The acceptable probabilities for the various limit states have not yet been defined or quantified [3, 4], but the acceptance of probabilistic concepts marks an important step forward in design thinking and should stimulate further research and study in the right direction.

Limit state design philosophy is partly based on the **classical reliability theory**, which is briefly described below [12]. Figure 1.6–1 illustrates the classical reliability theory. Figure 1.6–1(a) shows the probability distribution curves for the random variables F (the load) and f (the strength). The probability of failure at a particular load range F_i to $F_i + dF_i$ is given by

$$\text{probability of failure} = A_f \times dA_F \tag{1.6–1}$$

where dA_F is the area in Fig. 1.6–1(a) representing the probability of the load F falling within the small range F_i to $F_i + dF_i$, and A_f is that representing the probability of the strength f falling below F_i. Hence

$$\text{overall probability of failure} = \int_0^1 A_f \, dA_F \tag{1.6–2}$$

A more convenient way of approaching the problem is to plot the probability distribution curve for the quantity 'strength–load', as shown in Fig. 1.6–1(b); the probability of failure is then given by the area under that part of the curve where $(f - F)$ is negative. There are important reasons

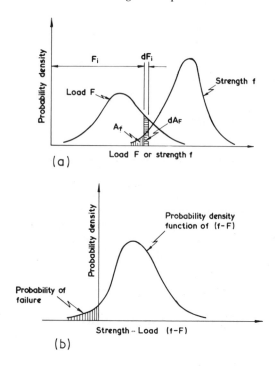

Fig. 1.6–1   **Classical structural reliability theory [12]**

why, in practice, the classical reliability theory cannot be applied in its purest form [12]. One of these is that for structures such as those covered by BS 8110 (or ACI 318–83) [16, 17], the probabilities of failure that are socially acceptable must be kept very low (e.g. 1 in 10^6). At such low levels, the probability of failure is very sensitive to the exact shapes of the distribution curves for strength and load. To determine these exact shapes would require very large numbers of statistical data, and such data are not yet available. In particular, sufficient numbers of extreme values of the strengths of complete structures (to define accurately the shapes of the tails of the distribution curves) may never be available. Also in the simple example illustrated in Fig. 1.6–1, only one load and one strength variable are considered. For a real structure, there will in general be many loads and many modes of failure, usually with complex correlations between them, making it very difficult to calculate the probability of failure. In limit state design [3, 4], our engineering experience and judgement have been used to modify, and to remedy the inadequacies of, the pure probabilistic approach to structural design [15].

Further information on structural design and safety concepts may be found in an interesting book [18] by the late Professor Lord Baker of Cambridge University.

References

1   Harris, Sir Alan. Design philosophy and structural decisions. In *Handbook of Structural Concrete*, edited by Kong, F. K., Evans, R. H., Cohen, E. and Roll, F. Pitman, London and McGraw-Hill, New York, Chapter 3, 1983.

2   CEB. *Recommendations for an International Code of Practice for Reinforced Concrete*. English edition. American Concrete Institute and Cement and Concrete Association, 1964.

3   CEB–FIP. *Model Code for Concrete Structures*, English edition. Cement and Concrete Association, London, 1978.

4   BS 8110:1985. *Structural Use of Concrete—Part 1: Code of Practice for Design and Construction*. British Standards Institution, London, 1985.

5   Rowe, R. E. *et al. Handbook to BS 8110*. Cement and Concrete Association, Slough, 1987.

6   Rowe, R. E. Euro-Codes. *The Structural Engineer*, **63A**, No. 10, Oct. 1985, p. 317.

7   FIP. *Practical Design of Reinforced and Prestressed Concrete Structures*. Thomas Telford, London, 1984.

8   Neville, A. M. and Kennedy, J. B. *Basic Statistical Methods for Engineers and Scientists*. 3rd ed. International Textbook Co., Scranton, 1986.

9   Kreyszig, E. *Advanced Engineering Mathematics*, 5th ed. Wiley, New York, 1983.

10 Jeffrey, Sir Harold. *Theory of Probability*. Oxford University Press, 1961.

11 Heyman, J. Plastic design and limit state design. *The Structural Engineer*, **51**, No. 4, April 1973, pp. 127–31.

12 Kemp, K. O. Concepts of structural safety. *Civil Engineering and Public Works Review*, **68**, No. 799, Feb. 1973, pp. 132–41.

13 Thoft-Christensen, P. and Baker, M. J. *Structural Reliability Theory and its Applications*. Springer-Verlag, New York, 1982.

14 Blockley, D. *The Nature of Structural Design and Safety*. Ellis Horwood, Chichester, 1980.

15 CIRIA Report 63: *The Rationalization of Safety and Serviceability Factors in Structural Codes*. Construction Industry Research and Information Association, London, 1977.

16 ACI Committee 318. *Building Code Requirements for Reinforced Concrete* (ACI 318–83). American Concrete Institute, Detroit, 1983.

17 ACI Committee 318. *Commentary on Building Code Requirements for Reinforced Concrete* (ACI 318R–83). American Concrete Institute, Detroit, 1983.

18 Baker, Lord. *Enterprise v. Bureaucracy—the Development of Structural Air-Raid Shelters during the Second World War*. Pergamon Press, Oxford, 1978.

Chapter 2
Properties of structural concrete

Preliminary note: Readers interested only in structural design to BS 8110 may concentrate on the following sections:

(a) *Section 2.5(a): Strength of concrete.*
(b) *Section 2.5(e): Durability of concrete.*

2.1 Introduction

Concrete is a composite material which consists essentially of:

(a) a binding medium of cement and water, called the cement–water paste, or simply the **cement paste**; and
(b) particles of a relatively inert filler called aggregate.

The selection of the relative proportions of cement, water and aggregate is called **mix design**, which will be dealt with later in this chapter. At this stage it is sufficient to note that in mix design the most important requirements are:

(a) The fresh concrete must be workable, or placeable.
(b) The hardened concrete must be strong enough to carry the load for which it has been designed.
(c) The hardened concrete must be able to withstand the conditions to which it will be exposed in service.
(d) It must be capable of being produced economically.

These requirements may be summed up as workability, strength, durability and economy.

2.2 Cement

The chemistry of cement is rather complicated [1, 2], but for our purpose it is sufficient to know its main properties and recognize its main constituents. **Portland cement** is a finely pulverized clinker produced by burning, at about 1450 °C in a kiln, mixtures containing lime (CaO), silica (SiO_2), alumina (Al_2O_3) and iron oxide (Fe_2O_3). The main oxide composition of Portland cement is, typically: lime 60–65%, silica 18–25%, alumina 3–8%, iron oxide (Fe_2O_3) 0.5–5%. Different types of Portland cement are obtained by varying the relative proportions of these four predominant chemical compounds, and by grinding the clinker to different degrees of fineness. The two most common types of cement [3] are **ordinary Portland cement** and **rapid-hardening Portland cement**, which are both covered by BS 12.

When cement in mixed with water, the various compounds of the cement begin to react chemically with the water. For a short time, this cement–water paste remains plastic, and it is possible to disturb it and remix without harmful effects, but as the chemical reactions continue, the paste begins to stiffen, or **set**. The arbitrary beginning and the ending of the period of setting are called the **initial set** and the **final set**. The definition of the stiffness of the paste which is to be regarded as set is necessarily vague. In BS 12, for example, the **initial setting time** is defined as the interval between the time when water is added to the cement and the time when the paste will just withstand a prescribed pressure. Similarly, the **final setting time** is defined as the interval between the time when water is first added and that when the paste has further stiffened to be able to withstand a higher prescribed pressure. In the test for setting times, a mix of **standard consistence** is used: BS 12 defines standard consistence as that of a cement paste to which such an amount of water has been added that, at a prescribed time after adding water, the paste is just able to withstand a prescribed pressure. The water content of such a paste, expressed as a percentage of the dry weight of the cement, is usually from 26 to 33%.

If the cement sets too rapidly, the concrete will stiffen too quickly for it to be transported and properly placed in the moulds or formwork; if it sets too slowly, it might delay the use of the structure through insufficient strength. BS 12 specifies that the initial setting time must be at least 45 minutes and the final setting time at most 10 hours. In practice, the rate of setting is controlled by adding about 7–8% of gypsum to the clinker as it is being ground. This gypsum is sometimes called a **retarder**, because it retards the setting of the cement.

After the paste has attained final set, it continues to hydrate and increase in rigidity and strength; this process is called **hardening**. It is important to understand that setting and hardening are the result of chemical reactions, in which water plays an important part; it is not just a matter of the paste drying out. In the absence of moisture, these chemical reactions stop; in the presence of moisture they may continue for many years so that the hardened paste continues to gain strength. The setting and hardening of the cement paste is accompanied by the liberation of heat, called the **heat of hydration**, which for ordinary Portland cement averages about 300 kJ/kg at 7 days, rising to about 340 kJ/kg at 28 days. In mass concrete construction such as in dams, where it is difficult for this heat to escape, it may become necessary to control the rise in temperature of the concrete by using low-heat Portland cement (BS 1370), for which the heat of hydration does not exceed 250 kJ/kg at 7 days and 300 kJ/kg at 28 days.

The rate of hardening increases with the **fineness** to which the cement has been ground. In practice, fineness is defined by the specific surface of the cement particles. For example, BS 12 stipulates that the specific surface of ordinary Portland cement must not be less than 225 000 mm^2/g. Cement particles are angular in shape, and it is usual to state the nominal particle size in terms of standard sieve numbers. For ordinary Portland cement, practically all particles will pass a No. 100 sieve (BS 410, 150 μm aperture width) and over 95% will pass a No. 200 sieve (75 μm).

The fineness to which the raw materials are ground affects the **soundness** of the cement. A cement is said to be unsound if excessive expansion of some of the constituents occurs after the cement has set; such expansion causes cracking, disruption and disintegration of the mass, and hence threatens the security of any concrete structure in which such cement has been used. An important cause of unsoundness is the amount of free lime (CaO) encased in the cement particles. Such hard-burnt lime hydrates very slowly and the expansive reactions may continue for months or even years after the cement has set. Fine grinding of the raw materials brings them into close contact when burned, thus reducing the chance of free lime existing in the clinker. In BS 12, an accelerated test for soundness is used; this consists essentially in measuring the expansion of a cement paste of standard consistence at prescribed times.

All specifications for cement prescribe some form of tests for its **strength** which are usually carried out on mortar or on concrete made with the cement. According to BS 12, compression tests are carried out on 70.7 mm mortar cubes or on 100 mm concrete cubes. In the mortar test, cement is gauged with a standard sand (called Leighton Buzzard sand) in the proportions by weight of one part cement to three parts sand, and the water/cement (w/c) ratio is 0.40 by weight. In the concrete test, the water/cement ratio is 0.60 and the aggregate/cement ratio is to be adjusted by trial and error such that certain prescribed workability requirements are met. The cubes are made and stored in a prescribed manner and the minimum strength requirements for ordinary Portland cement are:

Mortar cubes
3-day compressive strength $\geq$ 15 N/mm^2
7-day compressive strength $\geq$ 23 N/mm^2

Concrete cubes
3-day compressive strength $\geq$ 8 N/mm^2
7-day compressive strength $\geq$ 14 N/mm^2

Rapid-hardening Portland cement

Rapid-hardening Portland cement often has a higher lime content than ordinary Portland cement and is more finely ground; in other respects the two cements are very similar. BS 12 specifies that the specific surface must not be less than 325 000 mm^2/g. Practically all particles of this cement will pass a No. 200 sieve (BS 410, 75 μm aperture width) and about 99% will pass a No. 350 sieve (45 μm).

Rapid-hardening Portland cement is capable of developing as great strength in three days as ordinary Portland cement does in seven days. The strength requirements in BS 12 are:

Mortar cubes
3-day compressive strength ≥ 21 N/mm^2
7-day compressive strength ≥ 28 N/mm^2

Concrete cubes
3-day compressive strength ≥ 12 N/mm^2
7-day compressive strength ≥ 17 N/mm^2

Though rapid-hardening Portland cement has a higher rate of hardening and a higher rate of heat development than ordinary Portland cement, the setting times of the two cements are similar. In BS 12, the setting-time requirements are the same for both; similarly, the requirements for soundness are the same.

Unit weight of cement
The specific gravity of the particles of cement is generally within the range of 3.1–3.2; for most calculations, it is taken as 3.15. The unit weight of bulk cement depends, obviously, on the degree of compactness; as a rough guide, it may be taken as 1450 kg/m^3. In the UK, one bag of cement weighs about 50 kg.

2.3 Aggregates

Aggregates used in concrete making are divided into two categories: (a) the **coarse aggregate** such as crushed stone, crushed gravel or uncrushed gravel, which consist of particles that are mainly retained on a 5 mm sieve, and (b) the **fine aggregate** such as natural sand, or crushed stone sand or crushed gravel sand, which consist of particles mainly passing a 5 mm sieve.

It is a most important requirement that the aggregate should be durable and chemically inert under the conditions to which it will be exposed. Other important requirements concern the size, the shape, the surface texture, and the grading, which are discussed below. Surprisingly enough, the compressive strength of concrete is not much affected by the strength of the aggregates unless they are very weak; the flexural strength of concrete, however, is more affected by weak aggregates.

In this section we shall restrict our discussions to aggregates from natural sources such as those referred to above; these aggregates are covered by BS 882. However, it is worth noting that, despite the present-day engineer's preference for natural aggregates, he will find these in increasingly short supply in the not too distant future. To quote an ACI report: 'In many areas, supplies and reserves of naturally occurring aggregates will become depleted and increased emphasis will therefore be placed on manufactured aggregates, many of which are lightweight' [4]. In the UK, **lightweight aggregates** for concrete should comply with BS 3797. See also BS 8110 : Part 2 : Clause 5.

Size of aggregates
In reinforced and prestressed concrete construction, nominal maximum sizes of the coarse aggregates are usually 40, 20, 14 or 10 mm. The nominal

maximum size actually used in a job depends on the dimensions of the concrete member. Particles too large in relation to these dimensions may affect the strength adversely, particularly the flexural strength; a good practice is to ensure that the maximum size does not exceed 25% of the minimum thickness of the member nor exceed the concrete **cover** to the reinforcement (i.e. the clear distance between the reinforcement bar and the formwork). There is also the important requirement that the aggregate should be small enough for the concrete to flow around the reinforcement bars so that it can be adequately compacted. For example, BS 8110 [3] recommends that

(a)   the nominal maximum size of the coarse aggregate should be less than the horizontal clear distance between the reinforcement bars by at least 5 mm; and

(b)   it should not exceed $1\frac{1}{2}$ times the clear vertical distance between the bars.

Apart from the above considerations, it is advantageous to use the higher maximum sizes because, in general, as the maximum size of the aggregate increases, a lower water/cement ratio can be used for a given workability, and a higher strength is obtained. This applies for a nominal maximum size of up to 40 mm; above this, the gain in strength due to the reduced water/cement ratio is offset by the adverse effects of the lower bond area between the cement paste and the aggregate and of the discontinuities caused by the large particles.

Shape and surface texture of aggregates
In practice, the shapes of aggregates are usually described by terms such as rounded, irregular or angular, which are necessarily imprecise; similarly terms such as smooth or rough are used to describe surface textures.

The **particle shapes** affect the strength of the concrete mainly by affecting the cement-paste content required for a given workability. If the cement content is the same, then an angular aggregate would require a higher w/c ratio than an irregular one, which in turn will require a higher w/c ratio than a rounded one.

The **surface texture** affects concrete strength in two ways. First, it affects the bond between the cement paste and the aggregate particles; second, it affects the cement-paste content required to achieve a given workability. On balance a rough surface results in a higher concrete strength, particularly flexural strength.

The above remarks apply to both coarse and fine aggregates.

Grading of aggregates
For a concrete to be durable, it has to be dense and, when fresh, it should be sufficiently workable for it to be properly compacted. The mortar, i.e. the mixture of cement, water and fine aggregate, should be slightly more than sufficient to fill the voids in the coarse aggregate; in turn the cement paste should be slightly more than sufficient to fill the voids in the fine aggregate. The voids in an aggregate depend on its particle-size distribution, or **grading**. The grading of the aggregates affects the strength

of concrete mainly indirectly, through its important effect on the water/cement ratio required for a specified workability. A badly graded aggregate requires a higher water/cement ratio and hence results in a weaker concrete.

In practice, the grading of an aggregate is determined by **sieve analysis**, in which the percentage of the aggregate passing through each of a series of standard sieves is determined. For the grading to comply with BS 882, the percentages for the coarse aggregate should fall within the limits in Table 2.3–1; for the fine aggregate, the limits should fall within the overall limits in Table 2.3–2 and, additionally, not more than one in ten consecutive samples shall have a grading outside the limits for any of the gradings C, M or F in Table 2.3–2. In Table 2.3–2, the gradings of the fine aggregates are divided into three zones, ranging from the coarser zone C through the medium zone M to the finer zone F. The division into zones is largely based on the percentage passing the 600 μm (No. 25) sieve. It has been found that in a fresh concrete mix, the content of the fine aggregate passing the 600 μm sieve has an important effect on the workability. Furthermore, many naturally occurring sands divide themselves at this particular size.

Table 2.3–1   Grading limits for coarse aggregate

| | *Percentage by weight passing the standard sieves* | | |
| | *Nominal size of aggregate* | | |
Standard sieve (mm)	40 mm *to* 5 mm	20 mm *to* 5 mm	14 mm *to* 5 mm
50.0	100		
37.5	90–100	100	—
20	35–70	90–100	100
14	—	—	90–100
10	10–40	30–60	50–85
5	0–5	0–10	0–10

Table 2.3–2   Grading limits for fine aggregate

| | *Percentage by mass passing BS sieve* | | | |
| | | *Additional limits for grading* | | |
Sieve size	*Overall limits*	*C*	*M*	*F*
10.00 mm	100	—	—	—
5.00 mm	89–100	—	—	—
2.36 mm	60–100	60–100	65–100	80–100
1.18 mm	30–100	30–90	45–100	70–100
600 μm	15–100	15–54	25–80	55–100
300 μm	5–70	5–40	5–48	5–70
150 μm	0–15	—	—	—

Strength of aggregates

As stated earlier, within fairly wide limits the crushing **strength of the aggregate** has little effect on the compressive strength of the concrete. Thus, for aggregates complying with BS 882, their strengths are not an important consideration in general structural concrete construction. However, if a concrete of high characteristic strength is required, say above 60 N/mm^2, it may be necessary to use crushed-rock coarse aggregate with a natural-sand or crushed-sand fine aggregate. Broadly speaking, the compressive strength of a practical structural concrete cannot exceed that of the aggregate used in making it, but the 'ceiling' strength of the concrete using a particular type of aggregate depends not only on the strength but also on the surface characteristics of the aggregate. Indeed the crushing strengths of aggregates complying with BS 882 are generally well above 100 N/mm^2; strengths of above 200 N/mm^2 are not uncommon.

Unit weight of aggregates

The specific gravity of aggregate particles depends on the mineral contents; for sand and gravel, it is usually taken to be 2.60. The bulk density, that is the unit weight per cubic metre, further depends on the moisture content and the degree of compaction. As a rough guide, the bulk density of sand and gravel is about 1 700 kg/m^3.

2.4   Water

BS 3148 gives the requirements for the testing of water for its suitability for use in concrete making.

If the water is suitable for drinking it is generally suitable for concrete making. If the water is suspected to be unsuitable, two series of test cubes may be made: one series with the suspected water and another with drinking water. The strengths and general appearances at 7 and 28 days will provide useful information.

2.5   Properties of concrete

Strength and durability are generally considered the most important qualities of concrete. Other important properties include creep and shrinkage characteristics, and the elastic modulus. Fire resistance, resistance to abrasion and thermal conductivity are sometimes important considerations, but these will not be discussed here and the reader is referred to specialist texts [5].

Unit weight of concrete

For structural design purposes, the unit weight of concrete made with normal aggregates covered by BS 882 is usually taken as 24 kN/m^3. This unit weight includes an allowance for the weight of reinforcement bars.

2.5(a)   Strength of concrete

The **compressive strength** of concrete is the most common measure for judging the quality of concrete. In the UK, the characteristic strength of concrete (see Section 1.4) is based on the 28-day **cube strength**; that is, the crushing strength of standard 150 mm cubes at an age of 28 days after mixing; 100 mm cubes may be used if the nominal maximum size of the aggregate does not exceed 25 mm. Procedures for making and testing the cubes are given in BS 1881:Parts 108 and 116. In the USA, 6 × 12 in (150 × 300 mm nominal) cylinder specimens are tested, in accordance with ASTM standard C39, to determine the **cylinder strength**. The range of cube strengths for reinforced and prestressed concree work is usually from 25 to 60 N/mm^2.

The cylinder strength is usually only about 70–90% of the cube strength. The difference is due to the frictional forces which develop between the platen plates of the testing machine and the contact faces of the test specimen. These end forces produce a multiaxial stress state which increases the compressive strength of the concrete. The multiaxial stress effects are significant throughout the cube; in the cylinder, however, the height/width ratio is sufficiently large for the mid-height region to be reasonably free from these effects. For practical purposes the cylinder strength may be taken as the **uniaxial compressive strength** of concrete.

Nowadays, tests are rarely carried out to measure directly the **tensile strength** of concrete, mainly because of the difficulty of applying a truly concentric pull. A method for measuring the **indirect tensile strength**, sometimes referred to as the **splitting tensile strength**, is described in BS 1881 and ASTM standard C496. The test, which was developed independently in Brazil and in Japan, consists essentially in loading a standard concrete cylinder (300 × 150 mm diameter in the UK; 12 × 6 in diameter in the USA) across a diameter until failure occurs, by splitting across a vertical plane (Fig. 2.5–1(a)). If the concrete cylinder behaves as a linearly elastic body, the distribution of the horizontal stresses along the vertical diameter would be as shown in Fig. 2.5–1(b), with a uniform tensile stress of 2 $F/\pi dl$ over most of the diametrical plane. Of course, the concrete cylinder does not behave exactly as a linearly elastic body, but both BS 1881 and ASTM C496 state that the splitting tensile strength f_t may be taken as

$$f_t = \frac{2F}{\pi dl} \ (\text{N/mm}^2) \tag{2.5–1}$$

where F = maximum applied force (N);
$\quad\quad d$ = cylinder diameter (mm);
$\quad\quad l$ = cylinder length (mm).

For practical concrete mixes such as those used in reinforced and prestressed concrete construction, the splitting tensile strength generally varies from about 1/8 to 1/12 of the cube strength.

The determination of the **flexural strength** of concrete, as described in BS 1881 and ASTM standard C78, consists essentially of testing a plain

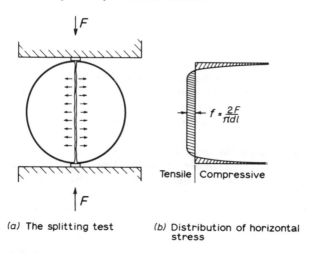

(a) The splitting test

(b) Distribution of horizontal stress

Fig. 2.5–1   Splitting tensile test

concrete beam under symmetrical two-point loading applied at one-third-span points. According to the above British and American standards, the flexural strength is to be calculated as M/Z, where M is the bending moment at the section where rupture occurs and Z is the elastic section modulus of the beam. The flexural strength so calculated is often referred to as the **modulus of rupture**, and is only a hypothetical stress based on the assumption of linear-elastic behaviour up to the instant of rupture. The modulus of rupture overestimates the true flexural strength of the concrete, and is for use as a comparative measure for practical purposes only. The modulus of rupture is usually about $1\frac{1}{2}$ times the splitting cylinder strength.

The **nature of strength** of concrete is complex and not yet fully understood [5]. However, it can be said that strength is primarily dependent on the free **water/cement ratio**, that is the ratio of the weight of the free mixing water to that of the cement in the mixture—the free mixing water being understood to include the surface water of the aggregates but not their absorbed water. Other things being equal, the lower the w/c ratio the higher is the strength (Fig. 2.5–2). To hydrate completely, cement needs to combine with about 38% by weight of water [1], but in practice complete hydration of the cement does not take place and is not aimed at. Under practical conditions, irrespective of the w/c ratio used, the cement combines with only about 23% water by weight. However, with a w/c ratio as low as 0.23, the concrete is very dry and difficult to place and compact; additional water is therefore required to lubricate the mix, that is to provide the necessary workability. In simple terms, and with some loss of accuracy, it can be said that the **non-evaporable water** is that part of the mixing water which actually combines with the cement in the formation of the various products of hydration, referred to collectively as the **gel**. In other words, the non-evaporable water is 23% of the cement by weight. Water in excess of this amount evaporates when the concrete dries and is

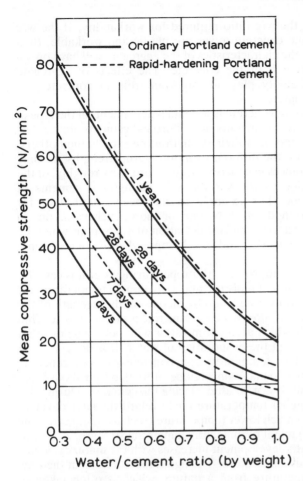

Fig. 2.5–2   Relation between water/cement ratio and mean compressive strength—100 mm cubes (after Road Note No. 4: Reference 6)

called the **evaporable water**. The spaces left after the loss of the evaporable water and the spaces occupied by any bubbles of air trapped in the concrete mix are collectively known as the **voids**, the presence of which is a great source of weakness. It is thus seen that the relation between the w/c ratio and strength is intimately connected with the voids in the concrete and with the degree of hydration of the cement. Further discussion of this aspect of concrete strength would require a study of the physical structure and porosity of the hardened cement paste. It is sufficient for our purpose to note that concrete strength increases with the **gel/space ratio**, which is the ratio of the volume of the gel to the sum of the solid volume of the cement that has hydrated and the volume of the mixing water.

As stated in Section 2.3, the **characteristics of the aggregate** such as the size, shape, surface texture, and grading, affect the strength of concrete principally by affecting the w/c ratio required for workability. If the w/c ratio is the same and the concrete mixes are plastic and workable, then considerable changes in the characteristics of the aggregate will have only a small effect on the strength of the concrete. The effects of aggregate characteristics on workability play an important role in concrete mix design, as we shall see later.

The long-term strength of concrete is not much dependent on the **type of cement** used. Basically, the behaviours of all Portland cements are similar. It is true that some cements gain their strength more rapidly than others, but, for a given w/c ratio, the differences in long-term strengths are only about 10%. These differences in concrete strenght are, very roughly, of the same order as those due to the variations in strength of cements of nominally the same type. For example, it has been found that, for concretes of average strength of 35 N/mm^2 or more, such variations in cement quality correspond to a standard deviation of 3.6 N/mm^2 when the cement comes from many works, and of 3 N/mm^2 when it comes from a single works [7].

The strength of concrete improves with proper **curing** after it is cast. By curing is meant the provision of moisture and a favourable temperature for the cement to continue to hydrate, thereby increasing the strength of the concrete. In practice, shortly after the fresh concrete is placed, it is covered with absorbent materials which are kept moist or with polythene or other impervious sheets to prevent the evaporation of water. It is difficult to be precise about the effect of temperature on concrete [8], but generally speaking it is good practice to maintain a temperature between about 5 and 20 °C during the first half day or so after placing the concrete. During this initial period, a much higher temperature might retard the later development in strength, while a much lower temperature (such as would cause the fresh concrete to freeze) might permanently impair the strength. After this initial period, the strength development increases with the **maturity**, which is the product of **age** and temperature. The age is measured from the time of mixing and the temperature from a datum, which is often taken as −11.7 °C. Of course this assumes the presence of moisture for the cement to continue to hydrate. It should be pointed out that even when the concrete is not specially protected against drying, the evaporable water is not immediately lost under ordinary climatic conditions, and hence the strength will continue to increase for some time. However, in the absence of evaporable water in the concrete, there can be no strength increase.

2.5(b)   Creep and its prediction

When a stress is applied to a concrete specimen and kept constant, the specimen shows an immediate strain followed by a further deformation which, progressing at a diminishing rate, may become several times the original immediate strain (Fig. 2.5–3). The immediate strain is often referred to as the **elastic strain** and the subsequent time-dependent strain referred to as the creep strain, or simply the **creep**. That part of the strain which is immediately recoverable upon removal of the stress is called the

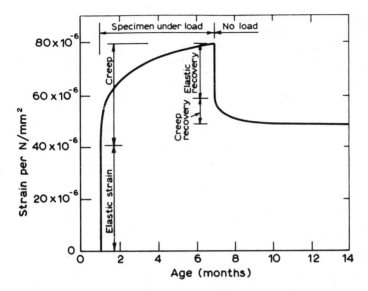

Fig. 2.5–3 **Typical strain/time curve for concrete subjected to constant load followed by load removal**

elastic recovery and the delayed recovery the **creep recovery**. The elastic recovery is less than the elastic strain; the creep recovery is much less than the creep.

The **mechanism of creep** is still subject to controversy [8–12]. Though it is likely that creep is related to the structure of the gel, it is hard to suggest definite conclusions on the mechanism of creep; the difficulty is that a satisfactory theory must explain in a unified way the behaviour of concrete under various environmental and stress conditions. Perhaps the only non-controversial statement that can be made is that the presence of some evaporable water is essential to creep [12]. Methods for estimating creep values have been suggested by various authors [8–12] including the CEB [13]. One thing is common to all these methods: the accuracy is not high and an error of ±30% or more is possible. An account is given below of an approximate but simple method previously proposed for engineers in practice [9, 10]. *To estimate the creep of a particular concrete, start from the creep values in Table 2.5–1 and then successively allow for the effects of various other factors as explained in the subsequent paragraphs.*

Table 2.5–1 shows some **creep values of specimen concrete mixes**, loaded to within one-third of the cube strength, assuming loading starts after 28 days wet curing and service condition of 70% relative humidity and 15 °C. These are **limiting-creep** values and are reached when the stresses are sustained for a very long period, say 30 years. Creeps corresponding to shorter periods of loading, as well as the effects of various factors are estimated as explained below. In practical structures, where the concrete is unlikely to be stressed beyond one-half of the cube strength, the creep of a given concrete under a given period of sustained stress may be

Table 2.5–1   Creep of specimen mixes

Mix Ref.	Mix proportions (by weight)	Creep (per N/mm^2)	Remarks
A	$1:2:4^a$ w/c = 0.65	120×10^{-6}	Cement: ordinary Portland
B	$1:1.5:3$ w/c = 0.55	100×10^{-6}	Fine aggregate: sand
C	$1:1:2$ w/c = 0.40	70×10^{-6}	Coarse aggregate: crushed gravel

[a] Cement : fine aggregate : coarse aggregate (by weight).

taken as roughly proportional to the **stress**. Hence Mix C, say, in Table 2.5–1 will have a limiting creep of $(70 \times 10^{-6}) \times 10 = 700 \times 10^{-6}$ if the sustained stress is 10 N/mm^2. For different concretes of the same cement-paste content, creep is approximately proportional to the **stress/strength ratio**, i.e. the ratio of the applied stress to the cube strength of the concrete, at the time of loading. Hence, a concrete of 45 N/mm^2 cube strength stressed to 15 N/mm^2 would have approximately the same creep as another concrete of 30 N/mm^2 stressed to 10 N/mm^2, provided the two concretes have the same cement-paste content.

Creep depends on the **duration of loading**, as shown in Table 2.5–2.

The effect on creep of the **age at loading** is mainly due to the increase in strength of concrete with age. Since, for a given stress, creep is inversely proportional to strength, the effect of age at loading can be estimated provided the strength–age relation is known. For example, if the one-year strength of a concrete is 1.5 times the 28-day strength, then by applying the load at age one year instead of at age 28 days, the long-term creep will be reduced by the ratio 1/1.5. As pointed out earlier, strength is related to the maturity, and a sample calculation is given in Reference 9.

If concrete is wet cured to an age of 28 days and then loaded, the limiting creeps in air at 50% relative humidity (RH) and at 100% RH are respectively about $1\frac{1}{3}$ and $\frac{1}{2}$ times that for storage in air at 70% RH. Creeps at 70% RH are given in Table 2.5–1; for other **relative humidities**, creeps can be estimated from the above general information.

If concrete is stored for a sufficiently long time, say a year, at a particular

Table 2.5–2   Effect of duration of loading

Duration of loading	Percentage of long-term creep
28 days	40
6 months	60
1 years	75
5 years	90
10 years	95
30 years	100

humidity and then loaded and stored under the same humidity, it will be found that the value of this humidity does not have a significant effect on the creep. This indicates that when the concrete has reached moisture equilibrium with the surrounding atmosphere creep becomes almost independent of the relative humidity of the surrounding air.

Up to about 100 °C the general shape of the creep/time curve is similar to that at normal temperatures, and the relation between creep and the stress/strength ratio remains linear. The rate of creep increases with an increase in **temperature** up to a maximum at about 70 °C, thereafter decreasing somewhat up to 100 °C. For general design purposes, it seems sufficient to assume that creep increases linearly with temperature at a rate of $1\frac{1}{4}\%$ of the 15 °C creep for each degree Celsius.

For reinforced and prestressed concrete work, the **cement-paste content** by volume, calculated by the procedure shown in Table 2.5–3, nearly always lies within the range 28–40%. Within this range, creep (for the same stress/strength ratio) can be assumed to increase at the approximate rate of 5% for each per cent increase in the cement-paste content by volume. In calculating this cement-paste content, the speciic gravity of Portland cements may be taken as 3.15 and that of common gravel and stone aggregates as 2.60. Table 2.5–3 shows sample calculations for two mixes.

Within the limits of concrete mixes used in reinforced and prestressed concrete work, the effect of the **w/c ratio** is indirect. Consider an increase in the w/c ratio. First, it causes a reduction in strength (Fig. 2.5–2), thereby increasing the stress/strength ratio for a given stress. Second, for a given aggregate/cement ratio, an increase in the w/c ratio increases the cement-paste content. By allowing for these two effects, the influence of a change in the w/c ratio is automatically taken care of [9, 10].

Within the range of practical mixes, the **aggregate/cement ratio**, **aggregate content**, **cement content** and **water content** do not by themselves have important effects on creep. Their effects are mainly due to their influence on the w/c ratio and the cement-paste content, and can be allowed for accordingly.

Table 2.5–3 Calculation for cement-paste content

	Mix			
	$1:1\frac{1}{2}:3$ w/c = 0.55		Cement/agg. = 1:5 w/c = 0.55	
Ingredient	Weights	Volumes	Weights	Volumes
Cement	1	1/3.15 = 0.31	1	1/3.15 = 0.31
Water	0.55	0.55/1 = 0.55	0.55	0.55/1 = 0.55
Aggregate	4.5	4.5/2.6 = 1.73	5.0	5.0/2.6 = 1.92
		2.59		2.78
	Cement-paste content: (0.31 + 0.55)/2.59 = 0.333		Cement-paste content: (0.31 + 0.55)/2.78 = 0.309	

The **type of cement** affects creep primarily through its effect on the rate of hardening of the concrete. For a given stress applied at a given age, creep occurs in an increasing order for concretes made with the following cements: high-alumina, rapid-hardening Portland, ordinary Portland, Portland blast furnace, low-heat Portland, and Portland-pozzolan. It appears that creep varies inversely with the rapidity of hardening of the cement. For a given stress, the creep of concretes of a given mix proportion but made with different types of cements is proportional to the stress/strength ratio. **Fineness of cement** itself does not seem to have important effects on creep, and is significant only in so far as it affects the rate of hardening of concrete.

The relative magnitudes of creep of concretes of the same mix proportion but made with different **types of aggregates** are illustrated in Fig. 2.5–4. In general, concretes made with aggregates of high moduli of elasticity and which are hard and dense have lower creeps. For a given stress, creep decreases as the maximum size of the coarse aggregate increases and when well-graded and well-shaped aggregates are used. It should be noted that the effects of the **size**, **shape**, **surface texture**, and **grading** of the aggregate on creep are largely due to their effects on the amount of water required for workability. For comparable workability, concretes of a given mix proportion made with aggregates of larger maximum size which are well shaped and well graded have lower water/cement ratios and lower cement-paste contents than concretes made with aggregates of smaller maximum size, or with badly graded, flaky and elongated aggregates.

The foregoing paragraphs give mainly factual statements on creep and creep prediction. For a brief explanation of the mechanics of creep and of the reasons why the various factors influence creep the way they do, the

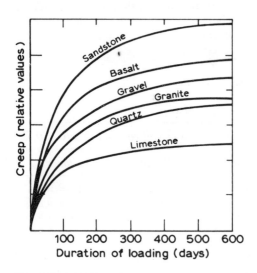

Fig. 2.5–4   Effect of type of aggregate on creep

reader is referred to Reference 10. An authoritative and fuller account has been given by Neville [11].

2.5(c)   Shrinkage and its prediction

The volume contraction which occurs as the concrete hardens and dries out is called the drying shrinkage, or simply the **shrinkage**. Shrinkage is thought to be due mainly to the loss of adsorbed water in the gel; its nature, like that of creep, is thought to be primarily related to the physical structure of the gel rather than its chemical composition [10, 12].

The shrinkage on first drying is partly irreversible and is called the **initial drying shrinkage**. If dry concrete is resaturated with water, an expansion, sometimes referred to as the **moisture movement**, of about 60% of the initial drying shrinkage will occur. If this saturated concrete is redried, it will shrink again, this shrinkage being substantially equal to the expansion. The process of saturation and drying may be repeated indefinitely, and for every cycle the expansion and shrinkage are very nearly equal, and will be equal if the hydration of the cement is complete before the drying shrinkage occurs. As hydration in concrete is rarely ever complete, the shrinkage in each cycle is nearly always slightly larger than the expansion.

If, after casting, the concrete is continuously cured in saturated atmosphere or under water, it undergoes an expansion. Such expansion is mostly due to the adsorption of water by the gel, and is about 150×10^{-6} for concrete of aggregate/cement ratio $6:1$ and w/c ratio 0.6. This expansion increases with the cement content; for neat cement paste the expansion will be 10 times as large. About 80% of this expansion occurs in the first month after casting, 90% in 6 months and practically the full amount in one year.

For practical design purposes, we are less concerned with the nature of shrinkage than with the prediction of shrinkage values. An approximate but simple procedure has been proposed for such purposes [14]. *To estimate the shrinkage of a particular concrete, start from the values in Table 2.5−4 and then successively allow for the effects of the various other factors as explained below.*

Table 2.5−4 shows some **shrinkage values of specimen concrete mixes**, made of ordinary Portland cement and crushed-gravel aggregate and sand. The concretes are wet cured for 28 days after casting and then exposed to 70% RH at 15 °C. These shrinkage values are the ultimate values reached

Table 2.5−4   Shrinkage of specimen mixes

Mix Ref.	Mix proportions (by weight)	Shrinkage	Remarks
A	$1:2:4$ w/c = 0.65	400×10^{-6}	Volume/surface
B	$1:1.5:3$ w/c = 0.55	500×10^{-6}	ratio of
C	$1:1:2$ w/c = 0.40	600×10^{-6}	specimens = 60 mm

Table 2.5–5   Effect of duration of exposure

Duration of exposure	Percentage of long-term shrinkage	Duration of exposure	Percentage of long-term shrinkage
28 days	40	5 years	90
6 months	60	10 years	95
1 year	75	30 years	100

after a long time of exposure, say 30 years. Effects of the **duration of exposure** can be allowed for using Table 2.5–5.

Using the shrinkage at 70% RH as the reference magnitude, shrinkage can be assumed to increase at the approximate rate of 2% for each per cent decrease in **relative humidity** down to about 40% RH, and decrease at the approximate rate of 3% for each per cent increase in RH up to about 90% RH [13, 14]. At 100% RH there is an expansion equal numerically to about 20% of the shrinkage at 70% RH. Shrinkages at 70% RH are given in Table 2.5–4; for other humidities, shrinkage can be estimated from the above general information.

There is insufficient information [8, 13] on the variation of shrinkage with **temperature**. As an approximate guidance, it is suggested [14] that, within the range of atmospheric temperatures in Great Britain, shrinkage can be assumed to increase at the rate of 1% of the 15 °C shrinkage for each degree Celsius rise in temperature and decrease at the same rate with fall in temperature.

For reinforced and prestressed concrete work, the **cement-paste content** by volume, calculated by the procedure shown in Table 2.5–3, generally lies within the range 28–40%. Within this range shrinkage can be assumed to increase at the approximate rate of 7% for each per cent increase in the cement-paste content by volume, calculated as in Table 2.5–3.

The mix proportions of a concrete are completely defined by the cement-paste content, the w/c ratio and the ratio of fine aggregate to coarse aggregate. The effect of mix proportions on shrinkage is largely due to their effect on the cement-paste content. Having allowed for the effect of the cement-paste content, the additional effect of the **w/c ratio** is less important (Fig. 2.5–5) while the effect of the **ratio of fine aggregate to coarse aggregate** is in itself insignificant.

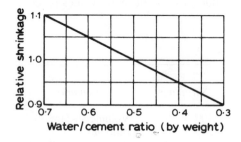

Fig. 2.5–5   Effect of water/cement ratio on shrinkage of concrete of given cement-paste content

The **aggregate content** affects shrinkage partly because the aggregate restrains the amount of shrinkage that can occur and partly because the aggregate content determines the amount of cement paste, which causes the shrinkage. However, the aggregate content is completely defined by the cement-paste content; by allowing for the effect of the latter as previously explained, the effect of the aggregate content is automatically taken care of.

The cement-paste content and the w/c ratio completely define the **water content**; for the purpose of practical design, it is only necessary to allow for the effects of the cement-paste content and the w/c ratio as explained above; the effect of the water content is then automatically taken care of. Similarly, for practical design purposes, there is no need to allow specially for the effects of the **cement content**, the **aggregate/cement ratio** or the **ratio of fine aggregate to coarse aggregate**.

For ordinary and rapid-hardening Portland cements, the variations of the **chemical composition** within the permitted ranges of BS 12 have negligible effect on concrete shrinkage. Where the mix proportions are kept constant, increasing the **fineness of cement** tends to increase shrinkage slightly. Using a fineness of $225\,000$ mm^2/g (which is the minimum specific surface of ordinary Portland cement complying with BS 12) as the reference, concrete shrinkage can be assumed to increase at the approximate rate of $\frac{1}{3}\%$ for each per cent increase in fineness. However, increasing the fineness of cement reduces the amount of mixing water required for a given workability and hence reduces the cement-paste content and the w/c ratio, so that in practice the increase in shrinkage due to the increase in cement fineness may be offset by the decrease due to the reduction in the cement-paste content and the w/c ratio.

For concretes of the same mix proportions but made of different **types of cements**, the shrinkages have the following approximate relative values: ordinary Portland (100), rapid-hardening Portland (110), low-heat Portland (115), high-alumina (100, but at a more rapid rate).

The relative shrinkage of concretes of the same mix proportions but made with different **types of aggregates** are illustrated in Fig. 2.5–6. Concretes having low shrinkages usually contain non-shrinking aggregates such as quartzite gravel, mountain limestone, blast furnace slag, dolomite, felspar, granite, etc.; high shrinkages will be caused by the use of aggregates which exhibit appreciable volume changes on wetting and drying, such as sandstone, slate, basalt and certain dolerites. Other conditions being equal, the effect of using a shrinkable aggregate is to increase the shrinkage of the concrete by an amount at least equal to the shrinkage of the aggregate itself. Both coarse and fine aggregates influence shrinkage but the influence of the coarse aggregate is more important. In general, shrinkage of concrete is reduced by using aggregates of high moduli of elasticity and which are hard and dense.

The **grading, size, shape** and **surface texture** do not have significant direct effects on concrete shrinkage. Their effects are largely indirect and due to their influence on the amount of mixing water required for a given workability.

The size and shape of the concrete member affect concrete shrinkage through their influence on the rate of loss of moisture; shrinkage varies

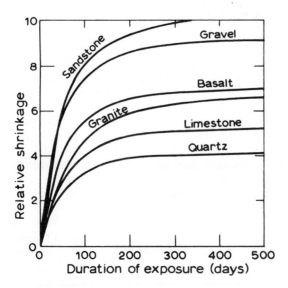

Fig. 2.5–6   Effect of type of aggregate on shrinkage [15]

inversely with the ratio of the volume to the surface area of the member. For the purposes of structural design, it may be assumed that, for a given **volume/surface ratio**, shrinkage is independent of the shape of the member. Figure 2.5–7 provides guidance on the effect of the volume/surface ratio on concrete shrinkage.

The length of the **moist-curing period** has no effect on the magnitude of the total long-term shrinkage, provided this period is not shorter than 2 or 3 days. However, a sufficiently long period of moist curing before drying enables the concrete to develop higher tensile strength before shrinkage takes place and hence reduces the incidence of cracking.

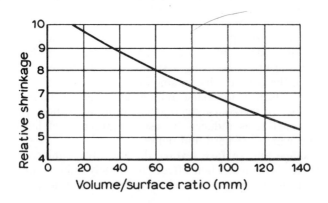

Fig. 2.5–7   Effect of volume/surface ratio on shrinkage [16]

2.5(d)   Elasticity and Poisson's ratio

Figure 2.5–8 shows a typical short-term stress/strain curve for concrete. Up to about one-third of the maximum stress, the curve is approximately straight; from that stress up to the maximum stress it is curved ascending, and beyond that it is curved descending.

The term **Young's modulus** is only applicable to the initial straight part of the curve. In fact, even the initial portion of the curve is slightly curved, and the slope of the line OA, which is the tangent to the curve at the origin, is called the **initial tangent modulus**. The slope of the tangent at an arbitrary point B is the **tangent modulus** at that point. The initial tangent modulus and the tangent modulus are not used in day-to-day structural design, where the interest lies in knowing the strain corresponding to a given stress. This information can be obtained from the **secant modulus**, which is the slope of the chord OB, and its value depends on the stress at B. As previously explained (Fig. 2.5–3), when a stress is applied to the concrete, the observed strain is made up of two parts: the elastic strain and the creep. Therefore, the secant modulus as determined from a laboratory test depends on the rate of application of the load. BS 1881 : Part 121 recommends that standard cylinder specimens should be tested and specifies that the stress should be applied at a rate of 0.6 ± 0.4 N/mm^2 per second; the secant modulus shall be that corresponding to a stress equal to one-third of the cylinder strength. These specifications may appear rather arbitrary, but they have been chosen to yield a value which is of practical use in structural design.

The secant modulus of concrete increases with the gel/space ratio and decreases with an increase in the voids (Section 2.5(a)). Broadly speaking, the various factors that affect concrete strength have similar effects on its modulus. It is therefore common practice to relate the secant modulus to the strength. For example, BS 8110 gives the relations between strength and modulus of elasticity shown in Table 2.5–6.

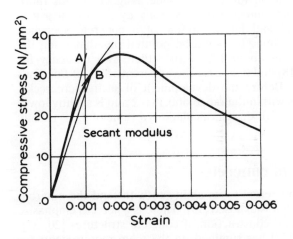

Fig. 2.5–8   Typical short-term stress/strain curve for concrete

Table 2.5–6   Relation between strength and modulus of elasticity
(BS 8110 : Part 2 : Clause 7.2)

| Compressive strength f_{cu} (N/mm²) | Static modulus E_c | | Dynamic modulus E_{cq} |
	Mean value (kN/mm²)	Typical range (kN/mm²)	
20	24	18–30	
25	25	19–31	
30	26	20–32	
40	28	22–34	See eqn (2.5–2)
50	30	24–36	
60	32	26–38	

In the table, the term **static modulus** refers to the secant modulus determined in accordance with BS 1881, as described above. The term **dynamic modulus** refers to the modulus of elasticity determined by an electrodynamic method described in BS 1881, in which the first natural mode frequency [17] of the longitudinal vibration of a standard test beam is measured. The longitudinal vibration subjects the test beam to very small stresses only; hence the dynamic modulus is often taken as being roughly equal to the initial tangent modulus and is therefore higher than the secant modulus. The dynamic modulus method of test is more convenient to carry out than the static method, and BS 8110 : Part 2 gives the following approximate formula for calculating the static (secant) modulus E_c from the dynamic modulus E_{cq}:

$$E_c(\text{kN/mm}^2) = 1.25 E_{cq}(\text{kN/mm}^2) - 19 \qquad (2.5-2)$$

Elastic analyses of concrete structures sometimes require a knowledge of the **Poisson's ratio** of concrete, although the final results of the analyses are seldom critically dependent on the precise value assigned to this ratio. Poisson's ratio may be determined experimentally by precise measurements of the longitudinal and lateral strains of a concrete specimen subjected to axial compression. It should be pointed out that when the lateral tensile strain reaches about 100×10^{-6} there is a tendency for microcracks to develop [18] and hence for the apparent value of Poisson's ratio to increase rapidly. Before the development of such microcracks, Poisson's ratio usually lies within the range of 0.1–0.2 and is slightly lower for high strength concretes. For design calculations a value of 0.2 is usually assumed.

2.5(e)   Durability of concrete

The **durability** of concrete refers to its ability to withstand the environmental conditions to which it is exposed [19]. There is a need to emphasize durability in the design and construction of concrete structures [20, 21]. The I.Struct.E. Manual [22] has summed up the main requirements for durability as:

(a) an upper limit to the w/c ratio (see Table 2.5–7);
(b) a lower limit to the cement content (see Table 2.5–7);
(c) a lower limit to the concrete cover to reinforcement (see Table 2.5–7);
(d) good compaction; and
(e) adequate curing.

The durability of concrete is intimately related to its **permeability**, which term refers to the ease with which water can pass through the concrete. A low permeability makes the concrete better able to withstand the effects of weathering, including the effects of driving rain and the disruptive action of freezing and thawing. It is stated in Section 2.5(a), in connection with the strength of concrete, that the evaporable water and the bubbles of trapped air occupy spaces called voids. The permeability of concrete (cf. its strength) increases rapidly with the amount of voids. As expected, therefore, the permeability increases rapidly with the w/c ratio and, broadly speaking, factors which increase the strength of the concrete would reduce its permeability and improve its durability: low w/c ratio, good compaction and adequate curing. Indeed, the authors' own experience with permeability is that when the absorption of water in a sample of concrete exceeds 7%, there is bound to be corrosion of the reinforcement.

Corrosion of reinforcement can seriously affect the service life of a concrete structure [20, 21]. The mechanisms of reinforcement corrosion are explained in an ACI Committee report [23], which also gives guidance on the protective measures for new concrete construction, on the procedures for identifying corrosive environments and active corrosion in concrete, and on remedial measures. Broadly speaking, the factors which reduce the permeability of the concrete (and protect the reinforcement from the ingress of external moisture) will help to inhibit reinforcement corrosion: low w/c ratio, good compaction, adequate curing and an adequate concrete cover. The chloride contents of the concrete must be held down, marine aggregate must be washed and sea-water should not be used in concrete making.

Experience has shown that too low a cement content makes it more difficult to obtain a durable concrete. Hence it is often considered desirable in practice to specify a **minimum cement content** (see Table 2.5–7). It

Table 2.5–7 Durability requirements (BS 8110 : Clause 3.3.3)

Exposure condition	*Nominal cover (mm)*			
Mild	25	20	20	20
Moderate	—	35	30	20
Severe	—	—	40	25
Very severe	—	—	50	30
Maximum free w/c ratio	0.65	0.60	0.55	0.45
Minimum cement content (kg/m^3)	275	300	325	400
Concrete f_{cu} (N/mm^2)	30	35	40	50

should also be pointed out that, while a higher cement content usually improves the quality of the concrete, this is no longer so if the cement content exceeds a certain limit. Too high a cement content increases the risk of cracking due to drying shrinkage in thin sections or to thermal stresses in thicker sections. It is therefore necessary to specify a **maximum cement content**; BS 8110: Clause 6.2.4.1 recommends that the cement content should not exceed 550 kg/m^3.

Comments
(a) In Table 2.5–7, **nominal cover** is the term used by BS 8110 to mean the design depth of cover, measured from the concrete surface to the outermost surface of ALL steel reinforcement, including links. It is the dimension used in the design and indicated on the drawings. *The nominal cover should not be less than:*
(1) the amount shown in Table 2.5–7;
(2) the nominal maximum size of the aggregate; and
(3) (for main bars) the bar size.
(b) The **exposure conditions** in Table 2.5–7 are defined in BS 8110: Clause 3.3.4.1. Broadly, the environment condition is:
(1) Mild: for concrete surfaces protected against weather, e.g. indoors.
(2) Moderate: for concrete surfaces sheltered from severe rain or freezing while wet, and for concrete surfaces in contact with non-aggressive soil.
(3) Severe: for concrete surfaces exposed to severe rain or alternating wetting and drying or occasional freezing.
(4) Very severe: for concrete surfaces exposed to sea-water spray, de-icing salts or freezing conditions while wet.

2.5(f)   Failure criteria for concrete

The concept of failure under a multiaxial stress state was introduced earlier when the difference between cube and cylinder strengths was discussed. A three-dimensional stress state, with any combination of normal and shear stresses, can always be reduced to an equivalent **triaxial stress state** of three principal stresses

$$f_1; f_2; f_3$$

as explained, for example, in Section 11.5 of Reference 17. A particular case of the general triaxial state is the **biaxial stress state**

$$f_1; f_2(f_3 = 0)$$

in which only two of the three principal stresses are non-zero. Another particular case is the **uniaxial stress state**

$$f_1(f_2 = f_3 = 0)$$

with only one non-zero principal stress. A universal **failure criterion for concrete** which allows for all possible states of triaxial stresses has not yet been found, though some proposals have been made for design purposes, e.g. by Hannant, Newman and others [24, 25]. It would seem that the failure of concrete cannot be related solely to the stress state; it is influenced also by factors such as the water/cement ratio and the moisture content of the concrete and by the method, the rate, and the sequence of applying the stresses. Fortunately, in many important applications simplified failure criteria are adequate, and examples include the following:

(a)   In estimating the ultimate strength in bending (Chapter 4) or that in combined bending and axial force (Chapter 7), the simple criterion of a limiting compressive strain gives sufficiently accurate results.
(b)   In studying the tensile cracking under combined bending and shear (Chapters 6 and 9), a simple criterion based on the principal tensile stress or the principal tensile strain gives reasonable results.

Often the actual stress state can be idealized as a biaxial state with the stress in the third principal direction equal to zero. The element in Fig. 2.5–9(a) is under biaxial stresses with the third principal stress (normal to the plane of the paper) equal to zero; similarly, the normal and shear stresses in Fig. 2.5–9(b) are equivalent to another biaxial state of principal stresses.

In the subsequent discussions on the failure criteria, we shall adopt the usual sign convention for stresses in concrete:

    compressive stresses: positive
    tensile stresses: negative

For practical purposes, the strength of concrete under **biaxial stresses** $(f_1, f_2; f_3 = 0)$ may be estimated [26] from Fig. 2.5–10 in which f_c' represents the uniaxial compressive strength, which is usually taken as equal to the cylinder strength. The figure is symmetrical about a 45° axis and it may be divided into three regions: biaxial compression (region I), biaxial tension (II) and combined compression and tension (III). Several observations can be made:

(a)   Biaxial compression (region I) increases the compressive stress at failure above the uniaxial compressive strength.
(b)   Biaxial tension (region II) has little effect on the tensile stress at failure.

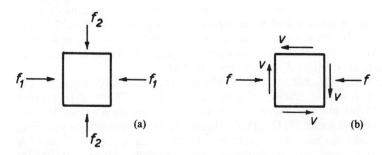

Fig. 2.5–9   Element subjected to (a) normal stresses; (b) normal and shear stresses

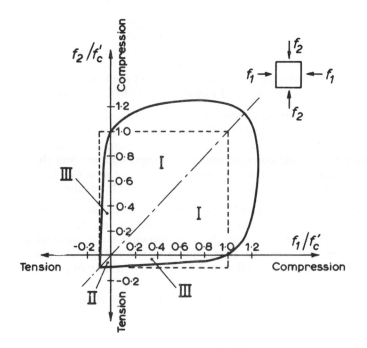

Fig. 2.5–10   Biaxial strength of concrete [26]

(c)   Combined compression and tension may appreciably reduce both the tensile and compressive stresses at failure.

Many designers neglect the interaction of f_1 and f_2 in a biaxial state, and assume for simplicity that failure occurs whenever a principal stress reaches the uniaxial compressive strength f'_c or the uniaxial tensile strength, which is often taken as $0.1f'_c$. This failure criterion, based on uniaxial strengths, is represented by the dotted square in Fig. 2.5–10: it can be unsafe for combined compression and tension (region III), although it is conservative for biaxial compression (region I)—see also Johnson and Lowe's research as described in the paragraph below eqn (2.5–5).

Using Fig. 2.5–10, a failure diagram may be derived for a plane element under **combined normal and shear stresses**; such a diagram (Fig. 2.5–11) shows that in the presence of a shear stress v, concrete will fail at a lower compressive stress f than the uniaxial compressive strength f'_c, or the uniaxial tensile strength f_t (taken here as $0.1f'_c$).

Today, in the absence of a generally acceptable failure criterion, the strength of concrete under triaxial stresses ($f_1 \geq f_2 \geq f_3$) is often investigated by making use of the following observation (after Hobbs, Pomeroy and Newman [25]): 'The influence of the intermediate principal stress f_2 on the failure of concrete has been found to be small and for practical purposes can be ignored.' Accordingly, Hobbs *et al.*[25] have made the following design recommendations for various stress combinations:

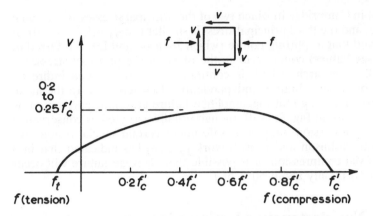

Fig. 2.5–11   Failure of concrete under combined direct and shear stresses

(a)   **Triaxial compression:** $f_1 \geq f_2 \geq f_3 > 0$
The maximum compressive stress that can be supported by concrete in a structure is given by

$$f_1 = 0.67f_{cu} + 3f_3 \qquad\qquad (2.5\text{--}3)$$

where f_{cu} is the characteristic cube strength, and should be divided by the usual partial safety factor γ_m when used in design: thus $f_1 = (0.67f_{cu}/\gamma_m) + 3f_3 = 0.45f_{cu} + 3f_3$ for $\gamma_m = 1.5$. Note that the stress $0.67f_{cu}/\gamma_m$ is identical to the peak stress in BS 8110's stress/strain curve (see Fig. 3.2–2(b)).

(b)   **Triaxial compression and tension:**

(1) $f_1 \geq f_2 \geq 0$; $f_3 < 0$ (i.e. only f_3 is tensile)
(2) $f_1 \geq f_2 \geq f_3 < 0$ (i.e. at least f_3 is tensile)

For either case (1) or case (2), the maximum compressive stress that can be supported is

$$f_1 = 0.67f_{cu} + 20f_3 \qquad\qquad (2.5\text{--}4)$$

and the maximum tensile stress that can be supported without cracks developing is given by

$$0 > f_3 > \frac{f_1 - 0.67f_{cu}}{20} \qquad\qquad (2.5\text{--}5)$$

Note that in eqns (2.5–4) and (2.5–5), the sign of f_3 is negative. Also, when the equations are used in design, the f_{cu} values should be divided by the partial safety factor γ_m, as in eqn (2.5–3).

The apparent clarity of the above summary account might well obscure the truth, stated earlier, that a universal failure criterion which allows for all possible stress states has not yet been found. Indeed, from time to time, observations are reported which may have far-reaching consequences. For example, Johnson and Lowe [27] have reported triaxial compression tests carried out in Cambridge in which two of the principal stresses are of equal magnitude, and the third principal stress is smaller (i.e. $f_1 = f_2 \geq f_3 \geq 0$). It was observed that a splitting failure (which Johnson and Lowe referred to as a **cleavage failure**) could occur in the plane of the principal stresses f_1 and f_2. Their research, which is continuing, has since been indirectly supported by results obtained independently elsewhere [28]. In the mean time it is worth noting that the simplified failure criterion represented by the dotted square in Fig. 2.5–10, though usually accepted as conservative for biaxial compression, may be unsafe under certain special conditions. For example, Johnson and Lowe's work [27, 28] has indicated that in a state of biaxial compression, it is possible that cleavage failure will occur when $f_1(= f_2)$ is only about 50% of f'_c.

2.5(g)   Non destructive testing of concrete

We have earlier referred to standard tests carried out on control specimens to determine the cube strength, the cylinder strength, the modulus of rupture, and so on. These may all be classified as destructive tests in the sense that the test specimens are destroyed in the course of the tests. There is another class of tests known as **non-destructive tests**, [29–31], which are particularly useful for assessing the quality of the concrete in the finished structure itself.

The **ultrasonic pulse velocity method** [29–31] is described in BS 4408: Part 5. It consists basically of measuring the velocity of ultrasonic pulses passing through the concrete from a transmitting transducer to a receiving transducer. The pulses are short trains of mechanical vibrations with frequencies within the range of about 10–200 kHz. They travel through concrete at velocities ranging from about 3 to 5 km/s. In general, the higher the velocity, the greater the strength of the concrete. Ultrasonic pulse velocity in concrete depends mainly on its elastic modulus and, since this is closely related to mechanical strength, the pulse velocity can be correlated to that strength. This correlation, however, is not unique but depends on the mix proportions and type of aggregate used, and also to a smaller extent on the moisture content of the concrete, its curing temperature and its age. The apparatus for this test generates ultrasonic pulses at regular intervals of time (usually from about 10 to 50 per second) and measures the time of flight between the transducers (that is the transit time) to an accuracy of better than ±1%. The distance between the transducers (that is the path length) must be measured to the same accuracy thus allowing the pulse velocity to be calculated to an accuracy of better than ±2%. Originally the apparatus incorporated a cathode-ray oscilloscope, but recently a portable form of the equipment has been developed (and marketed under the trade name *Pundit*) which indicates the transit time directly in digital form.

As mentioned above, estimation of strength from pulse velocity measurements requires a correlation curve. It has been found that such a correlation is of the form

$$\text{equivalent cube strength} = Ke^{kv} \tag{2.5–6}$$

where v is the pulse velocity, e the base of natural logarithms, and K and k constants which depend on the type of aggregate and the mix proportions. Thus the relationship may be conveniently plotted as log cube strength against pulse velocity. A similar correlation may be obtained for, say, modulus of rupture, but this would have different values of K and k. It is generally preferable to obtain the appropriate correlation curve for the particular concrete used by making ultrasonic tests on standard test specimens made with the concrete before subjecting them to destructive tests. Fig. 2.5–12 shows typical correlation curves.

The **rebound hardness test** [30] is also described in BS 4408 : Part 4. The **Schmit hammer** is the rebound instrument most widely used. It consists essentially of a metal plunger, one end of which is held against the concrete surface while the free end is struck by a spring-loaded mass which rebounds to a point on a graduated scale. This point is indicated by an index rider. The amount of rebound increases with increase in concrete strength for a particular concrete mix. The measurement taken is an arbitrary quantity referred to as the **rebound number**, and the correlation between this and the concrete strength clearly depends on the mechanical characteristics of the instrument. The test may be made on horizontal, vertical or inclined concrete surfaces, but a suitable correction must be made for the angle of inclination. The rebound test is relatively popular because of its simplicity of operation but it is generally considered to provide only an approximate indication of quality since it tests only the quality of the concrete near the surface (about 30 mm deep), and correlation between the rebound number and the concrete strength is rather scattered. BS 4408 : Part 4, gives details of the numerous factors which affect the results and the limitations within which the method is appropriate.

Of the two non-destructive tests mentioned, the ultrasonic method is the more reliable because it gives information about the state of the concrete throughout the width or depth of a structural member, whereas the rebound method indicates only the state of the concrete near the surface. Ultrasonic tests can be made on concrete from about 6 to 9 hours after

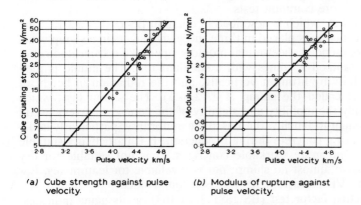

(a) Cube strength against pulse velocity.

(b) Modulus of rupture against pulse velocity.

Fig. 2.5–12   **Typical correlation curves for ultrasonic testing [30]**

placing, or at any age thereafter, but the rebound test is limited to the range of about 3 days to 3 months after placing. This later restriction applied because very green concrete could be damaged by the impact applied by the test, while the development of carbonation (which term refers to the reaction between the atmospheric carbon dioxide and the hydrated cement minerals) at the surface of the concrete after 3 months is likely to lead to considerable errors in the estimation of strength. The ultrasonic test gives a less accurate estimate for high-strength concrete (cube strength above 40 N/mm^2) than it does for medium to low-strength concrete, while the rebound test gives a less accurate estimate in the low-strength range (less than about 10 N/mm^2). Both methods are subject to certain errors due to particular effects. For example, the ultrasonic pulse velocity may be influenced by the presence of steel reinforcement running near to, and parallel to, the pulse path; the rebound test is influenced by the moisture condition of the concrete surface. Another criticism of the rebound test is that the hammer might strike a piece of aggregate, thereby giving quite misleading results.

2.6   Assessment of workability

Hitherto the term **workability** has been used qualitatively to describe the ease with which the concrete can be mixed, placed, compacted and finished. In fact, workability is rather difficult to define precisely; it is intimately related, among others, to (a) **compactibility**—that property of concrete which determines how easily it can be compacted to remove air voids, (b) **mobility**—that property which determines how easily the concrete can flow into the moulds and around the reinforcement, and (c) **stability**—that property which determines the ability of the concrete to remain a stable and coherent mass during handling and vibration. No single test has yet been devised which satisfactorily measures all the properties associated with workability. In practice, however, it is expedient to use some type of consistency measurement as an index to workability: the slump test, the compacting factor test, and the VB consistometer test are among the more common tests.

In the **slump test**, which is described in detail in BS 1881 : Part 102, the freshly made concrete is placed in three successive layers in a **slump cone**, which is an open-ended sheet-metal mould shaped like a truncated cone. Each layer is tamped 25 times with a standard rod. The mould is then removed carefully by lifting it vertically, and the reduction in height of the concrete is called the **slump** (Fig. 2.6–1). The slump test, which measures mainly the mobility and stability of the concrete, is very useful for detecting batch-to-batch variations in the uniformity (particularly in the water content) of a concrete during production. It is particularly suitable for the richer and more workable concretes. It is not suitable for dry mixes which have almost no slump; nor is it reliable for lean mixes. For such mixes the VB consistometer test is recommended.

The **compacting factor test** (BS 1881 : Part 103), as its name indicates, measures the compactibility of the concrete. The apparatus consists essentially of two truncated-cone hoppers with trap doors at the bottom

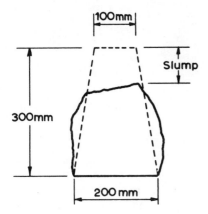

Fig. 2.6–1   Slump test

and a cylindrical container (Fig. 2.6–2). The upper hopper is first filled with a sample of the freshly made concrete, which is then dropped through the trap door into the second hopper. Again, by opening the trap door of this second hopper, the concrete is dropped into the cylindrical container below. Any surplus concrete above the top of the cylinder is then struck off flush, and the weight of the concrete in the cylinder is determined. The ratio of this weight to that of an equal volume of fully compacted concrete from the same batch is called the **compacting factor**. The latter weight is usually determined by filling the cylinder in several layers, each layer being vibrated to achieve full compaction. The compacting factor test provides a good indication of workability, and is particularly useful for the drier mixes of medium to low workabilities.

For measuring the workability of concretes which are so stiff that compaction by vibration is always necessary, the **VB consistometer test** (BS 1881:Part 104), developed by V. Bährner of Sweden, is more satisfactory than either the slump test or the compacting factor test. The VB apparatus (Fig. 2.6–3) consists essentially of a standard slump cone placed concentrically inside a cylinder of 240 mm internal diameter and 200 mm high, the cylinder being mounted on to a vibrating table. To carry out a VB test, the slump cone is filled with concrete as in a standard slump test. The slump cone is then removed, which allows the cone of concrete to subside. The swivel arm enables the transparent disc (230 mm diameter) to be swung into position over the subsided cone of concrete, as shown in Fig. 2.6–3. The screw on the swivel-arm holder is then tightened and vibration is started. The remoulding of the concrete in the cylinder is observed through the transparent disc. The **VB time** is the time in seconds between the start of virbration and the instant when the whole surface of the transparent disc is covered with cement paste, i.e. it is the time of vibration required to change the shape of the cone of concrete, left standing after lifting of the slump cone, into that of the cylinder with a level top surface. Thus, the stiffer and the less workable the concrete mix, the higher will be the VB time.

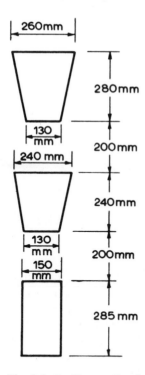

Fig. 2.6–2   Compacting factor apparatus

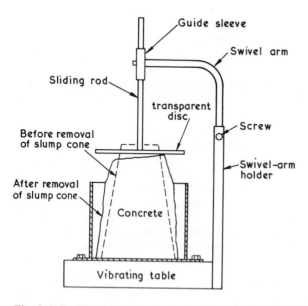

Fig. 2.6–3   The VB test

Table 2.6–1   Comparison of workability measurements

Slump *(mm)*	*VB time* *(seconds)*	*Compacting* *factor*
0	over 20	0.65–0.75
0–10	20–12	0.75–0.85
10–30	12–6	0.85–0.90
30–60	6–3	0.90–0.93
60–180	3–0	over 0.93

Table 2.6–1 gives some idea of the interrelationship of the three workability tests. The data apply to the mixes used in the comparisons of the tests and should not be assumed to be of general validity. Actually, accurate comparisons of the tests are not possible because each test measures the behaviour of concrete under a different set of conditions. The relationships between the slumps and VB times in Table 2.6–1 are reasonably reliable for a broad range of practical mixes, but the relationships between the compacting factors and slumps or VB times are much less consistent.

Table 2.6–2 defines arbitrarily four degrees of workability appropriate for the principal types of construction where high-frequency vibrators are used. The slumps given are those for concretes with aggregates having a nominal maximum size of about 10 mm. For concretes of comparable workability, the slump increases with the maximum size of the aggregates. Compacting factors are relatively little affected.

Table 2.6–2   Uses of concrete of different degrees of workability—maximum aggregate size 10 mm nominal (adapted from Reference 32)

Degree of *workability*	*Slump* *(mm)*	*Compacting* *factor*	*Type of construction*
Very low	0	0.75	Roads or other large sections
Low	0–5	0.85	Simple to normal reinforced concrete work
Medium	5–25	0.90	Normal to heavy reinforced concrete work
High	25–100	0.95	For sections with heavily congested reinforcement. Not normally suitable for vibration

2.7   Principles of concrete mix design

As stated at the beginning of this chapter, the aim in **mix design** is to select the optimum proportions of cement, water and aggregates to produce a concrete that satisfies the requirements of strength, workability, durability and economy. Mix design methods are useful as guides in the initial selection of these proportions, but it must be strongly emphasized that the final proportions to be adopted should be established by actual trials and adjustments on site.

It can be said that all practical mix design methods are based on the following two simple observations:

(a)   *The free w/c ratio is the single most important factor that influences the strength of the concrete.*
(b)   *The water content is the single most important factor that influences the workability of the fresh concrete mix.*

Note that in calculating the w/c ratio in (a) above, only the weight of the free water is used. The total water in the concrete mix consists of the water absorbed by the aggregate and the **free water**, which is the total water less the absorbed water and is available for the hydration and the lubrication of the mix. Similarly, the water content in (b) is expressed as the weight of the free water per unit volume of concrete. The water content required for a specified workability depends on the maximum size, the shape, grading and surface texture of the aggregate but is relatively independent of the cement content (i.e. the weight of cement per unit volume of concrete).

Mix design methods have been proposed by the Road Research Laboratory [6, 32], the American Concrete Institute [33, 34] and, more recently, the Department of the Environment [35]. Two of these methods are described below.

2.7(a)   Traditional mix design method

For several decades since the 1940s, mix design in the UK has been much influenced by a paper authored by Dr A. R. Collins which was published in *Concrete and Constructional Engineering* in October 1939 and which later became Road Note No. 4 [6]. According to the **Road Note No. 4 Method**, a w/c ratio is first chosen to satisfy the requirements of strength and durability; the aggregate/cement ratio is then chosen to satisfy the workability requirement. Note that specifying the w/c ratio and the aggregate/cement ratio is equivalent to specifying the w/c ratio and the water content; of the three variables, only two are independent.

Suppose it is required to design a mix to have a certain average strength at 28 days. This average strength is called the **target mean strength** and is statistically related to the required characteristic strength, as will be explained in Section 2.8. For the time being it is sufficient to note that the target mean strength will be some value which exceeds the characteristic strength by a suitable margin, called the **current margin**. Now, for properly compacted concretes, the strength depends primarily on the w/c ratio; in fact quantitative relationships are given in Fig. 2.5–2. Of course these relationships are no more than average values, and when trial mixes are actually carried out, the relationship may well be found to be somewhat different. Nevertheless, Fig. 2.5–2 provides a useful starting point. For example, with ordinary Portland cement, a w/c ratio of about 0.58 is required for a target mean strength of 30 N/mm². However, the durability requirement imposes another ceiling on the w/c ratio. For most structural concrete, it is advisable to keep the w/c ratio below, say, 0.60 irrespective of the strength requirement; where the concrete member will in service be exposed to freezing temperatures when wet, the w/c ratio should be kept below, say, 0.45.

The workability required for various types of construction is given in Table 2.6–2; Table 2.7–1 then gives the aggregate/cement ratio, which depends on the shape, grading and maximum size of the aggregate. The shapes of the aggregates referred to in Table 2.7–1 are (a) rounded, such as beach gravel, (b) irregular, such as water-worn river gravels, and (c) angular, such as crushed rocks. Corresponding to each degree of workability, the table gives aggregate/cement ratios for four gradings of aggregates which are shown in Fig. 2.7–1: grading No. 1 has the lowest proportion of fine material, while grading No. 4 has the highest proportion. The lower the proportion of fine material, the lower is the water content required for a given workability; in other words a higher aggregate/cement ratio can be used. On the other hand, where the reinforcement is congested or where it is desired to have concrete surfaces that present a good appearance on removal of the formwork, a higher proportion of 'fine material is advantageous. Note that Table 2.7–1 refers to aggregates of 10 mm maximum size. In general, as the maximum size of the aggregate increases, a higher aggregate/cement ratio may be used, and vice versa. As a very rough guide, the aggregate/cement ratios in Table 2.7–1 may be increased by about 10% if the maximum size of the

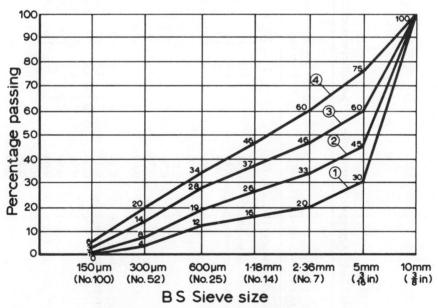

Fig. 2.7–1   Grading curves for 10 mm aggregate [32]

Table 2.7–1   Aggregate/cement ratios [32] for different workabilities—maximum aggregate size 10 mm

Degree of workability (Table 2.6–2)	Aggregate/cement ratio by weight															
	Very low				Low				Medium				High			
Grading number (Fig. 2.7–1)	1	2	3	4	1	2	3	4	1	2	3	4	1	2	3	4
w/c ratio by weight: rounded gravel aggregate																
0.40	5.6	5.0	4.2	3.2	3.9	3.9	3.3	2.6	3.9	3.5	3.0	2.4	3.5	3.2	2.8	2.3
0.45	7.2	6.4	5.3	4.1	4.7	4.9	4.1	3.2	4.7	4.3	3.7	3.0	4.2	3.9	3.4	2.9
0.50		7.8	6.4	4.9	5.4	5.8	4.9	3.8	5.4	5.0	4.3	3.5	4.8	4.5	4.0	3.4
0.55			7.5	5.7	6.1	6.7	5.7	4.4	6.1	5.7	4.9	4.0	5.3	5.1	4.5	3.9
0.60				6.5	6.7	7.5	6.4	5.0	6.7	6.3	5.5	4.5	5.8	5.6	5.0	4.3
0.65				7.2	7.3		7.1	5.6	7.3	6.9	6.1	5.0	S	6.1	5.5	4.7
0.70					7.9		7.7	6.2	7.9	7.5	6.7	5.5		6.6	6.0	5.1
0.75								6.7			7.2	5.9		7.1	6.5	5.5
0.80								7.2			7.7	6.3		7.6	6.9	5.9
w/c ratio by weight: irregular gravel aggregate																
0.40	4.1	3.8	3.3	2.8	3.3	3.1	2.8	2.3	3.5	3.4	3.2	2.8	3.2	3.1	3.0	2.7
0.45	5.1	4.8	4.3	3.6	4.1	3.9	3.5	3.0	4.2	4.1	3.8	3.4	S	3.8	3.6	3.2
0.50	6.1	5.8	5.2	4.4	4.8	4.6	4.2	3.7	S	4.7	4.4	4.0		4.4	4.2	3.7
0.55	7.0	6.7	6.1	5.2	5.5	5.3	4.9	4.3		5.3	5.0	4.5		4.9	4.7	4.2
0.60	7.9	7.6	7.0	6.0	S	6.0	5.6	4.9		5.9	5.6	5.0		5.4	5.2	4.6
0.65			7.8	6.8		6.6	6.2	5.5		6.4	6.1	5.5		5.9	5.7	5.0
0.70						7.2	6.8	6.1		6.9	6.6	6.0		6.4	6.1	5.4
0.75						7.8	7.4	6.7		7.4	7.1	6.4		6.8	6.5	5.8
0.80							8.0	7.3								

w/c ratio by weight: crushed rock aggregate																
0.40	3.7	3.3	2.8	2.0	3.8	3.6	3.0	2.2	3.3	3.1	2.7	2.1	S	3.2	2.9	2.4
0.45	4.5	4.1	3.5	2.6	4.4	4.2	3.6	2.7	3.8	3.7	3.2	2.6		3.7	3.4	2.8
0.50	5.2	4.9	4.2	3.2	4.9	4.8	4.2	3.2	S	4.2	3.7	3.0		4.2	3.8	3.2
0.55	5.9	5.6	4.9	3.8	S	5.3	4.7	3.7		4.7	4.2	3.4		4.6	4.2	3.6
0.60	6.6	6.3	5.5	4.3		5.8	5.2	4.2		5.1	4.6	3.8		5.0	4.6	4.0
0.65	7.3	7.0	6.1	4.8		6.3	5.7	4.6		5.6	5.1	4.2		5.4	5.0	4.4
0.70	7.9	7.6	6.7	5.3		6.8	6.2	5.0		6.0	5.5	4.6		5.8	5.4	4.7
0.75			7.3	5.8		7.2	6.6	5.5		6.4	5.9	5.0				
0.80			7.8	6.3												

With crushed aggregate of poorer shape than that tested, segregation may occur at a lower aggregate/cement ratio. S indicates that the mix would segregate.

aggregate is 20 mm; they should be increased by about 30% if the maximum size is 40 mm; more accurate values are given in References 6 and 32. Also, Table 2.7–1 refers only to the four gradings in Fig. 2.7–1. In practice, the aggregates available may not have natural gradings closely resembling any of these four. It is possible to combine two or more aggregates to give a grading approximating to the one required; the procedure is quite simple and is explained in References 6 and 36. Or trial mixes may be prepared with the available aggregates and adjustments made by systematic trial and error; indeed the results of the trial mixes may well show that the grading of an aggregate as supplied gives quite satisfactory results.

Example 2.7–1
Given the following data, design a mix if the target mean strength is 40 N/mm^2 at 28 days.

Cement:	ordinary Portland
Aggregate:	irregular, maximum size 10 mm, grading similar to No. 2 in Fig. 2.7–1
Type of construction:	normal reinforced concrete work using high-frequency vibrator
Condition of exposure:	exposed to climate of Great Britain

SOLUTION
From Fig. 2.5–2 the required w/c ratio is, say, 0.48. For exposure to climatic conditions in Great Britain, any w/c ratio below, say, 0.6 or 0.55 would usually be satisfactory (but see Section 2.5(e) on Durability).

From Table 2.6–2 a low workability is sufficient. Table 2.7–1 shows that, for an irregular aggregate with low workability, the aggregate/cement ratio for grading No. 2 is 3.9 for w/c ratio = 0.45 and 4.6 for w/c ratio = 0.50. By linear interpolation, that for a w/c ratio of 0.48 is

$$3.9 + \frac{0.48 - 0.45}{0.50 - 0.45} \times (4.6 - 3.9) = 4.3$$

Ans.   w/c ratio = 0.48; aggregate/cement ratio = 4.3.

2.7(b)   DoE mix design method

The Department of the Environment's *Design of Normal Concrete Mixes* [35], published in November 1975, originated from the long-established Road Note No. 4 [6] referred to in Section 2.7(a). The **DoE mix design method** [35] is intended to replace the traditional method based on Road Note No. 4, but the principal objectives remain unchanged: to obtain a preliminary estimate of the mix proportions as a basis to make trial mixes to arrive at the final mix proportions that satisfy the strength, workability and durability requirements.

Compared with the Road Note No. 4 method the DoE mix design method has several new features:

(a) The mixes may be designed either for the cube compressive strength or the indirect tensile strength, though we shall here restrict our discussions to cube strengths only.

(b) The data for workability include the slump and the VB time, but exclude the compacting factor. This is because, as explained in Section 2.6, it is difficult to establish a consistent relationship between compacting factors and slumps or VB times.

(c) Only two broad types of aggregates are considered: crushed aggregate and uncrushed aggregate. The relevant aggregate characteristics that affect the workability and strength of the concrete are the particle shape and the surface texture. The test data collected since the publication of Road Note No. 4 have shown that for mix design purpose, it is sufficient to classify aggregates into crushed and uncrushed. Thus, if an aggregate is uncrushed, then whether it happens to be rounded gravel or irregular gravel, for example, does not have the significant effect on mix design that it was previously thought to have.

(d) To conform to established American and continental European practice, the final mix properties are expressed in terms of weights of materials per unit volume of fully compacted fresh concrete—for example:

cement content	340 kg/m^3
water content	160 kg/m^3
fine aggregate content	515 kg/m^3
coarse aggregate content	1385 kg/m^3

The DoE mix design procedure is based on the data reproduced here in Tables 2.7–2 and 2.7–3 and Figs. 2.7–2 and 2.7–3; it may be summarized as follows.

Step 1   Determining the free w/c ratio
(1a) Given the required characteristic strength at a specified age, use eqn (2.8–1) to obtain the target mean strength at that age, which is the compressive strength to be used in the mix design.

Table 2.7–2   Approximate compressive strengths (N/mm^2) of concrete mixes made with a free-water/cement ratio of 0.5 (after DoE [35])

Type of cement	Type of coarse aggregate	Compressive strengths (N/mm^2) Age (days)			
		3	7	28	91
Ordinary Portland	Uncrushed	18	27	40	48
	Crushed	23	33	47	55
Rapid-hardening Portland	Uncrushed	25	34	46	53
	Crushed	30	40	53	60

Suppose the target mean strength so obtained is 43 N/mm² at 28 days.

(1b)   Given the type of cement and aggregate, use Table 2.7–2 to obtain the compressive strength, at the specified age, that corresponds to a free w/c ratio of 0.5.

Suppose ordinary Portland cement and uncrushed aggregate are used. Then Table 2.7–2 shows that the compressive strength is 40 N/mm² at 28 days (and 27 N/mm² at 7 days and so on). This pair of data (40 N/mm², w/c ratio = 0.5) will now be used to locate the appropriate strength–w/c ratio curve in Fig. 2.7–2, as explained below.

(1c)   In Fig. 2.7–2, follow the 'starting line' to locate the curve which passes through the point (40 N/mm², w/c ratio = 0.5); in this particular case, it is the fourth curve from the top of the figure. This curve shows that, to obtain our target mean strength of 43 N/mm², we need a w/c ratio of 0.47.

Note that in Fig. 2.7–2 a curve happens to pass almost exactly through our point (40 N/mm², w/c ratio = 0.5); this does not

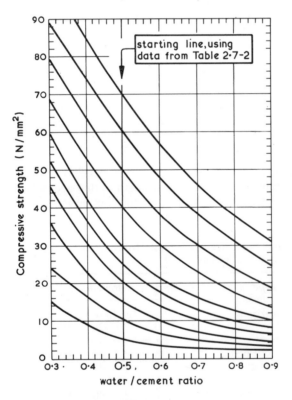

Fig. 2.7–2   Relationship between cube compressive strength and free-water/cement ratio (after DoE [35])

always happen, so that in practice it is usually necessary to interpolate between two curves in the figure.

(1d) If the w/c ratio as obtained in Step (1c) exceeds the maximum w/c ratio specified for durability (see Table 2.5–7), then adopt the lower value—resulting in a concrete having a higher strength than required.

Step 2 Determining the water content

Given the slump or VB time, determine the water content from Table 2.7–3.

In using Table 2.7–3, when coarse and fine aggregates of different types are used, the water content W is estimated as follows:

$$W = \tfrac{2}{3}W_f + \tfrac{1}{3}W_c \tag{2.7–1}$$

where W_f = water content appropriate to the type of fine aggregate;
W_c = water content appropriate to the type of coarse aggregate.

The aggregate type in Table 2.7–3 refers to all the aggregates used and not just the coarse aggregate. Indeed, the fine aggregate has a considerably greater specific surface area and hence a greater influence on the workability: this explains the greater weight, $\tfrac{2}{3}$, assigned to W_f in eqn (2.7–1).

Step 3 Determining the cement content

$$\text{Cement content (kg/m}^3) = \frac{\text{water content (from Step 2)}}{\text{w/c ratio (from Step 1)}}$$

$$\tag{2.7–2}$$

The value given by eqn (2.7–2) should be checked against any maximum or minimum cement contents that may have been specified, for durability for example; see Durability in Section 2.5(e). If the cement content calculated from eqn (2.7–2) is below a specified minimum, this

Table 2.7–3 Approximate free-water contents (kg/m³) required to give various levels of workability (after DoE [35])

Slump (mm): *VB time (seconds):*		0–10 > 12	10–30 12–6	30–60 6–3	60–180 3–0
Maximum size of aggregate (mm)	*Type of aggregate*				
10	Uncrushed	150	180	205	225
	Crushed	180	205	230	250
20	Uncrushed	135	160	180	195
	Crushed	170	190	210	225
40	Uncrushed	115	140	160	175
	Crushed	155	175	190	205

minimum must be used— resulting in a reduced w/c ratio and hence a higher strength than the target mean strength. If the calculated cement content is higher than a specified maximum, then the specified strength and workability cannot simultaneously be met with the selected materials; try changing the type of cement, the type and maximum size of the aggregate.

Step 4   Determining the aggregate content

Having calculated the water content and the cement content, the total aggregate content is in practice obtained quickly from a chart in the DoE document [35]. However, it can be easily calculated from first principles. For each cubic metre of fully compacted fresh concrete,

$$\begin{bmatrix} \text{volume occupied} \\ \text{by the aggregate} \end{bmatrix} = 1 - \frac{\text{cement content}}{\gamma_c} - \frac{\text{water content}}{\gamma_w}$$

(2.7–3)

where γ_c ($\doteqdot$ 3150 kg/m^3) is the density of the cement particles and γ_w (= 1000 kg/m^3) is that of water. Therefore

total aggregate content (kg/m^3)

= γ_a × [volume occupied by aggregates (eqn 2.7–3)]   (2.7–4)

where γ_a is the density of the aggregate particles. The DoE [35] recommends that if no information is available, γ_a should be taken as 2600 kg/m^3 for uncrushed aggregate and 2700 kg/m^3 for crushed aggregate.

Step 5   Determination of the fine and coarse aggregate contents

How much of the total aggregate content should consist of fine aggregate depends on the grading of the latter. Table 2.7–4 classifies fine aggregate into grading zones 1, 2, 3 and 4. The general principle in mix design is this: the finer the grading of the fine aggregate (i.e. the larger its surface area per unit weight) the lower will be the proportion, expressed as a percentage of the total aggregate, required to produce a concrete of otherwise similar properties.

Table 2.7–4   Grading limits for DoE mix design procedure

	Percentage by weight passing standard sieves			
Standard sieve	*Grading zone 1*	*Grading zone 2*	*Grading zone 3*	*Grading zone 4*
10 mm	100	100	100	100
5 mm	90–100	90–100	90–100	95–100
No. 7 (2.36 mm)	60–95	75–100	85–100	95–100
No. 14 (1.18 mm)	30–70	55–90	75–100	90–100
No. 25 (600 μm)	15–34	35–59	60–79	80–100
No. 52 (300 μm)	5–20	8–30	12–40	15–50
No. 100 (150 μm)	0–10	0–10	0–10	0–15

For a given slump and w/c ratio, the proportion of fine aggregate can be determined from Fig. 2.7–3 [35] in which the grading zones are those of Table 2.7–4. For example, suppose the specified slump is 10–30 mm, the w/c ratio is 0.47, and the fine aggregate is in grading zone 3, then Fig. 2.7–3 gives the proportion of fine aggregate as between 32 and 38% by weight, say 35%. Therefore, for this particular example,

fine aggregate content = 35% of total aggregate content
coarse aggregate content = (100–35)% of total aggregate content

Note that Fig. 2.7–3 is for use where the nominal maximum size of the coarse aggregate is 10 mm; the DoE document[35] contains similar design charts for 20 and 40 mm maximum sizes.

Example 2.7–2

Using the DoE mix design procedure, design a mix if the target mean strength is 43 N/mm^2 at 28 days and the required slump is 10–30 mm. The following data are given:

cement: ordinary Portland
aggregate type
  coarse: uncrushed, max. size 10 mm
  fine: (see Table 2.7–4)   uncrushed, grading zone 4
maximum w/c ratio: 0.60
maximum cement content: 550 kg/m^3 ⎫ see Table 2.5–7
Minimum cement content: 300 kg/m^3 ⎭

SOLUTION
The solution will follow the steps listed above.

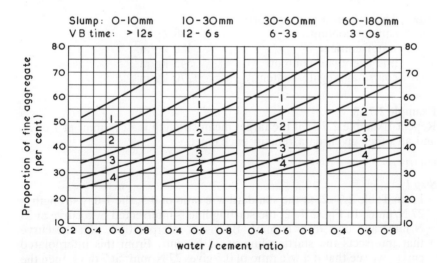

Fig. 2.7–3 Proportions of fine aggregate for grading zones 1, 2, 3, 4 (see Table 2.7–4)—for use with 10 mm nominal maximum size coarse aggregate (after DoE [35])

Step 1
   From Table 2.7–2, at the standard water/cement ratio of 0.5, the 28-day strength is 40 N/mm². Figure 2.7–2 then shows that if a w/c ratio of 0.5 gives 40 N/mm² then the w/c ratio that gives the target mean strength (43 N/mm²) is 0.47 approximately. This is inside the permitted maximum of 0.60. Therefore adopt a w/c ratio of 0.47.

Step 2
   From Table 2.7–3, for 10 mm nominal maximum size uncrushed aggregate, the water content to give a 10–30 mm slump is 180 kg/m³.

Step 3
   From eqn (2.7–2),

$$\text{cement content} = \frac{180}{0.47} = 385 \text{ kg/m}^3$$

$$> 300 \text{ kg/m}^3 \text{ and } < 550 \text{ kg/m}^3$$

Step 4
   From eqns (2.7–3) and (2.7–4), total aggregate content

$$= 2600 \left[1 - \frac{385}{3150} - \frac{180}{1000} \right] = 1815 \text{ kg/m}^3$$

Step 5
   From Fig. 2.7–3, for a slump of 10–30 mm, a w/c ratio of 0.47 and a fine aggregate in grading zone 4, the proportion of fine aggregate is 27.5 to 32% by weight, say 30%. Therefore

$$\text{fine aggregate content} = 0.30 \times 1815 = 545 \text{ kg/m}^3$$
$$\text{coarse aggregate content} = (1 - 0.30) \times 1815 = 1270 \text{ kg/m}^3$$

Ans.   The required mix proportions are:
cement content:	385 kg/m³
water content:	180 kg/m³
fine aggregate content:	545 kg/m³
coarse aggregate content:	1270 kg/m³

Example 2.7–3
Repeat Example 2.7–2 if the target mean strength is 32 N/mm² at 7 days, and all other given data remain unchanged.

SOLUTION

Step 1
   From Table 2.7–2, at the standard w/c ratio of 0.5, the 7-day strength is 27 N/mm². In Fig. 2.7–2, the starting line is intersected by a curve at 25 N/mm² and one at 30 N/mm². By visual interpolation, plot the curve that intersects the starting line at 27 N/mm². From this interpolated curve, we see that if a w/c ratio of 0.5 gives 27 N/mm² at 7 days, then the w/c ratio that gives the 7-day target mean strength (32 N/mm²) is 0.46 approximately.

Step 2
The water content is 180 kg/m^3 as in Example 2.7–2.

Step 3
The cement content is 180/0.46 = 390 kg/m^3.

Step 4

$$\text{Total aggregate content} = 2600 \left[1 - \frac{390}{3150} - \frac{180}{1000} \right] = 1810 \text{ kg/m}^3$$

Step 5
From Fig. 2.7–3, for a slump of 10–30 mm, a w/c ratio of 0.46 and a fine aggregate in grading zone 4, a suitable proportion of fine aggregate is, say, 31% by weight. Therefore

$$\text{fine aggregate content} = 0.31 \times 1810 = 560 \text{ kg/m}^3$$
$$\text{coarse aggregate content} = (1 - 0.31) \times 1810 = 1250 \text{ kg/m}^3$$

Ans. The required mix proportions are:

cement content:	390 kg/m^3
water content:	180 kg/m^3
fine aggregate content:	560 kg/m^3
coarse aggregate content:	1250 kg/m^3

Note that the answers in the above examples, and also that in Example 2.7–1, are only preliminary estimates. We began this section by saying that the final proportions should be established by trials and site adjustments. To stress this point once again, we conclude this section by stating that *the so-called mix design represents no more than an attempt to make a step in the right direction; adjustments should always be expected after experience with the actual materials and site conditions.*

2.8 Statistics and target mean strength in mix design

Analysis of numerous test results from a wide range of projects has demonstrated that the strength of concrete falls into some pattern of the normal frequency distribution curve (Fig. 1.3–3), symmetrical about the average with most of the test results falling close to the average. It is therefore possible to relate the required characteristic strength of a concrete to the **target mean strength** to be used in the mix design. Recalling (Section 1.4) that the characteristic strength is the cube strength below which not more than 5% of the test results may fall, it is immediately seen from eqn (1.4–1) that

$$\text{target mean strength} = \text{characteristic strength} + 1.64\sigma \quad (2.8\text{–}1)$$

where σ is the standard deviation of the strength tests. The quantity 1.64σ here represents the margin by which the target mean strength must exceed the required characteristic strength, and is called the **current margin**.

At the initial mix design stage the standard deviation is not accurately known and it is prudent to use a higher margin than 1.64 times the estimated standard deviation. For example, where the required characteristic strength is 20 N/mm^2 or above, a margin of about 15 N/mm^2 should be used in the initial mix design [37]. When a reasonably large number of test results becomes available, the current margin is 1.64σ. The use of small samples in statistical analysis introduces undesirable unknowns. Forty tests may be regarded as an approximate dividing line between large samples and small samples. BS 8110 does not give much guidance on this point. For characteristic strengths exceeding 20 N/mm^2, the old code CP 110 recommends that where 40 test results are available, the current margin is to be taken as 1.64σ or 7.5 N/mm^2, whichever is greater; where 100 test results are available, it may be taken as 1.64σ or 3.75 N/mm^2, whichever is greater.

From the above discussions it is clear that the target mean strength to be used in the mix design increases with the standard deviation. The poorer the quality control, the higher the standard deviation, and the higher will be the necessary target mean strength. A higher target mean strength will increase the cost of manufacture. On the other hand, to reduce the standard deviation will require better quality control and, therefore, higher cost. In practice a compromise is necessary.

Table 2.8–1 gives some idea of the standard deviations that might be expected under different conditions. The fact that Table 2.8–1 shows standard deviations rather than coefficients of variation (see eqn 1.3–6) might give the impression that the standard deviation is independent of the mean strength level. Whether this is so, or whether it is the coefficient of variation that is independent of the mean strength level, has led to some controversy. Experience indicates that, above a strength level of about 20 N/mm^2, the standard deviation seems to be fairly independent of the strength level [37]. Below this level, it is more reasonable to assume that the coefficient of variation is independent of the strength level.

The standard deviation [37] on about 60% of the sites in the UK is between 4.5 and 7.0 N/mm^2. Values much lower than 4N/mm^2 can seldom be achieved in practice, because the variability due to sampling and testing alone corresponds to a standard deviation of the order of 2.3N/mm^2; also, the variation due to the cement (of a nominally specified type) can sometimes correspond to a standard deviation of 3–3.6 N/mm^2.

Table 2.8–1   Standard deviations under different conditions

Conditions	Standard deviation (N/mm^2)
Good control with weight batching, use of graded aggregates, etc. Constant supervision	4–5
Fair control with weight batching. Use of two sizes of aggregates. Occasional supervision	5–7
Poor control. Inaccurate volume batching of all-in aggregates. No supervision	7–8 and above

Example 2.8–1
Determine the target mean strength and the current margin to be used in the mix design if the standard deviation is 6 N/mm^2 and if the characteristic strength is to be 30 N/mm^2.

SOLUTION
From eqn (2.8–1),

$$\text{target mean strength} = 30 + 1.64 \times 6 = \underline{39.8 \text{ N/mm}^2}$$
$$\text{current margin} \quad\quad = 1.64 \times 6 \quad\quad = \underline{9.8 \text{ N/mm}^2}$$

Example 2.8–2
A concrete mix is to be designed to give a 1% probability that an individual strength test result will fall below a certain specified value, f_{spec}, by more than $f \text{ N/mm}^2$. Determine the target mean strength.

SOLUTION
There is to be a 1% probability that a test result will fall below $(f_{\text{spec}} - f)$.

Referring to Fig. 1.3–4, we now want the shaded area of the tail of the normal distribution curve to be 0.01, i.e. we want the area OABC to be 0.5 − 0.01 = 0.49. From Table 1.3–3, $z = 2.33$. Therefore

$$f_{\text{spec}} - f = \text{target mean strength} - 2.33\sigma$$

That is

$$\text{target mean strength} = f_{\text{spec}} + 2.33\sigma - f \quad\quad\quad (2.8\text{–}2)$$

Example 2.3–3
A concrete mix is to be designed to give a 1% probability that the average of n consecutive test results will fall below a certain specified value f_{spec}. Determine the target mean strength.

SOLUTION
In statistical analysis [38, 39], it is known that when individual samples are taken n at a time from a normal distribution with a standard deviation σ and a mean $\bar{x}$, the average values calculated from the sets of n samples also have a mean $\bar{x}$ but a standard deviation σ_n which is equal to $\sigma/\sqrt{n}$. Therefore, if we wish to design a mix to give a 1% probability that the average of n consecutive test results will fall below the specified value f_{spec} then we know (from the arguments that lead to $z = 2.33$ in eqn 2.8–2) that the corresponding z (n) value must be 2.33. Therefore

$$\text{target mean strength} = f_{\text{spec}} + 2.33\sigma_n$$

Since $\sigma_n = \sigma/\sqrt{n}$, we have

$$\text{target mean strength} = f_{\text{spec}} + \frac{2.33}{\sqrt{n}} \sigma \qu\quad\quad (2.8\text{–}3)$$

Example 2.8–4
The ACI Building Code (ACI 318–83) [40] uses the concept of a specified strength f'_c, which is the 28-day cylinder compressive strength used in

current American design practice. The concrete mix must be designed to meet both of the following requirements:

(a) A 1% probability that an individual test result will fall below f'_c by more than 500 lbf/in² (3.44 N/mm²).
(b) A 1% probability that an average of three consecutive test results will fall below f'_c.

For each of the above two criteria, express the target mean strength in terms of the cylinder strength f'_c and the standard deviation σ. Determine the target mean strength to be used in the mix design if f'_c is 5500 lbf/in² and the standard deviation is 800 lbf/in².

SOLUTION
Criterion (a): From eqn (2.8–2),

$$\text{target mean strength} = f'_c + 2.33\sigma - 500 \qquad (2.8\text{–}4)$$

Criterion (b): From eqn (2.8–3),

$$\text{target mean strength} = f'_c + 2.33\sigma/\sqrt{3} = f'_c + 1.34\sigma \qquad (2.8\text{–}5)$$

For $f'_c = 5500$ lbf/in² and $\sigma = 800$ lbf/in², we have

target mean strength (eqn 2.8–4)

$$= 5500 + (2.33)\,(800) - 500 = 6865 \text{ lbf/in}^2$$

target mean strength (eqn 2.8–5)

$$= 5500 + (1.34)\,(800) = 6575 \text{ lbf/in}^2$$

Therefore, to satisfy both criteria, a target mean strength of 6865 lbf/in² should be used in the mix design.

Comments
Equations (2.8–4) and (2.8–5) are in fact those given in Clause 4.3.2.1 of the current American Code (ACI 318–83) [40].

Example 2.8–5
Concrete specifications sometimes refer to a mean-of-four compliance rule, which states that the average strength determined from any group of four consecutive test cubes should exceed the specified characteristic strength by not less than 0.5 times the current margin. If a concrete has been properly designed to give a specified characteristic strength, what is the probability of its failing to comply with the mean-of-four rule?

SOLUTION
According to the mean-of-four rule, the minimum value of the average of four test results must be at least that given by

$$\text{mean of four} = \text{characteristic strength} + \frac{\text{current margin}}{2}$$

$$= f_k + \frac{1.64\sigma}{2}$$

(where 1.64σ is the current margin given by eqn 2.8–1)

$$= (f_k + 1.64\sigma) - \frac{1.64\sigma}{2}$$

$$= \text{target mean strength} - \frac{1.64\sigma}{2}$$

$$= \text{target mean strength} - \frac{1.64\sigma}{\sqrt{4}}$$

As explained in the derivation of eqn (2.8–3), the probability of the mean of four tests falling below the target mean strength by $1.64\sigma/\sqrt{4}$ is the same as the probability of a single test falling below it by 1.64σ—that is, the probability is 5%. In other words, a concrete which actually has the specified characteristic strength will still risk a 5% probability of failing to comply with the rule.

Readers interested in the application of statistical concepts to mix design and quality control are urged to study the papers by Gregory and others [41–44].

2.9 Computer programs

(in collaboration with **Dr H. H. A. Wong**, University of Newcastle upon Tyne)

The FORTRAN programs for this chapter are listed in Section 12.2. See also Section 12.1 for 'Notes on the computer programs'.

References

1 Lea, F. M. *The Chemistry of Cement and Concrete*. Edward Arnold, London, 1970.
2 Neville, A. M. Properties of concrete—an overview. *Concrete International*, **8**, Feb. 1986, pp. 20–3; March 1986, pp. 60–3; April 1986, pp. 53–7.
3 BS 8110:1985. *Structural Use of Concrete— Part 1: Code of Practice for Design and Construction. Part 2: Code of Practice for Special Circumstances*. British Standards Institution, London, 1985.
4 ACI *ad hoc* Board Committee. Concrete-Year 2000. *Proc. ACI*, **68**, No. 8, Aug. 1971, pp. 581–9.
5 Neville, A. M. and Brooks, J. J. *Concrete Technology*. Longman, London, 1987.
6 DSIR. Road Research Laboratory. *Road Note No. 4: Design of Concrete Mixes*. HMSO, London, 1950.
7 Erntroy, H. C. *The Variation of Works Test Cubes*. Cement and Concrete Association, Slough, 1960.
8 ACI Committee 209. *Prediction of Creep, Shrinkage and Temperature Effects in Concrete Structures*. American Concrete Institute, Detroit, 1982.
9 Evans, R. H. and Kong, F. K. Estimation of creep of concrete in reinforced and prestressed concrete design. *Civil Engineering and Public Works Review*, **61**, No. 718, May 1966, pp. 593–6.
10 Evans, R. H. and Kong, F. K. Creep of prestressed concrete. In *Developments*

in Prestressed Concrete, edited by Sawko, F. Applied Science Publishers Ltd, London, 1978, Vol. 1, pp. 95–123.

11 Neville, A. M. *Creep of Concrete: Plain, Reinforced and Prestressed*. North-Holland Co., Amsterdam, 1970.

12 ACI Committee 209. *Shrinkage and Creep in Concrete: 1966–70* (ACI Bibliography No. 10). American Concrete Institute, Detroit, 1972.

13 CEB–FIP. *Model Code for Concrete Structures*, English Edition. Cement and Concrete Association, Slough, 1978.

14 Evans, R. H. and Kong, F. K. Estimation of shrinkage of concrete in reinforced and prestressed concrete design. *Civil Engineering and Public Works Review*, **62**, No. 730, May 1967, pp. 559–61.

15 Troxell, G. E., Raphael, J. M. and Davies, R. E. Long-time creep and shrinkage tests of plain and reinforced concrete. *Proc. ASTM*, **58**, 1958, pp. 1101–20.

16 Hansen, T. C. and Mattock, A. H. Influence of size and shape of member on the shrinkage and creep of concrete. *Proc. ACI*, **63**, No. 2, Feb. 1966, pp. 267–90.

17 Coates, R. C., Coutie, M. G. and Kong, F. K. *Structural Analysis*. 3rd edn. Van Nostrand Reinhold, Wokingham. (Tentatively scheduled for 1987/8).

18 Evans, R. H. and Kong, F. K. The extensibility and microcracking of in-situ concrete in composite prestressed beams. *The Structural Engineer*, **42**, No. 6, June 1964, pp. 181–9.

19 ACI Committee 201. *Guide to Durable Concrete*. American Concrete Institute, Detroit, 1982.

20 Somerville, G. The design life of concrete structures. *The Structural Engineer*, **64A**, No. 2, Feb. 1986, pp. 60–71.

21 Hognestad, E. Design of concrete for service life. *Concrete International*, **8**, No. 6, June 1986, pp. 63–7.

22 I.Struct.E/ICE Joint Committee. *Manual for the Design of Reinforced Concrete Building Structures*. Institution of Structural Engineers, London, 1985.

23 ACI Committee 222. *Corrosion of Metals in Concrete*. American Concrete Institute, Detroit, 1985.

24 Hannant, D. J. Nomograms for the failure of plain concrete subjected to short-term multiaxial stresses. *The Structural Engineer*, **52**, May 1974, pp. 151–65.

25 Hobbs, D. W., Pomeroy, C. D. and Newman, J. B. Design stresses for concrete structures subject to multiaxial stresses. *The Structural Engineer*, **55**, No. 4, April 1977, pp. 151–65.

26 Kupfer, H., Hilsdorf, H. K. and Rusch, H. Behaviour of concrete under biaxial stresses. *Proc. ACI*, **66**, No. 8, Aug. 1969, pp. 656–66.

27 Johnson, R. P. and Lowe, P. G. Behaviour of concrete under biaxial and triaxial stresses. In *Structures, Solid Mechanics and Engineering Design: Proceedings of the Southampton 1969 Civil Engineering Materials Conference* edited by Te'eni, M. Wiley-Interscience, 1971, Part 2, pp. 1039–51.

28 Lowe, P. G. Discussion of 'The Mohr envelope of failure for concrete: a study of its tension–compression part'. *Magazine of Concrete Research*, **27**, No. 91, June 1975, pp. 121–2.

29 *In situ/Non-destructive Testing of Concrete* (ACI SP–82). American Concrete Institute, Detroit, 1984.

30 Elvery, R. H. Estimating strength of concrete in structures. *Concrete*, **7**, No. 11, Nov. 1973, pp. 49–51.

31 Chung, H. W. An appraisal of the ultrasonic pulse technique for detecting voids in concrete. *Concrete*, **12**, No. 11, Nov. 1978, pp. 25–8.

32 Shacklock, B. W. *Concrete Constituents and Mix Proportions.* Cement and Concrete Association, Slough, 1974.

33 ACI Committee 211. *Standard Practice for Selecting Proportions for Normal, Heavyweight and Mass Concrete.* American Concrete Institute, Detroit, 1984.

34 ACI Committee 211. *Standard Practice for Selecting Proportions for Structural Lightweight Concrete.* American Concrete Institute, Detroit, 1981.

35 Teychenné, D. C., Franklin, R. E. and Erntroy, H. C. *Design of Normal Concrete Mixes.* Department of the Environment, HMSO, London, 1975.

36 *The Vibration of Concrete.* Institution of Civil Engineers and Institution of Structural Engineers, London, 1956.

37 Teychenné, D. C. The variability of the strength of concrete and its treatment in codes of practice. *Structural Concrete,* **3**, No. 1, Jan./Feb. 1966, pp. 33–47.

38 Neville, A. M. and Kennedy, J. B. *Basic Statistical Methods for Engineers and Scientists.* 3rd edn. International Textbook Co., Scranton, 1986.

39 Kreyszig, E. *Advanced Engineering Mathematics.* 5th edn. Wiley, New York, 1983.

40 ACI Committee 318. *Building Code Requirements for Reinforced Concrete* (ACI 318–83). American Concrete Institute, 1983.

41 Gregory, M. S. Strength specification of concrete. *Commonwealth Engineer,* June 1958, pp. 57–60; July 1958, pp. 61–4; Aug. 1958, pp. 76–7.

42 Gregory, M. S. Sequential analysis applied to quality control of concrete. *Commonwealth Engineer,* April 1958, pp. 51–4.

43 Chung, H. W. How good is good enough—a dilemma in acceptance testing of concrete. *Proc. ACI,* **75**, No. 8, Aug. 1978, pp. 374–80.

44 Philleo, R. E. Concrete production, quality control, and evaluation in service. In *Handbook of Structural Concrete,* edited by Kong, F. K., Evans, R. H., Cohen, E. and Roll, F. Pitman, London and McGraw-Hill, New York, 1983, Chapter 27.

Chapter 3
Axially loaded reinforced concrete columns

Preliminary note: Readers interested only in structural design to BS 8110 may concentrate on the following sections:

(a) *Section 3.2: Stress/strain characteristics.*
(b) *Section 3.4: Design to BS 8110.*
(c) *Section 3.5: Design details (BS 8110).*
(d) *Section 3.6: Design and standard method of detailing—examples.*

3.1 Introduction

Structural concrete members may be subjected to axial load, flexure, shear or torsion, or a combination of these; they may be prestressed or non-prestressed. The principles underlying their analysis and design are basically the same. However, the authors' experience is that a teaching text may with advantage begin with something simple—the axially loaded column; this provides a good opportunity to introduce the concepts of elastic and ultimate-strength behaviour and to demonstrate the important effects of shrinkage and creep. To avoid unnecessary distractions at this stage, the discussion will further be restricted to short columns in this chapter. The distinction between a short column and a slender column will be explained in Chapter 7; it is sufficient for the time being to define a **short column** as one in which the length is not more than about 15 times the minimum lateral dimension.

3.2 Stress/strain characteristics of steel and concrete

The behaviour of reinforced concrete columns is intimately related to the stress/strain characteristics of the reinforcement bars and the concrete. Figure 3.2–1(a) shows typical stress/strain curves for the reinforcement; these may be considered applicable for both tension and compression. Both mild steel bars and hot-rolled high yield bars have definite yield points. For mild steel, the plastic range (that is, the horizontal plateau of the curve) may extend up to a strain of about 0.015; for hot-rolled high yield steel, the plastic range extends up to about 0.005 strain when the curve rises again as a result of strain hardening. Cold-worked high yield bars do not have a definite yield point; for practical purposes, BS 4461 defines the yield stress of such bars as the stress at 0.43% strain (point P in Fig. 3.2–1 (a)).

For design purpose, BS 8110 idealizes the stress/strain curves for reinforcement to that shown in Fig. 3.2–1(b), which applies to both tension and compression. The partial safety factor γ_m for the ultimate limit state is taken as 1.15 (see Table 1.5–2). Therefore the **design strengths**, in tension and compression, are

$$\frac{f_y}{1.15} = 0.87f_y \tag{3.2–1}$$

The **design yield strains** are the strains at $0.87f_y$ and are hence calculated as $0.87f_y/E_s = 0.002$ for $f_y = 460$ N/mm^2 and 0.0011 for $f_y = 250$ N/mm^2. (*Note*: $E_s = 200$ kN/mm^2 from Fig. 3.2–1(b).)

The stress/strain characteristics of concrete were briefly referred to in Section 2.5(d). The exact shape of the stress/strain curve is much dependent on the concrete strength. Figure 3.2–2(a) shows typical curves

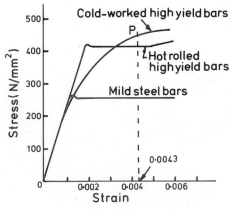

(a) Actual stress/strain curves

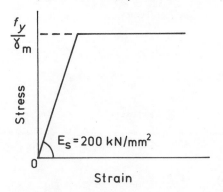

(b) Design stress/strain curves (BS 8110)

Fig. 3.2–1   Stress/strain curves for reinforcement

for short-term loading. Within the range of concrete mixes used in practical design, the following general statements may be made:

(a)   Up to about 50% of the maximum stress, the stress/strain curve may be approximated by a straight line.
(b)   The peak stress is reached at a strain of about 0.002.
(c)   Visible cracking and disintegration of the concrete does not occur until the strain reaches about 0.0035.

For design purposes, BS 8110 uses the idealized curve in Fig. 3.2–2(b),

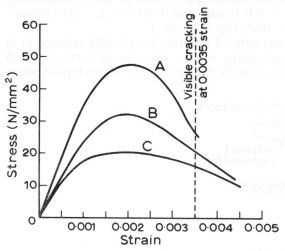

(a) Actual stress/strain curves

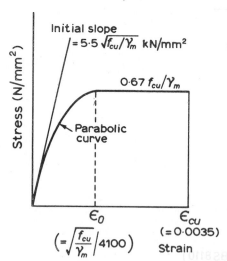

(b) Design stress/strain curves (BS 8110)

Fig. 3.2–2   **Stress/strain curves for concrete in compression**

with a maximum stress of $0.67 f_{cu}/\gamma_m$; concrete is assumed to fail at an **ultimate strain**, ε_{cu}, of 0.0035.

3.3   Real behaviour of columns

Figure 3.3–1 shows schematically a reinforced concrete column subjected to an axial load N. The four longitudinal bars are enclosed by lateral ties, or links as they are often called, the function of which will become clear later on. We shall consider the response of the column as the load N is progressively increased to the ultimate value.

Elastic behaviour
When the stresses in the concrete and the reinforcement are sufficiently low, the stress/strain relations may be considered linear (Figs 3.2–1(a) and 3.2–2(a)); therefore the usual elastic theory applies. From the **condition of equilibrium**,

$$N = f_c A_c + f_s A_{sc} \qquad (3.3\text{--}1)$$

where N = applied axial load;
$\quad f_c$ = compressive stress in the concrete;
$\quad f_s$ = compressive stress in the longitudinal reinforcement;
$\quad A_c$ = cross-sectional area of the concrete;
$\quad A_{sc}$ = cross-sectional area of the longitudinal reinforcement.

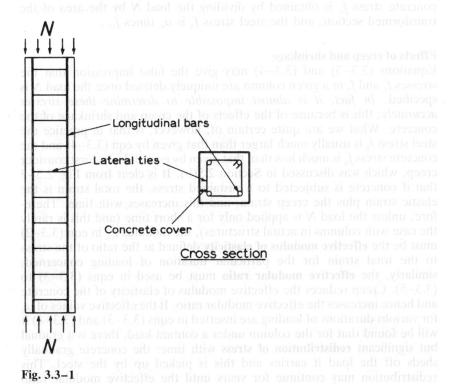

Longitudinal bars

Lateral ties

Concrete cover

Cross section

Fig. 3.3–1

Because of the bond between the reinforcement and the concrete, they will have equal strains under load (bond will be discussed in Chapter 6). Therefore, from the **condition of compatibility**,

$$f_c/E_c = f_s/E_s$$
(concrete strain = steel strain)   (3.3–2)

where E_c and E_s are respectively the modulus of elasticity of the concrete (see Table 2.5–6) and the steel; the **modulus of elasticity of steel** is usually taken as 200 kN/mm^2 in design.

From eqns (3.3–1) and (3.3–2),

$$f_c = \frac{N}{A_c + \alpha_e A_{sc}}$$ (3.3–3)

$$f_s = \frac{\alpha_e N}{A_c + \alpha_e A_{sc}}$$ (3.3–4)

i.e.

$$f_s = \alpha_e f_c$$ (3.3–5)

where $\alpha_e = E_s/E_c$ is called the **modular ratio**.

In eqns (3.3–3) and (3.3–4), the quantity $A_c + \alpha_e A_{sc}$ represents the concrete area plus α_e times the steel area, and is often referred to as the area of the **transformed section** or **equivalent section**. Therefore, when the stresses are within the linear ranges in Figs 3.2–1(a) and 3.2–2(a), the concrete stress f_c is obtained by dividing the load N by the area of the transformed section, and the steel stress f_s is α_e times f_c.

Effects of creep and shrinkage
Equations (3.3–3) and (3.3–4) may give the false impression that the stresses f_c and f_s in a given column are uniquely defined once the load N is specified. *In fact, it is almost impossible to determine these stresses accurately*; this is because of the effects of the creep and shrinkage of the concrete. What we are quite certain of, however, is that in practice the steel stress f_s is usually much larger than that given by eqn (3.3–4) and the concrete stress f_c is much less than that given by eqn (3.3–3). First consider creep, which was discussed in Section 2.5(b). It is clear from Fig. 2.5–3 that if concrete is subjected to a sustained stress, the total strain is the elastic strain plus the creep strain, and this increases with time. Therefore, unless the load N is applied only for a short time (and this is rarely the case with columns in actual structures), the modulus E_c in eqn (3.3–2) must be the **effective modulus of elasticity** defined as the ratio of the stress to the total strain for the particular duration of loading concerned; similarly, the **effective modular ratio** must be used in eqns (3.3–3) to (3.3–5). Creep reduces the effective modulus of elasticity of the concrete and hence increases the effective modular ratio. If the effective values of α_e for various durations of loading are inserted in eqns (3.3–3) and (3.3–4) it will be found that for the column under a contant load, there is a gradual but significant **redistribution of stress** with time: the concrete gradually sheds off the load it carries and this is picked up by the steel. This redistribution may continue for years until the effective modular ratio

settles down to an approximately steady value, which unfortunately cannot be determined with precision (see Table 2.5–2). For practical concrete mixes, the effective modular ratio for long-term loading may be two to three times the short-term modular ratio; BS 8110: Clause 2.5.2 recommends that the effective modular ratio should be taken as 15.

If the sustained load on a column is removed, there is an immediate elastic recovery (see Fig. 2.5–3), and **residual stresses** are set up. The reinforcement ends up in compression and the concrete in tension, as demonstrated in Example 3.3–1 below. The residual tensile stress in the concrete may sometimes be high enough to cause cracking.

The effect of shrinkage of concrete (see Section 2.5(c)) causes further redistribution of stresses. A plain concrete column undergoing a (hypothetical) uniform shrinkage will experience no stresses; but in a reinforced concrete column, the reinforcement bars resist the shrinkage and set up tensile stresses in the concrete and compressive stresses in the steel itself (See Example 3.3–2).

From the above discussion it is clear that the actual stresses in a column may be quite different from the values predicted by eqns (3.3–3) and (3.3–4). The difficulty of determining the creep and shrinkage with precision means that the actual stresses in a column cannot be determined precisely. If the column is subjected to a history of load applications and removals, the residual stresses cconstitute further complications.

Example 3.3–1
A 400 × 200 mm rectangular concrete column is reinforced with six bars of size 25 (*Note:* **bar size** refers to the nominal diameter of the bar in millimetres; see 'detailing notation' in Example 3.6–3.) An axial load of 1000 kN is applied and sustained for a long time. For the level of concrete stresses in question, the short-term modular ratio is 7.5 and the effective modular ratio for the long-term loading is 15. Determine:

(a) the concrete and steel stresses immediately upon application of the load;
(b) the long-term stresses; and
(c) the residual stresses when the sustained load is removed.

SOLUTION

$$A_{sc} = \text{area of six size 25 bars} = 2945 \text{ mm}^2 \text{ (see Table A2–1)}$$

$$A_c = 400 \times 200 \text{ mm}^2 - 2945 \text{ mm}^2 = 77000 \text{ mm}^2$$

(*Note:* In practical design, A_c is usually taken as the nominal concrete area, 400 × 200 mm², without allowing for the area taken up by the reinforcement.)

$$\alpha_e(\text{short term}) = 7.5; \qquad \alpha'_e(\text{long term}) = 15$$

$$A_c + \alpha_e A_{sc} = 77000 + 7.5 \times 2945$$

$$= 99100 \text{ mm}^2$$

$$A_c + \alpha'_e A_{sc} = 77000 + 15 \times 2945$$

$$= 121000 \text{ mm}^2$$

(a) *Short-term stresses.* From eqns (3.3–3) and (3.3–4),

$$f_c = \frac{1000 \times 10^3}{99\,100} = \underline{10.1\ \text{N/mm}^2}\ (\text{compression})$$

$$f_s = 7.5 \times 10.1 = \underline{75.8\ \text{N/mm}^2}\ (\text{compression})$$

(b) *Long-term stresses*

$$f_c = \frac{1000 \times 10^3}{121\,000} = \underline{8.26\ \text{N/mm}^2}\ (\text{compression})$$

$$f_s = 15 \times 8.3 = \underline{124\ \text{N/mm}^2}\ (\text{compression})$$

(*Note:* The above results show that during the period of sustained loading the concrete stress is reduced from 10.1 to 8.26 N/mm^2 while the steel stress is increased from 75.8 to 124 N/mm^2.)

(c) *Residual stresses.* The immediate reductions in stresses upon load removal are given by eqns (3.3–3) and (3.3–4) using $\alpha_e = 7.5$. These are, from (a) above, 10.1 and 75.8 N/mm^2. Therefore

$$\text{residual concrete stress} = 8.26\ \text{N/mm}^2 - 10.1\ \text{N/mm}^2$$
$$= \underline{1.84\ \text{N/mm}^2}\ (\text{tension})$$

$$\text{residual steel stress} = 124\ \text{N/mm}^2 - 75.8\ \text{N/mm}^2$$
$$= \underline{48.2\ \text{N/mm}^2}\ (\text{compression})$$

(As an exercise, the reader should verify that the residual tensile force in the concrete is equal to the residual compressive force in the steel.)

Example 3.3–2
Derive expressions for the stresses in the concrete and the reinforcement of a column due to a shrinkage strain ε_{cs}.

SOLUTION
Use the usual notation, but let f_c be the tensile stress in the concrete and f_s the compressive stress in the steel. From the conditions of equilibrium and compatibility, we have, respectively:

$$f_c A_c = f_s A_{sc} \tag{3.3–6}$$

$$\varepsilon_{cs} - \frac{f_c}{E_c} = \frac{f_s}{E_s} \tag{3.3–7}$$

Solving, the shrinkage stresses are

$$f_c = \varepsilon_{cs} E_s \frac{A_{sc}}{A_c + \alpha_e A_{sc}}\ (\text{tension}) \tag{3.3–8}$$

$$f_s = \varepsilon_{cs} E_s \frac{A_c}{A_c + \alpha_e A_{sc}}\ (\text{compression}) \tag{3.3–9}$$

Ultimate strength behaviour
Returning to Fig. 3.3–1, if the load is increased until failure of the column occurs, it will be found that the maximum value of the load N (which is the

ultimate strength of the column) is practically independent of the load history or of any creep and shrinkage effects. The ultimate limit state of collapse is reached at a load given very nearly by

$$N = 0.67f_{cu}A_c + f_yA_{sc} \tag{3.3-10}$$

where f_{cu} is the cube strength of the concrete and f_y the yield strength of the reinforcement; for cold-worked high yield bars (which do not have a distinct yield point), BS 4461 defines their yield stress as the stress at 0.43% strain.

The coefficient 0.67 in eqn (3.3–10) is due to a number of possible factors:

(a) columns have a much larger height/width ratio than standard cubes and hence the effect of end restraints is almost insignificant compared with that in a cube test;
(b) the apparent concrete strength increases with the rate of loading [1, 2], and, since the rate of loading in a column test is much slower than that in a cube test, the apparent strength is reduced;
(c) the compaction of the concrete in a reinforced concrete column is likely to be less complete than in a cube.

In terms of the cylinder strength f'_c, eqn (3.3–10) becomes

$$N = 0.85f'_cA_c + f_yA_{sc} \tag{3.3-11}$$

Figure 3.3–2(a) shows a typical mode of failure, with the reinforcement buckling after the ultimate strength of the column is reached. It is important, therefore, that the ties or links should be sufficiently closely spaced to prevent premature buckling; also, if the size of the links is inadequate, premature buckling of the type in Fig. 3.3–2(b) may occur. Design rules governing the provision of links are given in Section 3.5.

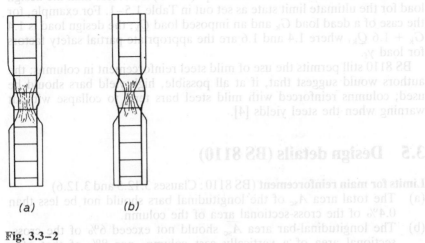

(a)   (b)

Fig. 3.3–2

3.4   Design of axially loaded short columns (BS 8110)

The design of reinforced concrete columns is based on the empirical equation (3.3–10). This equation, and indeed the whole range of our knowledge of column behaviour, are based on the many tests [3–7] carried out during the past half a century or so. For design, eqn (3.3–10) is modified by the introduction of partial safety factors γ_m, which for the ultimate limit state of collapse are 1.5 and 1.15 respectively for concrete and reinforcement

$$N = 0.67 \frac{f_{cu}}{\gamma_m} A_c + \frac{f_y}{\gamma_m} A_{sc} \qquad (3.4–1)$$

For the ultimate limit state, therefore, the equation becomes

$$N = 0.45 f_{cu} A_c + 0.87 f_y A_{sc}$$

To allow for eccentricity of loading due to construction tolerances, BS 8110 further limits the ultimate axial load to about 90% of this, so that in design

$$N = 0.4 f_{cu} A_c + 0.75\, f_y A_{sc} \qquad \textbf{(3.4–2)}$$

where f_{cu} = the characteristic strength of the concrete;
   f_y = the characteristic strength of the reinforcement;
   A_c = the area of concrete (in design it is usual to take A_c as the **nominal area** without deduction for the area of the reinforcement); and
   A_{sc} = the area of longitudinal reinforcement.

In limit state design a structural member is, as explained in Section 1.2, usually designed for the ultimate limit state and checked for the serviceability limit states of cracking and excessive deflection. For the particular case of short **braced columns**, that is columns which are restrained in position at both ends, it is generally not necessary to check serviceability in design. It is therefore only necessary to proportion the member so that the value of N from eqn (3.4–2) is not less than the design load for the ultimate limit state as set out in Table 1.5–1. For example, for the case of a dead load G_k and an imposed load Q_k, the design load is 1.4 G_k + 1.6 Q_k, where 1.4 and 1.6 are the appropriate partial safety factors for load γ_f.

BS 8110 still permits the use of mild steel reinforcement in columns; the authors would suggest that, if at all possible, high yield bars should be used; columns reinforced with mild steel bars tend to collapse without warning when the steel yields [4].

3.5   Design details (BS 8110)

Limits for main reinforcement (BS 8110: Clauses 3.12.5 and 3.12.6)
(a)   The total area A_{sc} of the longitudinal bars should not be less than 0.4% of the cross-sectional area of the column.
(b)   The longitudinal-bar area A_{sc} should not exceed 6% of the cross-sectional area of a vertically cast column, nor 8% of that of a

horizontally cast column, except that at laps of reinforcement bars (in both types of columns) the limit may be increased to 10%.

Comments
(a) Tests have shown that where the **steel ratio** A_{sc}/A_c is too low, eqn (3.3–10) and hence eqn (3.4–2) are not applicable. Also, reinforcement is required to resist bending moments which may exist irrespective of whether the design calculations show that they exist. We have seen in Section 3.3 that shrinkage and creep cause a redistribution of load from the concrete to the reinforcement. Unless a lower limit is placed on the steel ratio, the steel stress may reach the yield level (see Example 3.3–2) even when under service load. Also, the 0.4% lower limit helps to protect columns in structural frames against failure in tension when, for example, the surrounding floors near the column are unloaded above but heavily loaded below, or when the structural frame is subjected to unequal foundation settlements.
(b) The upper limits on the steel ratios are to avoid congestion and hence unsatisfactory compaction of the concrete. Indeed, the I.Struct.E. Manual [8] recommends an upper limit of only 4%.
(c) Designers normally consider that a minimum of four longitudinal bars should be used in a rectangular column and six bars in a circular column. The I.Struct.E. Manual [8] further recommends that longitudinal bars should not be smaller than size 12 and their spacing should not exceed 250 mm.

Lateral ties or links (BS 8110 : Clause 3.12.7)
(a) All longitudinal bars should be enclosed by **links** (sometimes called ties or stirrups), which should be so arranged that every corner and alternate bar shall have lateral support provided by the corner of a link having an included angle of not more than 135°. No bar shall be further than 150 mm from a bar restrained by a link.
(b) For circular columns, where the longitudinal reinforcement is located round the periphery of a circle, adequate lateral support is provided by a circular link passing round the bars.
(c) Links should have a minimum diameter of at least one-quarter of that of the largest longitudinal bar, and the maximum link spacing should not exceed 12 times the diameter of the smallest longitudinal bar. (The I.Struct.E. Manual [8] further recommends that the link spacing should not exceed the smallest cross-sectional dimension of the column.)

Comments
The purpose of links is primarily to prevent the outward buckling of the longitudinal bars, as illustrated in Fig. 3.3–2. The diameter and spacing of the links are therefore related to the diameter of the longitudinal bars. The minimum size is one-quarter of that of the largest longitudinal bar, but the minimum size of any reinforcement bar in British practice is size 6. For practical reasons, most designers regard size 8 bars as the minimum for use

as column links; smaller-size links often fail to hold the main longitudinal bars securely, and may themselves be pushed out of shape in the concreting process.

Links also provide lateral restraint to the concrete, resulting in an enhanced axial-load capacity (see 'Failure criteria for concrete' in Section 2.5 (f)).

Concrete cover for durability (BS 8110 : Clause 3.3)

Table 2.5–7 in Section 2.5(e) gives the concrete covers to meet the durability requirements for the columns and other structural members.

Note that in Table 2.5–7 the meaning of **nominal cover** is as defined in BS 8110 : Clause 3.3.1.1, namely: the nominal cover is the design depth of concrete to all steel reinforcement, including links. It is the dimension used in design and indicated on the drawings.

Fire resistance (BS 8110 : Clause 3.3.6)

The fire resistance of a column depends on its minimum dimension and the concrete cover, as shown in Table 3.5–1. Note that the fire resistance requirements may in practice dictate the size of the column and the concrete cover.

Table 3.5–1 Fire resistance requirements for columns
(BS 8110 : Part 2 : Clause 4.3.1)

Fire rating (hours)	Minimum dimension (mm)		Concrete cover to MAIN reinforcement (mm)
	Fully exposed	*50% exposed*	
1	200	160	25
2	300	200	35
3	400	300	35
4	450	350	35

3.6 Design and detailing—illustrative examples

Example 3.6–1

Calculate the ultimate axial load of a 300 mm square column section having six size 20 bars, if $f_{cu} = 40$ N/mm^2 and $f_y = 460$ N/mm^2.

SOLUTION

From eqn (3.4–2):

$$N = 0.4 f_{cu} A_c + 0.75 f_y A_{sc}$$
$$= (0.4)\,(40)\,(300^2 - 1885) + (0.75)\,(460)\,(1885) \text{ N}$$
$$(\text{where } A_{sc} = 1885 \text{ mm}^2 \text{ from Table A2–1})$$
$$= \underline{2060 \text{ kN}}$$

In practice, A_c is usually taken for simplicity as the nominal cross-sectional area without deduction for the area of the steel. In this case,

$$N = (0.4) (40) (300^2) + (0.75) (460) (1885) \text{ N}$$
$$= \underline{2090 \text{ kN}}$$

Example 3.6–2
Design a short, braced reinforced concrete column for an ultimate axial load of 2000 kN. Given: $f_{cu} = 40$ N/mm^2, $f_y = 460$ N/mm^2.

SOLUTION
From eqn (3.4–2),

$$N = 0.4f_{cu}A_c + 0.75f_yA_{sc}$$

Assume a 2% steel ratio, i.e. $A_{sc} = 0.02A_c$, say,

$$(2000) (10^3) = (0.4) (40)A_c + (0.75) (460) (0.02A_c)$$

$$A_c = 87340 \text{ mm}^2, \text{ say } 300 \text{ mm square}$$

If a 300 mm square section satisfies the fire-resistance and architectural requirements, it is only necessary to calculate A_{sc} from eqn (3.4–2).

$$(2000) (10^3) = (0.4) (40) (300^2) + (0.75) (460)A_{sc}$$

$$A_{sc} = 1623 \text{ mm}^2$$

The area of six size 20 bars is 1885 mm^2 from Table A2–1. Therefore adopt the following values:

column section	300 mm square
main bars	6—size 20
links	size 8 at 200 mm centres
nominal cover	30 mm to links (see Tables 2.5–7 and 3.5–1)

Figure 3.6–1 shows the reinforcement arrangement for a typical length of column between floors. (See Comment (b) below on standard method of detailing. See also Fig. 3.6–2.)

Comments
(a) The reinforcement details in Fig. 3.6–1 satisfy BS 8110's requirements as listed in Section 3.5.

Main steel ratio = $\dfrac{1885}{300^2}$ = 2.1% > 0.4% and < 6%

Link size	8 mm > ¼ of 20 = 5 mm
Link spacing	200 mm < 12 times 20 = 240 mm
Nominal cover	30 mm to links, which is suitable for moderate exposure (Table 2.5–7) and adequate for a 4-hour fire resistance (Table 3.5–1)

(b) Figure 3.6–1 serves the purpose of showing the beginner a fairly clear picture of how the reinforcement bars are arranged. However, the way in which the reinforcement has been drawn and described is not suitable for use in detail drawings for actual construction purpose.

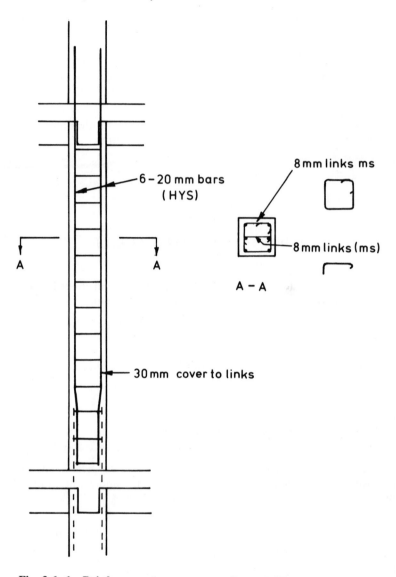

6 − 20 mm bars
(HYS)

30 mm cover to links

8 mm links ms

8 mm links (ms)

A − A

Fig. 3.6−1   Reinforcement arrangement: Example 3.6−2
(see also standard detailing in Fig. 3.6−2)

Example 3.6−3 explains the standard method of detailing reinforced
concrete [9] in current British practice.

Example 3.6−3
Revise the reinforcement drawings in Fig. 3.6−1 to conform to the British
standard method of detailing structural concrete [9].

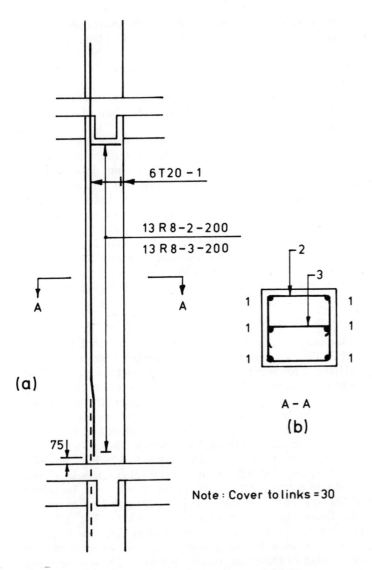

(a)

6 T 20 – 1

13 R 8–2–200
13 R 8–3–200

A ← A →

75

A – A

(b)

Note: Cover to links = 30

Fig. 3.6–2   Fig. 3.6–1 redrawn to conform to standard method of detailing [9]

SOLUTION
The revised drawing is shown in Fig. 3.6–2. Note that the cross-section has for clarity been drawn to a larger scale than the elevation; this conforms to current practice. [9].

Comments
(a)   According to the standard method of detailing [9], only one bar of each type is shown in full in **column** elevations, **slab** plans and **wall**

elevations, the remaining bars being indicated by short lines. Thus in Fig. 3.6–2(a), only one of the six size 20 longitudinal bars is drawn in full.

(*Note:* In **beams**, all longitudinal bars are shown in full on the elevation.)

(b)   Only one link or set of links in each column (or beam) is drawn in full; a short line is used to indicate the first or the last link of a group, as shown in Fig. 3.6–2(a).

(c)   The positions of bars are established by dimensioning to the faces of the existing concrete or the formwork, thus enabling the steel fixer to work from these faces. See, for example, the 75 mm dimension that establishes the position of the longitudinal bars in Fig. 3.6–2(a).

(d)   All bars that need to be fixed in a certain part before it can be concreted must be detailed with that part; these bars are then shown in broken lines in succeeding portions. Thus, in Fig. 3.6–2(a), the broken lines represent the longitudinal bars that are separately detailed with the column length underneath the lower floor level.

(e)   In Fig. 3.6–2(a), current British **detailing notation** [9] has been used to describe the reinforcement. Briefly, the sequence of description is as follows:

Number, Type, Size, Mark, Centres, Location

Consider, for example, the longitudinal bars; in the label '6T20-1' in Fig. 3.6–2(a), the first figure denotes the number of bars, the letter the **type** of bar—T for high yield deformed bars and R for plain round bars—the figure after the letter denotes the **bar size** (i.e. the nominal diameter of the bar in millimetres) and the number after the hyphen is the identification **bar mark**. Thus 6T20-1 represents six high yield deformed bars of size 20 mm, the bars being identified by the bar mark 1; in this example, bar mark 1 refers to a cranked bar bent to the dimensions A, B, C, D and radius r specified in Fig. 3.6–3(a).

Similarly, the label 13R8-2-200 in Fig. 3.6–2(a) refers to 13 mild steel bars of size 8 mm identified by the bar mark 2, the bars being spaced at 200 mm centres. In this example, bar mark 2 refers to a link bent to the dimensions A, B and radius r specified in Fig. 3.6–3(b); bar mark 3 refers to a link bent as specified in Fig. 3.6–3(c).

(f)   The full notation of a bar or group of bars is given once only, preferably in plan or elevation, though it may be given in section in exceptional cases. Thus in Fig. 3.6–2, the full notations are all given in the elevation in Fig. (a); the section in Fig. (b) indicates the position and bar mark of every bar included in that section.

(g)   In Fig. 3.6–3 the bending details are specified in accordance with BS 4466 [10]—see Figs A2–1, A2–2 and A2–3.

(h)   So far we have concentrated our attention on how to detail structural concrete, so as to communicate effectively with the men on the construction site. In fact, the performance of the structure itself is significantly affected by the reinforcement details. Readers interested in the effect of detailing on structural behaviour are referred to the *Handbook of Structural Concrete* [11].

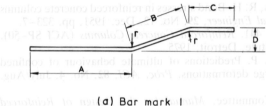

(a) Bar mark 1

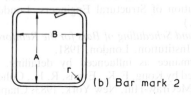

(b) Bar mark 2

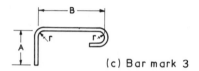

(c) Bar mark 3

Fig. 3.6–3   Bending dimensions for reinforcement bars of Fig. 3.6–2
(see also Appendix 2: Figs. A2–1, A2–2, and A2–3)

3.7   Computer programs

(in collaboration with **Dr H. H. A. Wong**, University of Newcastle upon Tyne)

The FORTRAN programs for this chapter are listed in Section 12.3. See also Section 12.1 for 'Notes on the computer programs'.

References

1   Evans, R. H. Effect of rate of loading on the mechanical properties of some materials. *Journal ICE*, **18**, June 1942, p. 296.

2   Evans, R. H. Effect of rate of loading on some mechanical properties of concrete. *Proceedings* of a Conference on Mechanical Properties of Non-Metallic Brittle Materials. Butterworths Scientific Publications, London, 1958, pp. 175–92.

3   ACI–ASCE Committee 441. *Reinforced Concrete Columns* (ACI Bibliography No. 5). American Concrete Institute, 1965, 122pp.

4   Evans, R. H. and Lawson, K. T. Ultimate strength of axially loaded columns reinforced with square twisted steel and mild steel. *The Structural Engineer*, **33**, No. 11, Nov. 1955, pp. 335–43.

5   Hawkes, J. M. and Evans, R. H. Bond stresses in reinforced concrete columns and beams. *The Structural Engineer*, **29**, No. 12, Dec. 1951, pp. 323–7.
6   ACI–ASCE Committee 441. *Reinforced Concrete Columns* (ACI SP–50). American Concrete Institute, Detroit, 1975.
7   Fafitis, A. and Shah, S. P. Predictions of ultimate behaviour of confined columns subjected to large deformations. *Proc. ACI*, **82**, No. 4, July/Aug. 1985, pp. 423–33.
8   I.Struct.E./ICE Joint Committee. *Manual for the Design of Reinforced Concrete Building Structures*. Institution of Structural Engineers, London, 1985.
9   Concrete Society and I.Struct.E. Joint Committee. *Standard Method of Detailing Structural Concrete*. Institution of Structural Engineers, London. (Scheduled for publication in 1987/8.)
10 BS 4466:1981. *Bending Dimensions and Scheduling of Bars for the Reinforcement of Concrete*. British Standards Institution, London, 1981.
11 Taylor, H. P. J. Structural performance as influenced by detailing. In *Handbook of Structural Concrete* edited by Kong, F. K., Evans, R. H., Cohen, E. and Roll, F. Pitman, London and McGraw-Hill, New York, 1983, Chapter 13.

Chapter 4
Reinforced concrete beams—
the ultimate limit state

Preliminary note: Readers interested only in structural design to BS 8110 may concentrate on the following sections:

(a) *Section 4.4(c) and (d): BS 8110 stress blocks.*
(b) *Section 4.5: Use of BS 8110 design charts.*
(c) *Section 4.6: Use of BS 8110 simplified stress block.*
(d) *Section 4.8: Flanged beams.*
(e) *Section 4.10: Design details.*
(f) *Section 4.11: Design example.*

4.1 Introduction

It was pointed out in Section 3.3 that the actual stresses in a reinforced concrete column may bear little resemblance to the values calculated on the basis of the elastic theory. In reinforced concrete beams, in addition to the effects of shrinkage and creep and of loading history, there are the uncertain effects of the cracking of the concrete in the tension zone; as in columns, conventional calculations for the stresses in reinforced concrete beams do not give a clear indication of their potential strengths. Therefore, during the past several decades there has been a gradual move in design from elastic stress calculations to ultimate strength methods [1, 2]. For example, ultimate strength design for beams was introduced into both the American and British design codes in the 1950s, and the limit state design procedures in current British practice make specific requirements for ultimate strength calculations.

Ultimate strength design is basically a return to forgotten fundamentals. Though pronounced interest in the ultimate strength of structural members dates back only 40 or 50 years, its beginnings may be traced further back than the concepts of elasticity. In Europe the origin of systematic thought regarding ultimate flexural strength of beams was due to G. Galilei [3]. His work, devoted exclusively to ultimate strength, was published as early as 1638, 40 years before Robert Hooke made the statement 'ut tensio sic vis' [4], which is now known as Hooke's law and which enabled Navier to develop the fundamental theorems of the theory of elasticity some one and a half centuries later.

4.2 A general theory for ultimate flexural strengths

As a result of the extensive research work in the past several decades [1, 5], the ultimate load behaviour of reinforced concrete beams is now quite well understood. Current design methods in American and British codes are based on the general theory described below. The following assumptions are made:

(a) The strains in the concrete and the reinforcing steel are directly proportional to the distances from the neutral axis, at which the strain is zero.
(b) The ultimate limit state of collapse is reached when the concrete strain at the extreme compression fibre reaches a specified value ε_{cu}.
(c) At failure, the distribution of concrete compressive stresses is defined by an idealized stress/strain curve.
(d) The tensile strength of the concrete is ignored.
(e) The stresses in the reinforcement are derived from the appropriate stress/strain curve.

Figure 4.2–1(a) shows a beam cross-section having an area A_s of longitudinal **tension reinforcement** and an area A'_s of longitudinal **compression reinforcement**; the distance d from the top face to the centroid of the tension reinforcement is called the **effective depth** of the beam. Figure 4.2–1(b), in which x denotes the **neutral axis depth**, shows the strains distributed in accordance with assumption (a). The assumption of linear strain distribution, which implies that plane sections remain plane, is not exactly correct but is justifiable for practical purposes. A critical review of the research on this subject is given in Reference 6.

From assumption (b), the maximum concrete compressive strain has a specified value ε_{cu} at the instant of collapse. Therefore, the concrete strains

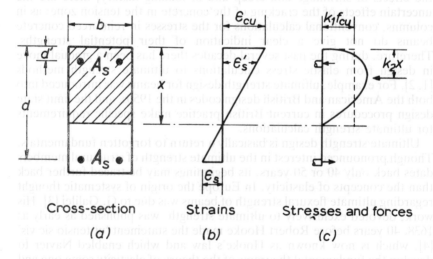

Cross-section	Strains	Stresses and forces
(a)	(b)	(c)

Fig. 4.2–1   Strain and stress distributions at failure

at distance d and d' from the top of the beam can be obtained immediately from the geometry of Fig. 4.2–1(b). The relationship between the strain in a reinforcement bar and that in the adjacent concrete depends on the bond (see Chapter 6) between the concrete and the steel, but it is accurate enough [1, 2] to assume here that they are equal. Therefore, the strains ε_s in the tension reinforcement and ε_s' in the compression reinforcement are given by the condition of compatibility as

$$\varepsilon_s = \frac{d - x}{x}\varepsilon_{cu} \tag{4.2–1}$$

$$\varepsilon_s' = \frac{x - d'}{x}\varepsilon_{cu} \tag{4.2–2}$$

Assumption (c) refers to an idealized stress distribution for the concrete in compression, i.e. for the concrete above the neutral axis (shaded portion in Fig. 4.2–1(a)). The stress distribution diagram (Fig. 4.2–1(c)) is generally referred to as the **stress block**. A comparison of Fig. 4.2–1(c) with Fig. 3.2–2(a) shows that the stress block in ultimate flexural strength analysis is the stress/strain curve drawn with a horizontal axis for stress and a vertical axis for strain. Since the beginning of this century, a large number of ultimate strength theories have been proposed, but essentially they differed only in the shape assumed for the stress block [5]. Hence if the characteristics of the stress block are expressed in general terms, ultimate strength equations can then be derived from the principles of mechanics. The two relevant **characteristics of the stress block** (further discussed in Section 4.4) are the ratio k_1 of the average compressive stress to the characteristic concrete strength f_{cu}, and the ratio k_2 of the depth of the centroid of the stress block to the neutral axis depth. The forces on the beam section can be expressed in terms of these characteristics:

concrete compression	$= k_1 f_{cu} bx$
concrete tension	$=$ ignored (assumption (d))
reinforcement compression	$= A_s' f_s'$
reinforcement tension	$= A_s f_s$

where the steel tensile stress f_s and the steel compressive stress f_s' are related to the strains ε_s and ε_s' by the respective stress/strain curves for the reinforcement. From the condition of equilibrium,

$$k_1 f_{cu} bx = A_s f_s - A_s' f_s' \tag{4.2–3}$$

In eqns (4.2–1) to (4.2–3), the neutral axis depth x is in effect the only unknown. For an arbitrary value of x, the steel strains ε_s and ε_s' are given by eqns (4.2–1) and (4.2–2), and the corresponding stresses f_s and f_s' by the stress/strain curves. However, such a set of (x, f_s, f_s') values will not in general satisfy eqn (4.2–3). In practice, a trial and error procedure is usually adopted: a value of x is assumed, the steel strains (and hence stresses) are then determined. If eqn (4.2–3) is not satisfied an adjustment is made to x by inspection, and the procedure repeated (several times) until eqn (4.2–3) is sufficiently closely satisfied. The **ultimate flexural strength** M_u (often called the **ultimate moment of resistance**) of the beam is then obtained by taking moments about a convenient horizontal axis. For example, taking moments about the level of the tension reinforcement gives

$$M_u = (k_1 f_{cu} bx)(d - k_2 x) + A_s' f_s'(d - d') \tag{4.2-4}$$

Or, by taking moments about the centroid of the concrete stress block,

$$M_u = A_s f_s(d - k_2 x) + A_s' f_s'(k_2 x - d') \tag{4.2-5}$$

The M_u values from these equations are of course the same.

In Fig. 4.2–1(a) a rectangular section is shown, but the above theory is of general validity, being equally applicable to the arbitrary cross-section (provided it is symmetrical about a vertical axis).

Special case: A_s only

In the particular case of a **singly reinforced beam**, i.e. a beam with no compression reinforcement, a graphical solution may conveniently be used. Equation (4.2–3) now becomes

$$k_1 f_{cu} bx = A_s f_s$$

or

$$f_s = \frac{k_1 f_{cu} b}{A_s} x \tag{4.2-6}$$

From eqn (4.2–1),

$$x = \frac{\varepsilon_{cu}}{\varepsilon_{cu} + \varepsilon_s} d \tag{4.2-7}$$

Combining eqns (4.2–6) and (4.2–7),

$$f_s = \frac{k_1 f_{cu}}{\varrho} \frac{\varepsilon_{cu}}{\varepsilon_{cu} + \varepsilon_s} \tag{4.2-8}$$

where ϱ is the **steel ratio** A_s/bd.

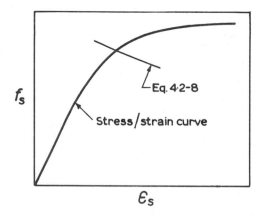

Fig. 4.2–2   **Graphical solution for f_s and ε_s at failure**

In the beam at the ultimate limit state of collapse the values of f_s and ε_s must satisfy eqn (4.2–8); they must also satisfy the stress/strain curve for the steel. Therefore the required value of f_s can be determined graphically by solving eqn (4.2–8) simultaneously with the stress/strain curve, as illustrated in Fig. 4.2–2. The ultimate moment of resistance is then

$$M_u = A_s f_s (d - k_2 x)$$

$$= A_s f_s \left(1 - \varrho \frac{k_2 f_s}{k_1 f_{cu}} \right) d \tag{4.2–9}$$

from eqn (4.2–6) after simplification.

In eqn (4.2–9), the quantity $[1 - \varrho k_2 f_s / k_1 f_{cu}]d$ is the **lever arm** for the ultimate resistance moment. The ratio of the lever arm to the effective depth d is sometimes referred to as the **lever arm factor**. Similarly, the ratio x/d is sometimes referred to as the **neutral axis factor**.

4.3   Beams with reinforcement having a definite yield point

Figure 4.3–1(a) shows the cross-section of a beam with reinforcement bars, such as mild steel or hot-rolled high yield steel, which have a definite yield point f_y. If the steel ratio $\varrho \ (= A_s/bd)$ is below a certain value to be defined later, it will be found that as the bending moment is increased the steel strain ε_s reaches the yield value ε_y while the concrete strain ε_c is still below the ultimate value ε_{cu} (Fig. 4.3–1(b)). Such a beam is said to be **under-reinforced**; in an under-reinforced beam, the steel yields before the concrete crushes in compression. Since the concrete does not crush (and hence the beam does not collapse) until the extreme compression fibre strain reaches ε_{cu}, the beam will continue to resist the increasing applied moment; this it does by an upward movement of the neutral axis, resulting in a somewhat increased lever arm while the total compression force in the concrete remains unchanged. At collapse, the strain distribution is as in Fig. 4.3–1(c); since the steel has a definite yield point, the steel stress is equal to the yield stress. The ultimate resistance moment of an under-reinforced section is therefore given by eqn (4.2–9) with f_s replaced by f_y:

$$M_u = A_s f_y \left(1 - \varrho \frac{k_2 f_y}{k_1 f_{cu}} \right) d \tag{4.3–1}$$

The failure of an under-reinforced beam is characterized by large steel strains, and hence by extensive cracking of the concrete and by substantial deflection. The ductility of such a beam provides ample warning of impending failure; for this reason, and for economy, designers usually aim at under-reinforcement.

If the steel ratio ϱ is above a certain value, the concrete strain will reach the ultimate value ε_{cu} (and hence the beam will fail) before the steel strain reaches the yield value ε_y, and the strain distribution at collapse is as shown in Fig. 4.3–2. Such a section is said to be **over-reinforced**. From eqn (4.2–6)

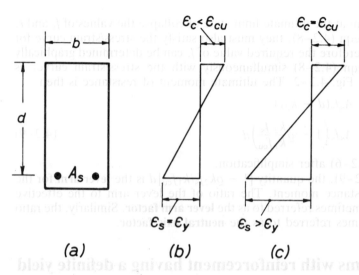

(a)   (b)   (c)

Fig. 4.3–1

$$f_s = \frac{x}{d} \frac{k_1 f_{cu}}{\varrho}$$

where $\varrho = A_s/bd$. Since the steel stress is below the yield point, $f_s = E_s \varepsilon_s$ and the above equation can be written as

$$\varepsilon_s = \frac{x}{d} \frac{k_1 f_{cu}}{\varrho E_s} \tag{4.3–2}$$

From eqn (4.2–7),

$$\frac{x}{d} = \frac{\varepsilon_{cu}}{\varepsilon_{cu} + \varepsilon_s}$$

Fig. 4.3–2

Combining with eqn (4.3–2),,

$$\left(\frac{x}{d}\right) = \frac{\varepsilon_{cu}}{\varepsilon_{cu} + \left(\frac{x}{d}\right)\dfrac{k_1 f_{cu}}{\varrho E_s}}$$

or

$$\frac{k_1 f_{cu}}{\varrho E_s}\left(\frac{x}{d}\right)^2 + \varepsilon_{cu}\left(\frac{x}{d}\right) - \varepsilon_{cu} = 0 \qquad (4.3\text{–}3)$$

which may be solved for the neutral axis factor x/d (and hence for x) since all the other quantities are known. The ultimate moment of resistance of the over-reinforced beam may then be obtained by taking moments about the tension reinforcement:

$$M_u = k_1 f_{cu} bx(d - k_2 x) \qquad (4.3\text{–}4)$$

The failure of an over-reinforced beam is initiated by the crushing of the concrete, while the steel strain is still relatively low. The failure is therefore characterized by a small deflection and by the absence of extensive cracking in the tension zone. The failure, often explosive, occurs with little warning.

A section is said to be **balanced** if the concrete strain reaches ε_{cu} simultaneously as the steel strain reaches ε_y; that is, if the strain distribution at collapse is as shown in Fig. 4.3–3. The neutral axis depth factor x/d of a balanced section has a unique value, which is in fact given by eqn (4.2–7) with ε_s replaced by ε_y:

$$\frac{x}{d} = \frac{\varepsilon_{cu}}{\varepsilon_{cu} + \varepsilon_y} \quad \text{(for balanced section)} \qquad (4.3\text{–}5)$$

The steel ratio also has a unique value, given by eqn (4.2–8) with f_s replaced by f_y and ε_s by ε_y:

$$\varrho = k_1 \frac{f_{cu}}{f_y} \frac{\varepsilon_{cu}}{\varepsilon_{cu} + \varepsilon_y} \quad \text{(for balanced section)} \qquad (4.3\text{–}6)$$

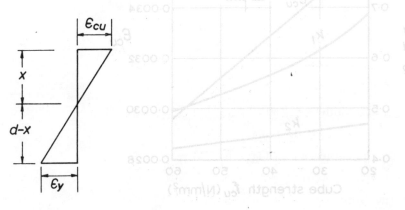

Fig. 4.3–3

The ultimate moment of resistance of a balanced section may be obtained either from eqn (4.3–1) in which ϱ now satisfies eqn (4.3–6), or from eqn (4.3–4) in which x now satisfies eqn (4.3–5).

As explained above, *the immediate cause of failure of all three types of beams is the crushing of the concrete when the compressive strain reaches the ultimate value* ε_{cu}. However, in an under-reinforced beam, the failure is initiated by the large strain increase in the tension reinforcement at yield. For this reason, the failure of an under-reinforced beam is sometimes referred to as a **primary tension failure**; that of an over-reinforced beam is referred to as a **primary compression failure**.

4.4   Characteristics of some proposed stress blocks

(a)   Hognestad *et al*

In the general flexural theory in Section 4.2 and in the more restricted theory in Section 4.3, the properties of the concrete stress block have been expressed in terms of the characteristic ratios k_1 and k_2. Much research has been carried out to study the characteristics of the stress block [1, 2, 5, 7–9]. In particular, the tests by Hognestad *et al.* [7, 8] had a considerable influence on American and, indirectly, British design thinking. Their results are summarized in Fig. 4.4–1, in which the concrete cube strengths f_{cu} have been obtained from the cylinder strengths f'_c using a conversion factor of 0.8. The figure shows that the ultimate strain ε_{cu} varies with the concrete strength; however, current American and British design codes

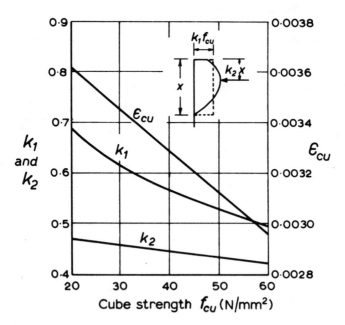

Fig. 4.4–1   Characteristics of Hognestad *et al.*'s stress block

assume for simplicity that ε_{cu} has a definite value irrespective of the concrete strength.

(b)   Whitney's equivalent rectangular block

Both BS 8110 and the ACI Building Code [10] make use of the concept of an equivalent rectangular stress block, which was pioneered by Whitney [11]. Whitney found that if the actual stress block was replaced by a fictitious rectangular block of intensity 0.85 times the cylinder strength f_c' and of such a depth x_w that the area of $0.85f_c'x_w$ was equal to that of the actual block, then the centroids of the two blocks were very nearly at the same level (Fig. 4.4–2). The depth x_w of Whitney's block is not directly related to the neutral axis depth x; x is determined by the strain distribution (eqns 4.2–1 and 4.2–7), but x_w is to be determined from the condition of equilibrium. For a rectangular beam of width b having tension reinforcement A_s,

$$0.85f_c'bx_w = A_sf_s$$

where f_s is the steel stress at collapse. For a primary tension failure, f_s is equal to the yield stress f_y; therefore

$$x_w = \frac{A_sf_y}{0.85f_c'b} = \frac{f_y}{0.85f_c'}\varrho d \tag{4.4–1}$$

where ϱ is the steel ratio A_s/bd. The ultimate moment of resistance may be obtained by taking moments about the centroid of the stress block or about that of the tension reinforcement:

$$M_u = A_sf_y\left(d - \frac{x_w}{2}\right) \tag{4.4–2(a)}$$

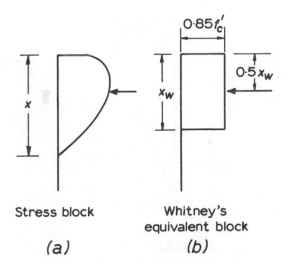

Stress block

(a)

Whitney's
equivalent block

(b)

Fig. 4.4–2   **Whitney's equivalent rectangular stress block**

or $M_u = 0.85 f_c' b x_w \left(d - \dfrac{x_w}{2} \right)$ (4.4–2(b))

If the reinforcement does not have a definite yield point Whitney suggested that, in eqns (4.4–1) and (4.4–2), f_y may be taken as the stress corresponding to a strain of 0.004.

Equation (4.4–1) shows that x_w increases with ϱ. However, when ϱ exceeds the balanced value (eqn 4.3–6) a primary compression failure occurs and eqn (4.4–2(a)) is not applicable. Whitney has proposed that, where x_w as given by eqn (4.4–1) exceeds $0.536d$ (which is his experimentally determined value for a balanced failure), then for design purposes x_w should be taken as $0.536d$ and M_u calculated from eqn (4.4–2(b)).

Whitney's method is rarely applied directly to present-day design; it is included here because of its historical significance.

(c) BS 8110 stress block

Figure 4.4–3(a) shows the idealized stress block adopted in BS 8110 for ultimate strength calculations in design; it is derived from the stress/strain curve in Fig. 3.2–2(b) with the concrete partial safety factor γ_m taken as 1.5 (see Table 1.5–2). BS 8110 further assumes that the ultimate concrete strain is constant at $\varepsilon_{cu} = 0.0035$, and that the parabolic part of the stress block ends at a strain $\varepsilon_0 = \sqrt{f_{cu}}/5000$.

As an exercise, the reader should verify that eqns (4.4–3) and (4.4–4) below are correct (*Hint*: useful properties of the parabola are given in textbooks on structural mechanics, e.g. Reference 12; see also Problem 4.1.)

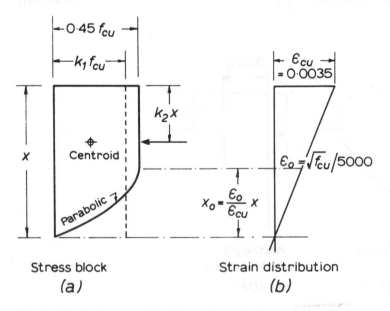

Fig. 4.4–3 **Design stress block for ultimate limit state—BS 8110**

area of stress block, $k_1 f_{cu} x = 0.45[1 - \sqrt{f_{cu}/52.5}]f_{cu}x$

therefore

$$k_1 = 0.45[1 - \sqrt{f_{cu}/52.5}] \qquad (4.4\text{-}3)$$

Taking moments about the top of the block,

$$(\text{area of block}) \cdot k_2 x = 0.45 f_{cu} x \left(\frac{x}{2}\right) - \frac{0.45 f_{cu} x_0}{3}\left(x - \frac{x_0}{4}\right)$$

where

$$x_0 = \frac{\varepsilon_0}{\varepsilon_{cu}} x \quad (\text{from Fig. 4.4-3(b)})$$

therefore

$$k_2 = \frac{\left[2 - \dfrac{\sqrt{f_{cu}}}{17.5}\right]^2 + 2}{4\left[3 - \dfrac{\sqrt{f_{cu}}}{17.5}\right]} \qquad (4.4\text{-}4)$$

Graphs of k_1 and k_2 values are plotted in Fig. 4.4-4.

(d)   BS 8110 simplified rectangular stress block
As an alternative to the stress block in Fig. 4.4-3, BS 8110 states that the moment of resistance may be determined from a simplified **rectangular stress block** of intensity $0.45f_{cu}$ extending from the compression face to a depth of $0.9x$ (Fig. 4.4-5). Note that the stress intensity of $0.45f_{cu}$ *already includes an allowance for the partial safety factor* γ_m. The characteristic ratios k_1 and k_2 were defined earlier for a general stress block in Section 4.2. The ratio k_1 is defined by concrete compression $= k_1 f_{cu} bx$.

Hence, for the BS 8110 rectangular block,

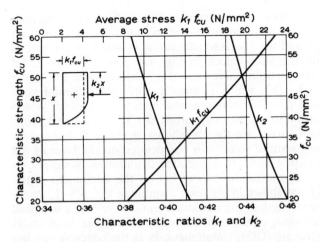

Fig. 4.4-4   Characteristics of BS 8110 stress block—ultimate limit state [13]

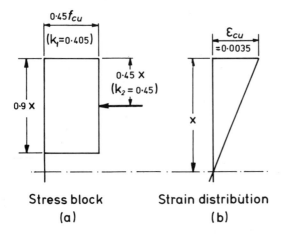

Stress block (a)   Strain distribution (b)

Fig. 4.4–5   Simplified design stress block for ultimate limit state—BS 8110

$$k_1 f_{cu} bx = (0.45 f_{cu})(b)(0.9x)$$
$$= 0.405 f_{cu} bx$$

i.e.

$$k_1 = 0.405$$

Clearly

$$k_2 = 0.45$$

4.5   BS 8110 design charts—their construction and use

The general theory in Section 4.2 and the equations in Section 4.3 become directly applicable to British design practice if the ultimate concrete strain ε_{cu} is taken as 0.0035 and the characteristic ratios k_1 and k_2 are taken as those associated with the stress block in Fig. 4.4–3 or Fig. 4.4–5. Of course, the design stress/strain curves for the reinforcement are to be derived from Fig. 3.2–1(b), which is equally applicable to mild steel (characteristic strength $f_y = 250$ N/mm^2), and high yield steel ($f_y = 460$ N/mm^2). With the partial safety factor γ_m taken as 1.15 (see Table 1.5–2), the design stress/strain curves are then as shown in Fig. 4.5–1.

In Section 4.3, the balanced steel ratio was defined on the basis of the yielding of the steel (eqn 4.3–6). In design, the definition is based not on the actual yield stress of the reinforcement but on the attainment of the design strength $0.87 f_y$, as shown in the inset diagram in Fig. 4.5–1, where f_y is the characteristic strength of the reinforcement. Therefore, a **balanced section** is defined for design purposes as one in which the steel stress reaches the design strength $0.87 f_y$ simultaneously as the concrete reaches the strain 0.0035. The **under-reinforced section** and the **over-reinforced**

section are likewise defined on the basis of the 0.0035 concrete strain and the $0.87f_y$ steel stress.

The steel ratio $\varrho (= A_s/bd)$ for the balanced section is, from eqn (4.3–6),

$$\varrho(\text{balanced}) = \frac{k_1 f_{cu}}{0.87 f_y} \frac{0.0035}{0.0035 + \varepsilon_y} \qquad (4.5\text{–}1)$$

where ε_y is the design yield strain $0.87f_y/E_s$. Suppose, for example, $f_{cu} = 40$ N/mm^2 and $f_y = 460$ N/mm^2; from Fig. 4.4–4, $k_1 f_{cu} = (0.396)(40) = 15.84$ N/mm^2; from Fig. 4.5–1, $0.87f_y = 400$ N/mm^2, $\varepsilon_y = 0.002$. Therefore

$$\varrho(\text{balanced}) = \left(\frac{15.84}{400}\right)\left(\frac{0.0035}{0.0035 + 0.0020}\right) = 0.0252$$

For ϱ below the balanced value the beam is under-reinforced, and the ultimate moment of resistance is given by eqn (4.3–1):

$$M_u = A_s(0.87f_y)\left[1 - \varrho\frac{k_2(0.87f_y)}{k_1 f_{cu}}\right]d$$

i.e.

$$\frac{M_u}{bd_2} = 0.87f_y\varrho\left[1 - \frac{0.87f_y k_2}{k_1 f_{cu}}\varrho\right] \qquad (4.5\text{–}2)$$

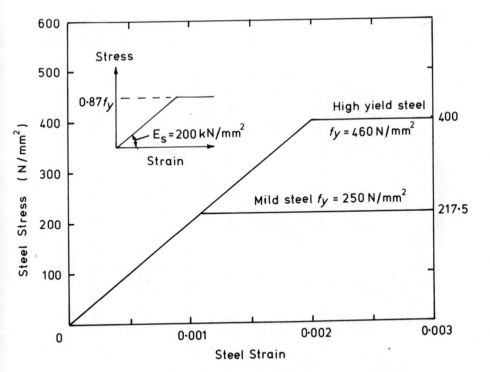

Fig. 4.5–1   **Design stress/strain curves for ultimate limit state—BS 8110**

where k_1 and k_2 values are given in Fig. 4.4–4.

Thus M_u/bd^2 values may be computed for various values of the steel ratio ϱ, up to the balanced value given by eqn (4.5–1). For the particular case of $f_{cu} = 40$ N/mm^2 and $f_y = 460$ N/mm^2, we saw above that ϱ(balanced) $= 0.0252$. As an exercise, the reader should show that the graph of M_u/bd^2 against ϱ is the first portion of the bottom curve of the **beam design chart** (Fig. 4.5–2); that is, up to the kink at $A_s/bd = 2.52\%$.

For ϱ exceeding the balanced value, the beam is over-reinforced, and eqn (4.3–4) applies:

$$M_u = k_1 f_{cu} bx(d - k_2 x)$$

i.e.

$$\frac{M_u}{bd^2} = k_1 f_{cu}\left(\frac{x}{d}\right)\left[1 - k_2\left(\frac{x}{d}\right)\right] \qquad (4.5-3)$$

where k_1 and k_2 values are given in Fig. 4.4–4, and the neutral axis factor x/d is given by eqn (4.3–3):

$$\frac{k_1 f_{cu}}{\varrho E_s}\left(\frac{x}{d}\right)^2 + 0.0035\left(\frac{x}{d}\right) - 0.0035 = 0 \qquad (4.5-4)$$

in which E_s is 200 kN/mm^2, since ϱ now exceeds the balanced value.

As an exercise, the reader should choose some values of ϱ between 2.52% and 3.50%, and then use eqns (4.5–3) and (4.5–4) to complete the verification of the design curve (the bottom one, for $\varrho' = 0$) in Fig. 4.5–2.

For a **doubly reinforced beam**, i.e. a beam having both tension

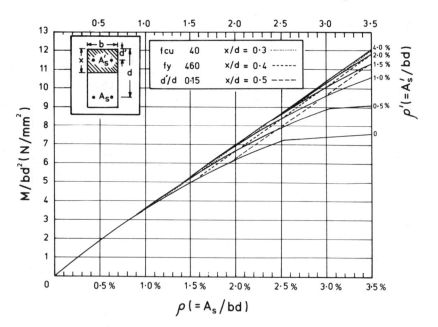

Fig. 4.5–2   Beam design chart—ultimate limit state (BS 8110)

reinforcement A_s and compression reinforcement A'_s, the trial and error procedure in Section 4.2 may be used. A trial neutral axis depth is first assumed and the steel strains ε_s and ε'_s, in the tension and compression reinforcement respectively, are then worked out from the compatibility condition (eqns (4.2–1) and (4.2–2)):

$$\varepsilon_s = \frac{d - x}{x}(0.0035) \tag{4.5–5}$$

$$\varepsilon'_s = \frac{x - d'}{x}(0.0035) \tag{4.5–6}$$

where d is the effective depth and d' the depth to the centroid of the compression reinforcement (see Fig. 4.2–1). The steel stresses f_s and f'_s are next obtained from the appropriate stress/strain curve in Fig. 4.5–1. The equilibrium condition (eqn 4.2–3) is then checked:

$$k_1 f_{cu} bx = A_s f_s - A'_s f'_s$$

i.e.

$$k_1 f_{cu}\left(\frac{x}{d}\right) = \varrho f_s - \varrho' f'_s \tag{4.5–7}$$

where $\varrho = A_s/bd$, $\varrho' = A'_s/bd$, and k_1 is as read off from Fig. 4.4–4. If eqn (4.5–7) is not satisfied, the trial value of x is increased or decreased by inspection and the process repeated until it is reasonably well satisfied. The ultimate moment of resistance of the doubly reinforced beam is then given by eqn (4.2–5):

$$M_u = A_s f_s(d - k_2 x) + A'_s f'_s(k_2 x - d')$$

i.e.

$$\frac{M_u}{bd^2} = \varrho f_s\left[1 - k_2\left(\frac{x}{d}\right)\right] + \varrho' f'_s\left[k_2\left(\frac{x}{d}\right) - \left(\frac{d'}{d}\right)\right] \tag{4.5–8}$$

Thus, for a specified value of ϱ' and a given combination of $(f_{cu}, f_y, d'/d)$ values, the curves of M_u/bd^2 against ϱ may be constructed (see Example 4.5–1). Figure 4.5–2 shows a set of such curves for $f_{cu} = 40 \text{ N/mm}^2$, $f_y = 460 \text{ N/mm}^2$ and $d'/d = 0.15$.

BS 8110 gives a comprehensive set of design charts covering different combinations of f_{cu}, f_y and d'/d values.

Design for the ultimate limit state (BS 8110)
The design use of the BS 8110 design charts will be illustrated by the worked examples below. The first two examples deal with the **construction of the BS 8110 charts** and serve as an introduction to the later ones.

Example 4.5–1
With reference to the BS 8110 design chart in Fig. 4.5–2, verify the M/bd^2 **value** for a beam with $\varrho(= A_s/bd) = 3\%$ and $\varrho'(= A'_s/bd) = 1\%$ for $f_{cu} = 40 \text{ N/mm}^2$, $f_y = 460 \text{ N/mm}^2$. Work from first principles and conform to BS 8110.

SOLUTION

Try $x/d = 0.7$. From eqns (4.5–5) and (4.5–6),

$$\varepsilon_s = \left(\frac{1 - x/d}{x/d}\right)(0.0035) = 0.0015$$

$$\varepsilon_s' = \left(\frac{x/d - d'/d}{x/d}\right)(0.0035) = 0.00275$$

From Fig. 4.5–1 (for $f_y = 460 \text{ N/mm}^2$),

$$f_s = 300 \text{ N/mm}^2; \qquad f_s' = 400 \text{ N/mm}^2$$

substituting into eqn (4.5–7) (with k_1 from Fig. 4.4–4),

$$k_1 f_{cu}\left(\frac{x}{d}\right) = \varrho f_s - \varrho' f_s'$$

$$\text{LHS} = (0.396)(40)(0.7) = 11.09$$

$$\text{RHS} = (0.03)(300) - (0.01)(400) = 5 < \text{LHS}$$

Therefore the assumed x/d is too high.

Try $x/d = 0.4$. Repeat the above process;

$$\varepsilon_s = \left(\frac{1 - 0.4}{0.4}\right)(0.0035) = 0.00525$$

$$\varepsilon_s' = \left(\frac{0.4 - 0.15}{0.4}\right)(0.0035) = 0.00245$$

whence

$$f_s = 400 \text{ N/mm}^2; \qquad f_s' = 400 \text{ N/mm}^2$$

$$\text{LHS of eqn (4.5–7)} = (0.396)(40)(0.4) = 6.34$$

$$\text{RHS} = (0.03)(400) - (0.01)(400) = 8.00 > \text{LHS}$$

Therefore the assumed x/d is too low. By inspection the correct x/d value must be between 0.4 and 0.7.

Try $x/d = 0.5$. Repeat the above process:

$$\varepsilon_s = \left(\frac{1 - 0.5}{0.5}\right)(0.0035) = 0.0035$$

$$\varepsilon_s' = \left(\frac{0.5 - 0.15}{0.5}\right)(0.0035) = 0.00245$$

whence

$$f_s = 400 \text{ N/mm}^2; \qquad f_s' = 400 \text{ N/mm}^2$$

$$\text{LHS of eqn (4.5–7)} = (0.396)(40)(0.5) = 7.92$$

$$\text{RHS} = (0.03)(400) - (0.01)(400) = 8.00 \doteq \text{LHS}$$

Hence $x/d = 0.5$ is sufficiently accurate.
  From eqn (4.5–8),

$$\frac{M_u}{bd^2} = \varrho f_s \left[1 - k_2 \left(\frac{x}{d} \right) \right] + \varrho' f_s' \left[k_2 \left(\frac{x}{d} \right) - \left(\frac{d'}{d} \right) \right]$$

(where $k_2 = 0.444$ from Fig. 4.4–4)

$= (0.03)(400)[1 - (0.444)(0.5)] + (0.01)(400)[(0.444)(0.5)$

$- (0.15)]$

$= 9.62$ (and $x/d = 0.5$ as found above)

This agrees with the M_u/bd^2 value given by the BS 8110 chart (Fig. 4.5–2).

Example 4.5–2

The beam design chart in Fig. 4.5–2 shows the neutral axis depth factors x/d for various steel ratios ϱ and ϱ'. For a given combination of f_{cu}, f_y, and d'/d, explain how such x/d **ratios** may be determined.

SOLUTION

For a general beam section, such as that in Fig. 4.2–1, the trial and error procedure in Example 4.5–1 above is appropriate.

In the particular case of a singly reinforced beam ($\varrho' = 0$), a more direct (but not necessarily quicker) approach may be used.

Step 1

Compare the actual steel ratio with the balanced ratio from eqn (4.5–1):

$$\varrho(\text{balanced}) = \frac{k_1 f_{cu}}{0.87 f_y} \frac{0.0035}{0.0035 + \varepsilon_y} \qquad (4.5\text{–}9)$$

where k_1 is determined from Fig. 4.4–4 and the strain ε_y at which the design strength $0.87 f_y$ is first reached is given by Fig. 4.5–1.

If the actual ϱ value does not exceed the balanced value, the steel stress reaches $0.87 f_y$ under the design ultimate moment. The neutral axis depth factor is therefore given by eqn (4.2–6) with f_s replaced by $0.87 f_y$:

$$0.87 f_y = \frac{k_1 f_{cu} b}{A_s} x$$

i.e.

$$\frac{x}{d} = \frac{0.87 f_y}{k_1 f_{cu}} \varrho \qquad (4.5\text{–}10)$$

Step 2

If the actual ϱ value exceeds the balanced value in Step 1, the beam is over-reinforced. The x/d ratio should then be calculated from eqn (4.5–4):

$$\frac{k_1 f_{cu}}{\varrho E_s} \left(\frac{x}{d} \right)^2 + 0.0035 \left(\frac{x}{d} \right) - 0.0035 = 0$$

where E_s is 200 kN/mm^2, since ϱ now exceeds the balanced value.

Example 4.5–3

The design ultimate moment for a beam of width 250 mm and effective depth 700 mm is 300 kNm. If $f_{cu} = 40$ N/mm^2 and $f_y = 460$ N/mm^2, design the reinforcement.

SOLUTION

$$\frac{M}{bd^2} = \frac{(300)(10^6)}{(250)(700^2)} = 2.45 \text{ N/mm}^2$$

From design chart (Fig. 4.5–2), $\varrho = 0.68\%$,

$$A_s = \left(\frac{0.68}{100}\right)(250)(700) = 1190 \text{ mm}^2$$

Provide four 20 mm bars   ($A_s = 1257$ mm^2 from Table A2–1)

Example 4.5–4
Repeat Example 4.5–3 if the design ultimate moment is 900 kNm. What then is the x/d ratio of the beam section so designed?

SOLUTION

$$\frac{M}{bd^2} = \frac{(900)(10^6)}{(250)(700^2)} = 7.35 \text{ N/mm}^2$$

If no compression steel is used, then the design chart (Fig. 4.5–2) shows that $\varrho = 2.9\%$. Moreover, x/d will be well over 0.5, which is not good practice (see comments below). Therefore use compression steel; let $\varrho' = 0.5\%$ say. Using the curve for $\varrho' = 0.5\%$ in Fig. 4.5–2, we have

$$\varrho = 2.26\%; \qquad \varrho' = 0.5\%; \qquad \frac{x}{d} = 0.45$$

$$A_s' = \left(\frac{0.5}{100}\right)(250)(700) = 875 \text{ mm}^2$$

Provide two 25 mm bars   (982 mm^2)

$$A_s' = \left(\frac{2.26}{100}\right)(250)(700) = 3955 \text{ mm}^2$$

Provide five 32 mm bars   (4021 mm^2)

The tension reinforcement would be arranged in two layers, such that 700 mm is the average effective depth.

Comments
The above solution shows that a singly reinforced section with $\varrho = 2.9\%$ would have satisfied the M/bd^2 requirement. However, such a beam section would have an x/d ratio well in excess of 0.5 and would fail in a brittle manner with inadequate warning before collapse. Section 4.9 gives further information on ductility and the significance of the x/d ratio.

Example 4.5–5
The effective depth d of a beam section is to be $2\frac{1}{2}$ times its width b. The design ultimate moment is 1600 kNm. If the tension steel ratio is not to exceed 3% and the compression steel ratio is not to exceed 1.5%,

determine the minimum required values of b and d. What is the neutral axis depth factor x/d for such a section?

SOLUTION
The dimensions will be a minimum if ϱ and ϱ' have the maximum values of 3 and 1.5% respectively. From the design chart (Fig. 4.5–2):

$$\frac{M}{bd^2} = 10.1, \quad \text{i.e.} \quad \frac{M}{b(2.5b)^2} = 10.1$$

Therefore

$$b^3 = \frac{(1600)(10^6)}{(2.5^2)(10.1)} = 25.35 \times 10^6$$

$$b = 293.7 \text{ mm}; \qquad d = 2.5b = 735 \text{ mm}$$

say

$$\underline{300 \text{ mm by } 750 \text{ mm}}$$

From Fig. 4.5–2: $x/d = \underline{0.4}$ approximately.

Example 4.5–6 (NOT to be attempted until after reading Chapter 7)
BS 8110: Clause 3.4.4.1 states that, for a beam section subjected to **combined bending and axial force**, the effect of the axial thrust may be ignored if it does not exceed $0.1f_{cu}$ times the cross-sectional area. Comment.

SOLUTION
Reference to Fig. 7.1–7 makes it clear that for small values of $\alpha(= N/f_{cu}bh)$, an increase in the **axial thrust** leads to an increase in $\beta(= M/f_{cu}bh^2)$. For the particular column section illustrated in Fig. 7.1–7, β increases with α up to $\alpha \doteq 0.2$. In design, it is fairly safe to say that, for α up to 0.1, β always increases with α. Hence where the axial thrust on a beam section does not exceed $0.1f_{cu}bh$, ignoring its presence will err on the safe side.

Limit on lever arm
BS 8110: Clause 3.4.4.1 states that the lever-arm factor z/d should not be taken as greater than 0.95. This restriction applies irrespective of whether the parabolic–rectangular stress block (Fig. 4.4–3) or the simplified rectangular stress block (Fig. 4.4–5) is used.

Effective span
It is convenient at this point to quote BS 8110's definitions of **effective span** and **effective length**:

(a) The **effective span** of a simply supported beam is the smaller of:
   (1) the distance between centres of supports; or
   (2) the clear distance between supports plus the effective depth.
(b) The **effective span** of a continuous beam is the distance between centres of supports.
(c) The **effective length** of a cantilever is:

(1) its length to the face of the support plus half its effective depth; or , where it forms the end of a continuous beam,
(2) the length to the centre of the support.

4.6   Design formulae and procedure— BS 8110 simplified stress block

As an alternative to the parabolic–rectangular stress block of Fig. 4.4–3, BS 8110 permits design calculations to be based on the simplified rectangular stress block of Fig. 4.4–5. In this section we shall:

(1) derive the basic design formulae from first principles (Section 4.6(a));
(2) explain how to design from first principles (Section 4.6(b));
(3) explain the design procedure of the I.Struct.E. Manual [14]—the manual's formulae will be derived and the technical background explained (Section 4.6(c)).

(*Note:* BS 8110 permits up to 30% **moment redistribution**. Design formulae and procedure allowing for moment redistribution will be given in Section 4.7, while the technical fundamentals of moment redistribution will be explained in Section 4.9. For the time being the reader need only note that *all the formulae and design procedures in this section can be used for up to 10% moment redistribution.*)

4.6(a)   Derivation of design formulae

BS 8110 intends that, where the simplified stress block is used, x/d *should not exceed 0.5*. Consider the beam section in Fig. 4.6–1(a). As Example 4.6–1 will show, for x/d values up to 0.5 the tension reinforcement is bound to reach the design strength of $0.87f_y$ at the ultimate limit state. Therefore, the forces are as shown in Fig. 4.6–1(b), from which

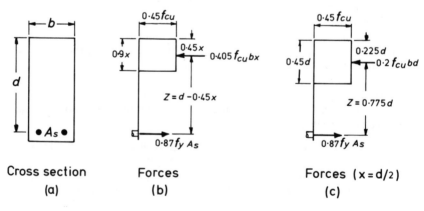

Cross section
(a)

Forces
(b)

Forces $(x = d/2)$
(c)

Fig. 4.6–1

concrete compression $= (0.45f_{cu})(0.9x)b$

$$= 0.405f_{cu}bx \qquad\qquad (4.6{-}1)$$

Equating this to the steel tension:

$$0.405f_{cu}bx = 0.87f_yA_s$$

$$\frac{x}{d} = 2.15\frac{f_y}{f_{cu}}\frac{A_s}{bd} \qquad\qquad (4.6{-}2)$$

The moment M corresponding to the forces in Fig. 4.6–1(b) is simply the concrete compression or the steel tension times the lever arm z, where

$$z = d - 0.45x \qquad\qquad (4.6{-}3)$$

Using the concrete compression, say,

$$M = (0.405f_{cu}bx)(d - 0.45x)$$
$$= (0.405x/d)(1 - 0.45x/d)f_{cu}bd^2$$
$$= Kf_{cu}bd^2 \qquad\qquad (4.6{-}4)$$

As expected, M increases with x/d and hence with A_s (see eqn 4.6–2). In design, BS 8110 limits x/d to not exceeding 0.5. When $x/d = 0.5$, the forces are as shown in Fig. 4.6–1(c); the moment M_u, which corresponds to these forces, represents the maximum **moment capacity** of a singly reinforced beam. From Fig. 4.6–1(c):

$$M_u = (0.2f_{cu}bd)z$$
$$= 0.156f_{cu}bd^2 \quad (\text{since } z = 0.775d)$$
$$= K'f_{cu}bd^2 \qquad\qquad (4.6{-}5)$$

where $K' = 0.156$. Of course, the same result of $0.156f_{cu}bd^2$ can be obtained by writing $x/d = 0.5$ in eqn (4.6–4).

Where the applied bending moment M exceeds M_u of eqn (4.6–5), the excess $(M - M_u)$ is to be resisted by using an area A'_s of **compression reinforcement** (Fig. 4.6–2(a)) such that the neutral axis depth remains at the maximum permitted value of $0.5d$ (i.e. depth of stress block $= 0.9x =$

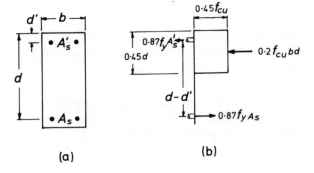

(a)   (b)

Fig. 4.6–2

0.45d). Example 4.6–2 shows that the compression reinforcement will reach the design strength of 0.87f_y provided d'/x does not exceed 0.43; that is (for $x/d = 0.5$), provided d'/d does not exceed 0.21. In other words, the force in the compression reinforcement can normally be taken as 0.87$f_y A_s'$, and this has a lever arm of $(d - d')$ about the tension reinforcement. Equating this additional resistance moment to the excess moment,

$$0.87f_y A_s'(d - d') = M - M_u \qquad\qquad \textbf{(4.6–6)}$$

where $M_u = K' f_{cu} bd^2$ (eqn 4.6–5). Equation (4.6–6) gives the required area A_s' of the compression reinforcement. An area A_s of tension reinforcement must then be provided to balance the total compressive force in the concrete and the compression reinforcement. Referring to Fig. 4.6–2(b),

$$0.87f_y A_s = 0.2f_{cu} bd + 0.87f_y A_s' \qquad\qquad \textbf{(4.6–7)}$$

Noting from eqn (4.6–5) that $0.2f_{cu} bd = M_u/z$, we can write eqn (4.6–7) as

$$A_s = \frac{M_u}{0.87f_y z} + A_s' \qquad\qquad \textbf{(4.6–8)}$$

where $z = 0.775d$ (from Fig. 4.6–1(c)); and
$\quad\quad\quad M_u = K' f_{cu} bd^2$ (from eqn 4.6–5).

The term 'balanced section' has been defined in Section 4.3, and again in a slightly different way in Section 4.5. In connection with the use of the simplified stress block here, a **balanced section** is defined as a singly reinforced section having such an area A_s of tension reinforcement that the x/d ratio is equal to 0.5. From eqn (4.6–2)

$$0.5 = 2.15\frac{f_y}{f_{cu}}\frac{A_s}{bd}$$

or

$$\varrho(\text{balanced}) = 0.233\frac{f_{cu}}{f_y} \qquad\qquad \textbf{(4.6–9)}$$

where $\varrho = A_s/bd$.

Example 4.6–1
Determine the limiting value of the neutral axis depth x for which tension reinforcement, of $f_y = 460$ N/mm^2, will reach the design strength of 0.87f_y at the ultimate limit state.

SOLUTION
The strain distribution at the ultimate limit state is shown in Fig. 4.6–3. For $f_y = 460$ N/mm^2, the design strength $0.87f_y = 400$ N/mm^2, and $\varepsilon_s = 400/E_s = 0.002$. Therefore, from the geometry of Fig. 4.6–3,

$$\frac{0.0035}{x} = \frac{0.002}{d - x}$$

$$x = 0.64d$$

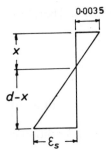

Fig. 4.6–3

Therefore, for x/d values not exceeding 0.64, tension reinforcement of $f_y = 460$ N/mm² (or less) will reach the design strength $0.87f_y$ at the ultimate limit state.

Example 4.6–2
The design formulae for doubly reinforced beams (eqns 4.6–6 to 4.6–8) assume that the compression reinforcement reaches the design strength $0.87f_y$ at the ultimate limit state. Determine the limiting value of the d'/x and d'/d ratios for this to be possible.

SOLUTION
Equations (4.6–6) to (4.6–8) are based on the condition that $x/d = 0.5$; that is, the provision of the area A'_s of compression steel to the balanced singly reinforced section is matched by the provision of an additional area $A_{s(add)}$ of tension reinforcement such that the tensile force $0.87f_yA_{s(add)}$ is equal to the compressive force $0.87f_yA'_s$, so that x/d remains at 0.5.

Figure 4.6–4 shows that for $x/d = 0.5$, the strain ε'_s in the compression reinforcement is

$$\varepsilon'_s = \frac{x - d'}{x}(0.0035)$$

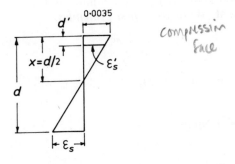

Fig. 4.6–4

For the design strength $0.87f_y$ to be reached, $\varepsilon_s' = 0.87f_y/E_s$. Therefore

$$\frac{0.87f_y}{E_s} = \frac{x - d'}{x}(0.0035)$$

where $E_s = 200$ kN/mm², i.e.

$$\frac{d'}{x} = 1 - \frac{f_y}{800} \qquad (4.6\text{--}10)$$

For $f_y = 460$ N/mm², this gives $d'/x = 0.43$. For the balanced condition of $x/d = 0.5$, we have

$$\frac{d'}{d} = \frac{1}{2}\left(\frac{d'}{x}\right) = 0.21$$

Therefore, as long as d'/d does not exceed 0.21 (i.e. d'/x does not exceed 0.43) the compression reinforcement can be assumed to reach the design strength of $0.87f_y$.

Referring to the more general eqn (4.6–10), if d'/x exceeds $1 - (f_y/800)$, a reduced stress f_s' should be used. From Fig. 4.6–4,

$$\varepsilon_s' = \frac{x - d'}{x}(0.0035)$$

Hence $f_s' = E_s\varepsilon_s'$ where $E_s = 200$ kN/mm², i.e.

$$f_s' = 700\left(1 - \frac{d'}{x}\right) \qquad (4.6\text{--}11)$$

4.6(b)   Designing from first principles

The I.Struct.E. Manual [14] gives a convenient procedure for practical design, which will be explained later—after Example 4.6–4. In the mean time, Examples 4.6–3 and 4.6–4 below will explain how designs can be carried out from **first principles**.

Example 4.6–3
The design ultimate moment M for a rectangular beam of width b 250 mm and effective depth d 700 mm is 300 kNm. If $f_{cu} = 40$ N/mm² and $f_y = 460$ N/mm², design the reinforcement. *Work from first principles.*

SOLUTION

Step (a)
   Check concrete capacity M_u. From eqn (4.6–5),

$$M_u = 0.156f_{cu}bd^2$$

$$= (0.156)(40)(250)(700^2)\ \text{Nmm}$$

$$= 764.4\ \text{kNm}$$

Since $M < M_u$, no compression steel is required.

Step (b)

Find lever arm z. From Fig. 4.6–1(b),

$$\text{concrete compression} = 0.405f_{cu}bx$$
$$\text{lever arm } z = d - 0.45x$$

Hence

$$M = 0.405f_{cu}bx(d - 0.45x)$$
$$(300)(10^6) = (0.405)(40)(250)(x)(700 - 0.45x)$$
$$x^2 - 1560x + 165\,000 = 0$$
$$x = 114 \text{ m}$$
$$\text{(or 1441 mm, which is inadmissible)}$$

Hence

$$z = 700 - 0.45x = 649 \text{ mm}$$

Step (c)

Find A_s. From Fig. 4.6–1(b),

$$M = 0.87f_yA_sz$$
$$\text{(where } z = 649 \text{ mm from Step (b) above)}$$
$$(300)(10^6) = (0.87)(460)A_s(649)$$
$$A_s = 1155 \text{ mm}^2$$

Provide four 20 mm bars   ($A_s = 1257 \text{ mm}^2$)

Comment

Example 4.5–3 solves the same problem using BS 8110's design chart. See also Example 4.6–5 which follows the I.Struct.E. Manual's procedure [14].

Example 4.6–4

Repeat Example 4.6–3 if M is 900 kNm. What is that x/d ratio of the beam section so designed?

SOLUTION

Step (a)

Check concrete capacity M_u.

$$M_u = 764.4 \text{ kNm}   \text{(from Step (a) of Example 4.6–3)}$$

Since $M > M_u$, compression steel is required.

Step (b)

Find compression steel area A_s'. From eqn (4.6–6),

$$0.87f_yA_s'(d - d') = M - M_u$$
$$(0.87)(460)A_s'(700 - 60 \text{ say}) = (900 - 764.4)(10^6)$$
$$A_s' = 530 \text{ mm}^2$$

Step (c)

Find tension steel area A_s. From eqn (4.6–7),

$$0.87f_yA_s = 0.2f_{cu}bd + 0.87f_yA_s'$$

$$(0.87)(460)A_s = (0.2)(40)(250)(700) + (0.87)(460)(530)$$

$$A_s = 4030 \text{ mm}^2$$

<u>Provide two 20 mm top bars</u> $(A_s' = 628 \text{ mm}^2)$

<u>Provide five 32 mm bottom bars</u> $(A_s = 4021 \text{ mm}^2)$

Step (d)

The x/d ratio. As explained in the paragraph preceding eqn (4.6–6), the x/d ratio is 0.5.

Comments

(a)   Example 4.5–4 solves the same problem using BS 8110's design chart. See also Example 4.6–6 which uses the I.Struct.E. Manual's procedure [14].

(b)   Examples 4.6–3 and 4.6–4 deal with design. For the use of BS 8110's simplified stress block in **analysis**, see Examples 4.6–8 and 4.6–9.

4.6(c)   Design procedure for rectangular beams (BS 8110/I.Struct.E. Manual)

The design procedure given below is that of the I.Struct. E. Manual [14]. Comments have been added to explain the derivation of the formulae and the technical background. As stated at the beginning of this section, the same procedure can be used for up to 10% moment redistribution. For **moment redistribution** exceeding 10%, see Section 4.7. For **flanged beams**, see Section 4.8. Consider the beam section in Fig. 4.6–1(a). Suppose the design bending moment is M. Proceed as follows.

Step 1

Calculate M_u for concrete.

$$M_u = K'f_{cu}bd^2$$

where $K' = 0.156$.

Comments

See eqn (4.6–5).

Step 2

If the design moment $M \le M_u$ of Step 1: the tension reinforcement A_s is given by

$$A_s = \frac{M}{(0.87f_y)z} \tag{4.6–12}$$

where the lever arm z is obtained from Table 4.6–1.

Table 4.6–1   Lever-arm and neutral axis depth factors [14]

$K = M/bd^2 f_{cu}$	0.05	0.06	0.07	0.08	0.09	0.100	0.104	0.110	0.119	0.130	0.132	0.140	0.144	0.150	0.156
(z/d)			0.94	0.93	0.91	0.90	0.89	0.87	0.87	0.86	0.84	0.82	0.82	0.81	0.80	0.79	0.775
(x/d)			0.13	0.16	0.19	0.22	0.25	0.29	0.30	0.32	0.35	0.39	0.40	0.43	0.45	0.47	0.50

Comments

(a)   The formula $A_s = M/(0.87f_y z)$ follows from Fig. 4.6–1(b), where it is seen that, by taking moments about the centroid of the concrete stress block, $M = (0.87f_y A_s)z$.

(b)   Table 4.6–1 is extracted from the I.Struct. E. Manual [14]. The lever-arm factors z/d in the table are given by the BS 8110 formula:

$$\frac{z}{d} = 0.5 + \sqrt{\left(0.25 - \frac{K}{0.9}\right)}$$

where $K = M/(f_{cu}bd^2)$. For the derivation of this formula, see Example 4.6–10.

(c)   Note that the z/d and x/d values given in Table 4.6–1 for $K = 0.156$ are in fact equally valid for $K > 0.156$ (though this is not pointed out in the I.Struct.E. Manual [14]). That is, for $K \geq 0.156$, $z/d = 0.775$ and $x/d = 0.50$. Study Example 4.6–11 for a full explanation.

Step 3

If the design moment $M > M_u$ of Step 1:

(3a)   Compression reinforcement is required and its area A_s' is given by

$$A_s' = \frac{M - M_u}{0.87f_y(d - d')} \tag{4.6–13}$$

where d' is the depth of the compression steel from the concrete compression face (Fig. 4.6–2(a)).

(3b)   If

$$\frac{d'}{x} > \left(1 - \frac{f_y}{800}\right), \quad \text{use } 700\left(1 - \frac{d'}{x}\right)$$

in lieu of $0.87f_y$ in the formula for A_s'.

(3c)   The area of tension reinforcement A_s is calculated from

$$A_s = \frac{M_u}{0.87f_y z} + A_s' \tag{4.6–14}$$

where $z = 0.775d$ (see Comment (c) below).

Comments

(a)   The formula for A_s' in Step (3a) was derived earlier as eqn (4.6–6).

(b)   The formulae in Step (3b) were derived earlier in Example 4.6–2. Example 4.6–2 shows that, where $d'/d > (1 - f_y/800)$, then the compression reinforcement A_s' does not reach the design stress $0.87f_y$ so that, in using the formula in Step (3a), a reduced stress $f_s' = 700(1 - d'/x)$ should be used.

(c)   The formula in Step (3c) was derived earlier as eqn (4.6–8). Equation (4.6–8) also makes it clear that z is to be taken as $0.775d$. See also Example 4.6–11.

Example 4.6–5
The design ultimate moment M for a rectangular beam of width 250 mm and effective depth 700 mm is 300 kNm. If $f_{cu} = 40$ N/mm² and $f_y = 460$ N/mm², design the reinforcement. *Follow the procedure of the I.Struct.E. Manual* [14].

SOLUTION

Step 1

$$M_u = 0.156f_{cu}bd^2$$

$$= (0.156)(40)(250)(700^2)$$

$$= 764.4 \text{ kNm}$$

Step 2
$M < M_u$. Calculate

$$K = \frac{M}{f_{cu}bd^2} = \frac{(300)(10^6)}{(40)(250)(700^2)}$$

$$= 0.061$$

From Table 4.6–1, $z = 0.93d = 651$ mm. From eqn (4.6–12),

$$A_s = \frac{M}{(0.87f_y)z} = \frac{(300)(10^6)}{(0.87)(460)(651)}$$

$$= 1152 \text{ mm}^2$$

Provide four 20 mm bars $(A_s = 1257 \text{ mm}^2)$

Comments
Example 4.5–3 solves the same problem using BS 8110's design chart, which gives $A_s = 1190$ mm².

Example 4.6–6
Repeat Example 4.6–5 if the design ultimate moment is 900 kNm. What is the neutral axis depth factor x/d of the beam section so designed?

SOLUTION

Step 1

$$M_u = 0.156f_{cu}bd^2$$

$$= (0.156)(40)(250)(700^2)$$

$$= 764.4 \text{ kNm}$$

Step 2
$M = 900$ kNm $> M_u$ of Step 1. Therefore move to Step 3.

Step 3
$M > M_u$ of Step 1.

(3a) Compression reinforcement is required. From eqn (4.6–13),

$$A'_s = \frac{M - M_u}{0.87f_y(d - d')}$$

where $d' = 60$ mm, say,

$$= \frac{(900 - 764.4)(10^6)}{(0.87)(460)(700-60)} = 530 \text{ mm}^2$$

(3b) Check d'/x. From Table 4.6–1,

$$\frac{x}{d} = 0.5 \quad \text{(see Comment (c) under Table 4.6–1)}$$

$$x = (0.5)(700) = \underline{350}$$

$$\frac{d'}{x} = \frac{60}{350} = 0.17 < \left(1 - \frac{f_y}{800}\right)$$

Therefore the compression steel reaches the design strength of $0.87f_y$ and the A'_s calculated in Step (3a) above is acceptable.

(3c) $A_s = \dfrac{M_u}{0.87f_yz} + A'_s$ (see eqn 4.6–14)

(where $z = 0.775d$ from Table 4.6–1)

$$= \frac{(764.4)(10^6)}{(0.87)(460)(0.775)(700)} + 530$$

$$= 4051 \text{ mm}^2$$

Provide two 20 mm top bars   (628 mm²)

Provide five 32 mm bottom bars   (4021 mm²)

Comments
Example 4.5–4 solves the same problem using BS 8110's design chart, which gives $A'_s = 981$ mm² and $A_s = 3955$ mm², so that $A'_s + A_s = 4936$ mm² which exceeds the $A'_s + A_s$ obtained using the simplified stress block here. The higher total steel area in Example 4.5–4 arises for two reasons:

(a) The design chart in Fig. 4.5–2 assumes $d'/d = 0.15$. In the present example, $d'/d = 60/700 = 0.09$.
(b) In Example 4.5–4, the compression steel ratio ϱ' is taken as 0.5%, which is the lowest value shown in the design chart (Fig. 4.5–2). If ϱ' had been taken as 0.3%, for example, a smaller total for $A'_s + A_s$ would have resulted (though in using Fig. 4.5–2 a curve for $\varrho' = 0.3\%$ would have to be added by interpolation).

Example 4.6–7
The effective depth d of a beam section is to be $2\frac{1}{2}$ times its width b. The design ultimate moment is 1600 kNm. If $\varrho\ (= A_s/bd)$ is not to exceed 3% and $\varrho'(= A'_s/bd)$ is not to exceed 1.5%, determine the minimum required

values of b and d, if $f_{cu} = 40$ N/mm^2 and $f_y = 460$ N/mm^2. What is the neutral axis depth factor x/d for such a section?

SOLUTION

The design formulae based on BS 8110's simplified stress block are derived on the assumption that, for any doubly reinforced beam, $x/d = 0.5$. Hence the steel areas A_s and A'_s must obey the definite relation given in the I.Struct.E. Manual [14], namely: $A_s = M_u/(0.87f_yz) + A'_s$. Equation (4.6–7) expresses this relation in the equivalent form

$$0.87f_y\varrho = 0.2f_{cu} + 0.87f_y\varrho'$$

Thus we can either specify that $\varrho = 3\%$ or that $\varrho' = 1.5\%$, but not both. Suppose we choose to specify $\varrho = 3\%$; then

$$(0.87)(460)(0.03) = 0.2(40) + (0.87)(460)\varrho'$$

whence $\varrho' = 1.0\%$. From eqn (4.6–6),

$$A'_s = \frac{M - M_u}{0.87f_y(d - d')}$$

Dividing by bd^2 and rearranging

$$0.87f_y\left(\frac{A'_s}{bd}\right)\left(1 - \frac{d'}{d}\right) = \frac{M}{bd^2} - \frac{M_u}{bd^2}$$

$$(0.87)(460)(0.01)(1 - 0.15 \text{ say}) = \frac{M}{bd^2} - \frac{M_u}{bd^2}$$

Substituting in $M = (1600)(10^6)$ and $M_u = 0.156f_{cu}bd^2$, we obtain

$$bd^2 = (165.9)(10^6)$$

For $d = 2.5b$, we have $b = 298$ mm, $d = 746$ mm <u>say 300 mm by 750 mm</u> ($x/d = 0.5$).

Comments

See also Example 4.5–5, in which the above problem is solved using BS 8110's design chart.

Examples 4.6–3 to 4.6–7 are **design examples**. Examples 4.6–8 and 4.6–9 below deal with the **analysis** of given beam sections. The reader should note that the formulae based on BS 8110's simplified stress block should not be used unless the following conditions are met:

(a) For a singly reinforced beam section, $x/d \leq 0.5$.
(b) For a doubly reinforced beam section, $x/d = 0.5$.

Example 4.6–8

Determine the ultimate moment of resistance M and the x/d ratio of the beam section in Fig. 4.6–5, using:

(a) design chart; and
(b) the BS 8110 simplified stress block.

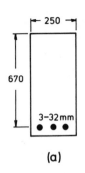

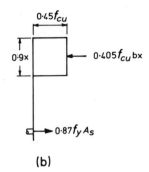

(a)   (b)

Fig. 4.6–5

Here, $f_{cu} = 40$ N/mm²; $f_y = 460$ N/mm².

SOLUTION

(a) **Using design chart** (Fig. 4.5–2).
A_s (three 32 mm bars) = 2412 mm²

$$\varrho = \frac{2412}{(250)(670)} = 1.44\%$$

From design chart (Fig. 4.5–2),

$M/bd^2 = 4.9$ N/mm²

$\quad M = (4.9)(250)(670^2)(10^{-6}) = \underline{550 \text{ kNm}}$

$\quad x/d = \underline{0.35}$ approximately (by interpolation)

(b) **Using BS 8110 simplified stress block** (Fig. 4.6–5 (b)). With reference to Fig. 4.6–5(b), the equilibrium condition is as given in eqn (4.6–2):

$$0.405 f_{cu} bx = 0.87 f_y A_s$$

$$(0.405)(40)(250)x = 0.87(460)(2412)$$

$$x = 238.3 \text{ mm}$$

$$x/d = \underline{0.36} < 0.5 \text{ OK}$$

From eqn (4.6–3) (or Fig. 4.6–5(b)),

$z = d - 0.45x = 670 - (0.45)(238.3)$

$z = 562.8$ mm

$M = 0.87 f_y A_s z$

$\quad = (0.87)(460)(2412)(562.8)(10^{-6}) = \underline{543 \text{ kNm}}$

Example 4.6–9
Determine the ultimate resistance moment M and the x/d ratio of the beam section in Fig. 4.6–6 using:

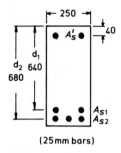

(25mm bars)

Fig. 4.6–6

(a)   design chart; and
(b)   the BS 8110 simplified stress block

Here, $f_{cu} = 40$ N/mm^2; $f_y = 460$ N/mm^2.

SOLUTION
(a)   **Using design chart.**

$$A_{s1} \text{ (two 25 mm)} = 982 \text{ mm}^2$$

$$A_{s2} \text{ (three 25 mm)} = 1473 \text{ mm}^2$$

$$A_s = A_{s1} + A_{s2} = 2455 \text{ mm}^2$$

$$A_s' \text{ (two 25 mm)} = 982 \text{ mm}^2$$

$$\varrho = \frac{A_s}{bd} = \frac{2455}{250 \times \frac{1}{2}(680 + 640)} = 1.49\%$$

$$\varrho' = \frac{A_s'}{bd} = \frac{982}{250 \times \frac{1}{2}(680 + 640)} = 0.59\%$$

From Fig. 4.5–2,

$$M = 5.3bd^2 = (5.3)(250)\left(\frac{680 + 640}{2}\right)^2 (10^{-6}) = \underline{577 \text{ kNm}}$$

$$\frac{x}{d} < \underline{0.3}$$

Comment
The d value used in the above calculations is taken as the mean of 640 mm for A_{s1} and 680 mm for A_{s2} and is hence only an approximation. The design chart gives curves for $x/d = 0.3, 0.4$ and 0.5 only; hence the precise x/d value cannot be determined.

(b)   **Using BS 8110 simplified stress block.** With reference to the forces in Fig. 4.6–7, let us assume at this stage that both the compression and tension steels reach the full design strength of $0.87f_y$. Therefore, the condition of equilibrium is (see Comment (a) below)

$$0.87f_y A_s = 0.405f_{cu}bx + 0.87f_y A_s'$$

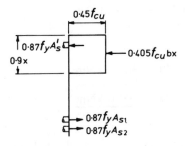

Fig. 4.6–7

$$(0.87)(460)(2455) = (0.405)(40)(250x) + 0.87(460)(982)$$
$$x = 145.5 \text{ mm}; \qquad x/d = \underline{0.22}$$

It is necessary to check that the steel design strengths $0.87f_y$ are in fact attained. From Fig. 4.6–8,

$$\varepsilon_s' = \left(\frac{105.5}{145.5}\right)(0.0035) = 0.003$$

$$\varepsilon_{s1} = \left(\frac{494.5}{145.5}\right)(0.0035) = 0.012$$

These strains exceed the design yield strains (see Fig. 4.5–1) and therefore the design strengths $0.87f_y$ are reached. The ultimate moment of resistance is conveniently determined by taking moments about the centroid of the concrete compression block in Fig. 4.6–7:

$$0.87f_yA_s'\left(\frac{0.9x}{2} - d'\right)$$
$$= (0.87)(460)(982)(65.48 - 40)(10^{-6}) \quad = \quad 10.0 \text{ kNm}$$
$$0.87f_yA_{s1}\left(d_1 - \frac{0.9x}{2}\right)$$

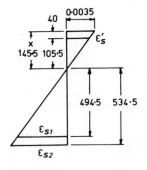

Fig. 4.6–8

$$= (0.87)(460)(982)(640 - 65.48)(10^{-6}) = 225.6 \text{ kNm}$$

$$0.87 f_y A_{s2} \left(d_2 - \frac{0.9x}{2} \right)$$

$$= (0.87)(460)(1473)(680 - 65.48)(10^{-6}) = 362.0 \text{ kNm}$$

$$\underline{M = 597.6 \text{ kNm}}$$

Comments
(a) Note that the condition of equilibrium has to be worked out from Fig. 4.6–7:

$$0.87 f_y A_s = 0.405 f_{cu} bx + 0.87 f_y A'_s \tag{4.6–15}$$

Equation (4.6–7) assumes that $x/d = 0.5$ and cannot therefore be used here.

(b) If the strain distribution in Fig. 4.6–8 had shown that the design strengths $0.87 f_y$ were not attained in some or all of the reinforcement, then a trial and error procedure would have to be used (see Example 4.5–1).

In Examples 4.6–10 and 4.6–11 below, we shall derive the formulae, given in BS 8110: Clause 3.4.4.4, for the z/d factors in Table 4.6–1 above (reproduced from Table 24 of the I.Struct.E. Manual [14]).

Example 4.6–10
A bending moment M is applied to a rectangular beam section. If $M \le M_u$ of eqn (4.6–5), show that z/d is given by the following BS 8110 formula:

$$\frac{z}{d} = 0.5 + \sqrt{\left(0.25 - \frac{K}{0.9} \right)} \tag{4.6–16}$$

Where $K = M/f_{cu} bd^2$;
z = lever-arm distance measured from the tension reinforcement to the centroid of the concrete compression block.

SOLUTION
For $M \le M_u$ of eqn (4.6–5) we have, from eqn (4.6–4),

$$K = (0.405x/d)(1 - 0.45x/d)$$

From Fig. 4.6–1,

$$z = d - 0.45x$$

Eliminating x from these two equations, we have

$$\left(\frac{z}{d} \right)^2 - \left(\frac{z}{d} \right) + 1.111K = 0$$

from which

$$\frac{z}{d} = 0.5 + \sqrt{\left(0.25 - \frac{K}{0.9} \right)}$$

Example 4.6–11

A bending moment M is applied to a beam section. If $M > M_u$ of eqn (4.6–5), show that z/d is given by the following BS 8110 formula:

$$\frac{z}{d} = 0.5 + \sqrt{\left(0.25 - \frac{K'}{0.9}\right)} \qquad (4.6\text{–}17)$$

where $K' = M_u/f_{cu}bd^2$ (see eqn 4.6–5);

$\quad\quad z$ = lever-arm distance measured from the tension reinforcement A_s to the centroid of the concrete compression block.

SOLUTION

When $M = M_u$ of eqn (4.6–5), x/d is equal to 0.5. When $M > M_u$ (i.e. $K > K'$) compression reinforcement is required, but x/d remains equal to 0.5. Therefore, with reference to eqn (4.6–4), the equation

$$K' = (0.405x/d)(1 - 0.45x/d)$$

will give $x/d = 0.5$ when K' is set equal to 0.156 of eqn (4.6–5). From Fig. 4.6–1(b),

$$z = d - 0.45x$$

This expression gives the lever-arm distance z measured from the tension steel A_s to the centroid of the concrete compression block, and remains valid for $M > M_u$.

Eliminating x from the above two equations,

$$\left(\frac{z}{d}\right)^2 - \left(\frac{z}{d}\right) + 1.111K' = 0$$

from which

$$\frac{z}{d} = 0.5 + \sqrt{\left(0.25 - \frac{K'}{0.9}\right)}$$

Of course, setting $K' = 0.156$ in this equation (see also eqn 4.6–5) gives $z = 0.775d$, which agrees with that in Fig. 4.6–1(c).

4.7 Design formulae and procedure—BS 8110 simplified stress block (up to 30% moment redistribution)*

As stated at the beginning of Section 4.6, the formulae and design procedure in that section are valid for up to 10% moment redistribution. We shall now explain how those formulae and the design procedure can be modified for application for *up to 30% moment redistribution*. Of course, the design formulae and procedure in this section, and those in Section 4.6, both

(a) conform to BS 8110, and

---

* Section 4.7 may be omitted on first reading; beginners should move on to Section 4.8.

(b)   are completely consistent with Section 4.4.5.1 of I.Struct.E. Manual [14].

The technical fundamentals of moment redistribution will be explained in Section 4.9, which will also explain why **moment redistribution** is carried out and how it is carried out. For our present purpose, it is sufficient to note that BS 8110 allows the bending moments obtained from the elastic analysis of a continuous structure (e.g. continuous beam) to be redistributed by up to 30% to make the detailing and the construction easier.

BS 8110 defines the **moment redistribution ratio** β_b as

$$\beta_b = \frac{\text{moment at the section after redistribution}}{\text{moment at the section before redistribution}} \leq 1 \qquad (4.7\text{--}1)$$

i.e.

$$\beta_b = \frac{100 - (\% \text{ of moment redistribution})}{100} \qquad (4.7\text{--}2)$$

For the reasons to be explained in Section 4.9, BS 8110 requires that the x/d ratio of a beam section should satisfy the condition

$$\frac{x}{d} \leq (\beta_b - 0.4) \qquad \textbf{(4.7--3)}$$

This condition will of course place restrictions on the amounts of longitudinal reinforcements to be used.

Equation (4.6–5) states that the maximum moment capacity of a singly reinforced beam section is

$$M_u = K' f_{cu} b d^2$$

where $K' = 0.156$. The value of $K' = 0.156$ assumes that $x/d = 0.5$. We can see from eqn (4.7–3) that for $x/d = 0.5$, $\beta_b \geq 0.9$; we can also see from eqn (4.7–2) that for $\beta_b \geq 0.9$, the moment redistribution must not exceed 10%. In other words, eqn (4.6–5), with $K' = 0.156$, applies only where the redistribution does not exceed 10%. Where there is more than 10% moment redistribution, eqn (4.7–3) states that $x/d \leq (\beta_b - 0.4)$; hence the maximum **moment capacity** M_u must be calculated afresh from eqn (4.6–4):

$$\begin{aligned} M_u &= (0.405 x/d)(1 - 0.45 x/d) f_{cu} b d^2 \\ &\quad (\text{where } x/d = (\beta_b - 0.4) \leq 0.5) \\ &= 0.405(\beta_b - 0.4)[1 - 0.45(\beta_b - 0.4)] f_{cu} b d^2 \\ &\quad (\text{where } \beta_b \leq 0.9 \text{ (since } x/d \leq 0.5)) \\ &= [0.40(\beta_b - 0.4) - 0.18(\beta_b - 0.4)^2] f_{cu} b d^2 \\ &= K' f_{cu} b d^2 \qquad\qquad\qquad\qquad\qquad\qquad\qquad (4.7\text{--}4) \end{aligned}$$

where

$$K' = 0.40(\beta_b - 0.4) - 0.18(\beta_b - 0.4)^2 \qquad (4.7\text{--}5)$$

where $\beta_b \leq 0.9$, so that

$$K' \leq 0.156 \quad \text{(see Comments below)}$$

Comments on eqn (4.7–5)
Equations (4.7–4) and (4.7–5) are based on the condition that $x/d \leq (\beta_b - 0.4)$, so that $x/d = 0.5$ when $\beta_b = 0.9$. When the simplified stress block is used, BS 8110 restricts x/d to not exceeding 0.5. This means that in using eqns (4.7–4) and (4.7–5), β_b is not to be taken as greater than 0.9 whatever the actual amount of moment redistribution. When β_b has the maximum permissible value of 0.9, eqn (4.7–5) gives $K' = 0.156$ (agreeing with eqn 4.6–4). *Therefore, provided β_b is taken as not exceeding 0.9, eqn (4.7–5) is valid for all % moment redistribution from 0 to 30%* (30% being the maximum permitted by BS 8110).

Example 4.7–1
Using eqn (4.7–5), calculate the value of K' for the following percentages of moment redistribution:

(a)   0–10%;
(b)   20%;
(c)   30%.

SOLUTION
For a given **percentage of moment redistribution**, it is seen from eqn (4.7–2) that

$$\beta_b = \frac{100 - (\% \text{ of moment redistribution})}{100}$$

(a)   0–10% moment redistribution. The corresponding β_b values are

$$\beta_b = 1 \text{ for zero moment redistribution}$$

$$\beta_b = 0.9 \text{ for 10\% moment redistribution}$$

In using eqn (4.7–5), β_b is not to be taken as exceeding 0.9. Substituting $\beta_b = 0.9$ into eqn (4.7–5) gives $K' = 0.156$.
(b)   20% moment redistribution. From eqn (4.7–2), $\beta_b = 0.8$. From eqn (4.7–5), $K' = 0.132$.
(c)   30% moment redistribution. From eqn (4.7–2), $\beta_b = 0.7$. From eqn (4.7–5), $K' = 0.104$.

Example 4.7–2
The bending moment at a beam section is M. Show that, *for the whole range of moment redistribution from 0 to 30%*, the lever arm z is given by the following BS 8110 formulae:*

$$\frac{z}{d} = 0.5 + \sqrt{\left(0.25 - \frac{K}{0.9}\right)} \quad \text{if } M \leq M_u \text{ of eqn (4.7–4)} \quad (4.7–6)$$

---

* The z/d factors in Table 24 of the I.Struct.E. Manual [14] are based on these equations, which are given in BS 8110: Clause 3.4.4.4.

$$\frac{z}{d} = 0.5 + \sqrt{\left(0.25 - \frac{K'}{0.9}\right)} \quad \text{if } M > M_u \text{ of eqn (4.7–4)} \quad (4.7–7)$$

where $K = M/f_{cu}bd^2$;
$\quad\quad\quad K' = M_u/f_{cu}bd^2$ (see eqn 4.7–4);
$\quad\quad\quad z$ = lever-arm distance measured from tension reinforcement A_s to the centroid of the concrete compression block.

SOLUTION

Case I: *0–10% moment redistribution*
For 0–10% moment redistribution, β_b in eqns (4.7–4) and (4.7–5) becomes equal to 0.9, so that eqn (4.7–5) reduces to eqn (4.6–5). In other words, for 0–10% moment redistribution, the moment equations revert back to eqns (4.6–4) and (4.6–5). We conclude from Examples 4.6–10 and 4.6–11 that:

$$\frac{z}{d} = 0.5 + \sqrt{\left(0.25 - \frac{K}{0.9}\right)} \quad \text{for } M \le M_u$$

(*Note:* eqn 4.7–4 ≡ eqn 4.6–5)

$$\frac{z}{d} = 0.5 + \sqrt{\left(0.25 - \frac{K'}{0.9}\right)} \quad \text{for } M > M_u$$

(*Note:* eqn 4.7–4 ≡ eqn 4.6–5)

Case II: *Moment redistribution > 10%*
When the moment redistribution > 10%, then eqn (4.7–4) sets a more severe limit on M_u than does eqn (4.6–5). However, if $M \le M_u$ of eqn (4.7–4), then we also have $M \le M_u$ of eqn (4.6–5), and the operating equation reverts back to eqn (4.6–4), i.e.

$$M = \left(0.405\frac{x}{d}\right)\left(1 - 0.45\frac{x}{d}\right)f_{cu}bd^2$$

Hence we conclude from Example 4.6–10 that

$$\frac{z}{d} = 0.5 + \sqrt{\left(0.25 - \frac{K}{0.9}\right)} \quad \text{for } M \le M_u \text{ of eqn (4.7–4)}$$

For $M > M_u$ of eqn (4.7–4), the capacity of the concrete section is given by eqn (4.7–4):

$$M_u = K'f_{cu}bd^2$$

where

$$K' = 0.4(\beta_b - 0.4) - 0.18(\beta_b - 0.4)^2$$

Of course, eqn (4.6–3) still holds:

$$z = d - 0.45x$$

where x is now governed by eqn (4.7–3):

$$x = (\beta_b - 0.4)d,$$

so that

$$z = d[1 - 0.45(\beta_b - 0.4)]$$

Eliminating $(\beta_b - 0.4)$ from the expressions for z and K', we have

$$\left(\frac{z}{d}\right)^2 - \left(\frac{z}{d}\right) + 1.111K' = 0$$

i.e.

$$\frac{z}{d} = 0.5 + \sqrt{\left(0.25 - \frac{K'}{0.9}\right)} \quad \text{for } M > M_u \text{ of eqn (4.7–4)}$$

where K' is given by eqn (4.7–5).

Note that BS 8110 restricts moment redistribution to not exceeding 30%.

Design procedure for rectangular beams (BS 8110/I.Struct.E. Manual)
The design procedure below is that of the I.Struct.E. Manual [14]. Comments have been added to explain the derivation of the formulae and the technical background. As explained at the beginning of this section, the procedure is valid for up to 30% moment redistribution. For **flanged beams**, see Section 4.8.

Consider again the beam section in Fig. 4.6–1(a). Suppose the design bending moment is M. Proceed as follows.

Step 1
Calculate M_u for concrete:

$$M_u = K'f_{cu}bd^2 \tag{4.7–8}$$

where K' is obtained from Table 4.7–1.

Table 4.7–1 K' factors for beams [14]

% moment redistribution	0–10	15	20	25	30
Values K'		0.156	0.144	0.132	0.119	0.104

Comments
Example 4.7–1 explains how the K' factor in Table 4.7–1 can be derived. See also eqn (4.7–5).

Step 2
If the design moment $M \leq M_u$ of Step 1, the tension reinforcement A_s is given by

$$A_s = \frac{M}{(0.87f_y)z} \tag{4.7–9}$$

where z is obtained from Table 4.7–2.

Comments
(a) The formula $A_s = M/(0.87f_y z)$ follows from Fig. 4.6–1(b); it is seen

Table 4.7–2   Lever-arm and neutral axis depth factors [14]

$K = M/bd^2 f_{cu}$	0.05	0.06	0.07	0.08	0.09	0.100	0.104	0.110	0.119	0.130	0.132	0.140	0.144	0.150	0.156
z/d		0.94	0.93	0.91	0.90	0.89	0.87	0.87	0.86	0.84	0.82	0.82	0.81	0.80	0.79	0.775
x/d		0.13	0.16	0.19	0.22	0.25	0.29	0.30	0.32	0.35	0.39	0.40	0.43	0.45	0.47	0.50

30%   25%   20%   15%   0–10%

Limit of table for various % of moment redistribution

that, by taking moments about the centroid of the concrete block: $M = (0.87f_y A_s)z$.
(b)   Table 4.7–2 is extracted from the I.Struct.E. Manual [14]. The lever-arm factors z/d are derived in Example 4.7–2.
(c)   In using Table 4.7–2, pay attention to the 'Limit of the table for various % of moment redistribution' as stated at the bottom of the table. Thus for 0–10% redistribution and $K > 0.156$, z/d and x/d are still to be taken as 0.775 and 0.50 respectively (as though $K = 0.156$). Similarly, for 15% redistribution and $K > 0.144$, z/d and x/d are still to be taken as 0.80 and 0.45 respectively. (Study Example 4.7–2 again, if necessary; note in particular the use of K and K' in the formulae for z/d.)

Step 3
If the design moment $M > M_u$ of Step 1:

(3a)   Compression reinforcement is required and its area A'_s is given by (see Comment (a) below)

$$A'_s = \frac{M - M_u}{0.87f_y(d - d')}$$

where d' is the depth of the compression steel from the concrete compression face.
(3b)   If

$$\frac{d'}{x} > \left(1 - \frac{f_y}{800}\right), \quad \text{use } f'_s = 700\left(1 - \frac{d'}{x}\right) \tag{4.7–10}$$

in lieu of $0.87f_y$ in the above formula for A'_s. See Comment (b) below.
(3c)   The area of the tension reinforcement A_s is given by (see Comment (c) below)

$$A_s = \frac{M_u}{0.87f_y z} + A'_s \tag{4.7–11}$$

where z is taken from Table 4.7–2.

Comments
(a)   The formula for A'_s in Step (3a) was derived in Section 4.6 as eqn (4.6–6). Of course, M_u is now calculated from K' in Table 4.7–1. See also eqn (4.7–5).

(b)   The formulae in Step (3b) were derived in Example 4.6–2; i.e. eqns (4.6–10) and (4.6–11).

(c)   To derive the formula in Step (3c), equate the forces in Fig. 4.7–1:

$$0.87f_y A_s = 0.405f_{cu}bx + 0.87f_y A'_s \qquad (4.7\text{–}12)$$

Noting from Fig. 4.7–1(b) that $M_u = (0.405f_{cu}bx)z$ we have

$$0.87f_y A_s = \frac{M_u}{z} + 0.87f_y A'_s \qquad (4.7\text{–}13)$$

$$A_s = \frac{M_u}{0.87f_y z} + A'_s$$

Of course, if the compression steel A'_s does not reach $0.87f_y$ then A_s should be calculated from:

$$0.87f_y A_s = \frac{M_u}{z} + f'_s A'_s \qquad (4.7\text{–}14)$$

where f'_s is as given in Step (3b) above.

Example 4.7–3
Repeat Example 4.6–5, assuming 15% moment redistribution.

SOLUTION

Step 1
$$M_u = K'f_{cu}bd^2 \quad \text{(where K' is 0.144 from Table 4.7–1)}$$
$$= (0.144)(40)(250)(700^2) \text{ Nmm}$$
$$= 705.6 \text{ kNm}$$

Step 2
$M(= 300 \text{ kNm}) < M_u$ of Step 1
Calculate

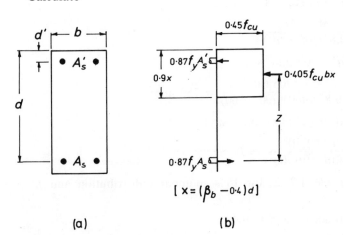

(a)   (b)

Fig. 4.7–1

$$K = \frac{M}{f_{cu}bd^2} = \frac{(300)(10^6)}{(40)(250)(700^2)} = 0.061$$

From Table 4.7–2, $z = 0.93d = 651$ mm.

$$A_s = \frac{M}{(0.87f_y)z} = \frac{(300)(10^6)}{(0.87)(460)(651)} = 1151 \text{ mm}^2$$

Comments

It can be seen that Step 2 here is the same as Step 2 of Example 4.6–5. This is because $M \leq M_u$ of Step 1. The solution (Case II) of Example 4.7–2 explains this point further.

Example 4.7–4

After 15% moment redistribution, the design moment M for a rectangular beam section of width 250 mm and effective depth 700 mm is 900 kNm. If $f_{cu} = 40$ N/mm^2 and $f_y = 460$ N/mm^2, design the reinforcement. What is the neutral axis depth factor x/d of the beam section so designed? (*Note:* This is a repeat of Example 4.6–6, apart from the 15% moment redistribution.)

SOLUTION

Step 1

$$M_u = K'f_{cu}bd^2 \quad \text{(where } K' \text{ is 0.144 from Table 4.7–1)}$$
$$= (0.144)(40)(250)(700^2) \text{ Nmm}$$
$$= 705.6 \text{ kNm}$$

Step 2

$M(= 900 \text{ kNm}) > M_u$ of Step 1
Therefore move to Step 3.

Step 3

$M > M_u$ of Step 1
(3a)   Compression steel is required:

$$A_s' = \frac{M - M_u}{0.87f_y(d - d')}$$

where $d' = 60$ mm, say,

$$= \frac{(900 - 705.6)(10^6)}{(0.87)(460)(700 - 60)} = 759 \text{ mm}^2$$

(3b)   Check d'/x ($d' = 60$ mm, say):

$$K = \frac{(900)(10^6)}{(40)(250)(700^2)} = 0.184$$

From Table 4.7–2, for 15% moment redistribution and $K \geq 0.144$,

$$z/d = 0.80; \qquad x/d = 0.45$$

(see Comment (c) under Table 4.7–2). Hence

$$\frac{d'}{x} = \frac{60}{(0.45)(700)}$$

$$= 0.19 < \left(1 - \frac{f_y}{800}\right)$$

Therefore the compression steel reaches $0.87f_y$ and A_s' as calculated in Step (3a) above is acceptable.

(3c) $A_s = \dfrac{M_u}{0.87f_y z} + A_s'$

z/d was found in Step (3b) to be 0.80.

$$A_s = \frac{(705.6)(10^6)}{(0.87)(460)(0.80)(700)} + 759 = 3907 \text{ mm}^2$$

Provide three 20 mm top bars (942 mm²)

Provide five 32 mm bottom bars (4021 mm²)

4.8   Flanged beams

The T-beam and the L-beam in Fig. 4.8–1 are examples of **flanged beams**. In practice the flange is often the floor slab and the question arises of what width of the slab is to be taken as the effective width; that is, the width b in Fig. 4.8–1. BS 8110 gives the following recommendations:

(a)   for a T-beam the **effective width** b should be taken as (1) $b_w + 0.2l_z$ or (2) the actual flange width, whichever is less;

(b)   for an L-beam the **effective width** b should be taken as (1) $b_w + 0.1l_z$ or (2) the actual flange width, whichever is less, where b_w is the web width (Fig. 4.8–1) and l_z is the distance between points of zero moment along the span of the beam. For a continuous beam, l_z may of course be determined from the bending moment diagram, but BS 8110 states that l_z may be taken as 0.7 times the effective span (as defined at the end of Section 4.5).

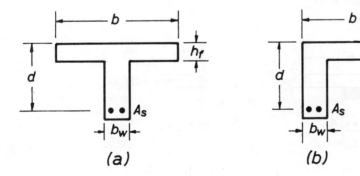

Fig. 4.8–1

For flanged beams, BS 8110: Clause 3.12.5.3 requires that **transverse reinforcement** should be provided near the top surface and across the full effective width of the flange. The area of such transverse reinforcement should not be less than 0.15% of that of the longitudinal cross-sectional area of the flange.

If the neutral axis is within the flange thickness, then a flanged beam may be analysed and designed as a rectangular beam of the same width b and effective depth d. The design chart in Fig. 4.5–2 becomes directly applicable. (*Note:* In Fig. 4.5–2, $\varrho = A_s/bd$ and $\varrho' = A_s'/bd$, where b is now the effective flange width.) The design procedure is straightforward. Work out M/bd^2 and if the design chart shows that $x \leq h_f$ (where h_f is the flange thickness) then the A_s value is read off the chart.

The I.Struct.E. Manual [14] gives a comprehensive procedure for practical design, using BS 8110's simplified rectangular stress block. Before explaining that procedure, we shall first describe a conventional quick design method.

Quick design method

In this approximate method, the x/d ratio is not checked explicitly. Instead, two simplifying assumptions are made:

(a) The depth of the BS 8110 rectangular stress block is not less than the flange thickness, i.e. $0.9x \geq h_f$. (If in fact $0.9x < h_f$, then the design errs slightly on the safe side.)

(b) The compressive force in the web below the flange (shaded area in Fig. 4.8–2(a)) is neglected.

The forces in the beam section are then as shown in Fig. 4.8–2(b), where the steel stress f_s depends on the x/d ratio. For $f_y = 460$ N/mm², Example 4.6–1 has shown that $f_s = 0.87f_y$ provided x/d does not exceed 0.64. Hence

$$M = 0.87f_y A_s \left[d - \frac{h_f}{2} \right] \qquad (4.8–1)$$

where $(d - h_f/2)$ is taken as the lever arm. If x/d exceeds 0.64, then the steel stress f_s is less than the design strength, and the **moment capacity** of the flange is given by

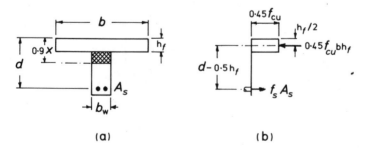

(a)   (b)

Fig. 4.8–2   Assumptions in quick design method

$$M_u = 0.45f_{cu}bh_f\left[d - \frac{h_f}{2}\right] \qquad (4.8-2)$$

When using this quick method, designers do not check whether $0.9x$ exceeds h_f. Instead, the design moment M is compared with M_u of eqn (4.8–2). If $M \le M_u$, then the steel area A_s is calculated from eqn (4.8–1). In the unlikely event that $M > M_u$ of eqn (4.8–2), it is usually simplest to increase the web dimensions. Otherwise, use the I.Struct.E. Manual's procedure [14] below.

I.Struct.E. Manual's design procedure (Case I: Moment redistribution not explicitly considered)

The procedure given below is that of the I.Struct.E. Manual [14]. Comments have been added to explain the derivation of the formulae and the technical background. Though moment redistribution is not explicitly considered, this 'Case I procedure' is in fact valid for *up to 10% moment redistribution*.

Step 1
Check x/d ratio. Calculate

$$K = \frac{M}{f_{cu}bd^2}$$

where M is the design moment and b the effective flange width. Obtain z/d and x/d from Table 4.6–1.

Comments
See comments on Table 4.6–1.

Step 2
Check whether $0.9x \le h_f$. If $0.9x \le h_f$ the BS 8110 rectangular stress block lies wholly within the flange thickness. The tension steel area A_s is determined as for a rectangular beam, using eqn (4.6–12):

$$A_s = \frac{M}{0.87f_y z} \qquad (4.8-3)$$

where z is obtained from Table 4.6–1 (see Step 1 above).

Comments
See eqn (4.6–12) and Table 4.6–1 and the comments that follow it.

Step 3
Check whether $0.9x > h_f$. If $0.9x > h_f$, the BS 8110 rectangular stress block lies partly outside the flange. Calculate the ultimate resistance moment of the flange M_{uf}:

$$M_{uf} = 0.45f_{cu}(b - b_w)h_f(d - 0.5h_f) \qquad (4.8-4)$$

Comments
The flanged section in Fig. 4.8–3(a) is considered to be made up of two components as shown in Fig. 4.8–3(b): the flange component and the web component. Equation (4.8–4) is obtained by considering the forces in Fig. 4.8–3(c) and taking moments about the tension steel A_s.

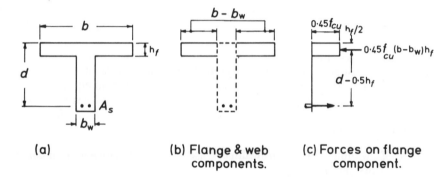

(a)

(b) Flange & web
components.

(c) Forces on flange
component.

Fig. 4.8–3

Step 4
    Calculate K_f.

$$K_f = \frac{M - M_{uf}}{f_{cu}b_w d^2}$$ (4.8–5)

where b_w is the web width. If $K_f \le 0.156$ (where 0.156 is K' as given by eqn 4.6–5), then calculate A_s from

$$A_s = \frac{M_{uf}}{0.87f_y(d - 0.5h_f)} + \frac{M - M_{uf}}{0.87f_y z}$$ (4.8–6)

where z is obtained from Table 4.6–1.

Comments
(a)   With reference to Fig. 4.8–4, the moment to be resisted by the web component (Fig. 4.8–4(c)) is the excess moment $M - M_{uf}$ (see eqn 4.8–4 for M_{uf}). Equation (4.8–5) treats the web component as a rectangular section, in the same way as eqns (4.6–4) and (4.6–5) do.

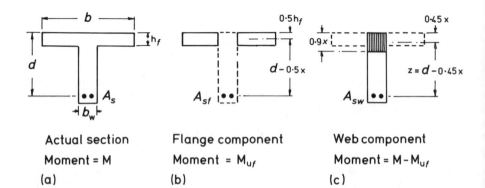

Actual section

Moment = M

(a)

Flange component

Moment = M_{uf}

(b)

Web component

Moment = $M-M_{uf}$

(c)

Fig. 4.8–4

(b) If K_f of eqn (4.8–5) does not exceed K' $(= 0.156)$ of eqn (4.6–5), we know that the moment capacity of the web component is not exceeded, i.e. compression steel is not required.

(c) To derive eqn (4.8–6), consider the actual A_s in Fig. 4.8–4(a) to be made up of A_{sf} of the flange component (Fig. 4.8–4(b)) and A_{sw} of the web component (Fig. 4.8–4(c)). By considering Fig. 4.8–4(b),

$$M_{uf} = 0.87f_y A_{sf}(d - 0.5h_f)$$

or

$$A_{sf} = \frac{M_{uf}}{0.87f_y(d - 0.5h_f)} \qquad (4.8–6(a))$$

By considering Fig. 4.8–4(c),

$$M - M_{uf} = 0.87f_y A_{sw} z$$

or

$$A_{sw} = \frac{M - M_{uf}}{0.87f_y z} \qquad (4.8–6(b))$$

where z is from Table 4.6–1. Equation (4.8–6) follows from $A_s = A_{sf} + A_{sw}$, i.e.

eqn (4.8–6) = eqn (4.8–6(a)) + eqn (4.8–6(b))

(d) In the unlikely event that $K_f > 0.156$, redesign the section or consult Example 4.8–1 for design of **compression steel**.

I.Struct.E. Manual's design procedure (Case II: 0–30% moment redistribution)*

The procedure below is that of the I.Struct. E. Manual [14]. Comments have been added to explain the derivation of the formulae and the technical background. The design steps are identical to those of Case I above, except as mentioned below.

Step 1

As Step 1 of Case I, except that the x/d and z/d ratios are now to be obtained from Table 4.7–2.

Comment

Comment (c) on below Table 4.7–2 explains how to use the table for various % moment redistribution.

Step 2

As Step 2 of Case I, except that in calculating A_s from eqn (4.8–3), the lever arm z is now obtained from Table 4.7–2. (Note the similarities between eqns 4.8–3 and 4.7–9.)

Step 3

As Step 3 of Case I.

---

* This procedure (Case II) may be omitted on first reading. Readers may move on to Section 4.9 and then read Section 4.7 before returning to this page.

Step 4

K_f is calculated as in Step 4 of Case I, using eqn (4.8–5).

If $K_f \leq K'$, obtained from Table 4.7–1, then obtain the lever arm z from Table 4.7–2 and calculate A_s from eqn (4.8–6).

Comments
(a) The equations in this step are derived in the comments below Step 4 of Case I.
(b) The essential differences between Step 4 here and Step 4 of Case I are:
  (1) K_f is now compared with K' of Table 4.7–1 (and not with $K' = 0.156$ of eqn 4.6–5);
  (2) in using eqn (4.8–6), z is now obtained from Table 4.7–2 (and not from Table 4.6–1).
(c) If $K_f > K'$ of Table 4.7–1, redesign the section or consult Example 4.8–1 for the design of the **compression steel**.

Example 4.8–1
With reference to Step 4 of the design procedure above, if $K_f > K'$, explain how to design the **compression steel**.

SOLUTION
The I.Struct.E. Manual's advice [14] on this point is: 'consult BS 8110 for design of compression steel'. With reference to Fig. 4.8–4(c), the moment capacity of the web component is

$$M_{uw} = K' f_{cu} b_w d^2 \qquad (4.8–7)$$

where K' is obtained from Table 4.7–1. The design moment M is partly resisted by M_{uf} (see eqn 4.8–4) and partly by M_{uw} (eqn 4.8–7). The difference $M - M_{uf} - M_{uw}$ will be resisted by compression steel:

$$0.87 f_y A_s'(d - d') = M - M_{uf} - M_{uw} \qquad (4.8–8)$$

Equation (4.8–8) gives the required area A_s' of the compression steel. The tension steel area A_s is then determined from the equilibrium of forces:

$$0.87 f_y A_s = \text{flange compression} + \text{web compression}$$

$$+ \text{ steel compression}$$

$$0.87 f_y A_s = 0.45 f_{cu}(b - b_w) h_f + 0.45 f_{cu} b_w (0.9x) + 0.87 f_y A_s'$$
$$(4.8–9)$$

where, in the second term on the right-hand side, the neutral axis depth x is obtained from Table 4.7–2.

Comment
Compare eqn (4.8–8) with eqn (4.7–10); compare eqn (4.8–9) with eqn (4.7–12).

4.9 Moment redistribution—the fundamental concepts

Before discussing the ultimate load behaviour of reinforced concrete **continuous beams**, we shall briefly refer to that of a continuous beam made of an ideally elastic-plastic material, that is, a material having the stress/strain relation in Fig. 4.9–1(a). An ideally elastic-plastic beam section will have the **moment/curvature characteristics** in Fig. 4.9–1(b); that is for a section of the beam subjected to an increasing moment M, the curvature $1/r$ (where r is the radius of curvature) at that section increases linearly with M until the value M_p, called the **plastic moment of resistance**, is reached; the curvature then increases indefinitely.

Figure 4.9–2(a) shows a two-span uniform beam made of such a material, subjected to midspan loads Q; Figure 4.9–2(b) shows the elastic bending moment diagram. Suppose the magnitude of Q is just large enough for the moment at section C to reach the value M_p. Then, from Fig. 4.9–1(b), it is seen that a further increase in the magnitude of Q, to Q' say, will not increase the bending moment at C. A **plastic hinge** is said to have developed at C, because after the moment at that section reaches M_p, the beam behaves as though it is hinged there. Thus, under the increased loads Q', the moment at B is

$$M_B = \tfrac{5}{32}Ql + \tfrac{1}{4}(Q' - Q)l$$

where $5Ql/32$ is from Fig 4.9–2(b) and the increase of moment of $(Q' - Q)l/4$ is the simple-beam bending moment corresponding to the load increment $(Q' - Q)$. Therefore, as Q is increased the moments at B and D will eventually reach the value M_p (Fig. 4.9–2(c)) and the beam will collapse in the mode in Fig. 4.9–2(d), where the beam is no longer a structure but a **mechanism**; the collapse mode is often referred to as the **collapse mechanism**. Let Q_u be the value of Q at collapse. From Fig. 4.9–2(c),

$$M_B = \frac{Q_u l}{4} - \frac{M_c}{2}$$

where now both M_B and M_C equal M_p; hence

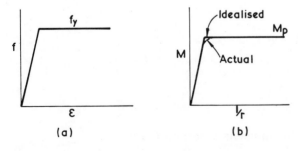

Fig. 4.9–1

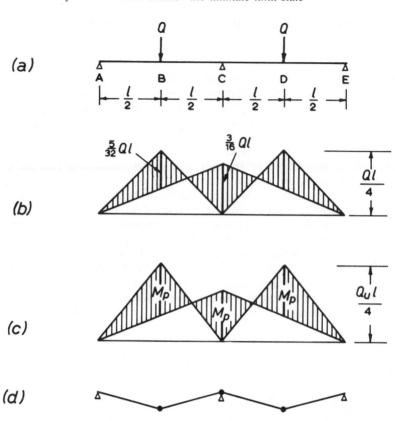

(a)

(b)

(c)

(d)

(e)

Fig. 4.9–2

$$M_p = \frac{Q_u l}{4} - \frac{M_p}{2}$$

or

$$Q_u = 6\left(\frac{M_p}{l}\right)$$

Therefore, at collapse, the moment at section C is

$$M_C = M_p = \tfrac{1}{6}Q_u l \qquad\qquad (4.9\text{–}1(a))$$

If the beam had remained elastic, the bending moment diagram at collapse would have been that of Fig. 4.9–2(b) with Q_u substituted for Q, and the (hypothetical) elastic moment at C would have been

$$(M_e)_C = \tfrac{3}{16} Q_u l \qquad\qquad\qquad (4.9-1(b))$$

Comparison of eqns (4.9–1(a), (b)) and of Figs 4.9–2(b) and (c), shows that the rotation at the plastic hinge at C has resulted in a **moment redistribution**; after the formation of the plastic hinge at C, the bending moment there is smaller (and that at B greater) than what it would have been if the beam had remained elastic. The bending moment at C after the moment redistribution is $Q_u l/6$. Let us, for the time being, define the **moment redistribution ratio** β_b as the ratio of the bending moment at a section after redistribution to that before redistribution (see formal definition in eqn 4.9–2). Then, for the beam in Fig. 4.9–2, the moment redistribution ratio at section C is

$$\beta_b = \frac{Q_u l/6}{3Q_u l/16} = 0.889$$

If, by design, the beam in Fig. 4.9–2(a) has a reduced moment of resistance of, say, $0.5\,M_p$ in the neighbourhood of C and an enhanced moment of resistance of, say, $1.25\,M_p$ in the neighbourhood of B and of D, then the bending moment diagram at collapse will be as in Fig. 4.9–2(e), from which

$$M_B = \frac{Q_u l}{4} - \frac{M_C}{2}$$

where M_B and M_C are now equal to $1.25 M_p$ and $0.5 M_p$ respectively. Whence

$$Q_u = 6\left(\frac{M_p}{l}\right)$$

as before.

It is thus seen that the designer has a choice: the loads Q_u may be resisted by a beam of uniform M_p or by one with $0.5 M_p$ at C and $1.25 M_p$ at B and D; in fact many other combinations are possible. For the combination of $0.5 M_p$ and $1.25 M_p$, the moment redistribution for section C is

$$\beta_b = \frac{Q_u l/12}{3Q_u l/16} = 0.444$$

where $Q_u l/12 = 0.5 M_p$ is the actual ultimate moment at C, and $3Q_u l/16$ is the elastic moment computed from the same loading. (*Note:* The value $3Q_u l/16$ may not be theoretically exact because of the local variations in resistance moment.)

Suppose now the uniform beam in Fig. 4.9–2(a) is made of a brittle elastic material so that it has the moment/curvature characteristics in Fig. 4.9–3; the curvature increases linearly with M until M_u is reached, when sudden rupture occurs. Referring to Fig. 4.9–2(b), the ultimate value of Q is given by

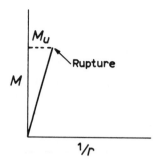

Fig. 4.9–3

$$\tfrac{3}{16}Q_u l = M_u, \quad \text{or} \quad Q_u = 5.33\left(\frac{M_u}{l}\right)$$

When Q is just under this value, the moments are as in Fig. 4.9–2(b). However, the instant Q exceeds Q_u, rupture at C occurs and the continuous beam splits into two simple beams, AC and CE. The simple-span moments at B and D are then

$$M_B = M_D = \tfrac{1}{4}Q_u l$$

Since this is greater than M_u of the beam ($= 3Q_u l/16$) rupture occurs at both B and D. In other words, as soon as the M_u value is exceeded at C, the whole structure collapses without warning. For a beam without ductility, such as this one, no moment redistribution is possible.

Structural design which takes account of moment redistributions, such as those in Fig. 4.9–2(c) and (e), and in which the cross-sections of the structural members are proportioned on the basis of their ultimate strengths, is called **plastic design** or **limit design** (which term should not be confused with limit state design). Strictly speaking, limit design refers only to design which takes account of moment redistribution, but leaves open the question of how the individual cross-sections of the members are to be proportioned after the design moments have been worked out. In structural steelwork, plastic design is an established design method and an authoritative account of the subject has been given by Baker, Heyman and Horne [15, 16] (see also References 17 and 18). In structural concrete [19, 20], ultimate strength procedures are adopted in BS 8110 for the design of individual member sections, but limit design as such is only partially recognized, and then only indirectly. Of course, whether limit design is accepted for concrete structures depends very much on adequate ductility in plastic hinge regions. Figure 4.9–4 shows that the moment/curvature relation for an under-reinforced beam resembles that of Fig. 4.9–1(b), while that for an over-reinforced beam resembles the one in Fig. 4.9–3. Of course, Fig. 4.9–4 only gives a simplified picture of the truth. The precise shape of the M against $1/r$ curve depends on the type of reinforcement steel, on the steel ratio ϱ, and, for doubly reinforced beams, on the difference in steel ratios $(\varrho - \varrho')$; in addition, the presence of an adequate amount of links or stirrups will increase ductility appreciably.

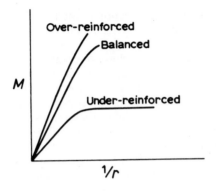

Fig. 4.9–4

In current British design practice, the bending moments in the members of a continuous structure are determined by elastic analysis. BS 8110 : Clause 3.2.2 then permits the elastic moments to be redistributed. The aim of such moment redistribution is to distribute bending moments away from peak moment regions, such as beam–column joints or supports of continuous members. This reduces the congestion of reinforcement bars at such regions and makes the structural members easier to detail and construct. Of course, certain conditions must be observed. Firstly, equilibrium must be maintained; this means that where the bending moments are reduced at some sections they will have to be appropriately increased at others. In other words, moment redistribution may lead either to an increase or a decrease in the design bending moment at a given section. For the purpose of certain design formulae (e.g. eqn 4.9–3 below), BS 8110 formally defines the **moment redistribution ratio** β_b as

$$\beta_b = \frac{\text{moment at a section after redistribution}}{\text{moment at the section before redistribution}} \leq 1 \qquad (4.9\text{--}2)$$

Where moment redistribution leads to a reduction in the design ultimate moment at a cross-section subjected to the largest moment within each hogging or sagging region, then BS 8110 (Amendment No. 1 of May 1986) requires that eqn (4.9–3) below should be used to check the x/d ratio of that section as finally designed:

$$\frac{x}{d} \leq (\beta_b - 0.4) \quad \text{or} \quad 0.6 \qquad\qquad \textbf{(4.9--3)}$$

whichever is the lesser.

For a singly reinforced beam, x/d increases with the steel ratio ϱ; for a doubly reinforced beam, it increases with the difference $(\varrho - \varrho')$. Therefore, in eqn (4.9–3), the β_b ratio is related indirectly to the ductility and rotation capacity of the member (see Fig. 4.9–4). To prevent an excessive demand on the ductility of a structural member, a 15% moment redistribution is normally to be taken as a reasonable limit [14], though certainly *BS 8110 permits up to 30% moment redistribution*. These limits apply where the moment redistribution leads to a reduction in the bending moment at a given section; where it leads to an increase in the bending moment, no restriction is necessary.

In the design of continuous beams it is easy to overlook one important point. Consider the fixed-end beam in Fig. 4.9–5(a). The elastic bending moment diagram corresponding to the ultimate loading is shown by the chain-dotted line in Fig. 4.9–5(b). The full line shows the redistributed moment diagram for use in the design of the individual cross-sections for the ultimate limit state. The dotted line is the elastic moment diagram corresponding to the service loading (see Table 1.5–1 for the γ_f factors for service loads and for ultimate loads). Figure 4.9–5(b) shows that the region ab, though under a sagging moment at the ultimate condition, is under a hogging moment at the service condition. The ultimate load condition requires no top reinforcement in the region ab, and consequently wide cracks would develop there at the service condition. To guard against such cracking, BS 8110 imposes the condition that

$$M_u \not< 0.7M_e \qquad\qquad (4.9\text{–}4)$$

where M_u is the ultimate resistance moment provided at any section of the member, and M_e is the moment at that section obtained from an elastic maximum-moments diagram covering all appropriate combinations of ultimate loads.

Equation (4.9–4) is compatible with the above-mentioned 30% limit on the moment reduction that is permitted in moment redistribution. Returning to Section 4.7, it is now clear why the design formulae and procedure there are stated as valid for only up to 30% moment redistribution. Note that the design formulae in Section 4.6 imply that x/d may reach 0.5. For x/d to reach 0.5, eqn (4.9–3) states that $\beta_b \geq 0.9$; in other words, there must be no more than 10% moment redistribution (see eqn 4.7–2). Hence the design formulae and procedure in Section 4.6 are valid only for up to 10% moment redistribution.

Example 4.9–1

The span lengths of a three-span continuous beam ABCD are: exterior spans AB and CD, 8 m each; interior span BC, 10 m. The characteristic

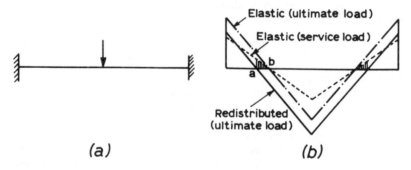

(a)

(b)

Fig. 4.9–5

dead load G_k (inclusive of self-weight) is 36 kN/m and the characteristic imposed load Q_k is 45 kN/m. Draw the **bending moment envelope** (or the maximum-moments diagram) for the ultimate condition. It is desired to take advantage of moment redistributions to equalize as far as possible the bending moments at the various critical sections. (See comments at the end on the loading arrangements.)

SOLUTION

Step 1

With reference to Table 1.5–1 and to the description of its use in Section 1.5, the design loads are

$$1.4G_k + 1.6Q_k = (1.4)(36) + (1.6)(45) = 122.4 \text{ kN/m}$$
$$1.0G_k = (1.0)(36) = 36 \text{ kN/m}$$

Step 2

There are three loading cases to consider (Fig. 4.9–6): Case 1 for maximum hogging moment (designated negative here) at B, Case 2 for maximum sagging moment at span BC and Case 3 for maximum sagging moment at spans AB and CD. (Strictly, there should be a Case 4, which is a mirror reflection of Case 1, but the reader should study the solution here and satisfy himself that that case is in fact covered.) See also the comments at the end of the solution.

Step 3

Consider Case 1. The reader should verify that the elastic bending moment diagram is that of the chain-dotted line in Fig. 4.9–6. (*Hint:* The support moments, −1097 and −667 kNm, may be determined by one of the methods in Reference 12, Chapters 4 and 6. The moment 506 kNm is obtained by considering AB as a simple beam supporting a uniformly distributed load of $1.4G_k + 1.6Q_k = 122.4$ kN/m and acted on by a hogging moment of 1097 kNm at B. Similarly for the moment 648 kNm.)

Step 4

Consider moment redistribution for Case 1. Equation (4.9–4) permits the hogging moment at B to be reduced, numerically, to 70% of the elastic value, provided of course that x/d of the section as finally designed does not exceed 0.3 (see eqn 4.9–3):

$$70\% \text{ of } (-1097 \text{ kNm}) = -768 \text{ kNm}$$

The redistributed moment M_B is −768 kNm; adjust M_C to the same value of −768 kNm. The reader should now verify that the redistributed moment diagram is the full line. (*Hint:* The moments 632 kNm in AB and 762 kNm in BC may be obtained using the hint in Step 3.)

Step 5

Consider, for Case 1, the condition $M_u \nleqslant 0.7M_e$ imposed by eqn (4.9–4). Strictly speaking, it is necessary to draw the complete moment diagram with ordinates equal to 70% of the chain-dotted curve. However, by inspection, only the portions represented by the dotted lines

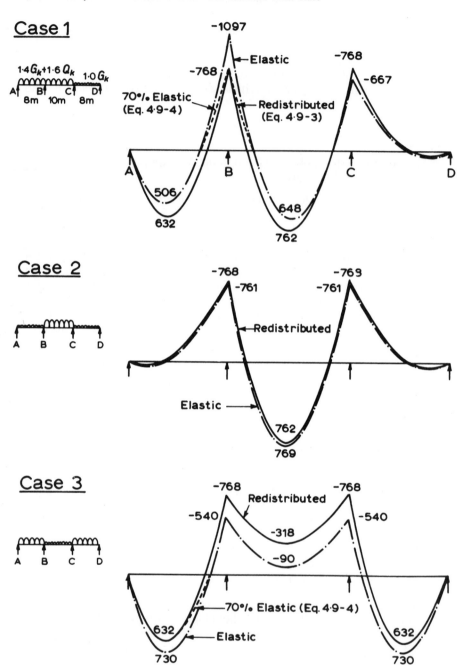

Fig. 4.9–6   Moment redistribution

need be drawn, because elsewhere eqn (4.9–4) is not critical (that is, elsewhere M_u exceeds 70%M_e).

Step 6

For Case 2, the reader should verify that the elastic moment diagram is shown by the chain-dotted line. (*Hint:* Use hint in Step 3.) Adjust M_B and M_C to −768 kNm for uniformity with Case 1. The maximum redistributed moment in span BC becomes 762 kNm. That this moment should be 762 kNm in both Case 1 and Case 2 is not accidental, but follows from the laws of equilibrium; in both Case 1 and Case 2 the span BC supports the load $1.4G_k + 1.6Q_k$ and is acted on by moments of −768 kNm at the ends.

Step 7

For Case 3, the elastic moments at B and C are again adjusted to −768 kNm for uniformity with Case 1. The maximum redistributed moments at AB and BC become 632 kNm as in Case 1.

Step 8

On the basis of the above-drawn moment diagram, the design moment envelope for the ultimate limit state is as shown in Fig. 4.9–7(a). Note that over the length ab the condition $M_u \nless 0.7M_e$ governs the design.

Step 9

For comparison, the elastic moment envelope is shown in Fig. 4.9–7(b), which is constructed from the chain-dotted curves in Fig. 4.9–6. Of course, this elastic moment envelope may legitimately be used to proportion the beam sections for the ultimate limit state; however, the

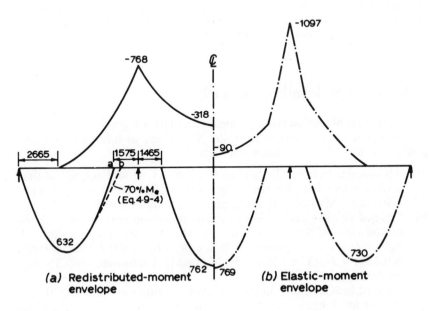

(a) Redistributed-moment envelope

(b) Elastic-moment envelope

Fig. 4.9–7

peak moment of -1097 kNm would have to be catered for and this will result in congestion of steel at section B and C. The use of the redistributed moments in Fig. 4.9–7(a) leads to a more even distribution of reinforcement throughout the beam and also to some overall saving in steel.

Step 10

When the beam is finally designed, a check should be made that the β_b ratios do not violate eqn (4.9–3).

$$\text{From Case 1:} \quad \beta_b = \frac{768}{1097} = 0.70 \text{ at B}$$

$$\text{From Case 2:} \quad \beta_b = \frac{762}{769} = 0.99 \text{ near midspan BC}$$

$$\text{From Case 3:} \quad \beta_b = \frac{632}{730} = 0.87 \text{ near midspans AB and CD}$$

Therefore the x/d ratios of the beam as finally designed must not exceed 0.3 at B, 0.59 near midspan BC, nor 0.47 near midspans AB and CD.

Comments on Step 2

As will be explained in Section 11.4(a), BS 8110: Clause 3.2.1.2.2 states that it is normally necessary to consider only two loading cases, as illustrated in Figs 11.4–1(c) and (d):

(a) the maximum load $(1.4G_k + 1.6Q_k)$ on all spans;
(b) $(1.4G_k + 1.6Q_k)$ on alternate spans, and the minimum load $1.0G_k$ on all other spans.

However, Example 4.9–1 is the first time we deal with a continuous beam, and we have therefore chosen to work from first principles as far as possible. Hence we have considered the loading cases as illustrated in Fig. 4.9–6.

4.10   Design details (BS 8110)

Minimum areas of tension reinforcement (BS 8110: Clause 3.12.5)
The requirements for steel of $f_y = 460$ N/mm^2 are as follows (main bars should not normally be smaller than size 16 [14]):

(a) **Rectangular beams**: The tension steel area A_s should not be less than 0.13% of bh where b is the beam width and h the overall depth. (The I.Struct.E. Manual [14] recommends a minimum of 0.2% bh.)
(b) **Flanged beams** (web in tension): The minimum percentages depend on the ratio of the web width b_w to the effective flange width b. If $b_w/b < 0.4$, the minimum is 0.18% b_wh; if $b_w/b \geq 0.4$, the minimum is 0.13% b_wh. (The I.Struct.E. Manual [14] recommends 0.2% b_wh for both cases.)
(c) **Flanged beams** (flange in tension over a continuous support): 0.26% of b_wh for T-beams; 0.2% of b_wh for L-beams.

(d) **Transverse reinforcement** in flanged beams: Transverse reinforcement shall be provided near the top surface of the flange, over the full effective width b. The area A_{st} of such reinforcement should not be less than 0.15% of $h_f l$, where h_f is the flange thickness and l the beam span.

Comments
(a) The minimum requirement for A_s will come into operation for those beams which, for architectural or other reasons, are made substantially larger than required by strength calculations. If the steel ratio is too low, the ultimate moment of resistance may become less than the cracking moment of the plain concrete section computed on the basis of its modulus of rupture. As soon as the load is sufficient to crack the beam, sudden collapse occurs without warning.
(b) As regards the **transverse reinforcement** in flanged beams, it can be shown from the principles of mechanics, and from laboratory experiments, that the compression force in the flange tends to induce longitudinal cracks at the flange/web junctions. The above-specified transverse reinforcement controls such longitudinal cracking as well as shrinkage and thermal cracking.

Minimum areas of compression reinforcement (BS 8110 : Clause 3.12.5)
Where compression reinforcement is required for the ultimate limit state, A_s' should not be less than 0.2% of bh for a rectangular beam or 0.2% of $b_w h$ for a flanged-beam web in compression. Use size 16 or larger bars [14].

Maximum areas of main reinforcement (BS 8110 : Clause 3.12.6)
Neither A_s nor A_s' should exceed 4% of bh (or 4% $b_w h$ for flange beams).

Links or stirrups (BS 8110 : Clauses 3.4.5 and 3.12.7)
Links or stirrups are required either to resist shear (see Chapter 6) or to contain the compression reinforcement against outward buckling. The **minimum practical link size** is size 8. The following requirements must be met:

(a) Where compression reinforcement is used in a beam, links of at least one-quarter of the diameter of the largest compression bar must be provided, at a spacing not exceeding 12 times the diameter of the smallest compression bar. These links should be so arranged that every corner and alternate bar in an outer layer of reinforcement is supported.
(b) Links for shear resistance, where required, must satisfy eqns (6.4–2) and (6.4–3).
(c) Except in beams of minor structural importance, such as lintels, **minimum links** are required along the entire span; see eqn (6.4–2).

Comments
Links provided to restrain the compression reinforcement can be considered fully effective in resisting shear, and vice versa. For **anchorage of links**, see Section 6.4, Step 6.

Slenderness limits (BS 8110: Clause 3.4.1.6)
To guard against lateral buckling, the **slenderness limits** in Table 4.10–1 should not be exceeded. In the table, the term slenderness limit is defined as the clear distance between lateral restraints, d is the effective depth and b_c the breadth of the compression face of the beam midway between restraints.

Table 4.10–1   Slenderness limits

Type of beam	Slenderness limit
Simply supported	$60b_c$ or $250b_c^2/d$ whichever is the lesser
Continuous	Same as above
Cantilever	$25b_c$ or $100b_c^2/d$ whichever is the lesser

Comments
The slenderness limits in Table 4.10–1 are intended for ordinary beams; design guidance on **deep beams**, in which the depth h is comparable to the span l, is given in the CIRIA deep-beam guide [21]. See also the recent References 22 and 23.

Minimum distance between bars (BS 8110: Clause 3.12.11.1)
(a)   The horizontal clear distance between bars should not normally be less than $h_{agg} + 5$ mm, or less than ϕ, whichever is greater, where h_{agg} is the nominal maximum size of the coarse aggregate and ϕ is the bar size (or the size of the larger bars if they are unequal).
(b)   Where the bars are arranged in two or more rows, the gaps between the corresponding bars in each row should be vertically in line and the vertical clear distance between the bars should not be less than $\frac{2}{3}h_{agg}$ or ϕ, whichever is greater.
(c)   Where an internal vibrator is intended to be used, sufficient space should be left between bars to enable the vibrator to be inserted.

Maximum distance between bars (BS 8110: Clause 3.12.11.2)
The restrictions on maximum distance are intended for controlling crack widths. They therefore apply to tension bars only and are summarized in Fig. 5.4–1. Detailed explanations are given in Section 5.4.

Minimum lap length (BS 8110: Clause 3.12.8)
The minimum lap should not in any case be less than 15 times the bar size or 300 mm whichever is the greater; for fabric reinforcement it should not be less than 250 mm. BS 8110's further requirements are as follows.

Tension laps
The lap length should be at least equal to the anchorage length (See eqns 6.6–3(a), (b)) required to develop the stress in the smaller of the two bars lapped.

(a)   Where a lap occurs at the top of a section as cast, and the minimum cover is less than twice the bar size, the lap length should be multiplied by a factor of 1.4.
(b)   Where a lap occurs at the corner of a section and the minimum cover to either face is less than twice the bar size, or where the clear distance between adjacent laps is less than 75 mm or six times the bar size, whichever is greater, the lap length should be increased by a factor of 1.4.
(c)   In cases where both conditions (a) and (b) apply, the lap length should be increased by a factor of 2.

Compression laps
The lap length should be at least 25% greater than the compression anchorage length (see eqns 6.6–3 (a), (b)). Table 4.10–2 gives values of ultimate anchorage bond lengths and lap lengths for Type 2 deformed bars ($f_y = 460$ N/mm^2) and BS 4483 Fabrics, for $f_{cu} = 40$ N/mm^2 and over, as calculated from eqn (6.6–3(b)).

Table 4.10–2   Ultimate anchorage lengths and lap lengths for $f_{cu} \geq 40$ N/mm^2 ($\phi =$ bar size)

Reinforcement type	$f_y = 460$ N/mm^2	Fabric
Tension anchorage and lap lengths	32ϕ	25ϕ
Compression anchorage length	26ϕ	20ϕ
Compression lap length	32ϕ	25ϕ

Curtailment and anchorage of bars (BS 8110 : Clause 3.12.9)
(a)   Except at an end support, every bar should extend beyond the theoretical cut-off point for a distance not less than the effective depth of the member or 12 bar sizes, whichever is greater. The **theoretical cut-off point** is defined as the location where the resistance moment of the section, considering only the continuing bars, is equal to the required moment.
(b)   A bar stopped in a tension zone should satisfy the additional requirement that it extends a full anchorage length l_u (eqn 6.6–3(b)) from the theoretical cut-off point, unless other conditions detailed in Clause 3.12.9.1 of BS 8110 are satisfied.
(c)   At a simple end support, each tension bar should have an effective anchorage of 12 bar sizes beyond the centre line of the support unless other conditions detailed in Clause 3.12.9.4 of BS 8110 are satisfied. The effective anchorage lengths of hooks and bends are explained in Fig. 6.6–1.

Comments
For anchorage of links, see Section 6.6.

Simplified rules for curtailment of bars (BS 8110 : Clause 3.12.10.2)
The recommendations above relate the curtailment of bars to the theoretical cut-off points. In practical design, bending moment diagrams are often

not drawn for members of secondary importance, and the theoretical cut-off points are not then known without further calculation. BS 8110 permits the following simplified rules to be applied, where the beam supports substantially uniformly distributed loads. In these rules, l refers to the effective span length, and ϕ the bar size.

(a) **Simply supported beams**: All the tension bars should extend to within $0.08l$ of the centres of supports. At least 50% of these bars should further extend for at least 12ϕ (or its equivalent in hooks or bends) beyond centres of supports.

(b) **Cantilevers**: All the tension bars at the support should extend a distance of $l/2$ or 45ϕ, whichever is greater. At least 50% of these bars should extend to the end of the cantilever.

(c) **Continuous beams of approximately equal spans**
   (1) All the tension bars at the support should extend $0.15l$ or 45ϕ from the face of support, whichever is greater. At least 60% of these bars should extend $0.25l$ and at least 20% should continue through the spans.
   (2) All the tension bars at midspan should extend to within $0.15l$ of interior supports and $0.1l$ of exterior supports. At least 30% of these bars should extend to the centre of supports.
   (3) At a simply supported end, the detailing should be as given in (a) for a simply supported beam.

Concrete cover for durability (BS 8110 : Clause 3.3)
Table 2.5–7 in Section 2.5(e) gives the concrete covers to meet the durability requirements for beams and other structural members. Note that in Table 2.5–7, the meaning of nominal cover is as defined in BS 8110 : Clause 3.3.1.1, namely: the **nominal cover** is the design depth of the concrete cover to all steel reinforcement, including links. It is the dimension used in the design and indicated on the drawings.

Fire resistance (BS 8110 : Clause 3.3.6)
The fire resistance of a beam depends on its width and the concrete cover, as shown in Table 4.10–3. Note that, in practice, the fire resistance requirements may dictate the size of the beam and the concrete cover.

Table 4.10–3   Fire resistance requirements for beams
(BS 8110 : Part 2 : Clause 4.3.1)

Fire rating (hours)	Minimum width (mm)		Concrete cover to MAIN reinforcement (mm)	
	Simply supported	Continuous	Simply supported	Continuous
1	120	80	30	20
2	200	150	50	50
3	240	200	70[a]	60[a]
4	280	240	80[a]	70[a]

[a] See BS 8110 : Part 2 : Clause 4 for protection against spalling.

4.11   Design and detailing—illustrative example

The design example below explains how to determine member details to satisfy strength and serviceability requirements. At this stage, the emphasis is on the detailed design calculations. Preliminary analysis and **member sizing**, together with other topics of practical importance, will be dealt with in Chapter 11.

Example 4.11–1

Figure 4.11–1 shows the L-shaped cross-section of an edge beam in a typical floor of a multistorey framed building. The dimensions 375 by 325 mm in Fig. 4.11–1(a) and the effective depths 320 and 325 mm in Fig. 4.11–1(b) have been obtained from a preliminary analysis and member sizing. The beam is continuous over many equal spans of 5500 mm each. Check the adequacy of the concrete section and design the reinforcement for an interior span. The design information is given below.

Exposure condition	moderate
Fire resistance	1 hour
Characteristic dead load g_k (inclusive of self-weight)	15.62 kN/m
Characteristic imposed load q_k	4.13 kN/m
Characteristic strengths	
Concrete f_{cu}	40 N/mm^2
Reinforcement f_y (main bars)	460 N/mm^2
f_y (links)	250 N/mm^2

SOLUTION
(See also the comments at the end.)

Step 1   Durability and fire resistance

From Table 2.5–7, the nominal cover for moderate exposure is 30 mm, i.e. 40 mm to main bars assuming 10 mm links.

From Table 4.10–3, the fire resistance for a 325 mm beam width and 40 mm cover to main bars exceeds one hour.

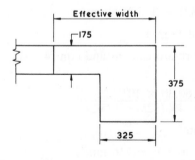

(a) Gross dimensions

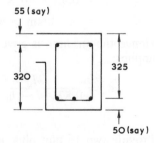

(b) Estimated effective depths

Fig. 4.11–1

Step 2   Design load

$$G_k = lg_k = 5.5 \times 15.62 = 85.91 \text{ kN}$$

$$Q_k = lg_k = 5.5 \times 4.13 = 22.72 \text{ kN}$$

From Table 1.5–1,

design load $F = 1.4G_k + 1.6Q_k$

$$= 120.27 + 36.36 = 156.63 \text{ kN}$$

Step 3   Ultimate moments

From Table 11.4–1, the ultimate moments are

M at supports: $0.08Fl = (0.08)(156.63)(5.5) = 68.92$ kNm

M at midspan: $0.07Fl = (0.07)(156.63)(5.5) = 60.30$ kNm

Step 4   Support main reinforcement

$$\frac{M}{bd^2} = \frac{68.92 \times 10^6}{325 \times 320^2} = 2.07 \text{ N/mm}^2$$

From Fig. 4.5–2,

$$A_s = 0.55\% \ bd = 0.0055 \times 325 \times 320 = 572 \text{ mm}^2$$

Provide two 20 mm bars ($A_s = 628 \text{ mm}^2$)

Step 5   Midspan main reinforcement

The midspan section is that of an L-beam. From Section 4.8,

effective width of flange $= b_w + \dfrac{l_z}{10}$

$$= 325 + \frac{0.7 \times 5500}{10}$$

$$= 710 \text{ mm}$$

From eqn (4.8–2),

$M_u = 0.45f_{cu}bh_f(d - h_f/2)$

$$= 0.45 \times 40 \times 710 \times 175(325 - 175/2) \text{ Nmm}$$

$$= 531 \text{ kNm} > 60.30 \text{ kNm of Step 3}$$

Hence the strength is governed by the reinforcement and eqn (4.8–1) applies:

$$A_s = \frac{M_u}{0.87f_y(d - h_f/2)} = \frac{60.30 \times 10^6}{0.87 \times 460(325 - 175/2)}$$

$$= 634 \text{ mm}^2$$

Provide two 16 mm plus one 20 mm bars ($A_s = 716 \text{ mm}^2$)

Step 6   Ultimate shear forces

From Table 11.4–1, the shear force at an interior support is

$$V = 0.55F = 0.55 \times 156.63 \text{ (Step 2)} = 86.15 \text{ kN}$$

Step 7   Shear reinforcement
See Example 6.12–1. Provide 10 mm mild steel links at 200 mm centres.

Step 8   Deflection
See Example 5.7–1.

Step 9   Cracking
See Example 5.7–2.

Step 10   Summary of outputs

Cross-sectional dimensions	as in Fig. 4.11–1(a)
Cover to longitudinal bars (Step 1)	40 mm
Support main bars (Step 4)	two 20 mm
Midspan main bars (Step 5)	two 16 mm plus one 20 mm
Links (Step 7)	10 mm ms at 200 mm

The reinforcement details are shown in Fig. 4.11–2, which conforms to the standard method of detailing, as explained in Example 3.6–3.

Comments on Step 1
Preliminary member sizing is discussed in Sections 11.1, 11.2 and 11.3.

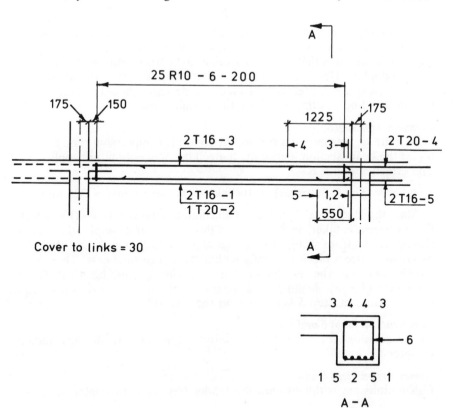

Fig. 4.11–2

Comments on Step 3

The approximate moment (and shear) coefficients in Table 11.4–1 are taken from BS 8110 : Clause 3.4.3. More detailed calculations that take advantage of moment redistribution, as illustrated in Example 4.9–1, usually lead to a more economical design. In practice Table 11.4–1 is used for the less important structural members, particularly when they are not to be constructed identically in large numbers.

Note that when Table 11.4–1 is used, moment redistribution is not permitted.

Comments on Step 4

At an interior support, the beam section resists a hogging bending moment; hence the concrete compression zone is at the bottom. The beam section is therefore designed as though it is rectangular. With reference to Fig. 4.11–1(b), the effective depth d has been calculated as follows:

$$\begin{array}{ll} \text{concrete cover} & = 30 \text{ mm} \\ \text{link diameter} & = 10 \text{ mm} \\ \text{main bar radius} & = \underline{10 \text{ mm}} \\ & \overline{50 \text{ mm}} \end{array}$$

$$d = 375 - 50 = 325 \text{ mm, say}$$

Admittedly, the diameter of the main bars, or even that of the links, is not precisely known in advance. It turns out that the estimated effective depth is slightly less than the actual value that corresponds to the bar sizes finally adopted. Hence the design calculations have erred slightly on the safe side, and may be left unamended; the designer should use his discretion in deciding whether to revise the calculations.

Comments on Step 5

Within the sagging moment region, the concrete compression zone is at the top of the beam section; hence the flanged beam equations are applicable. The effective depth is

$$d = 375 - 30(\text{cover}) - 10(\text{link}) - 10(\text{bar radius}) = 325 \text{ mm}$$

Also, the steel area A_s has been calculated from eqn (4.8–1) in which the lever arm is taken as $d - h_f/2$. In this particular example, the reader should verify from Table 4.6–1 that $0.9x$ is less than h_f, so that the rectangular stress block is wholly within the flange thickness. Therefore, strictly speaking, the tension steel area A_s should have been calculated from eqn (4.8–3), leading to a saving of about 20%. Of course eqn (4.8–1), used in Step 5 here, errs on the safe side.

Comments on Steps 6 and 7

The calculations for these steps are withheld until the reader has studied Chapter 6.

Comments on Steps 8 and 9

Calculations are withheld until the reader has studied Chapter 5.

Comments on Step 10

In the design calculations that lead to the output listed in Step 10, the effects of **torsion** have been ignored. It is reasonable to assume that the slab (see Fig. 4.11–1(a)) is supported by at least one other beam on the left-hand side. Hence there is only compatibility torsion, and this need not be explicitly considered in design, as explained later in Section 6.7.

After reading Section 6.6, it will become clear that explicit calculations for **local bond and anchorage** are not required for this particular case. Example 6.6–1 shows calculations for anchorage.

As explained under the heading 'Minimum areas of tension reinforcement' in Section 4.10, BS 8110 requires **transverse reinforcement** across the full effective width of the flange, as illustrated in Fig. 4.11–3. The area of this reinforcement should not be less than

$$0.0015 \times 175 \times 1000 = 265 \text{ mm}^2 \text{ per metre run}$$

Thus, 10 mm bars at 150 mm centres (523 mm^2 per metre run; see Table A2–2) will be satisfactory. Note that such transverse reinforcement bars have not been shown in Fig. 4.11–2; conventionally they are detailed with the slab, as explained in Comment (d) at the end of Example 3.6–3.

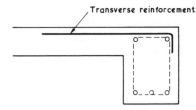

Fig. 4.11–3

4.12 Computer programs

(in collaboration with **Dr H. H. A. Wong**, University of Newcastle upon Tyne)

The FORTRAN programs for this chapter are listed in Section 12.4. See also Section 12.1 for 'Notes on the computer programs'.

Problems

4.1 Derive expressions for the shaded areas of the figures below and the locations of their centroids C.

Ans. (a) Area = $2x_0x_1/3$; $x_c = 5x_0/8$.
      (b) Area = $x_0x_1/3$; $x_c = x_0/4$.

4.2 Using the *properties of the parabola* obtained in Problem 4.1, derive eqns (4.4–3) and (4.4–4).

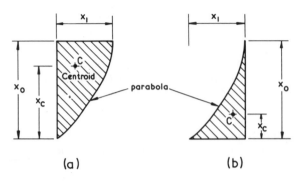

(a)   (b)

Problem 4.1

4.3 Briefly describe the modes of failure of (a) an under-reinforced beam and (b) an over-reinforced beam.

Ans. See paragraphs following: (a) eqn (4.3–1); and (b) eqn (4.3–4).

4.4 Figure 4.4–5 shows that, in BS 8110's simplified rectangular stress block, the stress intensity is $0.45f_{cu}$. Does this value of $0.45f_{cu}$ already include the allowance for the partial safety factor γ_m?

Ans. Yes (see Section 4.4(d)).

4.5 Figure 4.5–2 shows a beam design chart reproduced from BS 8110. Explain how you can construct such a design chart from *first principles*.

Ans. See Section 4.5, particularly Examples 4.5–1 and 4.5–2.

4.6 BS 8110: Clause 3.4.4.1. states that, for a beam section subjected to combined bending and axial force, the effect of the axial force may be ignored if it does not exceed $0.1f_{cu}$ times the cross-sectional area. Comment.

Ans. See Example 4.5–6.

4.7 Moment redistribution is not explicitly considered in Section 4.6. Nevertheless, all the design formulae and procedures in that Section are valid for up to 10% moment redistribution. Explain.

Ans. This is because eqn (4.7–3) is implicitly satisfied.

4.8 Clause 3.4.4.4 of BS 8110 and Clause 4.4.5.1 of the I.Struct.E. Manual [14] give the following formula (reproduced here as eqn 4.6–12 and 4.7–9) for the tension steel area A_s:

$$A_s = \frac{M}{0.87f_y z}$$

where z values are given here in Tables 4.6–1 and 4.7–2. Use these tables to obtain z for the following cases: Case (a): $K = 0.180$ and moment redistribution $\leq 10\%$; Case (b): $K = 0.150$ and moment redistribution $= 15\%$.

Ans. (a) $z/d = 0.775$; see Comment (c) below Table 4.6–1.
(b) $z/d = 0.80$; see Comment (c) below Table 4.7–2.

4.9 BS 8110 : Clause 3.4.4.4 gives the following design formulae:

(1) $A'_s = (K - K')f_{cu}bd^2/0.87f_y(d - d')$

(2) $A_s = K'f_{cu}bd^2/0.87f_yz + A'_s$

(a) State the equivalent forms in which these equations appear in the I.Struct.E. Manual [14].
(b) Derive these equations from first principles.

Ans. See eqns (4.6–13) and (4.6–14) (and also eqns 4.7–10 and 4.7–11) and their derivations.

4.10 In using the design formula in Problem 4.8, can it be assumed that the stress in the tension steel A_s will always reach $0.87f_y$?

Ans. Yes (see Example 4.6–1).

4.11 BS 8110's design formulae, as quoted in Problem 4.9, assume that the stress in the compression steel A'_s reaches $0.87f_y$. Is this assumption always correct?

Ans. No; see Example 4.6–2 and also Comment (c) below eqn (4.7–11).

4.12 Table 24 of the I.Struct.E. Manual [14], reproduced here as Table 4.6–1, gives z/d for various values of $K(= M/f_{cu}bd^2)$.

(a) State and derive the BS 8110 formulae for z/d.
(b) Knowing z/d, how can you find x/d from first principles?

Ans. (a) See Comment (b) below Table 4.6–1 and also Example 4.7–2.
(b) From eqn (4.6–3), $z/d = 1 - 0.45x/d$.

4.13 The I.Struct.E. Manual [14] gives a practical procedure for the rapid design of rectangular beams. Outline the procedure and derive the main design formulae from first principles.

Ans. See 'Design formulae and procedure—BS 8110 simplified stress block' in Sections 4.6 and 4.7 and the comments on each step of the procedure.

4.14 The I.Struct.E. Manual [14] gives a practical procedure for the rapid design of flanged beams. Outline the procedure and derive the main design formulae from first principles.

Ans.   See Section 4.8.

4.15   Explain how Examples 4.6–8(b) and 4.6–9(b) should be modified if the aim is to obtain **actual resistance moments**, for comparison with laboratory experiments, for example.

Ans.   To calculate the actual resistance moments, it is necessary to remove the allowances for the partial safety factors γ_m (= 1.5 for concrete and 1.15 for steel). This can be conveniently done by writing $1.5f_{cu}$ for f_{cu} and $1.15f_y$ for f_y in the equations in Examples 4.6–8(b) and 4.6–9(b).

References

1   ACI–ASCE Committee 327. Ultimate strength design. *Proc. ACI*, **52**, No. 5, Jan. 1956, pp. 505–24.
2   Hognestad, E. Fundamental concepts in ultimate load design of reinforced concrete members. *Proc. ACI*, **48**, No. 10, June 1952, pp. 809–32.
3   Galilei, G. *Dialogues Concerning Two New Sciences* (translated by H. Crew and A. de Salvio.) Macmillan, New York, 1914.
4   Pippard, A. J. S. Elastic theory and engineering structures. *Proc. ICE*, **19**, June 1961, pp. 129–56.
5   Evans, R. H. The plastic theories for the ultimate strength of reinforced concrete beams. *Journal ICE*, **21**, Dec. 1943, pp. 98–121.
6   Evans, R. H. and Kong, F. K. Strain distribution in composite prestressed concrete beams. *Civil Engineering and Public Works Review*, **58**, No. 684, July 1963, pp. 871–2 and No. 685, Aug. 1963, pp. 1003–5.
7   Hognestad, E., Hanson, N. W. and McHenry, D. Concrete stress distribution in ultimate strength design. *Proc. ACI*, **52**, No. 4, Dec. 1955, pp. 455–80.
8   Mattock, A. H., Kriz, L. B. and Hognestad, E. Rectangular concrete stress distribution in ultimate strength design. *Proc. ACI*, **57**, No. 8, Feb. 1961, pp. 875–928.
9   Rusch, H. Research towards a general flexural theory for structural concrete. *Proc. ACI*, **57**, No. 1, July 1960, pp. 1–28.
10   ACI Committee 318. *Building Code Requirements for Reinforced Concrete (ACI 318–83)*. American Concrete Institute, Detroit, 1983.
11   Whitney, C. S. Plastic theory of reinforced concrete design. *Trans. ASCE*, **107**, 1942, pp. 251–326.
12   Coates, R. C., Coutie, M. G. and Kong, F. K. *Structural Analysis*, 2nd edn. Nelson, London, 1980, pp. 149, 187.
13   Allen, A. H. *Reinforced Concrete Design to CP 110*. Cement and Concrete Association, London, 1974.
14   I.Struct.E./ICE Joint Committee. *Manual for the Design of Reinforced Concrete Building Structures*. Institution of Structural Engineers, London, 1985.
15   Baker, Lord and Heyman, J. *Plastic Design of Frames 1: Fundamentals*. Cambridge University Press, 1969.
16   Horne, M. R. *Plastic Theory of Structures*. Pergamon Press, Oxford, 1979.
17   Kong, F.K. and Charlton, T. M. The fundamental theorems of the plastic

theory of structures. *Proceedings of the M. R. Horne Conference on Instability and Plastic Collapse of Steel Structures, Manchester, 1983.* Granada Publishing, London, 1983, pp. 9–15.

18   Kong, F. K. Verulam letter on design of multistorey steel framed buildings. *The Structural Engineer*, **62A**, No. 11, Nov. 1984, p. 355.

19   Braestrup M. W. and Nielsen M. P., Plastic design methods. In *Handbook of Structural Concrete* edited by Kong, F. K., Evans, R. H., Cohen, E. and Roll, F. McGraw-Hill, New York and Pitman, London, 1983, Chapter 20.

20   Kong, F. K. Discussion of: Why not WL/8? by A. W. Beeby. *The Structural Engineer*, **64A**, No. 7, July 1986, p. 184.

21   Ove Arup and Partners. *CIRIA GUIDE 2: The Design of Deep Beams in Reinforced Concrete.* Construction Industry Research and Information Association, London, 1984, 131pp.

22   Kong, F. K., Tang, C. W. J., Wong, H. H. A. and Chemrouk, M. Diagonal cracking of slender concrete deep beams. *Proceedings of a Seminar on Behaviour of Concrete Structures.* Cement and Concrete Association, Slough, 1986, pp. 213–17.

23   Kong, F. K., Garcia, R. C., Paine, J. M., Wong, H. H. A., Tang, C. W. J. and Chemrouk, M. Strength and stability of slender concrete deep beams. *The Structural Engineer*, **64B**, No. 3, Sept. 1986, pp. 49–56.

Chapter 5
Reinforced concrete beams—
the serviceability limit states

Preliminary note: Readers interested only in structural design to BS 8110 may concentrate on the following sections:

(a) *Section 5.3: Deflection control in design (BS 8110).*
(b) *Section 5.4: Crack control in design (BS 8110).*

5.1 The serviceability limit states of deflection and cracking

The deflection of a structure or any part of a structure must not adversely affect the appearance or efficiency of the structure; similarly, any cracking of the concrete must not adversely affect its appearance or durability. Lately the serviceability of concrete structures has become a much more important design consideration than in the past, mainly because more efficient design procedures have enabled engineers to satisfy the ultimate limit state requirements with lighter but more highly stressed structural members. For example, during the past few decades, successive British codes have allowed the maximum service stress in the reinforcement to be approximately doubled in design.

Of the serviceability limit states, those of excessive flexural deflection and of excessive flexural cracking are currently the two that normally must be considered in design [1–4]. *In day-to-day practical design, the serviceability limit state requirements are met by the following straight-forward procedures*:

(a) *Deflections are controlled by simply limiting the span/depth ratios, as explained in Section 5.3.*
(b) *Crack widths are controlled by simply limiting the maximum spacings of the tension reinforcement, as explained in Section 5.4.*

However, an engineer's work is not confined to the simple task of complying with code requirements. There are times when he needs to estimate how a particular structure will behave; there are times when he needs to predict the deflections of a structure, for comparison with site measurements, for example. Therefore, in this chapter we shall also explain the analytical methods for calculating the magnitudes of deflections and crack widths, should special circumstances warrant such calculations to be done. Serviceability is concerned with structural behaviour under service loading, and service loading is sufficiently low for the results of an elastic analysis to be relevant. Therefore in the next section, we shall give an account of the elastic theory for reinforced concrete beams, leading to concepts and results which have applications in deflection and crack-width calculations.

5.2 Elastic theory: cracked, uncracked and partially cracked sections

In this section we shall describe the elastic theory for reinforced concrete for three types of member sections: the cracked section (Case 1), the uncracked section (Case 2) and the partially cracked section (Case 3). Case 1 is the **classical elastic theory** for reinforced concrete, which once occupied a central position in design but which has little direct application today; it is, however, still of some use in crack-width calculations (see Section 5.6). Case 2 is important in prestressed concrete design (see Chapters 9 and 10), and Case 3 is currently used for calculating deflections (see Section 5.5).

Case 1:   The cracked section
Figure 5.2–1(a) shows the cross-section of a beam subjected to a bending moment M. The following simplifying assumptions are made:

(a)   Plane sections remain plane after bending. In other words, the strains vary linearly with distances from the neutral axis. (For a critical review of the research on strain distribution, see Reference 5.)
(b)   Stresses in the steel and concrete are proportional to the strains.
(c)   The concrete is cracked up to the neutral axis, and no tensile stress exists in the concrete below it. (For this reason, the section in Fig. 5.2–1(a) is referred to as a **cracked section**.)

From assumption (a), the steel strains can be expressed in terms of the concrete strain ε_c on the compression face (Fig. 5.2–1(b)):

$$\varepsilon_s' = \frac{x - d'}{x}\varepsilon_c; \qquad \varepsilon_s = \frac{d - x}{x}\,\varepsilon_c \qquad (5.2\text{–}1)$$

From assumption (b) the concrete stress f_c on the compression face, the tension steel stress f_s and the compression steel stress f_s' are

$$f_c = E_c\varepsilon_c \qquad (5.2\text{–}2(a))$$
$$f_s' = E_s\varepsilon_s' = \alpha_c E_c\varepsilon_s' \qquad (5.2\text{–}2(b))$$
$$f_s = E_s\varepsilon_s = \alpha_c E_c\varepsilon_s \qquad (5.2\text{–}2(c))$$

where E_s and E_c are the moduli of elasticity of the steel and concrete respectively and α_c is the modular ratio E_s/E_c.

Since in Fig. 5.2–1(b) the concrete below the neutral axis is to be ignored (assumption (c)), the effective cross-section is that of Fig. 5.2–2(a). From the condition of equilibrium of forces,

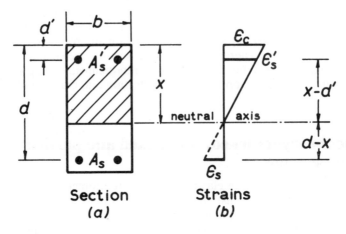

Fig. 5.2–1

$$\tfrac{1}{2}A_c f_c + A_s' f_s' = A_s f_s$$

where A_c is the area of the concrete in compression and A_s' and A_s are respectively the area of the compression steel and that of the tension steel. Using eqns (5.2–1) and (5.2–2) to express all stresses in terms of ε_c, we have

$$\tfrac{1}{2}A_c E_c \varepsilon_c + A_s' \alpha_e E_c \frac{x - d'}{x} \varepsilon_c = A_s \alpha_e E_c \frac{d - x}{x} \varepsilon_c$$

which simplifies to

$$A_c\!\left(\frac{x}{2}\right) + (\alpha_e A_s')(x - d') = (\alpha_e A_s)(d - x) \qquad (5.2\text{–}3)$$

Equation (5.2–3) states that the neutral axis of the cracked section passes through the centroid of the **transformed section** *or* **equivalent section**, obtained by replacing the areas A_s' and A_s by their respective equivalent concrete areas $\alpha_e A_s'$ and $\alpha_e A_s$ (Fig. 5.2–2(b)). In the figure, the areas of concrete displaced by the compression bars are indicated by voids; however, in practice such voids are usually ignored* and the concrete compression area A_c is taken as the nominal area bx, where b is the beam width. On writing bx for A_c, $\varrho'bd$ for A_s' and ϱbd for A_s, eqn (5.2–3) becomes

$$\tfrac{1}{2}bx^2 + \alpha_e\varrho'bd(x - d') = \alpha_e\varrho bd(d - x)$$

from which (and the reader should verify this) the neutral axis depth factor x/d is

$$\frac{x}{d} = -\alpha_e(\varrho + \varrho') + \sqrt{\left\{\alpha_e^2(\varrho + \varrho')^2 + 2\alpha_e\!\left(\varrho + \frac{d'}{d}\varrho'\right)\right\}} \qquad (5.2\text{–}4)$$

---

* The voids can be allowed for by writing $(\alpha_e - 1)\varrho'$ for $\alpha_e\varrho'$ in eqns (5.2–4) and (5.2–9).

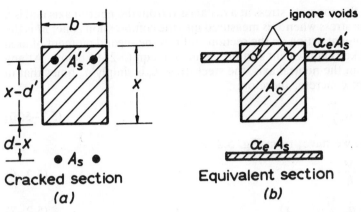

Fig. 5.2–2

where $\alpha_e = E_s/E_c$ ($E_s = 200$ kN/mm^2; E_c from Table 2.5–6), $\varrho = A_s/bd$, $\varrho' = A_s'/bd$ and the other symbols are as defined in Fig. 5.2–1(a). Equation (5.2–4) is plotted in Fig. 5.2–3.

With reference to the equivalent section, the well-known bending formula is applicable:

$$f_{ci} = \frac{M}{I_c} x_i \qquad\qquad\qquad (5.2\text{–}5(a))$$

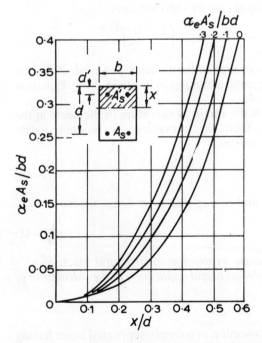

Fig. 5.2–3   Neutral axis depth of cracked section [6] ($d'/d = 0.1$)

where f_{ci} is the concrete stress at a distance x_i from the neutral axis and is a compressive stress when x_i is measured into the compression zone; M is the bending moment acting on the section and I_c is the second moment of area of the (cracked) equivalent section, defined in eqn (5.2–9) below. At any distance from the neutral axis, the steel stress f_{si} is simply α_e times that in the adjacent concrete; therefore

$$f_{si} = \alpha_e \frac{M}{I_c} x_i \qquad\qquad (5.2\text{–}5(\text{b}))$$

Specifically, we have, from Fig. 5.2–2,

$$f_c = \frac{M}{I_c} x \qquad\qquad (5.2\text{–}6)$$

$$f_s = \alpha_e \frac{M}{I_c}(d - x) \qquad\qquad (5.2\text{–}7)$$

$$f_s' = \alpha_e \frac{M}{I_c}(x - d') \qquad\qquad (5.2\text{–}8)$$

where f_c is the maximum compressive stress in the concrete, f_s is the stress in the tension reinforcement and f_s' that in the compression reinforcement. Referring again to Fig. 5.2–2(b),

$$I_c = \tfrac{1}{3}bx^3 + \alpha_e A_s'(x - d')^2 + \alpha_e A_s(d - x)^2$$
$$= \tfrac{1}{3}bx^3 + \alpha_e \varrho'bd(x - d')^2 + \alpha_e \varrho bd(d - x)^2$$

from which (and the reader should verify this) we have

$$\frac{I_c}{bd^3} = \frac{1}{3}\left(\frac{x}{d}\right)^3 + \alpha_e\varrho\left(1 - \frac{x}{d}\right)^2 + \alpha_e\varrho'\left(\frac{x}{d} - \frac{d'}{d}\right)^2 \qquad (5.2\text{–}9)$$

where I_c is the second moment of area of the (cracked) equivalent section and the other symbols have the same meanings as in eqn (5.2–4). Equation (5.2–9) is plotted in Fig. 5.2–4.

Referring to Fig. 5.2–1, if r is the **radius of curvature** of the beam at the section under consideration, then the **curvature** $1/r$ is immediately obtained from the strain diagram as

$$\frac{1}{r} = \frac{\varepsilon_c}{x} \qquad\qquad (5.2\text{–}10)$$

Substituting into eqn (5.2–6) and noting that $\varepsilon_c = f_c/E_c$, we have

$$\frac{1}{r} = \frac{M}{E_c I_c} \qquad\qquad (5.2\text{–}11)$$

which is the well-known curvature expression in structural mechanics.

The following worked example has useful application in the calculation of crack widths (see Section 5.6).

Example 5.2–1

Figure 5.2–5(a) shows the cross-section of a simply supported beam having a 10 m span and supporting a dead load g_k of 24 kN/m and an imposed load

$$\alpha_e$$
$$= \frac{E_s}{E_c}$$

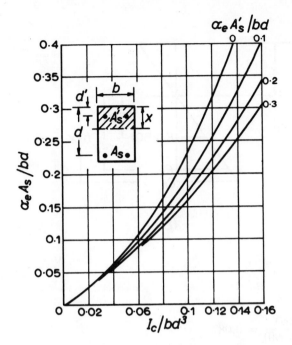

Fig. 5.2–4   **Second moment of area of cracked section [6]** $(d'/d = 0.1)$

q_k of 31.5 kN/m. The characteristic concrete and steel strengths are $f_{cu} = 40 \text{ N/mm}^2$ and $f_y = 460 \text{ N/mm}^2$ respectively; $E_s = 200 \text{ kN/mm}^2$ and E_c is given in Table 2.5–6.

(a) Determine the midspan service-load concrete strains at the level of the tension reinforcement, at the tension face (i.e. the soffit) of the beam, and at 250 mm below the neutral axis.

(b) If, because of creep, the value of E_c becomes half that in Table 2.5–6, repeat the calculations for the long-term service-load strains in (a) above.

SOLUTION

(a) **Short-term strains due to service loading.** For serviceability calculations, BS 8110: Part 2: Clauses 3.3.2 and 3.3.3 recommend that the partial safety factor $\gamma_f = 1$ for both g_k and q_k. Therefore

$$\text{service load moment} = \frac{(1.0 \times 24 + 1.0 \times 31.5) \times 10^2}{8}$$

$$= 694 \text{ kNm}$$

From Table 2.5–6,

$$E_c = 28 \text{ kN/mm}^2 \text{ for } f_{cu} = 40 \text{ N/mm}^2$$

therefore

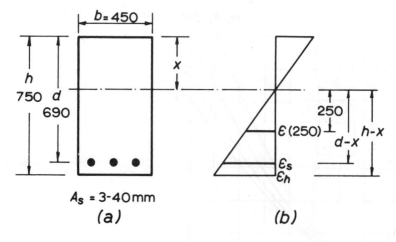

$A_s = 3\text{-}40$mm

(a)   (b)

Fig. 5.2–5

$\alpha_e = 200/28 = 7.14$

$\alpha_e\varrho = (7.14)\dfrac{(3769)}{(450)(690)} = 0.087$

From Figs 5.2–3 and 5.2–4,

$x/d = 0.35; \qquad I_c/bd^3 = 0.054$

therefore

$x = (0.35)(690) = 241.5$ mm
$I_c = (0.054)(450)(690^3) = (7982)(10^6)$ mm^4

From eqn (5.2–7),

$\varepsilon_s = \dfrac{\alpha_e}{E_s}\dfrac{M}{I_c}(d - x)$

$= \dfrac{(7.14)}{(200)(10^3)}\dfrac{(694)(10^6)}{(7982)(10^6)}(690 - 241.5)$

$= \underline{0.00139}$

(This is also the concrete strain at the level of the tension steel.)
From Fig. 5.2–5(b), the concrete strain at the tension face is

$\varepsilon_h = \dfrac{h - x}{d - x}\varepsilon_s = \left(\dfrac{750 - 241.5}{690 - 241.5}\right)(0.00139) = \underline{0.00157}$

At 250 mm below the neutral axis, the concrete strain is

$\varepsilon(250) = \left(\dfrac{250}{690 - 241.5}\right)(0.00139) = \underline{0.00077}$

(b) **Long-term strains due to service loading**

$E_c = (\tfrac{1}{2})(28)$ kN/mm^2 $= 14$ kN/mm^2

$$\alpha_e = 200/14 = 14.29$$

$$\alpha_e\varrho = (14.29)\frac{(3769)}{(450)(690)} = 0.173$$

From Figs 5.2–3 and 5.2–4,

$$x/d = 0.44; \qquad I_c/bd^3 = 0.085$$

whence

$$x = 303.6 \text{ mm}; \qquad I_c = (12565)(10^6) \text{ mm}^4$$

From eqn (5.2–7),

$$\varepsilon_s = \frac{(14.29)}{(200)(10^3)}\frac{(694)(10^6)}{(12565)(10^6)}(690 - 303.6)$$

$$= \underline{0.00152}$$

$$\varepsilon_h = \frac{(750 - 303.6)}{(690 - 303.6)}(0.00152) = \underline{0.00176}$$

$$\varepsilon(250) = \frac{(250)}{(690 - 303.6)}(0.00152) = \underline{0.00098}$$

Example 5.2–1 refers to a cracked section. Therefore the strain values calculated for the concrete below the neutral axis are only **average strains** (or apparent strains); that is, they are the average strains measured over a fairly long gauge length of, say, 150 mm or more. When measured over much shorter gauge lengths the concrete strains are likely to be quite erratically distributed; the subject is discussed in detail in Reference 5.

Case 2: The uncracked section

When the applied bending moment M is small enough for the maximum concrete tensile stress not to exceed the tensile strength or the modulus of rupture of the concrete, an analysis based on an **uncracked section** becomes relevant. The effective concrete section is then the full section bh, as shown in Fig. 5.2–6(a) and the equivalent section is as in Fig. 5.2–6(b).

As in eqn (5.2–3), the neutral axis of the uncracked section passes through the centroid of the equivalent section; the neutral axis depth x is therefore given by

$$A_c\left(x - \frac{h}{2}\right) + \alpha_e A_s'(x - d') = \alpha_e A_s(d - x) \qquad (5.2\text{–}12)$$

where A_c is now the entire concrete area bh. Referring again to Fig. 5.2–6(b), the second moment of area of the (uncracked) equivalent section is

$$I_u = \tfrac{1}{12}bh^3 + bh\left(x - \frac{h}{2}\right)^2 + \alpha_e A_s'(x - d')^2$$

$$+ \alpha_e A_s(d - x)^2 \qquad (5.2\text{–}13)$$

At any distance x_i from the neutral axis, the concrete stress f_{ci} and the steel stress f_{si} are given by

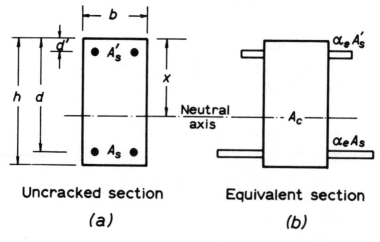

Uncracked section   **Equivalent section**

(a)   (b)

Fig. 5.2–6

$$f_{ci} = \frac{M}{I_u}x_i; \qquad f_{si} = \alpha_e\frac{M}{I_u}x_i \tag{5.2–14}$$

where stresses are compressive when x_i is measured into the compression zone, and tensile when x_i is measured into the tension zone. The curvature is given by

$$\frac{1}{r} = \frac{M}{E_cI_u} \tag{5.2–15}$$

which can be derived in the same manner as eqn (5.2–11).

Case 3: The partially cracked section

The partially cracked section is a device introduced by CP 110 in 1972 and retained in BS 8110 : Part 2 : Clause 3.6. Figure 5.2–7 shows a beam section in which, as usual, strains are assumed to be linearly distributed. However, in the tension zone (that is, below the neutral axis) some concrete tension still exists as represented by the triangular stress distribution in Fig. 5.2–7(c), in which the concrete tensile stress has a specified value f_{ct} at the level of the tension reinforcement. Note particularly that the concrete stresses above the neutral axis (but not those below it) and the reinforcement stresses are related to the strains in Fig. 5.2–7(b) by the usual equations:

$$f_c = E_c\varepsilon_c; \qquad f_s' = E_s\varepsilon_s'; \qquad f_s = E_s\varepsilon_s \tag{5.2–16}$$

Below the neutral axis, however, the concrete tensile stresses are not to be determined from the strain diagram, but from the specified value f_{ct}. For example, at the tension face or soffit of the beam, the concrete stress is

$$\text{concrete stress at tension face} = \frac{h-x}{d-x}f_{ct} \tag{5.2–17}$$

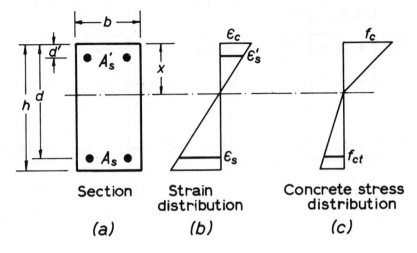

Fig. 5.2–7

The value f_{ct} is further assumed to be independent of the magnitude of the bending moment M acting on the section. In other words, as M varies, the concrete compressive stresses and the reinforcement stresses would vary, but the concrete tensile stresses remain unchanged, being pegged to the value assigned to f_{ct}. BS 8110: Part 2: Clause 3.6 recommends that:

(a) $f_{ct} = 1$ N/mm² for short-term loading;
(b) $f_{ct} = 0.55$ N/mm² for long-term loading.

A beam section having such characteristics certainly sounds rather artificial, but, as we shall see in Section 5.5, it has important applications in deflection calculations. For convenience of reference, we shall call the section a **partially cracked section**.

Strictly speaking, the neutral axis depth x of a partially cracked section should be computed by equating the tensile and compressive forces. Since the concrete tensile stress distribution is assumed to be independent of the moment M acting on the section, while other stresses vary with M, the value of x would vary with M as well. *For practical applications (see Section 5.5) the simplifying assumption is made that the neutral axis depth is the same as that in a cracked section; namely, x may be calculated from eqn (5.2–4)*. Figure 5.2–8 shows the partially cracked section incorporating this assumption: The strain distribution (Fig. 5.2–8(b)) and the steel and concrete compressive stresses (Fig. 5.2–8(c)) are determined as for a cracked section—after the moment M has been adjusted to allow for the contribution of the concrete tension in Fig. 5.2–8(d). Referring to Fig. 5.2–8(d), the average tensile stress is given by eqn (5.2–17) as $\frac{1}{2}(h - x)f_{ct}/(d - x)$; therefore

$$\text{tension force in concrete} = \frac{1}{2}\frac{b(h - x)^2}{(d - x)}f_{ct} \qquad (5.2–18)$$

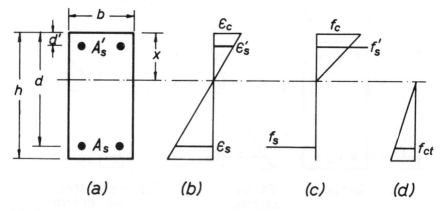

Fig. 5.2–8

The lever arm of this force about the neutral axis is $\frac{2}{3}(h - x)$; therefore

$$\text{moment due to concrete tension} = \frac{1}{3}\frac{b(h - x)^3}{(d - x)}f_{ct} \qquad (5.2\text{–}19)$$

It follows that when a moment M is applied to the partially cracked section, part of it (eqn 5.2–19) is resisted by the concrete tension; the net moment to be resisted by the concrete compression and by the forces in the reinforcement is

$$M_{net} = M - \frac{1}{3}\frac{b(h - x)^3}{(d - x)}f_{ct} \qquad (5.2\text{–}20)$$

Whence, in Fig. 5.2–8(c), the stresses are

$$f_c = \frac{M_{net}}{I_c}x \qquad (5.2\text{–}21)$$

$$f_s = \alpha_e\frac{M_{net}}{I_c}(d - x); \qquad f_s' = \alpha_e\frac{M_{net}}{I_c}(x - d') \qquad (5.2\text{–}22)$$

where α_e is the modular ratio E_s/E_c, and the second moment of area I_c is to be calculated from eqn (5.2–9).
   Similarly, the curvature produced by the moment M_{net} is

$$\frac{1}{r} = \frac{M_{net}}{E_cI_c} \qquad (5.2\text{–}23)$$

where M_{net} is obtained from eqn (5.2–20) and I_c is from eqn (5.2–9).
   The following worked example has useful application in the calculation of deflections (see Section 5.5).

Example 5.2–2

Figure 5.2–9 shows the cross-section of a beam, acted on by a load, part of which is permanent. The bending moment M_t due to the total load is 48 kNm and the bending moment M_p due to the permanent load is

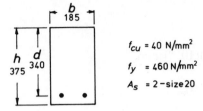

Fig. 5.2–9

36 kNm. Using the assumptions appropriate to a partially cracked section, determine:

(a) the long-term curvature of the beam under the permanent load, if f_{ct} (see Fig. 5.2–8(d)) has the specified value of 0.55 N/mm² appropriate to long-term loading;

(b) the instantaneous curvatures under the total load and the permanent load, if $f_{ct} = 1$ N/mm² for short-term loading; and

(c) the difference between the instantaneous curvatures under the total and permanent loads.

Given: $E_s = 200$ kN/mm², E_c from Table 2.5–6, E_c (long term) = $E_c/(1 + \phi)$ where the creep coefficient ϕ may in this example be taken as 2.5.

SOLUTION

(a) **Long-term curvature $1/r_{1p}$ due to M_p.** From Table 2.5–6, $E_c = 28$ kN/mm². Therefore

$$E_c \text{ (long term)} = \frac{28}{1 + \phi} = \frac{28}{1 + 2.5} = 8 \text{ kN/mm}^2$$

$$\alpha_e = 200/8 = 25$$

$$\alpha_e \varrho = (25)\frac{628}{(185)(340)} = 0.25$$

From Figs. 5.2–3 and 5.2–4,

$$x/d = 0.505 \qquad I_c/bd^3 = 0.105$$

Therefore

$$x = 0.505d = 171 \text{ mm}$$

$$I_c = 0.105bd^3 = (763)(10^6) \text{ mm}^4$$

From eqn (5.2–20),

$$M_p(\text{net}) = (36)(10^6) \text{ Nmm} - \frac{1}{3}\left\{\frac{185(375 - 171)^3}{(340 - 171)}\right\}(0.55)$$

$$= (36)(10^6) \text{ Nmm} - (2)(10^6) \text{ Nmm}$$

$$= (34)(10^6) \text{ Nmm}$$

Using eqn (5.2–23), the long-term curvature is

$$\frac{1}{r_{1p}} = \frac{M_p(\text{net})}{E_c I_c}$$

$$= \frac{(34)(10^6)}{(8)(10^3)(763)(10^6)} = \underline{(5.57)(10^{-6})} \text{ mm}^{-1}$$

(b) **Instantaneous curvatures $1/r_{it}$ and $1/r_{ip}$**

$$\alpha_e = 200/28 = 7.14$$

$$\alpha_e \varrho = (7.14)\frac{628}{(185)(340)} = 0.071$$

From Figs 5.2–3 and 5.2–4,

$$x/d = 0.32 \qquad I_c/bd^3 = 0.045$$

whence

$$x = 108.8 \text{ mm} \qquad I_c = (327)(10^6) \text{ mm}^4$$

From eqn (5.2–20),

$$M_t(\text{net}) = (48)(10^6) - \frac{1}{3}\left\{\frac{185(375 - 108.8)^3}{(340 - 108.8)}\right\}(1)$$

$$= (48)(10^6) - (5.03)(10^6) = (42.97)(10^6) \text{ Nmm}$$

$$M_p(\text{net}) = (36)(10^6) - (5.03)(10^6) = (30.97)(10^6) \text{ Nmm}$$

Using eqn (5.2–23), the instantaneous curvatures $1/r_{it}$ and $1/r_{ip}$, due respectively to the total and the permanent load, are

$$\frac{1}{r_{it}} = \frac{(42.97)(10^6)}{(28)(10^3)(327)(10^6)} = \underline{(4.69)(10^{-6})} \text{ mm}^{-1}$$

$$\frac{1}{r_{ip}} = \frac{(30.97)(10^6) \text{ mm}^{-1}}{(28)(10^3)(327)(10^6)} = \underline{(3.39)(10^{-6})} \text{ mm}^{-1}$$

(c) **Difference in instantaneous curvatures.**

$$\frac{1}{r_{it}} - \frac{1}{r_{ip}} = (4.69)(10^{-6}) - (3.39)(10^{-6}) = \underline{(1.30)(10^{-6})} \text{ mm}^{-1}$$

An examination of the calculations in (b) above shows that the difference in instantaneous curvatures may be obtained directly as

$$\frac{1}{r_{it}} - \frac{1}{r_{ip}} = \frac{M_t - M_p}{E_c I_c} \qquad (5.2\text{–}24)$$

Note that $M_t - M_p = M_t(\text{net}) - M_p(\text{net})$; that is the terms involving f_{ct} in eqn (5.2–20) cancel out.

5.3   Deflection control in design (BS 8110)

Excessive deflections may lead to sagging floors, to roofs that do not drain properly, to damaged partitions and finishes, and to other associated troubles [1, 2]. BS 8110 states that the final deflection, including the effects of creep and shrinkage, should not exceed either of the following limits:

(a) span/250;
(b) span/500 or 20 mm, whichever is the lesser, after the construction of the partitions or the application of finishes (BS 8110: Amendment No. 1, 1986).

These deflection limits are given as being reasonable values for use in practical design. The limit (a) of span/250 is considered to be that beyond which the deflection will be noticed by the user of the structure. The limit (b) is to prevent damage to partitions and finishes. Both limits are intended for general guidance only; where, for example, a special type of partition is used, the manufacturer's advice should be sought.

In design, it is usual to comply with the above deflection limits by a straight-forward procedure of limiting the ratio of the span to the effective depth [7–10]; it is only in exceptional cases that deflections are actually calculated (see Section 5.5) and compared with the limiting values. The practical procedure recommended by BS 8110: Clause 3.4.6 may conveniently be summarized as follows (see also the Comments at the end).

Step 1   Basic span/depth ratios

Select the basic span/effective depth ratios (usually referred to as the basic **span/depth** ratios) in Table 5.3–1. For **flanged sections** with $b_w/b >$ 0.3, obtain the span/depth ratio by linear interpolation between the values given in Table 5.3–1 for rectangular sections and for flanged sections with $b_w/b = 0.3$. (*Note:* For a flanged section, b is the effective flange width.)

Step 2   Long spans

For spans exceeding 10 m, there are three cases to consider, depending on whether it is necessary to limit the increase in deflection (to span/500 or 20 mm as stated above) after the construction of the partitions or finishes:

(a) If it is not necessary to limit such an increase in deflection, then the basis span/depth ratio obtained (in Step 1) from Table 5.3–1 remains valid.
(b) If it is necessary to limit such an increase, and the structural member is not a cantilever, then the basic span/depth ratio obtained from Table 5.3–1 should be multiplied by a modification factor equal to 10/span.

Table 5.3–1   Basic span/effective depth ratios (BS 8110: Clause 3.4.6.3)

Support condition	Rectangular sections	Flanged sections $\frac{b_w}{b} \leq 0.3$
Cantilever	7	5.6
Simply supported	20	16.0
Continuous	26	20.8

(c) If it is necessary to limit the increase in deflection, and the structural member is a **cantilever**, then the design must be justified by deflection calculation (see Section 5.5).

Step 3   Modification factor for tension reinforcement
The span/depth ratio is now multipled by the modification factor, obtained from Table 5.3–2, to allow for the effect of the **tension reinforcement**.

Step 4   Modification factor for compression reinforcement
If the beam is doubly reinforced, the span/depth ratio may be further multiplied by a modification factor, obtained from Table 5.3–3, to allow for the effect of the **compression reinforcement**.

Comments on Step 1
Limiting the span/depth ratio is an effective and convenient method of deflection control in practical design [7–10].

In Table 5.3–1, the span/depth ratios for flanged sections are smaller than those for rectangular sections. This is because a flanged section has a smaller area of concrete in the tension zone than that in a rectangular section of the same width b. As we shall see in Section 5.5 on deflection calculations, the concrete in the tension zone does contribute to the stiffness of the member even after cracking.

Comments on Step 2
For spans exceeding 10 m, the use of the ratios in Table 5.3–1 may lead to deflections exceeding span/500 or 20 mm after the construction of partitions and finishes. Hence the modification factor of 10/span is recommended. Long cantilevers are notorious as a source of troubles associated with excessive deflections; hence BS 8110 requires their design to be justified by deflection calculations.

Table 5.3–2   Modification factor for tension reinforcement
(BS 8110 : Clause 3.4.6.5)

Service stress		M/bd^2 (N/mm^2)								
f_s	(N/mm^2)	0.50	0.75	1.00	1.50	2.00	3.00	4.00	5.00	6.00
($f_y = 250$)	156	2.00	2.00	1.96	1.66	1.47	1.24	1.10	1.00	0.94
($f_y = 460$)	288	1.68	1.50	1.38	1.21	1.09	0.95	0.87	0.82	0.78

Table 5.3–3   Modification factor for compression reinforcement
(BS 8110 : Clause 3.4.6.6)

$\dfrac{100A'_{s,\,prov}}{bd}$	0	0.15	0.25	0.35	0.50	0.75	1.00	1.50	2.00	2.50	≥ 3.00
Factor	1.00	1.05	1.08	1.10	1.14	1.20	1.25	1.33	1.40	1.45	1.50

Comments on Step 3

(a)   Table 5.3–2 shows only the modification factors for the two commonly used values of reinforcement service stress f_s. For other f_s values, the reader may refer to the full table in BS 8110: Clause 3.4.6.5, or else use the following formulae given by the above-mentioned BS clause:

$$\text{Modification factor} = 0.55 + \frac{477 - f_s}{120\left[0.9 + \dfrac{M}{bd^2}\right]} \leq 2.0 \quad (5.3{-}1(a))$$

$$f_s = \frac{5}{8}f_y\left[\frac{A_{s,\text{req}}}{A_{s,\text{prov}}}\right]\left[\frac{1}{\beta_b}\right] \quad\quad (5.3{-}1\ (b))$$

where $A_{s,\text{req}}$ is the area of the tension reinforcement required at midspan to resist the moment M due to the design ultimate loads (at support for a cantilever), $A_{s,\text{prov}}$ the area of the tension reinforcement actually provided at midspan (at support for a cantilever) and β_b the ratio

$$\frac{\text{moment at the section after redistribution}}{\text{moment at the section before redistribution}}$$

from the respective maximum moments diagram. Note that the condition $\beta_b \leq 1$ does not apply here (as it does in eqn 4.7–1).

(b)   Deflection is influenced by the amount of the tension reinforcement and its stress f_s [3, 11]. In Table 5.3–2, the M/bd^2 value is used as a convenient measure of the tension steel ratio—the higher the M/bd^2 value, the higher the steel ratio. To study the effect on deflection of the steel stress and the steel ratio, consider the beam section in Fig. 5.2–5. By geometric reasoning, we see that the curvature at the section is

$$\frac{1}{r} = \frac{\varepsilon_s}{d - x} = \frac{f_s/E_s}{d - x} \quad\quad (5.3{-}2)$$

Equation (5.3–2) shows that, for a given value of f_s, the curvature $1/r$ increases with the neutral axis depth x. Since x increases with the steel ratio (see Fig. 5.2–3), this means that the curvature (and hence the deflection) increases with the steel ratio. Also, as x increases, the area of the concrete compression zone becomes larger, and consequently the effect of creep on deflection becomes more pronounced. In other words, for a given f_s value, an increase in the steel ratio leads to an increase in deflection; hence a lower limit must be placed on the span/depth ratio. This explains why the modification factors in Table 5.3–2 become smaller as the M/bd^2 value (and hence the steel ratio) is increased.

Equation (5.3–2) also shows that for a given value of x (i.e. for a given steel ratio), the curvature and hence the deflection increase with f_s. That is, if f_s is increased, then a lower limit must be placed on the span/depth ratio. This explains why the modification factors in Table 5.3–2 become smaller as f_s is increased.

Comments on Step 4
Figure 5.2–3 shows that the neutral axis depth x decreases with the compression steel ratio A_s'/bd. Therefore, for a given value of the service stress f_s in the tension steel, an increase in the compression steel ratio will increase the quantity $(d - x)$ in eqn (5.3–2) and consequently will reduce the curvature, and hence the deflection of the beam. Therefore, if the compression steel ratio is increased, the allowable span/depth ratio may also be increased.

Note also that the final span/depth ratios as determined from Steps 1 to 4 above will take account of normal creep and shrinkage deflections.

Example 5.3–1
The design ultimate moment for a rectangular beam of 11 m simple span is 900 kNm. If $f_{cu} = 40$ N/mm^2 and $f_y = 460$ N/mm^2, design the cross-section for the ultimate limit state and check that the span/effective depth ratio is within the allowable limit in BS 8110: Clause 3.4.6.

SOLUTION
Example 4.5–4 shows that a beam section having the following properties are appropriate for the ultimate limit state:

$$b = 250 \text{ mm}; \quad d = 700 \text{ mm}; \quad A_s = 4021 \text{ mm}^2;$$

$$A_s' = 982 \text{ mm}^2; \quad \frac{A_s}{bd} = 2.30\%; \quad \frac{A_s'}{bd} = 0.56\%$$

Hence the allowable span/depth ratio may be determined by the step-to-step procedure described on pp. 169–170.

Step 1
From Table 5.3–1, the basic span/depth ratio is 20.

Step 2
We shall assume that it is necessary to restrict the increase in deflection, after construction of the partitions and finishes, to the BS 8110 limits. Hence the modified factor for long span is

$$\frac{10}{\text{span}} = \frac{10}{11} = 0.91$$

Step 3
$$\frac{M}{bd^2} = \frac{(900)(10^6)}{(250)(700^2)} = 7.35 \text{ N/mm}^2$$

From Table 5.3–2, the modification factor for tension reinforcement is 0.78. (More accurately, eqn 5.3–1(a) gives 0.75.)

Step 4
$$\frac{100 A_{s,\text{prov}}'}{bd} = 0.56$$

From Table 5.3–3, the modification factor for compression reinforcement is 1.15. Therefore the allowable span/depth ratio is

$$(20)(0.91)(0.78)(1.15) = 16.33$$

The actual span/depth ratio = (11 m)/(700 mm)

$$= 15.71 < 16.33$$

Hence BS 8110's deflection limits are unlikely to be exceeded.

Example 5.3–2
If the allowable span/depth ratio determined in accordance with the steps above turns out to be smaller than the actual span/depth ratios, what remedial actions may be taken?

SOLUTION
Possible remedial actions include:

(a) The effective depth d may be increased to bring the actual span/depth ratio down to the allowable value.

(b) Additional tension steel can be provided over and above that required by the M/bd^2 value. This will reduce the service stress f_s (see eqn 5.3–1(b)) and increase the modification factor (see eqn 5.3–1(a)).

(c) A third possibility is to carry out a full deflection calculation, using the BS 8110 procedure as explained in Section 5.5. It will usually show that the span/depth ratio procedure is conservative.

5.4   Crack control in design (BS 8110)

BS 8110 : Clause 3.12.11.2.1 states that surface crack widths should not, in general, exceed 0.3 mm. Excessively wide cracks are objectionable mainly because they affect the appearance of the structure. The corrosion of the reinforcement depends mainly on the concrete cover and the porosity of the concrete; research on the effects of crack width on corrosion has not been conclusive [12–14], but it is prudent to limit crack widths in an aggressive environment, even when the structural member cannot be seen.

In practical design, it is usual to comply with the 0.3 mm crack-width limit by a straightforward procedure of limiting the **maximum distance between bars in tension**, as recommended by BS 8110 : Clause 3.4.7. BS 8110's detailing rules for crack control are conveniently summarized in Fig. 5.4–1.

Comments on Fig. 5.4–1
(a) In measuring the clear distances a_b between tension bars, ignore any bar with a diameter less than 0.45 times that of the largest bar. (*Note:* 0.45, and not 0.5, is used so that, say size 12 bars may be used with size 25 bars).

(b) Similarly, in measuring the clear distance a_c to the corner, ignore any bar having a size less than 0.45 times that of the largest bar.

(c) The side bars should have a size not less than $\sqrt{(s_b b/f_y)}$ where s_b is the centre-to-centre bar spacing and b the beam width. (*Note:* b need not be taken as greater than 500 mm.) For $s_b = 250$ mm, which is the maximum permissible spacing, the minimum sizes of the side bars are

$$0.75\sqrt{b} \quad \text{(for } f_y = 460 \text{ N/mm}^2)$$
$$1.00\sqrt{b} \quad \text{(for } f_y = 250 \text{ N/mm}^2)$$

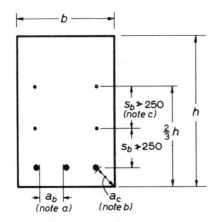

Notes :

(a) $a_b \not= $ value specified in Table 5·4-1

(b) $a_c \not= \frac{1}{2} \times $ value specified for a_b

(c) If (and only if) $h > 750\,\text{mm}$, side bars
     are required to a depth of $\frac{2}{3}h$.

Fig. 5.4−1   Reinforcement spacing rules for crack control

Note: the minimum sizes are to guard against the bar yielding locally
at a crack.

Table 5.4−1   Clear distance between bars (BS 8110 : Clause 3.12.11.2.3)

f_y (N/mm^2)	Redistribution to (+) or from (−) section considered							
	−30	−20	−10	0	+10	+20	+30	(%)
250	210	240	270	300	300	300	300	(mm)
460	115	130	145	160	180	195	210	(mm)

Comments on Table 5.4−1

(a) The values in Table 5.4−1 have been calculated from the following
equation, which may be more convenient to use than interpolation
from the table:

$$\text{clear spacing} = \frac{75\,000}{f_y}\left[\frac{100 - \beta\%}{100}\right] \leq 300 \tag{5.4−1}$$

where $\beta\%$ is the percentage of moment redistribution.

(b) Instead of using Table 5.4−1 or eqn (5.4−1), BS 8110 : Clause
3.12.11.2.4 states that the clear spacing may be calculated from

$$\text{clear spacing} = \frac{47\,000}{f_s} \leq 300 \tag{5.4−2}$$

where f_s is the steel service stress as defined in eqn (5.3–1(b)). Example 5.4–1 illustrates a useful application of eqn (5.4–2).
(c)   Equation (5.4–2) becomes identical with eqn (5.4–1) if $A_{s, req} = A_{s, prov}$ (see Problem 5.6).

Example 5.4–1
With reference to the detailing rules for crack control in Fig. 5.4–1, if the clear spacing a_b exceeds the value in Table 5.4–1, what remedial actions can be taken?

SOLUTION
Possible remedial actions include:

(a)   The three main bars may be replaced by, say, four or five smaller bars.
(b)   Additional small bars may be inserted between the three bars. According to the '0.45 rule' explained in Comment (a) to Fig. 5.4–1, size 12 bars may be inserted between size 25 bars and so on.
(c)   If the tension steel area actual provided is over and above that required for the ultimate limit state, this will have the effect of reducing the service stress f_s as given by eqn (5.3–1(b)). Then use eqn (5.4–2) to obtain a relaxed limit on the clear bar spacing.
(d)   A full crack width calculation may be carried out using BS 8110's procedure as explained in Section 5.6; this usually shows that the values in Table 5.4–1 are conservative.

5.5   Calculations of short-term and long-term deflections (BS 8110 : Part 2)

(*Note: In day-to-day design, deflections are controlled by a straightforward procedure of limiting the span/depth ratio—see Section 5.3*)
The difficulties concerning the calculation of deflections of concrete beams arise from the uncertainties regarding the flexural stiffness EI and the effects of creep and shrinkage. Before explaining how these uncertainties are dealt with in practice, we shall first refer to the well-known **moment–area theorems** [15], which express slopes and deflections in terms of the properties of the M/EI diagram. For elastic members, the quantity M/EI is equal to the curvature $1/r$ (see eqns 5.2–11, 5.2–15 and 5.2–23); therefore the moment–area theorems may be rephrased (more usefully) as the **curvature–area theorems** [16]:

(a)   The change in slope θ between two points A and B on a member is equal to the area of the curvature diagram between the two points:

$$\theta = \int_B^A \left(\frac{1}{r}\right) dx$$

where r is the radius of curvature of the typical element dx.
(b)   The deflection Δ of point B, measured from the tangent at point A, is equal to the moment of the curvature diagram between A and B, taken about the point B whose deflection is sought:

$$\Delta = \int_{B}^{A} x\left(\frac{1}{r}\right) dx$$

where x is measured from B.

Comments

The above results were first given the name curvature–area theorems in the first edition of the book, published in 1975; a formal proof using the principle of virtual work was given subsequently [16].

For estimating the deflections of concrete structures, the curvature–area theorems have distinct advantages over the conventional moment–area theorems:

(a)   Unlike the moment–area theorems, the curvature–area theorems express the purely geometrical relations between the slopes, θ, the deflections Δ and the curvatures $1/r$. Since the relations are purely geometrical, their validity is independent of the mechanical properties of the materials. That is, the curvature–area theorems are equally applicable irrespective of whether the structure is elastic or plastic or elasto-plastic.

(b)   Unlike the moment–area theorems, the curvature–area theorems can be used even where the deformations are caused by other effects than bending moments, e.g. by shrinkage and creep. Once the curvatures are known, the slopes and deflections are completely defined by the curvature–area theorems; whether the curvatures have been caused by bending moments or by shrinkage and creep does not affect the results.

Example 5.5–1

Figure 5.5–1 shows the curvature diagrams for a beam of uniform flexural rigidity EI, acted on by various loadings. Determine the midspan deflection a in each case.

SOLUTION

(a)   Referring to the curvature diagram for beam (a),

$$\text{shaded area} = \frac{1}{2}\left(\frac{l}{2}\right)\left(\frac{1}{r_m}\right) = \frac{l}{4}\left(\frac{1}{r_m}\right)$$

Moment of shaded area about left support

$$= \frac{l}{4}\left(\frac{1}{r_m}\right)\left(\frac{l}{3}\right) = \frac{l^2}{12}\left(\frac{1}{r_m}\right)$$

From the curvature–area theorem, this is the deflection of the left support from the tangent at midspan, and is numerically equal to the midspan deflection a. Therefore

$$a = \tfrac{1}{12}l^2\left(\frac{1}{r_m}\right)$$

(b)   Similarly,

$$a = \left(\tfrac{1}{2}l\frac{1}{r_m}\right)(\tfrac{1}{4}l) = \tfrac{1}{8}l^2\left(\frac{1}{r_m}\right)$$

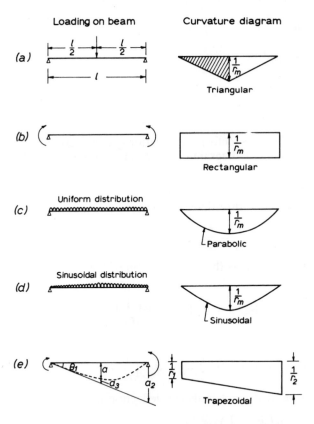

Fig. 5.5–1 Curvature diagrams for various loadings

(c)
$$a = \left(\frac{2}{3}\frac{l}{2}\frac{1}{r_m}\right)\left[\left(\frac{5}{8}\right)\left(\frac{l}{2}\right)\right] = \frac{1}{9.6}l^2\left(\frac{1}{r_m}\right)$$

Note: Useful properties of the parabola are given in books on structural mechanics, e.g. Reference 15. See also Problem 4.1 at the end of Chapter 4.

(d)   As an exercise, the reader should verify that

$$a = \frac{1}{\pi^2}l^2\left(\frac{1}{r_m}\right)$$

(e)   Referring to the deflected shape in Fig. 5.5–1(e), the reader should work out the solution in the following steps:

(1)   Determine the deflection a_2. This is the moment of the entire curvature diagram about the right support.

(2) $\theta_1 = a_2/l$.

(3)   Determine the deflection a_3. This is the moment of the left half of the curvature diagram taken about midspan.

(4)   Then

$$a = (\theta_1)\left(\frac{l}{2}\right) - a_3.$$

The answer should be

$$a = \tfrac{1}{16}l^2\left(\frac{1}{r_1} + \frac{1}{r_2}\right)$$

Example 5.5–2
Figures 5.5–2(a) and (b) show respectively an interior span of a continuous beam and the bending moment diagram. Derive an expression for the midspan deflection.

SOLUTION
Figure 5.5–2(c) shows the curvature diagram, where

$$\frac{1}{r_1} = \frac{M_1}{EI} \qquad \frac{1}{r_2} = \frac{M_2}{EI} \qquad \frac{1}{r_3} = \frac{M_3}{EI}$$

Figure 5.5–2(c) may be regarded as the superposition of Fig. 5.5–2(d) on Fig. 5.5–2(e). From Example 5.5–1(a),

$$\text{midspan deflection for Fg. 5.5–2(d)} = \tfrac{1}{12}l^2\left(\frac{1}{r_m}\right)$$

From Example 5.5–1(e),

$$\text{midspan deflection for Fig. 5.5–2(e)} = \tfrac{1}{16}l^2\left(\frac{1}{r_1} + \frac{1}{r_2}\right)$$

Therefore, by superposition, the net midspan deflection is

$$a = \tfrac{1}{12}l^2\left(\frac{1}{r_m}\right) - \tfrac{1}{16}l^2\left(\frac{1}{r_1} + \frac{1}{r_2}\right)$$

where, from the geometry of Fig. 5.5–2(c),

$$\frac{1}{r_m} = \frac{1}{2}\left(\frac{1}{r_1} + \frac{1}{r_2}\right) + \frac{1}{r_3}$$

Therefore

$$a = \tfrac{1}{12}l^2\left[\frac{1}{2}\left(\frac{1}{r_1} + \frac{1}{r_2}\right) + \frac{1}{r_3}\right] - \tfrac{1}{16}l^2\left(\frac{1}{r_1} + \frac{1}{r_2}\right)$$

which simplifies to

$$a = \tfrac{1}{12}l^2\frac{1}{r_3}\left[1 - \frac{\beta}{4}\right] \qquad\qquad (5.5–1)$$

where β is the ratio

$$\left(\frac{1}{r_1} + \frac{1}{r_2}\right)\Big/\left(\frac{1}{r_3}\right)$$

If the span in Example 5.5–2 supports a uniformly distributed load, it is necessary only to superimpose the results of Example 5.5–1(c) and (e). the reader should verify that in such a case the midspan deflection is

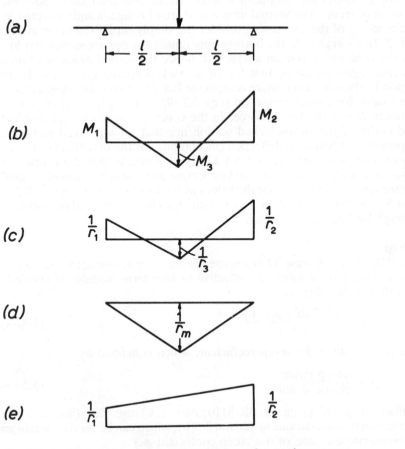

Fig. 5.5–2   **Superposition of curvature diagrams for continuous span**

$$a = \frac{1}{9.6}l^2\left(\frac{1}{r_3}\right)\left[1 - \frac{\beta}{10}\right] \tag{5.5–2}$$

where β is the ratio

$$\left(\frac{1}{r_1} + \frac{1}{r_2}\right)\Big/\left(\frac{1}{r_3}\right)$$

Before applying the curvature–area theorems to the determination of deflections, we shall first discuss how realistic assessments can be made of the flexural rigidity EI and of the curvatures due to creep and shrinkage:

Flexural stiffness EI

In design calculations, the modulus of elasticity of steel E_s is usually taken as 200 kN/mm^2. For a given concrete characteristic strength f_{cu}, the modulus of elasticity for short-term loading may be obtained from Table

2.5–6; E_c values for long-term loading will be discussed later under the heading of creep. The second moment of area I is significantly affected by the cracking of the concrete. Consider the simply supported beam in Fig. 5.5–3. In the region A, the bottom-fibre tensile stresses are sufficiently low for the concrete to remain uncracked; hence the second moment of area for this region would be that for an uncracked section (eqn 5.2–13). In region B, the situation is more complicated: at a section containing a crack, the I value for a cracked section (eqn 5.2–9) is appropriate. However, in between cracks the tensile forces in the concrete are not completely lost and neither I for an uncracked section nor that for a cracked section is appropriate. Clearly, in deflection calculations it is the sum effect of the EI values that is important, and BS 8110 recommends that the properties associated with the partially cracked section in Section 5.2 should be used for the entire beam, and that the values of the tensile stress f_{ct} in Figs 5.2–7 and 5.2–8 should be taken as 1 N/mm² for short-term loading and 0.55 N/mm² for long-term loading.

Creep
BS 8110: Part 2: Clause 3.6 recommends that, in calculating the curvatures due to long-term loading, the **effective** or **long-term modulus of elasticity** E_{eff} should be taken as

$$E_{eff} = \frac{E_c(\text{Table } 2.5\text{–}6)}{1 + \phi} \qquad (5.5\text{–}3)$$

where ϕ is called the **creep coefficient**, which is defined by

$$\phi = \frac{\text{creep strain}}{\text{elastic strain}} \qquad (5.5\text{–}4)$$

Values of ϕ are given in BS 8110: Part 2: Clause 7.3. (See also the prediction of creep strains in Section 2.5(b), which can be used to obtain an approximate estimate of the creep coefficient ϕ.)

Shrinkage
A plain concrete member undergoing a uniform shrinkage would shorten without warping. However, in a reinforced concrete beam, the reinforcement resists the shrinkage and produces a curvature. Consider the beam section in Fig. 5.5–4. A unit length of the beam is shown in Fig. 5.5–5. In Fig. 5.5–5, ε_{cs} represents the concrete shrinkage and is the uniform shortening which would occur over the unit length, had the beam been

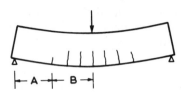

Fig. 5.5–3

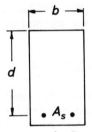

Fig. 5.5–4

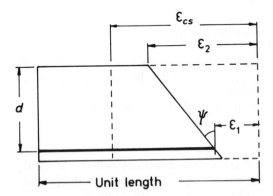

Fig. 5.5–5

unreinforced; ε_1 is the actual shortening over the unit length, at the level of the tension reinforcement; ε_2 is the actual shortening at the top. It is thus seen that the shrinkage curvature $1/r_{cs}$ of the beam at the section considered is equal to the angle ψ in Fig. 5.5–5. That is

$$\frac{1}{r_{cs}} = \frac{\varepsilon_2 - \varepsilon_1}{d} \tag{5.5-5}$$

Example 5.5–5 shows how eqn (5.5–5) can be used to derive eqn (5.5–6) below, which is the formula given by BS 8110:Part 2:Clause 3.6 for calculating the shrinkage curvature $1/r_{cs}$:

$$\frac{1}{r_{cs}} = \frac{\varepsilon_{cs}\alpha_e S_s}{I} \tag{5.5-6}$$

where ε_{cs} = concrete shrinkage;
       α_e = effective modular ratio E_s/E_{eff};
     E_{eff} = effective or long-term modulus of elasticity of the concrete (see eqn. 5.5–3);
       S_s = first moment of area of the reinforcement about the centroid of the beam section; and
        I = second moment of area of the beam section.

Values of the **concrete shrinkage** ε_{cs} are given in BS 8110 : Part 2 : Clause 7.4. An approximate estimate may also be made using the method of shrinkage prediction explained earlier in Section 2.5(c).

Loading for deflection calculations (BS 8110 : Part 2 : Clauses 3.3.2 and 3.3.3)
(a)   If the purpose of the calculation is to check the serviceability limit state of deflection, then the characteristic value should be used for both the dead load and the live load (i.e. $\gamma_f = 1$ for both).
(b)   If the purpose of the calculation is to obtain a best estimate of the deflection, then the characteristic value should be used for the dead load, but the expected or most likely value should be used for the live load.

Materials properties for deflection calculations
(BS 8110 : Part 2 : Clause 3.5)
(a)   To check the serviceability limit state of deflection, the characteristic strength of the concrete ($\gamma_m = 1$) should be used to obtain the modulus of elasticity in Table 2.5–6.
(b)   To obtain a best estimate of the deflection, the expected concrete strength should be used.

Short-term deflection
To calculate the **short-term deflection**, it is necessary only to apply the curvature–area theorems, using the EI value appropriate to the partially cracked section; creep and shrinkage effects do not come in.

Long-term deflection
In assessing the **long-term deflection**, the procedure of BS 8110 : Part 2 : Clauses 3.6 and 3.7 may conveniently be summarized as follows.

Step 1
Calculate the instantaneous curvature $1/r_{it}$ under the total load and the instantaneous curvature $1/r_{ip}$ due to the permanent load; form the difference $(1/r_{it} - 1/r_{ip})$.

Step 2
Calculate the long-term curvature $1/r_{lp}$ due to the permanent load.

Step 3
Add to the long-term curvature $1/r_{lp}$ the difference $(1/r_{it} - 1/r_{ip})$.

Step 4
Calculate the shrinkage curvature $1/r_{cs}$ from eqn (5.5–6). The required total curvature is then

$$\frac{1}{r} = \frac{1}{r_{lp}} + \left(\frac{1}{r_{it}} - \frac{1}{r_{ip}} \right) + \frac{1}{r_{cs}}$$

Step 5
From the total curvature so determined, deflections are readily calculated from the curvature–area theorem.

Comments
Step 4 of BS 8110's procedure states that the total long-term curvature $1/r$ is made up of three parts:

(a) The long-term curvature $1/r_{lp}$ due to the permanently applied load.
(b) The instantaneous curvature $(1/r_{it} - 1/r_{ip})$ due to the non-permanent load.
(c) The long-term curvature $1/r_{cs}$ due to the concrete shrinkage.

Since, by definition, a non-permanent load may or may not be acting at any given instant, we can say that the equation in Step 4 corresponds to the **maximum** long-term deflection, i.e. when the non-permanent load happens to be acting. The **minimum** long-term deflection is that when the non-permanent load happens not to be acting. This is calculated from a long-term curvature that excludes the instantaneous curvature $(1/r_{it} - 1/r_{ip})$ due to the non-permanent load; thus

$$\frac{1}{r} = \frac{1}{r_{lp}} + \frac{1}{r_{cs}}$$

Example 5.5–3
Figure 5.5–6 shows the cross-section of a beam which has a 5 m simple span and which supports a dead load G_k and an imposed load Q_k, both being uniformly distributed. It is known that half of the imposed load is permanent. Using the procedure recommended by BS 8110, calculate the total long-term deflection at midspan, for the purpose of checking the serviceability limit state. Given: $G_k = Q_k = 38.4$ kN; $f_{cu} = 40$ N/mm²; $f_y = 460$ N/mm²; $\phi = 2.5$; $\varepsilon_{cs} = 0.0003$.

SOLUTION
For the serviceability limit states, the partial safety factors γ_f and γ_m are taken as unity.

$$\text{Total load} = G_k + Q_k = 76.8 \text{ kN}$$

$$\text{Permanent load} = G_k + \tfrac{1}{2}Q_k = 57.6 \text{ kN}$$

M_t, the midspan moment due to the total load,

$$= \frac{(76.8)(5)}{8} = 48 \text{ kNm}$$

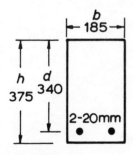

Fig. 5.5–6

M_p, the midspan moment due to the permanent load,

$$= \frac{(57.6)(5)}{8} = 36 \text{ kNm}$$

From Table 2.5–6,

$$E_c \text{ (short term)} = 28 \text{ kN/mm}^2$$

From eqn (5.5–3),

$$E_c \text{ (long term)} = \frac{28}{1 + 2.5} = 8 \text{ kN/mm}^2$$

The calculations for the deflection are carried out in the following steps.

Step 1
From Example 5.2–2(c),

$$\frac{1}{r_{it}} - \frac{1}{r_{ip}} = \frac{M_t - M_p}{E_c I_c}$$

and this was found to be $(1.30)(10^{-6})$.

Step 2
From Example 5.2–2(a),

$$\frac{1}{r_{lp}} = (5.57)(10^{-6})$$

Step 3

$$\frac{1}{r_{lp}} + \left(\frac{1}{r_{it}} - \frac{1}{r_{ip}}\right) = (5.57)(10^{-6}) + (1.30)(10^{-6})$$
$$= (6.87)(10^{-6})$$

Step 4
From eqn (5.5–6),

$$\frac{1}{r_{cs}} = \frac{\varepsilon_{cs} \alpha_e S_s}{I}$$

$$\varepsilon_{cs} = 0.0003$$

$$\alpha_e = E_s/E_{\text{eff}} = \frac{(200)(10^6)}{(8)(10^6)} = 25$$

From Example 5.2–2 (a),

$$x = 171 \text{ mm}; \qquad I_c = (763)(10^6) \text{ mm}^4$$

Hence

$$S_s = A_s(d - x)$$
$$= (628)(340 - 171) = (106.1)(10^3) \text{ mm}^3$$
$$\frac{1}{r_{cs}} = \frac{(0.0003)(25)(106.1)(10^3)}{(763)(10^6)} = (1.04)(10^{-6})$$

Step 5
Use the curvature–area theorems. From Example 5.5–1(c) and (b), the required long-term deflection is

$$a = \frac{1}{9.6}l^2(6.87 \times 10^{-6}) \quad \text{[Step 3]}$$

$$+ \frac{1}{8}l^2(1.04 \times 10^{-6}) \quad \text{[Step 4]}$$

$$= \underline{21.2 \text{ mm}}$$

Example 5.5–4
The structural member in Example 5.5–3 is a simply supported beam. If it had been the interior span of a continuous beam, explain how the solution to Example 5.5–3 would have to be modified.

SOLUTION
Calculate the total long-term curvatures $1/r_1$, $1/r_2$, and $1/1r_3$ at the left support, the right support and the midspan, respectively. Then use the method of superposition illustrated in Example 5.5–2 and eqn (5.5–2).

Example 5.5–5
BS 8110 : Part 2 : Clause 3.6 states that, in deflection calculations, the shrinkage curvature may be taken as

$$\frac{1}{r_{cs}} = \frac{\varepsilon_{cs}\alpha_e S_s}{I}$$

where the symbols are as defined earlier under eqn (5.5–6). Derive the equation.

SOLUTION
Consider again the typical beam section in Fig. 5.5–4 and the arguments that lead to eqn (5.5–5):

$$\text{shrinkage curvature } \frac{1}{r_{cs}} = \frac{\varepsilon_2 - \varepsilon_1}{d} \qquad (5.5\text{–}7)$$

where ε_2 is the concrete strain at the top level and ε_1 is that at the level of the tension steel, as shown in Fig. 5.5–5. Let

f_s = compressive stress in the steel due to the concrete shrinkage;

f_{c1} = concrete tensile stress at the level of the tension reinforcement, due to the concrete shrinkage; and

f_{c2} = concrete tensile stress at the top fibres of the beam section, due to the concrete shrinkage.

From the geometry of Fig. 5.5–5,

$$\varepsilon_2 = \varepsilon_{cs} - \frac{f_{c2}}{E_c} \qquad (5.5\text{–}8)$$

$$\varepsilon_1 = \varepsilon_{cs} - \frac{f_{c1}}{E_c} \qquad (5.5\text{–}9)$$

From the condition of equilibrium,

$$f_{c2} = \frac{(f_s A_s)}{A} + \frac{(f_s A_s) e_s}{I} e_s \qquad (5.5\text{--}10)$$

$$f_{c1} = \frac{f_s A_s}{A} - \frac{(f_s A_s) e_s}{I}(d - e_s) \qquad (5.5\text{--}11)$$

where e_s is the eccentricity of A_s from the centroid of the transformed section and A the cross-sectional area. From the compatibility condition,

$$\varepsilon_{cs} - \frac{f_{c1}}{E_c} = \frac{f_s}{E_s} \qquad (5.5\text{--}12)$$

Eliminating f_{c1} from eqns (5.5–11) and (5.5–12), we have

$$f_s = \frac{E_s \varepsilon_{cs}}{1 + \alpha_e \beta \left[1 + \dfrac{e_s^2}{k^2} \right]} \qquad (5.5\text{--}13)$$

where $\alpha_e = E_s/E_c$, $\beta = A_s/A$ and $k^2 = I/A$ (k is the radius of gyration). Substituting eqn (5.5–13) into eqns (5.5–10) and (5.5–11), we have

$$f_{c2} = \frac{E_s \beta \varepsilon_{cs}}{1 + \alpha_e \beta \left[1 + \dfrac{e_s^2}{k^2} \right]} \left[\left(1 + \frac{e_s^2}{k^2} \right) - \frac{e_s d}{k^2} \right] \qquad (5.5\text{--}14)$$

$$f_{c1} = \frac{E_s \beta \varepsilon_{cs}}{1 + \alpha_e \beta \left[1 + \dfrac{e_s^2}{k^2} \right]} \left[1 + \frac{e_s^2}{k^2} \right] \qquad (5.5\text{--}15)$$

From eqns (5.5–7) to (5.5–9),

$$\frac{1}{r_{cs}} = \frac{\varepsilon_2 - \varepsilon_1}{d} = \frac{f_{c1} - f_{c2}}{E_c d}$$

Using eqns (5.5–14) and (5.5–15),

$$\frac{1}{r_{cs}} = \frac{(\alpha_e \beta \varepsilon_{cs})}{1 + \alpha_e \beta \left[1 + \dfrac{e_s^2}{k^2} \right]} \left[\frac{e_s}{k^2} \right] \qquad (5.5\text{--}16)$$

As will be explained later in Example 9.4–1, the quantity $\alpha_e \beta (1 + e_s^2/k^2)$ is small compared with unit. Hence, eqn (5.5–16) becomes

$$\frac{1}{r_{cs}} = \frac{(\alpha_e \beta \varepsilon_{cs}) e_s}{k^2}$$

$$= \frac{\varepsilon_{cs} \alpha_e A_s e_s}{A k^2} \quad (\text{since } \beta = A_s/A)$$

$$= \frac{\varepsilon_{cs} \alpha_e S_s}{I}$$

where $S_s = A_s e_s$ and $I = A k^2$.

5.6   Calculation of crack widths (BS 8110 : Part 2)

(*Note: In day-to-day design, crack control is exercised by the straight-forward application of simple detailing rules—see Section 5.4*)

Much research has been done on cracks in concrete [17–20]. As the load on a beam increases, the number of cracks, and hence the crack spacing, rapidly reach a nearly constant value which does not change appreciably with further increase in the load. For a beam in such a condition, Beeby and his colleagues [17] have concluded that, directly over a reinforcement bar, the crack width increases with the concrete cover and with the average strain at the level at which cracking is being considered; with increasing distance from the bar, the crack width increases with the height to which the crack penetrates, i.e. the width increases with $(h - x)$ where h is the overall beam depth and x the neutral axis depth. Their research forms the basis of the following crack width formula in BS 8110:

$$\text{design surface crack width} = \frac{3a_{cr}\varepsilon_m}{1 + 2\left[\dfrac{a_{cr} - c_{min}}{h - x}\right]} \qquad (5.6\text{--}1)$$

where a_{cr} = the distance from the point considered to the surface of the nearest longitudinal bar;

ε_m = the average strain at the level where cracking is being considered, calculated allowing for the stiffening effect of the concrete in the tension zone, and is obtained from eqn (5.6–2);

c_{min} = the minimum cover to the tension steel;

h = the overall depth of the member; and

x = the neutral axis depth calculated on the assumption of a cracked section, i.e. using eqn (5.2–4) (or Fig. 5.2–3); this value of x is then used to obtain the strain ε_m in following the equation:

$$\varepsilon_m = \varepsilon_1 - \frac{b_t(h - x)(a' - x)}{3E_sA_s(d - x)} \qquad (5.6\text{--}2)$$

where ε_1 (see eqn 5.6–3) is the strain at the level considered, calculated on the assumption of a cracked section, with the concrete modulus E_c taken as half the value in Table 2.5–6 (to allow for creep effects), b_t is the width of the section at the centroid of the tension steel, a' is the distance from the compression face to the point at which the crack width is being calculated and A_s is the area of tension reinforcement. Using $(\frac{1}{2}E_c)$ with eqn (5.2–5(a)),

$$\varepsilon_1 = \frac{M}{(\frac{1}{2}E_c)I_c}x_1 \qquad (5.6\text{--}3)$$

where x_1 is the distance, measured from the neutral axis, to the point at which the strain ε_1 is sought.

A negative value for ε_m indicates that the section is uncracked. There are two special cases.

Case 1
Directly over a bar, the distance a_{cr} is equal to the concrete cover c_{min}, and eqn (5.6–1) reduces to

$$\text{width over a bar} = 3c_{min}\varepsilon_m \qquad (5.6\text{–}4)$$

Case 2
When the distance a_{cr} is large, eqn (5.6–1) approaches the following limit:

$$\text{width away from a bar} = 1.5(h - x)\varepsilon_m \qquad (5.6\text{–}5)$$

For a given member, ε_m is a maximum at the tension face; if $(h - x)$ is sufficiently small for the crack width at the tension face not to exceed the permissible limit of 0.3 mm, it will not exceed that limit anywhere. This explains why excessive crack widths rarely occur in slabs under service loading, provided the thickness does not exceed about 200 mm.

Example 5.6–1
Referring to the midspan section of the beam in Example 5.2–1 and Fig. 5.2–5, calculate the design surface crack width:

(a)   directly under a bar on the tension face;
(b)   at a bottom corner of the beam;
(c)   at a point on the tension face midway between two bars; and
(d)   at a point on a side face 250 mm below the neutral axis.

SOLUTION

(a)   ***Directly under a bar on tension face.*** Equation (5.6–4) applies. From Fig. 5.6–1(a),

$$a_{cr} = c_{min} = 40 \text{ mm}$$

From Example 5.2–1(b),

$$x = 303.6 \text{ mm}$$

and ε_1 (designated ε_h then) $= 0.00176$.
Substituting into eqn (5.6–2),

$$\varepsilon_m = 0.00176 - \frac{(450)(750-303.6)(750-303.6)}{3(200)(10^3)(3769)(690-303.6)}$$

$$= 0.001657$$

Using eqn (5.6–4),

$$\text{crack width} = (3)(40)(0.001657)$$

$$= \underline{0.20 \text{ mm}}$$

(b)   **Bottom corner.** From Fig. 5.6–1(b),

$$ij = (a_{cr} + 20) = \sqrt{(60^2 + 60^2)}$$

Therefore

$$a_{cr} = 64.85 \text{ mm}$$

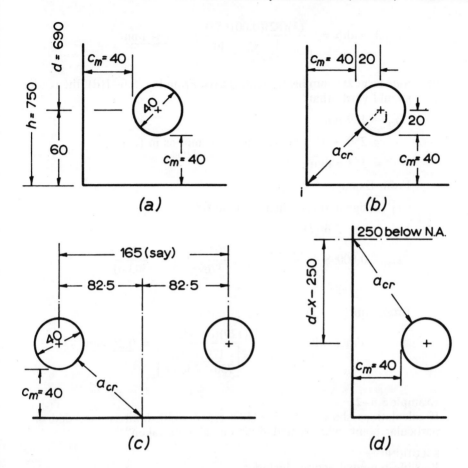

Fig. 5.6–1

$\varepsilon_m = 0.001657$ as in (a); $x = 303.6$ mm as in (a)

From eqn (5.6–1),

$$\text{crack width} = \frac{(3)(64.85)(0.001657)}{1 + 2\left[\dfrac{64.85 - 40}{750 - 303.6}\right]} = \underline{0.29\ \text{mm}}$$

(c) **Tension face: midway between two bars.** From Fig. 5.6–1(c), it is easy to show that

$$a_{cr} = \sqrt{(82.5^2 + 60^2)} - 20 = 82\ \text{mm}$$

$\varepsilon_m = 0.001657$ as in (a)

$x = 303.6$ mm as in (a)

From eqn (5.6–1),

$$\text{crack width} = \frac{(3)(82)(0.001657)}{1 + 2\left[\dfrac{82 - 40}{750 - 303.6}\right]} = \underline{0.34 \text{ mm}}$$

(d) **Side face: 250 mm below neutral axis.** From Fig. 5.6–1(d), the reader should verify that

$$a_{cr} = 129 \text{ mm}$$

$$a' = 250 + x \quad \text{(where } x \text{ is 303.6 mm as in (a))}$$

$$= 553.6 \text{ mm}$$

From Example 5.2–1(b),

$$\varepsilon_1 \text{ (designated } \varepsilon_{250} \text{ then) } = 0.00098$$

Using eqn (5.6–2),

$$\varepsilon_m = 0.00098 - \frac{(450)(750 - 303.6)(553.6 - 303.6)}{3(200)(10^3)(3769)(690 - 303.6)}$$

$$= 0.00092$$

From eqn (5.6–1),

$$\text{crack width} = \frac{(3)(129)(0.00092)}{1 + 2\left[\dfrac{129 - 40}{750 - 303.6}\right]} = \underline{0.25 \text{ mm}}$$

Example 5.6–2

If calculations show that BS 8110's crack width limit is exceeded at a particular point, what remedial actions may be taken?

SOLUTION

Possible remedial actions include:

(a) The area A_s of the tension reinforcement may be provided in the form of a larger number of smaller bars. This has the effect of reducing the distance a_{cr}.

(b) A larger A_s than that required for the ultimate limit state may be provided. This has the effect of increasing I_c and x (Figs 5.2–3 and 5.2–4) and hence reducing the strain ε_m.

(c) Or, an additional bar may be placed directly adjacent to the point where the crack width is excessive. From eqn (5.6–4) the crack width is now $3c_{min}\varepsilon_m$; thus if (say) $c_{min} = 50$ mm and $\varepsilon_m = (5f_y/8)/E_s$ (see eqn 5.3–1(b)), the crack width will be reduced to

$$\frac{(3)(50)(5/8)(460)}{(200)(10^3)} = 0.22 \text{ mm}$$

Finally it should be pointed out that the methods in Sections 5.4 and 5.6 are intended for flexural cracks in ordinary beams. They do not apply to shear cracks (Chapter 6) which fortunately are rarely a problem under service loading. Nor do these methods apply to deep beams where service crack widths may be a problem; some design guidance is given in Reference 21.

5.7 Design and detailing—illustrative examples

The following examples are a continuation of Example 4.11–1.

Example 5.7–1

Complete Step 8 of the solution to Example 4.11–1, by checking that the actual span/depth ratio is within the allowable limit. Dimensions and reinforcement details are given in Figs 4.11–1 and 4.11–2.

SOLUTION

Using the information in Figs 4.11–1 and 4.11–2, the midspan section is redrawn in Fig. 5.7–1. The calculations below follow the steps listed in Section 5.3; note, in particular, the provisions for flanged beams in Step 1 of Section 5.3.

Step 1   Basic span/depth ratio
From Table 5.3–1, the basic span/depth ratios are

> rectangular section: 26
>
> flanged section ($b_w/b \le 0.3$): 20.8

Actual $b_w/b = 325/710 = 0.46$
By interpolation, the basic span/depth ratio is

$$20.8 + \frac{(0.46 - 0.30)}{(1.00 - 0.30)}(26 - 20.8) = 22$$

Step 2   Modification factor for long span
Omitted, since span < 10 m.

Step 3   Modification factor for tension reinforcement
From Example 4.11–1, $M = 60.30$ kNm.

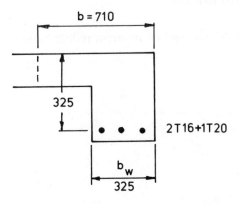

Fig. 5.7–1   Midspan section: dimensions and tension bars

$$\frac{M}{bd_2} = \frac{(60.30)(10^6)}{(710)(325^2)} = 0.8 \text{ N/mm}^2$$

From Table 5.3–2,
Modification factor = 1.48

Step 4   *Modification factor for compression reinforcement*
Omitted.

Step 5
The allowable span/depth ratio is

$$(22 \text{ of Step 1})(1.48 \text{ of Step 3}) = 32.6$$

$$\text{Actual span/depth ratio} = \frac{5500}{325}$$

$$= 16.9 < 32.6 \text{ OK}$$

Example 5.7–2
Complete Step 9 of the solution to Example 4.11–1, by checking that the clear distances between the tension bars are within the allowable limits.

SOLUTION
Using the information in Figs 4.11–1 and 4.11–2, the dimensions and longitudinal reinforcement details are redrawn in Figs 5.7–2 and 5.7–3.

Midspan section (Fig. 5.7–2)
From Table 5.4–1, for $f_y = 460$ N/mm^2 and zero redistribution, clear distance between bars = 160 mm.
From Fig. 5.7–2,

$$2a_b + 2(40 + 16) + 20 = 325$$

$$a_b = 96.5 \text{ mm} < 160 \text{ mm O.K.}$$

$$\left(a_c + \frac{16}{2}\right)^2 = \left(40 + \frac{16}{2}\right)^2 + \left(40 + \frac{16}{2}\right)^2$$

$$a_c = 59.88 \text{ mm} < \tfrac{1}{2} \text{ of 160 mm OK}$$

Support section (Fig. 5.7–3)
From Table 5.4–1, for $f_y = 460$ N/mm^2 and no moment redistribution,

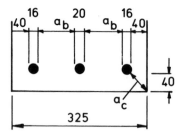

Fig. 5.7–2   Tension bars at midspan

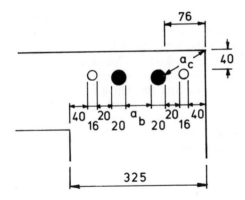

Fig. 5.7–3   Tension bars at support

clear distance between bars = 160 mm.
   From Fig. 5.7–3,

$$a_b + 2(40 + 16 + 20 + 20) = 325$$

$$a_b = 133 \text{ mm} < 160 \text{ mm OK}$$

$$\left(a_c + \frac{20}{2}\right)^2 = \left(40 + \frac{20}{2}\right)^2 + \left(76 + \frac{20}{2}\right)^2$$

$$a_c = 89.48 \text{ mm} > \tfrac{1}{2} \text{ of } 160 \text{ mm}\quad\text{(see Comment (a) below)}$$

Comments

(a)   The corner distance a_c is 89.48 mm and exceeds the BS 8110 limit in Fig. 5.4–1. One possible remedial action is, of course, to carry out a crack-width calculation (see Section 5.6). However, Example 5.4–1: solution part (c) explains an easier remedial action. That is, we shall take advantage of the fact that $A_{s,\text{prov}}$ is greater than $A_{s,\text{req}}$ and use the reduced f_s value to obtain a relaxed bar spacing from eqn (5.4–2).
   From Example 4.11–1, $A_{s,\text{req}} = 572 \text{ mm}^2$ and $A_{s,\text{prov}} = 628 \text{ mm}^2$. Hence, from eqn (5.3–1(b)),

$$f_s = \frac{5}{8}f_y\frac{A_{s,\text{req}}}{A_{s,\text{prov}}}\frac{1}{\beta_b}$$

$$= \frac{5}{8}(460)\left(\frac{572}{628}\right)\left(\frac{1}{1}\right)$$

$$= 261 \text{ N/mm}^2$$

Substituting into eqn (5.4–2),

$$\text{clear spacing} = \frac{47\,000}{f_s} = \frac{47\,000}{261} = 180 \text{ mm}$$

Therefore, we now have

$$a_c = 89.48 \text{ mm} < \tfrac{1}{2} \text{ of } 180 \text{ mm OK}$$

(b)   The corner distance a_c is well defined at the midspan section. At the support section, Fig. 5.7–3 shows that a_c is defined only for the right-hand side corner, but not for the left-hand side. Allen [22] has drawn attention to this ambiguity and asked the following question with reference to monolithic beam-and-slab construction: 'Where is the corner of the beam when considering the tension bars over a support?' It is reasonable to say that the 'corner' should not be literally interpreted as the point of intersection of the vertical face of the beam rib with the top face of the slab. Allen [22] has suggested that, in such circumstances, what needs checking is not a_c, but the clear distance between an outside tension bar and the adjacent slab bar near the top face of the slab.

5.8   Computer programs

(in collaboration with **Dr H. H. A. Wong**, University of Newcastle upon Tyne)

The FORTRAN programs for this chapter are listed in Section 12.5. See also Section 12.1 for 'Notes on the computer programs'.

Problems

5.1   Table 3.11 of BS 8110 (see Table 5.3–2 here) gives the modification factors for the tension reinforcement, for use in determining the allowable span/effective depth ratio. In the table, the quantity M/bd^2 is used as a measure of the amount of tension reinforcement. Comment on whether the modification factor depends on: (a) the tension steel area required for M/bd^2; or (b) the tension steel area actually provided.

Ans.   Strictly speaking, the modification factor depends on the amount of tension reinforcement actually provided. The values in Table 5.3–2, being based on M/bd^2 (and hence on the A_s required), are approximate and err on the safe side. For further information, see eqns (5.3–1(a) (b)) and the associated 'Comments on Step 3'.

5.2   Table 3.12 of BS 8110 (see Table 5.3–3 here) relates the modification factor to the amount of compression reinforcement. How does the compression steel affect the deflection?

Ans.   See 'Comments on Step 4' on p. 172.

5.3   In university courses, deflections are frequently calculated using Macaulay's method, which is based on the differential equation

$$\frac{d^2w}{dx^2} = -\frac{M}{EI}$$

where w here denotes the deflection and x the distance along the beam axis. Readers who use American textbooks will be familiar with the moment–area method, which is a more convenient tool favoured by Timoshenko and others. Section 5.5 here introduces a **curvature–area theorem**. Explain why, for concrete beams, this theorem is more convenient to use than either Macaulay's method or the moment–area method.

Ans.   For concrete beams, deflections are caused not only by bending moments, but also (and significantly) by creep and shrinkage. The deflection of a beam is completely defined by the curvature distribution; whether the curvature is caused by shrinkage or creep or bending moment is immaterial. The curvature–area theorem focuses attention on the purely geometrical relation between deflection and curvatures, and enables deflections to be calculated easily.

5.4   BS 8110 states that surface crack widths should not in general exceed 0.3 mm. Is this restriction on crack width to prevent the corrosion of the reinforcement?

Ans.   No; see first paragraph of Section 5.4.

5.5   In practical design, it is usual to comply with BS 8110's crack-width requirements by a straightforward procedure of (a) limiting the clear distance a_b between tension bars to within the value specified in Table 5.4–1 and (b) limiting the clear distance a_c from the corner of the beam to the nearest tension bar to not exceeding half the value given in that table.

The figure below shows the tension bars at the support section of a T-beam in a conventional monolithic beam-and-slab construction. Are the distances a_b and a_c to be measured as shown in the figure? (*Note:* Table 5.4–1 is extracted from BS 8110: Clause 3.12.11.2.3.)

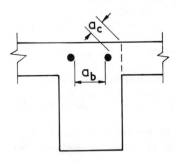

Ans. a_b, yes; a_c, no. (See Comment (b) at the end of Example 5.7–2.)

Problem 5.5

5.6   Equations (5.4–1) and (5.4–2) are taken from BS 8110: Clauses 3.12.11.2.3 and 3.12.11.2.4, respectively. Show that these two equations become identical when $A_{s,req} = A_{s,prov}$, where $A_{s,req}$ and $A_{s,prov}$ are as defined under eqn (5.3–1(b)). (*Hint:* Use eqn 5.3–1(b) to compute f_s. Then substitute this value of f_s into eqn 5.4–2. Next use eqn 4.7–2 to convert β_b into $\beta\%$).

References

1   *Deflection of Concrete Structures* (SP–86). American Concrete Institute, Detroit, 1985.
2   ACI Committee 435. *Allowable Deflections*. American Concrete Institute, Detroit, 1984.
3   Branson, D. E. *Deformation of Concrete Structures*. McGraw-Hill, New York, 1977.
4   Swamy, R. N. and Spanos, A. Deflection and cracking behaviour of ferrocement with grouped reinforcement and fiber reinforced matrix. *Proc. ACI*, **82**, No. 1, Jan./Feb. 1985, pp. 79–91.
5   Evans, R.H. and Kong, F. K. Strain distribution in composite prestressed concrete beams. *Civil Engineering and Public Works Review*, **58**, No. 684, July 1963, pp. 871–2 and No. 685, Aug. 1963, pp. 1003–5.
6   Bate, S. C. C. *et al. Handbook on the Unified Code for Structural Concrete*. Cement and Concrete Association, Slough, 1972.
7   Beeby, A. W. *Modified Proposals for Controlling Deflections by Means of Ratios of Span to Effective Depth*. Cement and Concrete Association, Slough, 1971.
8   Neville, A. M., Houghton-Evans, W. and Clark, C. V. Deflection control by span/depth ratio. *Magazine of Concrete Research*, **29**, No. 98, March 1977, pp. 31–41.
9   Rangan, B. V. Control of beam deflections by allowable span/depth ratios. *Proc. ACI*, **79**, No. 5, Sept./Oct. 1982, pp. 372–7.
10 Gilbert, R. I. Deflection control of slabs using allowable span to depth ratios. *Proc. ACI*, **82**, No. 1, Jan./Feb. 1985, pp. 67–72.
11 Pretorius, P. C. Deflections of reinforced concrete members: a simple approach. *Proc. ACI*, **82**, No. 6, Nov./Dec. 1985, pp. 805–12.
12 ACI Committee 222. Corrosion of metals in concrete. *Proc. ACI*, **82**, No. 1, Jan./Feb. 1985, pp. 3–32.
13 Hognestad, E. Design of concrete for service life. *Concrete International*, **8**, No. 6, June 1986, pp. 63–7.
14 Beeby, A. W. Corrosion of reinforcing steel in concrete and its relation to cracking. *The Structural Engineer*, **56A**, No. 3, March 1978, pp. 77–81.
15 Coates, R. C., Coutie, M. G. and Kong F. K. *Structural Analysis*, 2nd edn. Van Nostrand Reinhold (UK), 1980, p. 147.
16 Kong, F. K., Prentis, J. M. and Charlton, T. M. Principle of virtual work for a general deformable body—a simple proof. *The Structural Engineer*, **61A**, No. 6, June 1983, pp. 173–9.
17 Beeby, A. W. The prediction and control of flexural cracking in reinforced concrete members. *Proceedings* ACI Symposium on cracking, deflection and ultimate load of concrete slab systems. American Concrete Institute, Detroit, 1971, pp. 55–75.
18 Brondum-Nielsen, T. Serviceability limit state analysis of concrete sections under biaxial bending. *Proc. ACI*, **81**, No. 5, Sept./Oct. 1984, pp. 448–55.

19 ACI Committee 224. *Control of Cracking in Concrete Structures*. American Concrete Institute, Detroit, 1984.
20 Nawy, E. G. Structural elements—strength, serviceability and ductility. In *Handbook of Structural Concrete*, edited by Kong, F. K., Evans, R. H., Cohen, E. and Roll, F. Pitman, London, and McGraw-Hill, New York, 1983, Chapter 12.
21 Ove Arup and Partners. *CIRIA Guide 2: The Design of Deep Beams in Reinforced Concrete*. Concrete Industry Research and Information Association, London, 1984.
22 Allen, A. H. *Reinforced Concrete Design to CP110*. Cement and Concrete Association, Slough, 1974, p. 61.

Chapter 6
Shear, bond and torsion

Preliminary note: Readers interested only in structural design to BS 8110 may concentrate on the following sections:

(a) *Section 6.4: Shear design (BS 8110).*
(b) *Section 6.6: Bond and anchorage (BS 8110).*
(c) *Section 6.11: Torsion design (BS 8110).*
(d) *Section 6.12: Design and detailing example.*

6.1   Shear

Shear is an important but controversial topic in structural concrete [1–11]. In design, it is generally desirable to ensure that ultimate strengths are governed by flexure rather than by shear [1, 2]. Shear failures, which in reality are failures under combined shear forces and bending moments, are characterized by small deflections and lack of ductility. There is sometimes little warning before failure occurs, and this makes shear failures particularly objectionable.

It was stated in Section 3.4 that the effect of axial load on short columns need not be considered for the serviceability limit states; similarly, shear in beams is normally to be considered for the ultimate limit state only.

6.2   Shear failure of beams without shear reinforcement

Figure 6.2–1(a) shows half of a reinforced concrete beam acted on by a shear force V. An element in the beam would be subjected to shear stresses v, as in Fig. 6.2–1(b), and to horizontal normal stresses due to bending. If the element is near the neutral axis or within a flexurally cracked region, the bending stresses are comparatively small and may be neglected without serious loss in accuracy. The shear stresses in Fig. 6.2–1(b) are then equivalent to the principal stresses in Fig. 6.2–1(c), in which the principal tensile stresses are traditionally called the **diagonal-tension stresses**. It can be seen that (see *Failure criteria* in Section 2.5(f)) when the diagonal tension stresses reach the tensile strength of the concrete, a **diagonal crack** will develop. The preceding description, though convenient as an

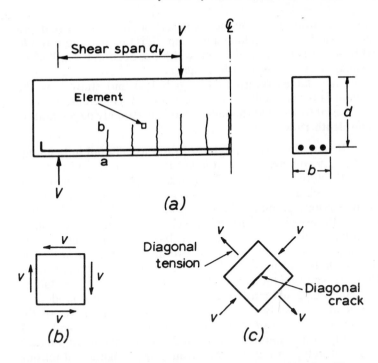

Fig. 6.2–1 Traditional concepts of shear and diagonal tension [1]

introduction to the concepts of diagonal tension and diagonal cracking, does not give a whole picture of the actual behaviour. In fact, the type of diagonal crack in Fig. 6.2–1(c), called a **web-shear crack**, occurs mainly in prestressed concrete beams (see Chapter 9) and only rarely in reinforced concrete beams. Of course, the behaviour of reinforced concrete beams is much influenced by the shear stresses, but the trouble is that we do not know how to calculate their values. In the earlier days it was usual to make various assumptions (which were not justified) and to prove that, below the neutral axis, v was everywhere equal to V/bz (b being the beam width and z the lever-arm distance) and that, above the neutral axis, v varied parabolically to zero at the compression face of the beam. It was realized some years ago that things were not so simple. Even today, 'the distribution of the shear stress across a flexurally cracked beam is not understood and an accurate determination of the magnitude of v is impossible' [1]; indeed, present-day designers no longer attempt to calculate the actual value of the shear stress v. However, there are advantages in retaining the concept of a **nominal shear stress** to be used as some sort of stress coefficient in design. In current British design practice, BS 8110 refers to this nominal shear stress as the **design shear stress** and defines it as

$$v = \frac{V}{b_v d} \tag{6.2–1}$$

where V is the shear force acting on the beam section, d the effective depth and b_v the beam width. For a rectangular beam, b_v is taken as the width b; for a **flanged beam**, it is taken as the web width b_w. Section 6.4 will show how design shear stresses are used in practice.

In the meantime, let us return to Fig. 6.2–1(a) and study how the rectangular beam fails eventually as the shear force V is increased. Many tests [1, 2] have established that the failure mode is strongly dependent on the **shear-span/depth ratio** a_v/d:

(a) $a_v/d > 6$: Beams with such a high a_v/d ratio usually fail in bending.
(b) $6 > a_v/d > 2.5$: Beams with a_v/d lower than about 6 tend to fail in shear. With reference to Fig. 6.2–1(a), as the force V is increased, the flexural crack a–b nearest the support would propagate towards the loading point, gradually becoming an inclined crack, which is known as a **flexure-shear crack** but which is often referred to simply as a diagonal crack (Fig. 6.2–2: crack a–b–c). With further increase in V, failure usually occurs in one of two modes. If the a_v/d ratio is relatively high, the diagonal crack would rapidly spread to e, resulting in collapse by splitting the beam into two pieces. This mode of failure is often called **diagonal-tension failure**; for such a failure mode, the ultimate load is sensibly the same as that at the formation of the diagonal crack. If the a_v/d ratio is relatively low, the diagonal crack tends to stop somewhere at j (Fig. 6.2–2); a number of random cracks may develop in the concrete around the longitudinal tension reinforcement. As V is further increased, the diagonal crack widens and propagates along the level of the tension reinforcement (Fig. 6.2–2: crack g–h). The increased shear force presses down the longitudinal steel and causes the destruction of the bond between the concrete and the steel, usually leading to the splitting of the concrete along g–h. If the longitudinal reinforcement is not hooked at the end, the destruction of bond and the concrete splitting will cause immediate collapse. If hooks are provided, the beam behaves like a two-hinge arch until the increasing force in the longitudinal reinforcement destroys the concrete surrounding the hooks, whence

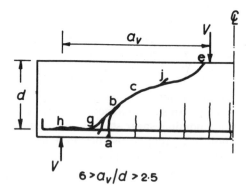

Fig. 6.2–2

collapse occurs. This failure mode is often called **shear-tension failure** or shear-bond failure; again the ultimate load is not much higher than the diagonal cracking load.

(c) $2.5 > a_v/d > 1$: For a_v/d lower than about 2.5 but greater than 1, the diagonal crack often forms independently and not as a development of a flexural crack (Fig. 6.2–3) [10]. The beam usually remains stable after such cracking. Further increase in the force V will cause the diagonal crack to penetrate into the concrete compression zone at the loading point, until eventually crushing failure of the concrete occurs there, sometimes explosively (Fig. 6.2–3; shaded portion). This failure mode is usually called **shear–compression failure**; for this mode, the ultimate load is sometimes more than twice that at diagonal cracking.

(d) $a_v/d < 1$: The behaviour of beams with such low a_v/d ratio approaches that of deep beams [12–21]. The diagonal crack forms approximately along a line joining the loading and support points (Fig. 6.2–4). It forms mainly as a result of the splitting action of the compression force that is transmitted directly from the loading point to the support; it initiates frequently at about $d/3$ above the bottom

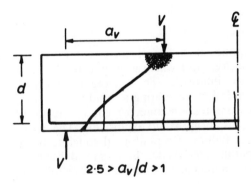

Fig. 6.2–3

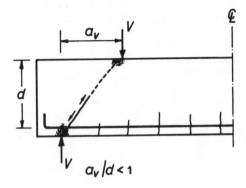

Fig. 6.2–4

face of the beam. As the force V is increased, the diagonal crack would propagate simultaneously towards the loading and support points. When the crack has penetrated sufficiently deeply into the concrete zone at the loading point, or, more frequently, at the support point, crushing failure of the concrete occurs. For a **deep beam failure mode**, the ultimate load is often several times that at diagonal cracking.

The **mechanisms of shear transfer** in a cracked concrete beam are illustrated in the free-body diagram in Fig. 6.2–5. The shear force V is resisted by the combined action of the shear V_{cz} in the uncracked concrete compression zone, the shear V_d from the **dowel action** of the longitudinal reinforcement, and the shear V_a which is the vertical component of the force due to **aggregate interlock** (sometimes called the **interface shear transfer**). Thus

$$V = V_{cz} + V_d + V_a \tag{6.2–2}$$

Quantitative evidence [10] is now available that for a typical reinforced concrete beam the shear force V is carried in the approximate proportions:

compression zone shear $V_{cz} = 20\text{–}40\%$

dowel action $\qquad\qquad V_d = 15\text{–}25\%$

aggregate interlock $\qquad V_a = 35\text{–}50\%$

According to Taylor [10], as the applied shear force is increased, the dowel action is the first to reach its capacity, after which a proportionally large shear is transferred to aggregate interlock. The aggregate interlock mechanism is probably the next to fail, necessitating a rapid transfer of a large shear force to the concrete compression zone, which, as a result of this sudden shear transfer, often fails abruptly and explosively. The above description suggests that the shear failure of a reinforced concrete beam is affected by a number of shear parameters besides the a_v/d ratio discussed earlier. The effects of the main parameters may be summarized as follows:

(a)   **concrete strength**. The dowel-action capacity, the aggregate-interlock capacity and the compression-zone capacity generally all increase with the concrete strength. Figure 6.2–6 shows the effect of concrete strength (and the steel ratio) on the nominal **ultimate shear stress**

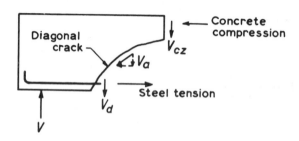

Fig. 6.2–5

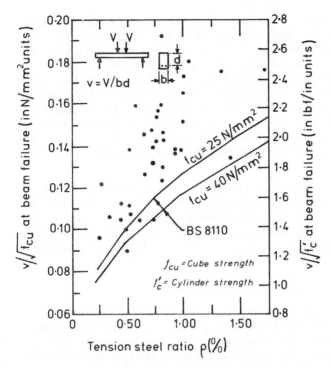

Fig. 6.2–6   Effect of f_{cu} and ϱ on nominal ultimate shear stress v [22]

which, in current British practice, is the design shear stress v calculated from eqn (6.2–1) with V taken as the shear force due to the ultimate load. In Fig. 6.2–6, the BS 8110 curves have been drawn using the values of the design concrete shear stress v_c, as given in Table 6.4–1, for $d = 400$ mm.

(b) **tension steel ratio**. The tension steel ratio [22] ϱ $(= A_s/bd)$ affects shear strength mainly because a low ϱ value reduces the dowel shear capacity and also leads to wider crack widths, which in turn reduces the aggregate-interlock capacity (Fig. 6.2–6).

(c) **strength of longitudinal reinforcement**. Provided the steel ratio is kept constant, the characteristic strength of the longitudinal reinforcement [2] has little effect on shear strength.

(d) **aggregate type**. The type of aggregate [10] affects shear strength mainly through its effect on the aggregate-interlock capacity. For this reason, though lightweight concrete can be made to have the same compressive strength as normal weight concrete, the shear stress (eqn 6.2–1) to be used in design has to be lower than for normal concrete. For example, BS 8110 : Part 2 : Clause 5.4 states that the shear stress values in Table 6.4–1 should be multiplied by 0.8 when applied to lightweight concrete.

(e) **beam size**. The ultimate shear stress reduces with the beam size [23, 24] particularly the beam depth; that is, larger beams are proportionately weaker than smaller beams. This is probably because in practice the aggregate-interlock capacity does not increase in the same proportion as the beam size. The design shear stress values in Table 6.4–1 allow for the influence of the effective depth d.

(f) **the effective shear-span/depth ratio** (M/Vd). The ultimate shear stress at a beam section increases rapidly as the M/Vd ratio [2] is reduced below about 2, where M is the bending moment, V the shear force, and d the effective depth; this is true for both distributed loading or concentrated loading. For two-point loading, as in Fig. 6.2–1(a), the critical (i.e. the largest) M/Vd ratio occurs at the loading point, where $M/Vd = a_v/d$.

A broad overview of the shear failure of reinforced concrete beams has been given above. As stated at the beginning of this chapter, shear is a controversial topic; new and potentially far-reaching concepts are continually being developed. Interested readers are recommended to study the latest papers by Kotsovos [3, 4], Regan [5] and Hsu [6], for example.

6.3   Effects of shear reinforcement

Referring to Fig. 6.2–1(a), the shear strength of the beam may be substantially increased by the suitable provision of **shear reinforcement**, or **web reinforcement** as it is often called; more important, such shear reinforcement increases the ductility of the beam and considerably reduces the likelihood of a sudden and catastrophic failure, which often occurs in beams without shear reinforcement.

Stirrups or **links** (Fig. 6.3–1) are the most common type of web reinforcement, though they are sometimes used in combination with **bent-up bars** (Fig. 6.3–2). Before diagonal cracking, the external shear force V produces practically no stress in the web reinforcement. When the diagonal crack forms, any web bar which intercepts the diagonal crack would suddenly carry a portion of the shear force V; web bars not intercepting the diagonal crack remain essentially unstressed. The mechanism of shear transfer is shown in Fig. 6.3–3, in which the meanings of the symbols are similar to those in Fig. 6.2–5, namely, V_{cz} is the shear carried by the uncracked concrete compression zone, V_a that by aggregate interlock across the diagonal crack, and V_d that carried by the dowel action of the

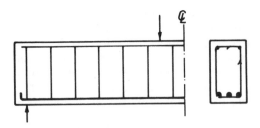

Fig. 6.3–1   Links or stirrups

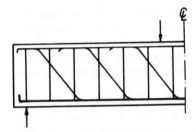

Fig. 6.3–2   **Combination system of links and bent-up bars**

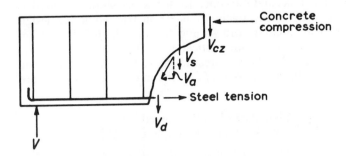

Fig. 6.3–3   **Shear transfer in beam with web reinforcement**

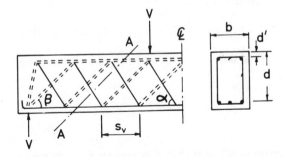

Fig. 6.3–4   **Truss analogy**

longitudinal reinforcement; V_s represents the shear force carried by the web bars crossed by the diagonal crack. Thus

$$V = V_{cz} + V_a + V_d + V_s$$
$$= V_c + V_s \tag{6.3–1}$$

where V_c represents $V_{cz} + V_a + V_d$. Conventionally V_c is referred to as the **shear carried by the concrete** and V_s as the **shear carried by the (web) steel**. In a typical beam, as the external shear V is increased, the web steel yields so that V_s remains stationary at the yield value, and subsequent increase in

V must be carried by V_{cz}, V_a and V_d. As the diagonal crack widens, the aggregate interlock becomes less effective and V_a decreases, forcing V_{cz} and V_d to increase rapidly. Failure of the beam finally occurs either by dowel splitting of the concrete along the longitudinal reinforcement or by crushing of the concrete compression zone resulting from the combined shear and direct stresses.

In current design practice, the stresses in the shear reinforcement are analysed by the **truss analogy**, illustrated in Fig. 6.3–4, in which the web bars are assumed to form the tension members of an imaginary truss, while the thrusts in the concrete constitute the compression members (shown dotted in the figure). The figure shows a general case of links at a longitudinal spacing s_v. The links and the concrete 'struts' are shown inclined at the general angles α and β, respectively, to the beam axis. To derive design equations using the truss analysis, draw a line A–A in Fig. 6.3–4, parallel to the concrete 'struts'. Consider the vertical equilibrium of the free body to the left of the line A–A. The web resistance V_s is contributed by the vertical components of the tensions $A_{sv}f_{yv}$ in the individual links that are crossed by A–A:

$$V_s = A_{sv}f_{yv} \sin \alpha \left[\begin{array}{l} \text{Number of links crossed} \\ \text{by A–A (see Example 6.3–1)} \end{array} \right]$$

$$= A_{sv}f_{yv} \sin \alpha \left[\frac{(d - d')\cot \alpha + (d - d')\cot \beta}{s_v} \right]$$

$$= A_{sv}f_{yv} [\cos \alpha + \sin \alpha \cot \beta] \left[\frac{d - d'}{s_v} \right] \qquad (6.3\text{–}2)$$

where A_{sv} is the area of *both* legs of each link and f_{yv} is the characteristic strength of the links.

In the particular case of **vertical links**, $\alpha = 90°$ and eqn (6.3–2) becomes

$$V_s = A_{sv}f_{ys} \left[\frac{d - d'}{s_v} \right] \cot \beta$$

$$\doteqdot A_{sv}f_{yv} \left[\frac{d}{s_v} \right] \cot \beta \qquad (6.3\text{–}3)$$

Tests [2] have led to the recommendation that β in eqn (6.3–3) could be taken as 45°, so that

$$V_s = A_{sv}f_{yv} \frac{d}{s_v} \qquad (6.3\text{–}4)$$

From eqn (6.3–1), $V_s = V - V_c$; and if we write $V = vbd$ and $V_c = v_c bd$, the above equation for vertical links becomes

$$(v - v_c) b = \frac{A_{sv}f_{yv}}{s_v} \qquad (6.3\text{–}5)$$

When the web reinforcement consists of a system of **bent-up bars**, it would be reasonable to use eqn (6.3–2) as it stands:

$$\left[\begin{array}{l} \text{shear carried by} \\ \text{bent-up bar system} \end{array} \right] = A_{sv}f_{yv} [\cos \alpha + \sin \alpha \cot \beta] \frac{d - d'}{s_v}$$

$$(6.3\text{–}6)$$

Again, test observations have led to the following recommendations on the use of eqn (6.3–6):

(a)   the angle α should not be less than 45°;
(b)   the spacing s_v should not exceed $1.5d$; otherwise the angle β tends to be less than 45°, with the consequence that the bent-up bar system becomes ineffective;
(c)   the shear force carried by the bent-up bars should not exceed 50% of the shear force V_s carried by the web steel. (That is, at least 50% of V_s should be provided by links.)

It should be noted once again that the truss analogy is no more than a design tool; though conceptually convenient, it presents an over-simplified model of the reinforced concrete beam in shear [1, 2]. For example, the truss analogy model completely ignores the favourable interaction between the web reinforcement and the aggregate-interlock capacity and the dowel force capacity; to this extent it tends to give conservative results, though the conservatism reduces as the amount of web steel increases. The truss analogy also assumes that the failure of the beam is initiated by the yielding or excessive deformation of the web reinforcement, but in very thin webbed reinforced or prestressed concrete beams (e.g. T-beams), failure may in fact be initiated by **web crushing**; in such a case the truss analogy would give unsafe results.

The above account of shear behaviour, together with that in Section 6.2 for beams without web reinforcement, may be amplified by the following summary statements:

(a)   For a beam without web reinforcement, Fig. 6.2–6 shows that, given the concrete strength f_{cu} and the longitudinal steel ratio ϱ, a safe lower bound value can be assigned to the nominal shear stress at collapse. Designating this nominal shear stress as v_c (where the suffix c will serve to remind us that we are referring to a beam without web steel), a safe estimate of the ultimate shear strength would be

$$V_c = v_c bd \qquad\qquad (6.3–7)$$

(b)   Where web reinforcement is used, it remains practically unstressed until diagonal cracking occurs, at which instant those web bars that intercept the diagonal crack will receive a sudden increase in stress [1]. If the amount of web steel is too small, the sudden stress increase may cause the instant yielding of the web bars. The authors [1] have drawn attention to the test observation that the web-steel ratio ϱ_v $(= A_{sv}/bS_v)$ should be such that the product $\varrho_v f_{yv}$ is not less than about 0.38 N/mm². For design purposes, we can round up 0.38 to 0.4, so that

$$\varrho_v f_{yv} \text{ (minimum)} \geq 0.4 \text{ N/mm}^2 \qquad\qquad (6.3–8)$$

(c)   Web reinforcement may confidently be assumed to be effective, only if every potential diagonal crack is intercepted by at least one web bar. Example 6.3–1 shows that, for this to be possible, the spacing of vertical links must not exceed

$$s_{v,max}\text{(vertical links)} = d$$

Tests [25] have shown that (1) links which intercept a diagonal crack near the top are relatively ineffective and that (2) a link, in addition to the one crossed by the diagonal crack and within a close distance to it, further increases the shear strength of the beam. Therefore, it is desirable in design to limit the maximum link spacing to, say, 75% of that above:

$$s_{v,max}(\text{vertical links}) = 0.75d \tag{6.3-9}$$

Example 6.3–1 also shows that, for bent-up bars (or inclined links),

$$s_{v,max}(\text{bent-up bars}) = (\cot \alpha + \cot \beta)d$$

It was explained earlier, in connection with eqn (6.3–6), that both α and β should not be less than 45°, so that

$$s_{v,max}(\text{bent-up bars}) = (\cot 45 + \cot 45)d$$
$$= 2d$$

Again, taking 75% of this value for design, we have

$$s_{v,max}(\text{bent-up bars}) = 1.5d \tag{6.3-10}$$

(d)   When ϱ_v and s_v satisfy the requirements in (b) and (c) above, the capacity of the web reinforcement may for design purposes be estimated by the truss analogy. The truss analogy assumes that the web steel can reach its yield stress before other failure modes occur, e.g. web crushing or compression zone failure. Tests [1, 2] have shown that such premature failures can be prevented by imposing a ceiling on the nominal ultimate shear stress. Denoting this nominal stress by v_u (to distinguish it from the v_c in eqn 6.3–7), the external shear force V must not be allowed to exceed

$$V = v_u bd \tag{6.3-11}$$

no matter how much web steel is used.

(e)   For a beam with web reinforcement, therefore, the shear resistance may be regarded as being made up of the sum of the concrete resistance and the web steel resistance:

$$V = V_c + V_s$$

or dividing by bd,

$$v = v_c + v_s \tag{6.3-12}$$

where v_c may conservatively be obtained from test results for beams without web reinforcement (e.g. Fig. 6.2–6). In fact the web steel does not only carry shear force itself, but it also increases the shear carrying capacity of the concrete.

(f)   The truss analogy does not differentiate between links and bent-up bars. When they are used in combination, the analogy gives their shear capacity as the sum of their capacities when used separately. The effects of links and bent-up bars used in combination are actually more than additive [1]. Bent-up bars are more effective than links in restricting the widening of the diagonal crack [1]; but links can perform the important function of preventing the pressing down of the longitudinal reinforcement and hence maintaining the dowel capacity.

Example 6.3–1
With reference to the truss analogy illustrated in Fig. 6.3–4, derive an expression for the number of links that will be crossed by the line A–A. Hence comment on the maximum permissible link spacing.

SOLUTION
With reference to Fig. 6.3–5, the horizontal projection of $A'-A' = (d - d')$ cot β. Similarly the horizontal projection of a link is $(d - d')$ cot α. Thus, the line A–A will intersect all the links within the horizontal distance $(\cot \alpha + \cot \beta)(d - d')$. That is,

$$\begin{bmatrix} \text{Number of links} \\ \text{crossed by A–A} \end{bmatrix} = \frac{(\cot \alpha + \cot \beta)(d - d')}{s_v}$$

$$\doteqdot \left[\frac{\cot \alpha + \cot \beta}{s_v} \right] d$$

The maximum permissible link spacing is that which will enable the line A–A to cross one link only, so that

$$s_{v,max} = (\cot \alpha + \cot \beta)d$$

For the particular case of vertical links, $\alpha = 90°$ and (see explanation following eqn 6.3–3) $\beta = 45°$, so that

$$s_{v,max}(\text{vertical links}) = d$$

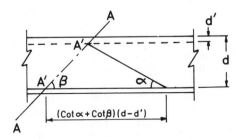

Fig. 6.3–5 Effective region for a web bar

6.4   Shear resistance in design calculations (BS 8110)

BS 8110's design procedure is based on the principles explained in Sections 6.2 and 6.3. The procedure is summarized below, with comments on the various steps listed at the end.

Step 1   The design shear stress

Calculate the **design shear stress** v from

$$v = \frac{V}{b_v d} \qquad\qquad (6.4-1)$$

where V = the ultimate shear force;

b_v = the beam width ($b_v = b$ for rectangular beam;

$b_v = b_w$ for **flanged beam**); and

d = the effective depth.

Step 2 $v > 0.8\sqrt{f_{cu}}$ *or* $5\,\text{N/mm}^2$

If v of Step 1 exceeds $0.8\sqrt{f_{cu}}$ or $5\,\text{N/mm}^2$, whichever is less, the product $b_v d$ must be increased to reduce v. The limits on v are therefore useful for preliminary calculations to check whether the overall dimensions of the cross-section are adequate. Having increased $b_v d$, if necessary, to ensure that v does not exceed $0.8\sqrt{f_{cu}}$ or $5\,\text{N/mm}^2$, move on to Step 3.

Step 3 $v < 0.5v_c$

Compare the design shear stress v with the **design concrete shear stress** v_c in Table 6.4–1. If $v < 0.5\,v_c$, then

(a)   No shear reinforcement is required for members of minor structural importance, such as lintels.

(b)   For all other structural members, provide **minimum links**, which

Table 6.4–1   Design concrete shear stress v_c—for $f_{cu} \geq 40\,\text{N/mm}^2$ (BS 8110 : Clause 3.4.5.4)[a,b,c]

$\dfrac{100A_s}{b_v d}$	Effective depth d (mm)						
	150	175	200	225	250	300	≥ 400
≤ 0.15	0.50	0.48	0.47	0.45	0.44	0.42	0.40
0.25	0.60	0.57	0.55	0.54	0.53	0.50	0.47
0.50	0.75	0.73	0.70	0.68	0.65	0.63	0.59
0.75	0.85	0.83	0.80	0.77	0.76	0.72	0.67
1.00	0.95	0.91	0.88	0.85	0.83	0.80	0.74
1.50	1.08	1.04	1.01	0.97	0.95	0.91	0.84
2.00	1.19	1.15	1.11	1.08	1.04	1.01	0.94
≥ 3.00	1.36	1.31	1.27	1.23	1.19	1.15	1.07

[a] The v_c values include an allowance for $\gamma_m = 1.25$ (see Table 1.5–2).
[b] The v_c values are for $f_{cu} \geq 40\,\text{N/mm}^2$. For f_{cu} values below $40\,\text{N/mm}^2$, the tabulated values should be multiplied by $(f_{cu}/40)^{1/3}$. thus:

(1)   for $f_{cu} = 35\,\text{N/mm}^2$, multiply by 0.956;
(2)   for $f_{cu} = 30\,\text{N/mm}^2$, multiply by 0.909;
(3)   for $f_{cu} = 25\,\text{N/mm}^2$, multiply by 0.855.

[c] The term A_s is that area of the longitudinal tension reinforcement which continues for at least a distance d beyond the section being considered, except at supports where the full area of the tension reinforcement may be used, provided the curtailment and anchorage requirements in Section 4.10 are met.

are defined as shear links that will provide a shear resistance of $0.4 \ \text{N/mm}^2$, i.e.

$$A_{sv}(\text{min. links}) \geq \frac{0.4 \ b_v s_v}{0.87 f_{yv}} \tag{6.4-2}$$

where A_{sv} is the area of the two legs of a link, f_{yv} the characteristic strength of the link (not to be taken as more than $460 \ \text{N/mm}^2$) and s_v the link spacing.

Note the following **link spacing requirements:**

(1)   In the direction of the span, the **link spacing** s_v should not exceed $0.75d$.
(2)   At right angles to the span, the horizontal spacing should be such that no longitudinal tension bar is more than 150 mm from a vertical leg of a link; this spacing should in any case not exceed d.

Step 4 $0.5 \ v_c \leq v \leq (v_c + 0.4)$

If v is between $0.5v_c$ and $(v_c + 0.4)$, provide minimum links as defined by eqn (6.4–2) for the whole length of the beam.

Step 5 $(v_c + 0.4) < v$

If v exceeds $v_c + 0.4$, provide links as follows:

$$A_{sv} \geq \frac{(v - v_c) \ b_v s_v}{0.87 f_{yv}} \tag{6.4-3}$$

where the symbols have the same meanings as in eqn (6.4–2). The link spacing requirements listed under eqn (6.4–2) apply equally to eqn (6.4–3) here.

For the use of **bent-up bars** see 'Comments on Step 5' at the end.

Step 6   *Anchorage of links*

BS 8110: Clause 3.12.8.6 lists the following requirements on the **anchorage of links**:

(a)   A link should pass round another bar of at least its own size, through an angle of 90°, and continue for a length of at least eight times its own size; or
(b)   It should pass round another bar of at least its own size, through an angle of 180°, and continue for a length of at least four times its own size.

Comments on Step 1
See eqn (6.2–1).

Comments on Step 2
BS 8110's limits of $0.8\sqrt{f_{cu}}$ and $5 \ \text{N/mm}^2$ are essentially the values assigned to the ultimate shear stress v_u of eqn (6.3–11), discussed earlier in Section 6.3. The limits $0.8\sqrt{f_{cu}}$ and $5 \ \text{N/mm}^2$ already include an allowance for the partial safety factor γ_m of 1.25.

Comments on Step 3
(a)   Longitudinal bars are customarily curtailed when they cease to be required to resist bending moment. It has been pointed out that such curtailment creates stress complications which may substantially

reduce the shear strength [1]. Therefore, in using Table 6.4–1, any longitudinal bars which are terminated within a distance d of the section concerned cannot be considered; indeed, near such termination, it is desirable (though not mandatory) to put in additional links locally [1].

(b) The use of v_c in shear design was explained in Section 6.3 in relation to eqn (6.3–7). Earlier, it was explained in Section 6.2 that the nominal shear stress at the collapse of a beam increases with the concrete strength f_{cu} and the tension steel ratio ϱ $(= A_s/bd)$. The v_c values in Table 6.4–1, which already allow for the partial safety factor γ_m, are plotted against experimental data in Fig. 6.2–6. This figure shows that some of the experimental values fall below the BS 8110 design values.

(c) Equation (6.4–2) gives the requirements for minimum links. The technical background to eqn (6.4–2) was given earlier in eqn (6.3–8):

$$\varrho_v f_{yv} \geq 0.4$$

That is,

$$\frac{A_{sv}}{b_v s_v} f_{yv} \geq 0.4$$

which becomes eqn (6.4–2), when the partial safety factor 0.87 is used for f_{yv}.

(d) The requirement that the link spacing s_v should not exceed $0.75d$ follows from eqn (6.3–9). At right angles to the span, the horizontal link spacing must also be kept sufficiently small to prevent the pressing down of any longitudinal reinforcement bar and hence maintain the dowel capacity.

Comments on Step 4
When $v > 0.5v_c$, BS 8110 allows for the possibility of diagonal cracking; if a diagonal crack does occur (because of accidental overloading, for example) there should be sufficient shear reinforcement to arrest its growth, and the minimum links specified by eqn (6.4–2) provide such a safeguard.

Comments on Step 5
(a) Equation (6.4–3) follows directly from eqn (6.3–5), if f_{yv} is multiplied by the coefficient 0.87 $(= 1/1.15$ where 1.15 is the partial safety factor $\gamma_m)$.

(b) *Note that irrespective of the amount of shear reinforcement used, the shear stress v from eqn (6.4–1) must not exceed $0.8\sqrt{f_{cu}}$ or 5 N/mm².* See Step 2 again, if necessary.

(c) BS 8110 does not permit shear reinforcement to be provided entirely in the form of bent-up bars, because there is insufficient evidence to show that bent-up bars on their own are satisfactory. Bent-up bars, however, may be used in combination with links (Fig. 6.3–2) but the links must contribute at least 50% of the total resistance of the shear reinforcement. The combined resistance of the links and bent-up bars is to be taken as the sum of their separate resistance; this point was

Table 6.4–2 Values of A_{sv}/s_v (mm) for various link-bar sizes ϕ and link spacings s_v

			ϕ	
s_v	8	10	12	16
100	1.00	1.57	2.26	4.02
150	0.67	1.05	1.51	2.68
200	0.50	0.79	1.13	2.01
250	0.40	0.63	0.90	1.61
300	0.33	0.52	0.75	1.34

discussed on paragraph (f) immediately preceding Example 6.3–1. BS 8110: Clause 3.4.5.6 gives the following equation for the shear resistance V_b of a system of **bent-up bars**:

$$V_b = A_{sb} \, (0.87 f_{yv}) \, (\cos \alpha + \sin \alpha \cot \beta) \, \frac{d - d'}{s_b} \qquad (6.4\text{–}4)$$

where A_{sb} is the cross-sectional area of the bent-up bar(s), f_{yv} the characteristic strength, α and β are the angles defined in Figs 6.3–4 and 6.3–5, d the effective depth, d' the concrete cover to the centres of the top reinforcement (see Fig. 6.3–4) and s_b is the spacing of the bent-up bars ($= s_v$ in Fig. 6.3–4). BS 8110 stipulates that bent-up bars should be so arranged that:
(1) The angles α and β are both greater than or equal to 45°;
(2) The spacing s_b should not exceed $1.5d$.
In practice, the angle α is specified by the designer. β and s_b are related by the following equation (see Example 6.3–1 and Fig. 6.3–5):

$$s_b = (\cot \alpha + \cot \beta) \, (d - d') \qquad (6.4\text{–}5)$$

Having specified a value for α not less than 45°, a value not exceeding $1.5d$ is specified for s_b, and β is then calculated from eqn (6.4–5). Equation (6.4–4) can then be used to calculate V_b. Equation (6.4–4) was derived in Section 6.3 as eqn (6.3–6).

Comments on Step 6
For further information on the subject of bond and anchorage, see Section 6.6.

The use of BS 8110's shear design procedure is illustrated in Example 6.4–1. As will be seen, Table 6.4–2 is a useful design aid.

Example 6.4–1
BS 8110's minimum links are defined by eqn (6.4–2).

(a) Briefly explain the technical background to the equation.
(b) Show that BS 8110's minimum link requirements are met by 'Grade 250 (mild steel) links equal to 0.18% of the horizontal section', as recommended by the I.Struct.E. Manual [26].

SOLUTION
(a) See the derivation of eqn (6.3–8) and the Comments (c) on Step 3 of BS 8110's shear design procedure.
(b) From eqn (6.4–2),

$$A_{sv} = \frac{0.4 \, b_v s_v}{0.87 f_{yv}}$$

$$= 0.18\% \text{ of } b_v s_v \text{ for } f_{yv} = 250 \text{ N/mm}^2$$

Example 6.4–2
The span lengths of a three-span continuous beam ABCD are: exterior spans AB and CD, 8 m each; interior span BC, 10 m. The characteristic dead load G_k (inclusive of self-weight) is 36 kN/m and the characteristic imposed load Q_k is 45 kN/m. The beam has a uniform rectangular section of width $b = 350$ mm, and the effective depth d may be taken as 800 mm. If $f_{cu} = 40 \text{ N/mm}^2$, $f_y = 460 \text{ N/mm}^2$, $f_{yv} = 250 \text{ N/mm}^2$, and if the longitudinal reinforcement is as in Fig. 6.4–1, design the shear reinforcement for span AB. Conform to BS 8110. (*Note:* The longitudinal reinforcement in Fig. 6.4–1 is described by the current British **detailing notation**: the first figure denotes the number of bars, the letter the type of steel—T for high yield steel and R for mild steel—the number after the hyphen is the identification **bar mark**. Thus, 2T32-1 represents two high yield bars of size 32 mm, the bars being identified by the bar mark 1; in this example, bar mark 1 refers to a straight bar of size 32 whose length is 3 m plus the projection into the span BC. For further information on detailing notation, see Example 3.6–3.)

SOLUTION
The design will be carried out in the steps listed previously; comments are given at the end of the solution.

Step 1   The design shear stress
First we draw the shear force envelope. For a continuous beam, the shear force is the algebraic sum of the simple-span shear and the shear due to the support bending moments. The beam was previously analysed in Example 4.9–1, and Fig. 4.9–6 shows that the redistributed support bending moment is 768 kNm for each loading case. Loading Cases 1 and 3 (Fig. 4.9–6):

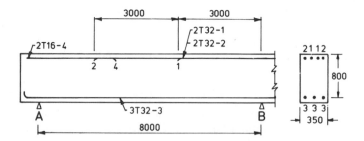

Fig. 6.4–1   Longitudinal steel details

design load $= 1.4G_k + 1.6Q_k = 122.4$ kN/m

$$V = \frac{(122.4 \text{ kN/m}) \ (8 \text{ m})}{2} + \frac{768 \text{ kNm}}{8 \text{ m}}$$

$= 585.6$ kN at B and 393.6 kN at A

Loading Case 2 (Fig. 4.9–6):

design load $= 1.0 \ G_k = 36$ kN/m

$$V = \frac{(36) \ (8)}{2} \pm \frac{768}{8}$$

$= 240$ kN at B and 48 kN at A

The **shear force envelope** is therefore as shown in Fig. 6.4–2(a). At the face of support B:

$$v \text{ (eqn 6.4–1)} = \frac{585.6}{(350) \ (800)} \ (10^3) = 2.09 \text{ N/mm}^2$$

Step 2   **The limits $0.8\sqrt{f_{cu}}$ and 5 N/mm^2**

$$0.8\sqrt{f_{cu}} = 0.8\sqrt{40} = 5.06 \text{ N/mm}^2$$

$$> 5 \text{ N/mm}^2$$

v at face of support $B = 2.09$ N/mm^2

$$< 5 \text{ N/mm}^2$$

Therefore the overall dimensions of the beam are adequate.

Step 3   **Comparison with $0.5v_c$**

This step will be omitted here because, in current design practice, minimum links are usually provided whether they are required by BS 8110 or not.

Step 4   **Values of $v_c + 0.4$**

The v_c values in Table 6.4–1 depend on the tension steel ratio. The following simplified interpretation of A_s values, which errs on the safe side, is adequate for normal design (see also the Comments at the end of this Example).

Within 3 m of B:

A_s (three size 32 bars) $= 2412$ mm^2

Therefore $100 \ A_s/b_v d = 0.86$; also

$$d > 400 \text{ mm}$$

By interpolation from Table 6.4–1:

$$v_c = 0.70 \text{ N/mm}^2$$

$$v_c + 0.4 = 1.10 \text{ N/mm}^2$$

Minimum links will be used when $v < v_c + 0.4$. That is,

$$v \text{ (minimum links)} = v_c + 0.4 = 1.10 \text{ N/mm}^2$$
$$V \text{ (minimum links)} = (1.10)\ (350)\ (800)\ (10^{-3}) = 308 \text{ kN}$$

Beyond 3 m from B:

$$A_s \text{ (two size 32 bars)} = 1608 \text{ mm}^2$$

Therefore

$$100\ A_s/b_v d = 0.57; \qquad d > 400 \text{ mm}$$
$$v_c \text{ (Table 6.4–1)} = 0.61 \text{ N/mm}^2 \text{ (by interpolation)}$$

As before,

$$v \text{ (minimum links)} = v_c + 0.4 = 1.01 \text{ N/mm}^2$$
$$V \text{ (minimum links)} = (1.01)\ (350)\ (800)\ (10^{-3}) = 282 \text{ kN}$$

The shear resistance of 308 and 282 kN are drawn as the chain-dot lines in Fig. 6.4–2(a); minimum links only are required where the shear force envelope lies inside these limits.

From Example 6.4–1(b):

Minimum link requirements are met by grade 250 (mild steel) links of 0.18% $b_v s_v$.

$$A_{sv} = 0.0018\ b_v s_v$$

so that

$$A_{sv}/s_v = (0.0018)\ (350) = 0.63 \text{ mm}$$

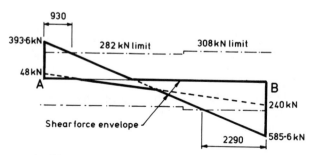

(a) Shear force envelope

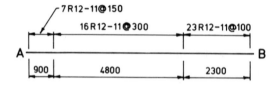

(b) Arrangement of links

Fig. 6.4–2

From Table 6.4–2, use

Size 12 links at 300 mm

Step 5 $v > (v_c + 0.4)$

Figure 6.4–2(a) shows that shear reinforcement exceeding minimum links is required for the regions within 2.29 m of B and within 0.93 m of A.

Within 2.29 m of B

$$v = 2.09 \text{ N/mm}^2 \quad \text{(see Step 1)}$$

$$v_c = 0.7 \text{ N/mm}^2 \quad \text{(see Step 4)}$$

From eqn (6.4–3),

$$\frac{A_{sv}}{s_v} = \frac{b_v(v - v_c)}{0.87 f_{yv}} = \frac{350(2.09 - 0.70)}{(0.87)\,(250)} = 2.24 \text{ mm}$$

From Table 6.4–2, use

Size 12 links at 100 mm

Within 0.93 m of A (see Comments at the end):

$$v = (393.6)\,(10^3)/(350)\,(800) = 1.41 \text{ N/mm}^2$$

$$v_c = 0.61 \text{ N/mm}^2 \quad \text{(see Step 4)}$$

$$\frac{A_{sv}}{s_v} = \frac{b_v(v - v_c)}{0.87 f_{yv}} = \frac{350(1.41 - 0.61)}{(0.87)\,(250)} = 1.29 \text{ mm}$$

From Table 6.4–2, use

Size 12 links at 150 mm

The shear reinforcement provided is summarized in Fig. 6.4–2(b), where the detailing notation is as explained in the question.

Comments on Step 1

The support moments are here taken as the redistributed moments, though the designer may, if he so wishes, base both the flexural and shear design on the elastic moments.

Comments on Step 2

Here $0.8 f_{cu}$ exceeds 5 N/mm². Hence the effective ceiling is 5 N/mm².

Comments on Step 3

Shear failures can be very dangerous, in that there may be no warning before disastrous collapse. Hence it is desirable to provide minimum links even when the calculations show that the shear stress v is well below $0.5 v_c$.

Comments on Step 4

When both top and bottom bars are present, as in Fig. 6.4–1, the area A_s should properly be interpreted from the bending moment envelope. Thus, the redistributed moment envelope in Fig. 4.9–7 shows that:

(a) Within 1575 mm of the support B, the bending moment is always a

hogging one; therefore, from Fig. 6.4–1, A_s = 4 top bars.
(b)   Within 2665 mm of support A, the moment is always a sagging one; therefore A_s = 3 bottom bars.
(c)   In between these regions, the moment is sometimes a hogging one and sometimes a sagging one, and it will be conservative to take A_s = two size 32 bars. In practical design, however, the procedure as explained here is considered too tedious; the simplified procedure as used in Step 4 is usually adopted.

Comments on Step 5

It has been pointed out that when the support reaction imposes a compression on the member, shear failure rarely initiates within a distance d of the support, where d is the effective depth [1]. Therefore, in a situation like this, the designer is justified in providing shear reinforcement on the basis of the shear force V at 800 mm from the support B, if he so wishes. Indeed, BS 8110: Clause 3.4.5.10 specifically gives the designer this freedom of choice. (Note that at the time when the authors called for the recognition of the enhanced shear resistance near supports, the then Code of Practice CP 114 did not recognize such enhanced resistance [1].)

Example 6.4–3

If the beam in Example 6.4–2 had been a flanged beam, having the cross-section in Fig. 6.4–3, how should the solution be modified?

SOLUTION
Since the solution to Example 6.4–2 is based on the dimensions of the rib, it applies equally to flanged beams. This is not to say that there is no difference in behaviour between rectangular beams and flanged beams [1, 25], but in current design practice such a difference is ignored.

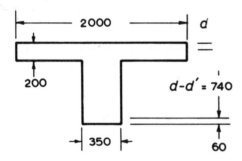

Fig. 6.4–3

6.5   Shear strength of deep beams

A beam having a depth comparable to the span length is called a **deep beam**. The design of deep beams is a topic which recurs in practice but which is not yet covered by BS 8110. Recently, the research at the Universities of Newcastle upon Tyne, Cambridge and Nottingham has led to the following design method, which also covers deep beams with web openings, [14–19].

The method is applicable where the span/depth ratio l/h does not exceed about 3; however, as will become clear later, the shear-span/depth ratio a_v/h (Fig. 6.5–1) is a more important parameter than the l/h ratio. The method, which has since been included in Clause 3.4.2 of CIRIA's *Deep Beam Design Guide* [12] and in Reynolds and Steedman's *Reinforced Concrete Designer's Handbook* [13] is based on the following formula for the ultimate shear strength V [14–19, 27, 28]:

$$V = C_1 \left[1 - 0.35\frac{a_v}{h}\right] f_t bh + C_2 \sum_{}^{n} A \frac{y}{h} \sin^2 \alpha$$

$$= \text{concrete resistance} + \text{steel resistance} \qquad (6.5-1)$$

where C_1 = a coefficient equal to 1.4 for normal weight concrete and 1.0 for lightweight concrete;

C_2 = a coefficient equal to 130 N/mm^2 for plain round bars and 300 N/mm^2 for deformed bars;

f_t = the cylinder splitting tensile strength of the concrete; if f_t is not available, it may be estimated from the cube strength f_{cu} by, say, f_t = 0.4 to 0.5$\sqrt{f_{cu}}$;

A = the area of a typical web bar—for the purpose of this equation, the longitudinal tension bars are also considered to be web bars;

y = the depth at which the typical bar intersects the critical

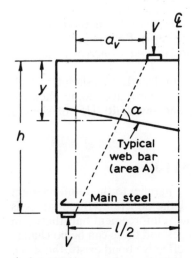

(a) Meanings of symbols

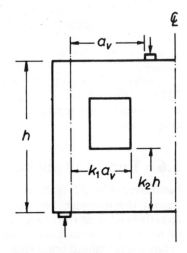

(b) Further symbols for beam with openings.

Fig. 6.5–1 Deep beam

diagonal crack, which is represented by the dotted line in Fig. 6.5–1;

α = the angle (Fig. 6.5–1) between the bar being considered and the diagonal crack ($\pi/2 > \alpha > 0$);

n = the total number of web bars, including the main longitudinal bars, that intercept the critical diagonal crack;

b = the beam width; and

a_v and h are as explained in Fig. 6.5–1.

As regards flexure, the design bending moment M should not exceed

$$M = 0.6A_s\frac{f_y}{\gamma_m}h \quad \text{or} \quad 0.6A_s\frac{f_y}{\gamma_m}l \tag{6.5–2}$$

whichever is less, where A_s is the tension steel area. Equation (6.5–2) assumes that the lever arm is $0.6h$ for $l/h > 1$ or $0.6l$ for $l/h < 1$. This is conservative because in deep beams the lever arm is unlikely to fall below $0.7h$ or $0.7l$ [18]. It should be noted that all the main longitudinal bars provided in accordance with eqn (6.5–2) also act as web bars; that is, the laws of equilibrium are unaware of the designer's discrimination between bars labelled as 'flexural' reinforcement' and bars labelled as 'shear reinforcement'.

The design method can be extended to **deep beams with openings** if the lever arm, $0.6h$ in eqn (6.5–2), is replaced by $\frac{3}{4}k_2h$, and the quantities a_v and h in the first term of eqn (6.5–1) are replaced by k_1a_v and k_2h respectively, where k_1a_v and k_2h are as defined in Fig. 6.5–1(b).

The above design method is based on a **structural idealization** explained in detail in Reference 16, which also gives a list of design hints. A worked example, in which the calculated collapse load for a large beam is compared with the actual load observed in a test to destruction, is given in Reference 17. Design examples illustrating the use of this method, both for deep beams with and without openings, are given in References 14 and 18, as well as in Reynolds and Steedman's *Reinforced Concrete Designer's Handbook* [13].

Readers interested in the practical design of deep beams should consult CIRIA's *Deep Beam Design Guide* [12], which also gives comprehensive recommendations regarding the **buckling and instability** [15] of slender deep beams.

6.6   Bond and anchorage (BS 8110)

Bond stress [29–32] is the shear stress acting parallel to the reinforcement bar on the interface between the bar and the concrete. Bond stress is directly related to the change of stress in the reinforcement bar; there can be no bond stress unless the bar stress changes and there can be no change in bar stress without bond stress. Where an effective bond exists, the strain in the reinforcement may for design purposes be assumed to be equal to that in the adjacent concrete. Effective bond exists if the relevant requirements in the code of practice are met; in BS 8110, these requirements are expressed in terms of certain nominal stresses, as we shall see later. Bond is due to the combined effects of adhesion, friction and (for deformed bars) bearing. In deformed bars with transverse ribs (or lugs) the concrete bearing stresses against the ribs contribute most of the bond. These bearing stresses, and to a lesser extent the frictional stresses, generate radial stresses and a circumferential tension in the concrete round the bar. As a result of this ring tension, the so-called bond failure is usually associated with the longitudinal splitting of the concrete along the bar, as, for example, in the shear-tension failures described in Section 6.2. Factors which help to prevent such splitting could be expected to increase the usable bond capacity: namely a higher concrete strength, heavier shear links and larger concrete cover to the reinforcement bars.

Ideally, design calculations should be related directly to the longitudinal splitting referred to above, but this is not yet possible. Instead, BS 8110 requires the checking of the **anchorage bond stress** f_b, which is the average bond stress calculated as the bar force F_s divided by the product of the anchorage length l and the nominal perimeter of the bar:

$$f_b = \frac{F_s}{\pi \phi l} \ngtr f_{bu} \qquad (6.6-1(a))$$

where ϕ is the bar size and f_{bu} is defined in eqn (6.6–2) below.

Since the bar force $F_s = f_s \, \pi\phi^2/4$, where f_s is the steel stress (which, of course, progressively diminishes to zero over the anchorage length l), eqn (6.6–1(a)) can also be expressed as

$$f_b = \frac{f_s \, \pi\phi^2/4}{\pi \phi l} = \frac{f_s \phi}{4l} \qquad (6.6-1(b))$$

$$\ngtr f_{bu}$$

The **ultimate anchorage bond stress** f_{bu} is given by BS 8110 as

$$f_{bu} = \beta \sqrt{f_{cu}} \qquad (6.6-2)$$

where β is a **bond coefficient** to be obtained as follows:

(a) For bars in tension in slabs, values of β are as given in Table 6.6–1.
(b) In beams, where minimum links have been provided in accordance with eqn (6.4–2), values of β are as given in Table 6.6–1. In beams where minimum links have not been so provided, the β values should be those listed in Table 6.6–1 for plain bars, irrespective of the type of bar actually used.

Table 6.6–1 Bond coefficient β (BS 8110 : Clause 3.12.8.4)

	β *(eqns 6.6–2 and 6.6–3)*	
Bar type	*Bars in tension*	*Bars in compression*
Plain bars	0.28	0.35
Deformed bars	0.50	0.63

(c)   The β values in Table 6.6–1 already include a partial safety factor γ_m of 1.4 (see Table 1.5–2).

The **anchorage bond length** l (often simply called the anchorage length) is the length of the reinforcement bar required to develop the stress f_s and is given by eqns (6.6–1(b)) and (6.6–2):

$$l = \frac{f_s \phi}{4\beta\sqrt{f_{cu}}} \qquad (6.6–3(a))$$

The **ultimate anchorage bond length** l_u (often referred to as the ultimate anchorage length or the full anchorage length) is the bar length required to develop the full design strength. Hence, by writing $f_s = 0.87f_y$ in eqn (6.6–3(a)),

$$l_u = \frac{0.87f_y\phi}{4\beta\sqrt{f_{cu}}} \qquad (6.6–3(b))$$

where ϕ is the bar size and β is as defined in eqn (6.6–2). Typically, for $f_{cu} = 40$ N/mm^2 and $f_y = 460$ N/mm^2 (deformed bars), the ultimate anchorage bond length is 32ϕ for bars in tension and 26ϕ for bars in compression (see also Table 4.10–2).

Where it is impracticable to provide the necessary anchorage length for bars in tension, the designer may use **hooks** or **bends** (Fig. 6.6–1), which should meet the detailing requirements of BS 4466, summarized here in Fig. A2–1. The equivalent or effective anchorage length of such a standard hook or bend may then be taken as multiples of the bar size ϕ shown in Table 6.6–2, adapted from the I.Struct.E. Manual [26].

For **anchorage of links**, see Section 6.4, Step 6.

Example 6.6–1
Referring to the beam in Example 6.4–2 (Fig. 6.4–1), check the tension and compression anchorage lengths required at support B.

SOLUTION
Where it is known that the bars are stressed to approximately their full design strengths, the anchorage lengths are calculated immediately from

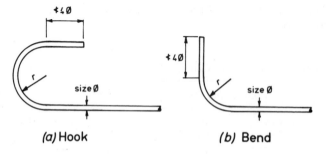

(a) Hook   *(b)* Bend

Fig. 6.6–1   Hooks and bends ($r \nless 2\phi$ for mild steel bars; $r \nless 3\phi$ for high yield bars up to size 20 mm and $\nless 4\phi$ or larger sizes)

Table 6.6–2   Effective anchorage lengths of standard hooks and bends (BS 4466)

Bar size ϕ	$f_y = 250$ N/mm²		$f_y = 460$ N/mm²	
	Bend	*Hook*	*Bend*	*Hook*
8	15ϕ		15ϕ	
10	13ϕ		15ϕ	
12	11ϕ	All	14ϕ	All
16	9ϕ	16ϕ	12ϕ	24ϕ
20	8ϕ		12ϕ	
25	8ϕ		12ϕ	
32	8ϕ		12ϕ	
40	8ϕ		12ϕ	

eqn (6.6–3(b)). For the purpose of illustration, we shall assume that the bar stresses are not known. From Fig. 6.4–1,

A_s (4 size 32 bars) = 3216 mm²

$\varrho = 3216/(350)\,(800) = 1.15\%$

A_s' (3 size 32 bars) = 2412 mm²

$\varrho' = 2412/(350)\,(800) = 0.86\%$

From the beam design chart (Fig. 4.5–2),

$$\frac{M_u}{bd^2} = 4 \text{ N/mm}^2$$

Actual $M/bd^2 = (768)\,(10^6)/(350)\,(800^2) = 3.43$ N/mm²

For anchorage length calculations, the bar stresses may be taken approximately as

$$f_s = \frac{3.43}{4}\,(0.87 f_y) = 343 \text{ N/mm}^2 \quad (\text{for } f_y = 460 \text{ N/mm}^2)$$

From eqn (6.6–3(a)) and Table 6.6–1, the anchorage bond lengths l are

$$l \text{ (tension bars)} = \frac{(343)\,(32)}{(4)\,(0.5)\,(\sqrt{40})} = \underline{868 \text{ mm}}$$

$$l \text{ (compression bars)} = \frac{(343)\,(32)}{(4)\,(0.63)\,(\sqrt{40})} = \underline{689 \text{ mm}}$$

Example 6.6–2
For many years, successive British Codes of Practice such as CP 110 and CP 114 required designers to check **local bond stresses**. BS 8110, however, does not require such checking. Comment.

SOLUTION
In the past, British designers check local bond stresses partly from habit and partly because successive codes of practice have referred to such

stresses. In the first and second editions (1975 and 1980 respectively) of this book, the authors wrote: 'Provided the anchorage length is sufficient, the local bond stress does not seem to have much significance.... It is desirable that (the requirement to check) local bond stresses will be dropped in future revisions of CP 110.'

6.7   Equilibrium torsion and compatibility torsion

BS 8110 implicitly differentiates between two types of torsion: **equilibrium torsion** (or **primary torsion**), which is required to maintain equilibrium in the structure, and **compatibility torsion** (or **secondary torsion**), which is required to maintain compatibility between members of the structure. To distinguish between the two types, it is helpful to note that (a) in a statically determinate structure, only equilibrium torsion can exist; (b) in an indeterminate structure both types may exist, but if the torsion can be eliminated by releasing redundant restraints then it is a compatibility torsion.

In general, where the torsional resistance or stiffness of members has not been explicitly taken into account in the analysis of the structure, no specific calculations for torsion will be necesary. Thus, BS 8110: Part 2: Clause 2.4.1 states that: 'In normal slab-and-beam or framed construction specific calculations are not usually necessary, torsional cracking being adequately controlled by shear reinforcement.' In other words, compatibility torsion may, at the discretion of the designer, be ignored in the design calculations; however, equilibrium torsion must be designed for.

6.8   Torsion in plain concrete beams

It is only recently that a substantial amount of experimental data has enabled engineers to obtain a reasonable working knowledge of torsion in structural concrete [33–37].

Figure 6.8–1 shows a plain concrete beam subjected to pure torsion. The torsional moment T induces shear stresses which produce principal tensile stresses at 45° to the longitudinal axis. When the maximum tensile stress reaches the tensile strength of the concrete, diagonal cracks form which tend to spiral round the beam [38]. For a plain concrete beam, failure immediately follows such diagonal cracking.

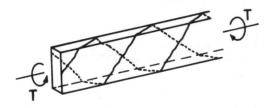

Fig. 6.8–1   Diagonal cracking due to torsion [38]

It can readily be shown (see, for example, Coates *et al.* [39]) that the problem of the torsion of an elastic beam of arbitrary cross-section can be reduced to the solution of the Poisson equation

$$\nabla^2 \phi_1 = -2 \qquad (6.8-1)$$

for a **torsion function** ϕ_1 which satisfies the boundary condition that $\phi_1 = 0$ along the lateral boundary of the cross-section. Equation (6.8–1) may conveniently be solved by, say, the finite difference technique [39]. Then, at any point (x, y) on the section, the torsional shear stresses v_t and the angle of rotation θ per unit length of the beam are given by [39]

$$\frac{v_t \ (x\text{-direction})}{\partial \ \phi_1/\partial y} = \frac{v_t \ (y\text{-direction})}{-\partial \phi_1/\partial x} = \frac{T}{2 \int \phi_1 \, dA} = G\theta \qquad (6.8-2)$$

where T is the torsional moment, G the shear modulus, A the cross-sectional area of the beam, and (x, y) are the coordinates referring to any arbitrarily chosen set of Cartesian axes. The quantity K, where

$$K = 2 \int \phi_1 \, dA \qquad (6.8-3)$$

is sometimes referred to as **St Venant's torsional constant**, which, for rectangular sections, has the values in Table 6.8–1.

Suppose we now construct a hill over the cross-section such that at any point the height of the hill is equal to the value of ϕ_1 at that point. The following numerical identities follow directly from eqns (6.8–2) and (6.8–3)

v_t (tangential to the contour line)

$$= \frac{T}{K} \text{ times the slope } \psi \text{ of the '}\phi_1\text{-hill'} \qquad (6.8-4(a))$$

$$K = \text{twice the volume of the '}\phi_1\text{-hill'} \qquad (6.8-4(b))$$

Now it is a well-known result in applied mechanics that, if a thin membrane is mounted over a cross-section and then inflated by a small pressure q, then at any point (x, y) the height z of the deflected surface of the membrane is given by the differential equation:

Table 6.8–1   Values of torsion constant K for rectangular section $(K = k h_{min}^3 h_{max})^a$

h_{max}/h_{min}	k	h_{max}/h_{min}	k
1.0	0.14	3.0	0.26
1.2	0.17	4.0	0.28
1.5	0.20	5.0	0.29
2.0	0.23	10.0	0.31
2.5	0.25	∞	0.33

[a] h_{max} = length of long side; h_{min} = length of short side; K = torsion constant (eqn 6.8–3).

$$\nabla^2 z = -\frac{q}{n} \qquad (6.8\text{–}5)$$

where n is the membrane tension per unit length. Equation (6.8–5) is of the same form as eqn (6.8–1). Therefore, the 'ϕ_1-hill' of eqns (6.8–4) must be geometrically similar to the surface of the deflected membrane. Hence we have the following conceptually useful **membrane analogy** for elastic torsion:

> If a thin membrane is mounted over the cross-section and inflated by pressure, then St Venant's torsion constant K is proportional to the volume under the membrane surface, and the torsional shear stress v_t in any specified direction at any point on the section is proportional to the slope of the membrane at that point, the slope being measured in a direction perpendicular to that of v_t.

The membrane analogy makes it clear, for example, that the maximum shear stress v_t occurs in the direction tangential to the contour (see eqn 6.8–4(a)) and that, normal to a contour line, the shear stress v_t must be zero, since that value of v_t is numerically equal to the slope (i.e. $= 0$) along a contour.

It is interesting and useful to consider how the membrane analogy can be extended to cover **plastic torsion**. If the torque T produces complete plasticity, then v_t is everywhere equal to the yield strength in shear (assuming, for the time being, that an elastic-perfectly plastic material is being considered); eqn (6.8–4(a)) is then interpreted as

$$\frac{T}{K}\psi = v_t \text{ at yield} = \text{constant}$$

That is, for a given section, the slope ψ of the inflated membrane must have a constant value everywhere equal to (K/T) times v_t at yield. Naturally, an inflated membrane cannot have such a constant slope; however, if dry sand is poured over the cross-section, the definite angle of repose of the sand will automatically lead to the formation of a heap having a constant slope on all faces. The membrane analogy is then modified as the **sand-heap analogy** (or **sand-hill analogy**), which states that the torsion constant K is twice the volume of the sand heap, provided that the yield shear stress v_t is interpreted at (T/K) times the slope ψ of the faces. In other words, eqns (6.8–4) for elastic torsion become applicable to plastic torsion when the words 'ϕ_1-hill' are replaced by 'sand-heap'.

Consider a rectangular section in plastic condition. The sand heap (or 'ϕ_1-hill') is then as shown in Fig. 6.8–2, in which ϕ_{1m} denotes the height at the ridge of the heap. The reader should verify that

$$\text{volume of sand heap} = \tfrac{1}{2}h_{min}\phi_{1m}(h_{max} - h_{min}) + \tfrac{1}{3}h_{min}^2\phi_{1m}$$

$$= K/2 \text{ (from eqn 6.8–4(b))} \qquad (6.8\text{–}6)$$

From Fig. 6.8–2, the slope of the faces of the sand heap is

$$\psi = 2\phi_{1m}/h_{min} \qquad (6.8\text{–}7)$$

Equation (6.8–4(a)) states that $v_t = (T/K)\psi$. Substituting in the values of K and ψ from eqns (6.8–6) and (6.8–7) and simplifying,

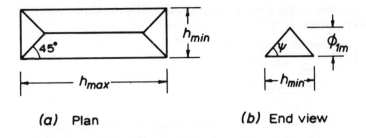

(a) Plan   (b) End view

Fig. 6.8–2   Sand heap for rectangular section

$$v_t = \frac{2T}{h_{min}^2[h_{max} - h_{min}/3]} \tag{6.8-8}$$

Marshall [40] has analysed the test results of many investigators and shown that a reasonable estimate of the cracking torque T could be obtained from eqn (6.8–8) by using the splitting cylinder strength f_t for v_t; this confirms similar conclusions drawn by others [33, 34]. It might appear surprising that eqn (6.8–8), which is based on a plastic stress distribution, should give better agreement with test results than does a strictly elastic analysis based on direct application of eqns (6.8–1) and (6.8–2). The reason is not quite understood, though it is possible that by the time the diagonal cracks become visible, the stress distribution has already been somewhat affected by microcracking. In this connection, it is appropriate to mention the authors' experience with modulus of rupture tests on plain concrete specimens: collapse, which occurs simultaneously with visible cracking, is always preceded by microcracking, and the microcracking load is only about 65% of the visible-cracking load [41].

The sand-heap analogy, being based on eqns (6.8–1) and (6.8–2), is applicable to a cross-section of arbitrary shape. For simple flanged sections, the sand heaps are approximately as shown in Fig. 6.8–3. Consider, for example, the T-section. Since the quantity v_t/ψ (where ψ is the slope of the faces of the sand heap) is the same for both the flange and the web, we have

$$\frac{v_t}{\psi} = \frac{T_f}{2 \times \text{volume of flange heap}} = \frac{T_w}{2 \times \text{volume of web heap}}$$

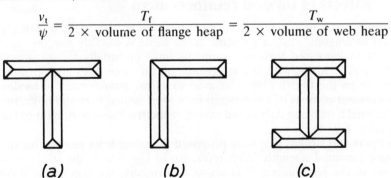

(a)   (b)   (c)

Fig. 6.8–3   Approximate sand heaps for simple flanged sections

where T_f and T_w denote the partial torques resisted by the flange and the web respectively. Since $T_f + T_w = T$, we have

$$T_f = T \frac{\text{volume of flange heap}}{\sum \text{volume}}$$

$$T_w = T \frac{\text{volume of web heap}}{\sum \text{volume}}$$

These relations can of course be extended to a cross-section composed of several rectangles, e.g. an I-section. Admittedly, the complexity of the sand heap at the intersection of the various components makes the evaluation of the volume rather tedious. However, if the sand heap for each component rectangle is assumed to have a uniform triangular cross-section (that is, if the local changes in shape near the ends and inter-sections are neglected) it is easy to show that the partial torque resisted by a typical component rectangle is

$$T_i = T \frac{(h_{min}^2 h_{max})_i}{\sum h_{min}^2 h_{max}} \tag{6.8–9}$$

where the suffix i refers to the typical component rectangle, and h_{min} and h_{max} are respectively the shorter and longer sides of the rectangle.

On the other hand, if a purely elastic analysis is used, then eqns (6.8–1) and (6.8–2) would lead to

$$T_i = T \frac{(k h_{min}^3 h_{max})_i}{\sum k h_{min}^3 h_{max}} \tag{6.8–10}$$

where the symbols have the same meanings as in eqn (6.8–9) except that the new coefficients k are from Table 6.8–1. Note that unless each component rectangle has the same aspect ratio h_{max}/h_{min}, the k coefficients in the numerator and denominator of eqn (6.8–10) do not cancel out.

For a more detailed explanation of the membrane analogy and the sand-heap analogy, and their application to torsion in structural concrete, the reader is referred to Sawko [42].

6.9   Effects of torsion reinforcement

As stated in the previous section, a plain concrete beam fails practically as soon as diagonal cracking occurs. If the beam is suitably reinforced, it will sustain increased torsional moments until eventually failure occurs by the steel yielding, with cracks opening up on three sides and crushing occurring on the fourth [38]. The most practical arrangement of **torsion reinforcement** consists of a combination of longitudinal bars and links, the longitudinal bars being distributed evenly round the inside perimeter of the links (Fig. 6.9–1).

Lampert and Collins [43] have proposed the **space truss analogy** for the ultimate torsional strength. With reference to Fig. 6.9–1, the space truss consists of the longitudinal bars acting as stringers, the legs of the links acting as posts, and the concrete between the cracks as the compression diagonals. With reference to Fig. 6.9–1, let

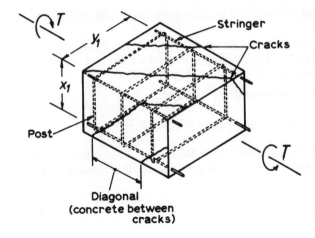

Fig. 6.9–1   Lampert and Collins's space truss analogy [43]

T = ultimate torsional moment of resistance;
A_s = total area of longitudinal reinforcement;
A_{sv} = area of the two legs of each link;
f_y = yield strength of the longitudinal reinforcement;
f_{yv} = yield strength of the links;
s_v = longitudinal spacing of the links;
x_1 = the smaller dimension between the corner bars, as labelled in Fig. 6.9–1;
y_1 = the larger dimension between the corner bars, as labelled in Fig. 6.9–1.

Considering a length s_v of the beam, the (steel volume) × (yield strength) products are respectively $A_s f_y s_v$ and $A_{sv} f_{yv}(x_1 + y_1)$. It is desirable that the longitudinal bars and the links should yield simultaneously; to achieve this condition, the volume–strength products should be made equal [43, 44], i.e.

$$A_s f_y s_v = A_{sv} f_{yv}(x_1 + y_1) \qquad (6.9\text{--}1)$$

Lampert and Collins have found that, if eqn (6.9–1) is satisfied, then the diagonal cracks (Fig. 6.9–1) can be assumed to be inclined at 45° to the axis of the member. We recall that in Section 6.3, the truss analogy was used to calculate the shear resistance of a beam. We shall now show that the ultimate torsional resistance may be calculated, in a similar way, from Lampert and Collins' space truss analogy.

With reference to Fig. 6.9–1, consider the intersection of the horizontal legs of the links by the diagonal cracks. Since each crack can be assumed to be inclined at 45° to the member axis, then on each horizontal face of the member,

$$\left[\begin{array}{c} \text{numer of horizontal legs} \\ \text{intersected by a crack} \end{array} \right] = \frac{y_1}{s_v} \qquad (6.9\text{--}2)$$

Similarly, on a vertical face of the member, each crack will intersect x_1/s_v vertical legs.

Assume that the member is torsionally under-reinforced, so that at failure the links intersected by the diagonal cracks yield in tension. First consider the horizontal legs of the links (Fig. 6.9–1):

$$\text{Tension in a horizontal leg} = \tfrac{1}{2}A_{sv}f_{yv}$$

$$\text{Moment about axis of member} = \tfrac{1}{2}A_{sv}f_{yv} \times \frac{x_1}{2}$$

From eqn (6.9–2), each crack on a horizontal face intersects y_1/s_v horizontal legs, and the torsional moment due to the tension in these legs is

$$\tfrac{1}{2}A_{sv}f_{yv} \times \frac{x_1}{2} \times \frac{y_1}{s_v}$$

Considering the two horizontal forces, the total torsional moment is twice this amount, i.e.

$$T \text{ (horizontal legs)} = \tfrac{1}{2}A_{sv}f_{yv}\frac{x_1 y_1}{s_v}$$

Similarly,

$$T \text{ (vertical legs)} = \tfrac{1}{2}A_{sv}f_{yv}\frac{y_1 x_1}{s_v}$$

Adding,

$$T = \frac{A_{sv}}{s_v}f_{yv}x_1 y_1 \qquad\qquad (6.9-3)$$

where the notation is as explained at the beginning of this section. Of course, the validity of eqn (6.9–3) is conditional upon eqn (6.9–1) being satisfied.

Equation (6.9–3) is a powerful tool in the hands of designers with a good understanding of structural behaviour. Note the following statements:

(a)   The Lampert and Collins study shows that the torsional strength of a properly reinforced beam is independent of the concrete strength, provided the beam is **torsionally under-reinforced**; that is, the longitudinal reinforcement and the links reach yield before the ultimate torque is reached; in fact, their equation applies only to under-reinforced beams. This restriction does not diminish the usefulness of eqn (6.9–3), because over-reinforced sections are avoided in design in any case. Not only are torsionally over-reinforced beams uneconomical, but they do not have the necessary ductility when subjected to overload. We shall see in Section 6.11 how over-reinforcement is guarded against in practice.

(b)   The torsional strength T is proportional to the area $x_1 y_1$ enclosed by the links; within reasonable limits, the ratio x_1/y_1 does not seem to be important.

(c)   The torsional strength of a box beam is sensibly the same as a solid beam (in fact eqn 6.9–3 does not differentiate between the two types of beams). Tests [44] have shown that this statement holds true provided the wall thickness of the box beam exceeds one-quarter of the overall thickness of the member in the direction of measurement. When the wall thickness is below this value, the strength of the box beam becomes less than that of the corresponding solid beam; for design purposes, this strength reduction may be conservatively estimated by the reduction factor of '4 times the ratio of wall thickness to member thickness' [44]. In the absence of more precise calculations, it is recommended that, to prevent excessive flexibility and possible buckling of the wall, the ratio of wall thickness to member thickness should not be less than 1/10 [44]. Conservatism is desirable in the design of box beams: under-reinforced solid beams exhibit a ductile failure mode in torsion, but the corresponding box beam with thin walls would fail in a brittle manner.

(d) The derivation of eqn (6.9–3) is based on several assumptions which cannot be rigorously justified:
(1) The diagonal cracks are inclined at 45° to the member axis.
(2) The links are uniformly stressed to f_{yv}.
(3) The dowel action of the reinforcement and the aggregate-interlock of the concrete do not contribute towards the torsional resistance.

Therefore such a derivation cannot, on its own, justify the acceptance of eqn (6.9–3) for design; there must be experimental evidence that the equation is acceptable. Such experimental evidence does exist; for example, the tests by Hsu and Kemp [44] have shown that the graph of T against $(A_{sv}/S_v)f_{yv}x_1y_1$ is approximately a straight line represented by the equation

$$T = a_c + a_s \frac{A_{sv}}{s_v} f_{yv} x_1 y_1 \qquad (6.9-4)$$

where the first term on the right-hand side, a_c, is conventionally referred to as the **concrete resistance torque** and the second term, $a_s(A_{sv}/s_v)f_{yv}x_1y_1$, **the steel resistance torque**. According to Hsu and Kemp [44], a_c is about 40% (note: not 100%) the cracking torque of the corresponding plain concrete member and

$$a_s = \frac{2}{3} + \frac{1}{3}\frac{y_1}{x_1}$$

Therefore the use of eqn (6.9–3) in design is equivalent to neglecting the concrete resistance and adopting for a_s its minimum possible value of unity.

6.10   Interaction of torsion, bending and shear

6.10(a)   Design practice (BS 8110)

Where a member is subjected to combined **torsion and bending**, BS 8110: Part 2: Clause 2.4.7 states that the reinforcements may be calculated separately for bending and for shear, and then added together. Similarly,

BS 8110: Part 2: Clause 2.4.6 states that for combined **torsion and shear**, the torsion and shear reinforcements are calculated separately, and then added together, in accordance with Table 6.10–1.

6.10(b)   Structural behaviour

The interaction of torsion, bending and shear has been studied by Hsu and others [44, 45]. For members in which the longitudinal reinforcement is symmetrical about both the vertical and horizontal axes of the cross-section, the **interaction of torsion and bending** is represented by curve (I) in Fig. 6.10–1(a), in which T and M are the torsional and bending moment combination that the beam is capable of resisting, T_0 is the ultimate strength in pure torsion, and M_0 that in pure bending. Curve (II) is the interaction curve for members with asymmetrical longitudinal reinforcement; for such members, the torsional capacity is increased by the application of limited amounts of bending. More recently, Lampert and Collins [43] have proposed the following interaction equations for rectangular beams:

$$\varrho_1\left(\frac{T}{T_0}\right)^2 + \frac{M}{M_0} = 1 \qquad\qquad (6.10\text{--}1(a))$$

$$\left(\frac{T}{T_0}\right)^2 - \frac{1}{\varrho_1}\left(\frac{M}{M_0}\right) = 1 \qquad\qquad (6.10\text{--}1(b))$$

where ϱ_1 is the ratio of the yield force ($A'_s f_y$) of the longitudinal steel at the top (i.e. the flexural compression zone) to the yield force ($A_s f_y$) of that at the bottom; the other symbols have the same meanings as explained earlier for Fig. 6.10–1(a). Equation (6.10–1(a)) applies for yielding of the bottom longitudinal steel and the links; eqn (6.10–1(b)) applies for yielding of the top longitudinal steel and the links. Lampert and Collins's interaction curve, therefore, consists of two parts, defined respectively by eqns (6.10–1(a), (b)); it is similar to Hsu's curve (II) in Fig. 6.10–1(a) except that there is now a kink at the intersection of eqns (6.10–1(a), (b)).

Table 6.10–1   Reinforcement for torsion and shear
(BS 8110: Part 2: Clause 2.4.6)

	$v_t \leq v_{tmin}$	$v_t > v_{tmin}$
$v \leq v_c$	Nominal shear reinforcement; no torsion reinforcement	Designed torsion reinforcement only
$v < v_c$	Designed shear reinforcement; no torsion reinforcement	Designed shear and torsion reinforcement

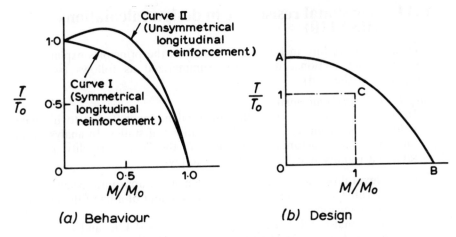

Fig. 6.10–1   Interaction of torsion and bending [44]

For the particular case of $\varrho_1 = 1$, their curve again resembles Hsu's curve (I).

In current design practice, the reinforcements for torsion and bending are calculated separately and then added together. The rationale for this procedure is illustrated in Fig. 6.10–1(b) [44]. In the figure, the ordinate $T/T_0 = 1$ on the vertical axis represents the strength in pure torsion of the member with torsional reinforcement only; the abscissa $M/M_0 = 1$ on the horizontal axis represents the strength in pure bending of the same member with flexural reinforcement only. Point C represents the design procedure of adding torsional and flexural reinforcements. As Hsu and Kemp [44] have pointed out, the addition of the torsional and flexural reinforcements increases both the pure torsion and the pure bending strengths, as illustrated by points A and B.

The **interaction of torsion and shear** is not quite understood [33–35]. For beams without links, a circular interaction curve has been suggested [44, 45]:

$$\left(\frac{T}{T_0}\right)^2 + \left(\frac{V}{V_0}\right)^2 = 1 \tag{6.10–2}$$

where T and V are the torsion and shear combination that can be resisted, T_0 is the strength in pure torsion and V_0 that in shear (and the accompanying bending). Insufficient test data are available for beams with links. In British design practice, as explained in Section 6.10(a), the reinforcements are calculated separately for torsion and for shear and are then added together.

The behaviour of members in combined **torsion, bending and shear** is even less understood [33–35]. For design purposes, it would seem to be conservative to calculate reinforcement requirements separately and then add them together [43, 47]; as Goode [38] has pointed out, the loading arrangement which gives the maximum torsional moment may not simultaneously give the maximum bending and shear.

6.11 Torsional resistance in design calculations (BS 8110)

The design procedure in BS 8110 is based on the principles explained in the preceding sections; it may be summarized as follows (see also comments at the end).

Step 1 The Torsional moment T

In a determinate structure, the torsional moment T due to ultimate loads will be given directly by the equations of statics. In analysing an indeterminate structure to determine T, the flexural rigidity and the **torsional rigidity** may be obtained as follows:

(a) Flexural rigidity EI: E may be taken as the appropriate value of E_c in Table 2.5–6 and I as the second moment of area of the nominal cross-section.

(b) Torsional rigidity GC: G may be taken as $0.42E_c$ and C as $0.5K$, where K is the St Venant value in Table 6.8–1.

Step 2 Torsional shear stress v_t

For a rectangular section, the **torsional shear stress** v_t may be calculated from the equation

$$v_t = \frac{2T}{h_{min}^2[h_{max} - h_{min}/3]} \qquad (6.11\text{–}1)$$

where T = the torsional moment due to ultimate loads;
h_{min} = the smaller dimension of the section; and
h_{max} = the larger dimension of the section.

Then proceed to Step 5, unless the member consists of a flanged section (or a box section), in which case proceed to Step 3 (or Step 4).

Step 3 Flanged sections

T-, L- or I-sections may be treated by dividing them into their component rectangles. This should normally be done so as to maximize the function $\sum (h_{min}^3 h_{max})$ which will generally be achieved if the widest rectangle is made as long as possible. The torsional shear stress v_t carried by each component rectangle may be calculated by treating it as a rectangular section subjected to a torsional moment of

$$T_i = T \frac{h_{min}^3 h_{max}}{\sum (h_{min}^3 h_{max})} \qquad (6.11\text{–}2)$$

where T_i is the torsional moment for a typical component rectangle (see also Step 8(f)).

Step 4 Box sections

Box and **hollow sections** in which wall thicknesses exceed one-quarter of the overall thickness of the member in the direction of measurement may be treated as solid rectangular sections.

Step 5 The torsional shear stress v_{tu}

In no case should the sum of the shear stresses resulting from shear force and torsion exceed the value v_{tu} from Table 6.11–1, nor in the case

Table 6.11–1 Torsional shear stresses
(BS 8110 : Part 2 : Clause 2.4.5)[a,b]

Concrete characteristic strength f_{cu} (N/mm^2)		
25	30	≥ 40
v_{tmin} 0.33	0.37	0.40
v_{tu} 4.00	4.38	5.00

[a] The v_{tmin} and v_{tu} values include an allowance for the partial
  safety factor γ_m.
[b] Values of v_{tmin} and v_{tu} are derived from the equations
   $v_{tmin} = 0.067 \sqrt{f_{cu}} \not> 0.4 \text{ N/mm}^2$
   $v_{tu} = 0.8\sqrt{f_{cu}} \not> 5 \text{ N/mm}^2$

of **small sections** ($y_1 < 550$ mm) should the torsional shear stress v_t
exceed $v_{tu}y_1/550$, where y_1 is the larger dimension of a link. In other
words, for all sections,

$$v_t + v \not> v_{tu} \text{ from Table 6.11–1} \tag{6.11–3}$$

where v_t is calculated from eqn (6.11–1) and v from eqn (6.4–1). For
small sections, there is the additional requirement that

$$v_t \not> v_{tu}y_1/550 \tag{6.11–4}$$

If either of eqns (6.11–3) and (6.11–4) cannot be satisfied, the overall
dimensions of the cross-section are inadequate. In the subsequent steps,
it will be assumed that these equations are satisfied (after revising cross-
sectional dimensions if necessary).

Step 6   *The torsion shear stress v_{tmin}*
   If v_t from eqn (6.11–1) exceeds the value v_{tmin} from Table 6.11–1,
torsion reinforcement should be provided, as explained in Step 7; if v_t
does not exceed v_{tmin}, the nominal links as defined by eqn (6.4–2) are
sufficient for resisting the torsional effects.

Step 7   *Torsion reinforcement*
This should consist of rectangular closed links together with longitudinal
reinforcement. This reinforcement is additional to any requirement for
shear and bending which occurs simultaneously with the torsion and
should be such that:

$$\frac{A_{sv}}{s_v} \geq \frac{T}{0.8x_1y_1(0.87f_{yv})} \tag{6.11–5}$$

$$A_s \geq \frac{A_{sv}}{s_v}\left(\frac{f_{yv}}{f_y}\right)(x_1 + y_1) \tag{6.11–6}$$

where T = the torsional moment due to ultimate loads;
         A_{sv} = the area of the legs of closed links at a section;
         A_s = the area of longitudinal torsion reinforcement;

f_{yv} = the characteristic strength of the links (but not to be taken as exceeding 460 N/mm²);

f_y = the characteristic strength of the longitudinal reinforcement (but not to be taken as exceeding 460 N/mm²);

s_v = the spacing of the links;

x_1 = the smaller centre-to-centre dimension of the links;

y_1 = the larger centre-to-centre dimension of the links.

Step 8   Detailing requirements

The detailing of torsion reinforcement should satisfy the following requirements:

(a) The links should be of a closed type similar to Code 74 of BS 4466 (see Fig. 6.11–1).

(b) The **link spacing** s_v should not exceed the least of x_1, $y_1/2$ or 200 mm, where symbols are as defined in Step 7.

(c) The longitudinal torsion reinforcement required by eqn (6.11–6) should be distributed evenly round the inside perimeter of the links. The clear distance between these bars should not exceed 300 mm and at least four bars, one in each corner of the links, should be used.

(d) Longitudinal torsion bars required at the level of the tension or compression reinforcement may be provided by using larger bars than those required for bending alone.

(e) All longitudinal torsion bars should extend a distance at least equal to the largest dimension of the cross-section beyond where it ceases to be required.

(f) In the component rectangles of T-, L- or I-sections, the reinforcement cages should be detailed so that they interlock and tie the component rectangles of the section together (see Fig. 6.11–2). Where the torsional shear stress v_t in a minor component rectangle is less than v_{tmin}, no torsion reinforcement need be provided in that rectangle.

Comments on Step 1

The relation $G = 0.42E_c$ may be derived from the elastic relationship $G = E/2(1 + v)$ by taking Poisson's ratio v as 0.19 (see Section 2.5(d) on

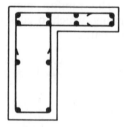

Fig. 6.11–1   Closed link for torsion

Fig. 6.11–2   Torsion reinforcement for flanged beam

Poisson's ratio). The relatively low value assigned to G, together with the fact that C is taken only as $\frac{1}{2}K$, has the effect of reducing the value of the torsional rigidity GC to be used in the structural analysis, and hence of reducing the magnitude of the compatibility torsion moment to be used in the design calculations; the use of $C = \frac{1}{2}K$ also allows for the reduction in GC due to flexural cracking.

Comments on Step 2
BS 8110's equation for v_t follows from eqn (6.8–8): the authors agree that it is difficult to accept the completely plastic stress distribution implied by this equation, but such acceptance is not necessary; the stress v_t may simply be regarded as a nominal stress, just as the stress v from eqn (6.4–1) is so regarded.

Comments on Step 3
BS 8110's intention is that the effective flange widths in Section 4.8 should apply to torsion as well. Note also that BS 8110's expression for the torsional moment for a component rectangle is neither that of eqn (6.8–9), which is based on a plastic stress distribution, nor that of eqn (6.8–10) which is based on an elastic stress distribution. A correct expression for structural concrete is not yet available; BS 8110's expression is conservative and may overestimate the torsion shear stress by 20% [40].

Recently, Goode [38] has drawn attention to tests on L-beams which have shown that if the flange is cracked in flexure its effectiveness in helping to resist torsion is small; he suggested that a flexurally cracked outstanding flange should be neglected and the beam designed as rectangular.

Comments on Step 4
See statement (c) following eqn (6.9–3); that statement also covers cases where the wall thickness is less than one-quarter the member size.

Comments on Step 5
As pointed out in statement (a) following eqn (6.9–3), it is important in practice to ensure that the section is torsionally under-reinforced. The v_{tu} values in Table 6.11–1 are intended to do just that. The additional restriction for small sections is to prevent the premature spalling of concrete at the corners which has been observed in tests; however, the factor $y_1/550$ would seem to be rather conservative.

Comments on Step 6
The v_{tmin} values in Table 6.11–1 are empirical. That is, the observed torsional moment at which diagonal cracking occurs in the test specimen is used in eqn (6.8–8) to compute the nominal stress v_t, which is then divided by a partial safety factor to give v_{tmin}. The v_{tmin} values in Table 6.11–1 are very conservative [38, 40].

Comments on Step 7
BS 8110's equation for A_{sv} follows directly from eqn (6.9–3) if the characteristic strength f_{yv} is replaced by the design strength $0.87f_{yv}$, and if the efficiency factor of 0.8 is inserted to allow for the fact that BS 8110 refers to the link dimensions and not to the dimensions between centres of

238   *Shear, bond and torsion*

longitudinal bars. The validity of eqn (6.9–3) depends on eqn (6.9–1) being satisfied, in which case the longitudinal steel A_s and the links A_{sv} would yield simultaneously. BS 8110's equation (6.11–6) permits A_s to exceed that required by eqn (6.9–1) and hence permits the links to yield first.

The reader should refer to Fig. 6.10–1(b) and note the implication of the BS 8110 procedure according to which the longitudinal bars for torsion and for bending are calculated separately without considering interaction effects, and then added together. Similarly, according to BS 8110, links for torsion and for shear are also calculated separately and added together.

Comments on Step 8
The important subject of detailing for torsion is discussed in an interesting paper by Mitchell and Collins [48].

Example 6.11–1
Explain, with reference to the concept of the **shear centre**, how to determine the design torsional moment for the section in Fig. 6.11–3.

SOLUTION
For the general case of a beam of arbitrary section, the torsional moment is obtained by taking moments, not about the centroid of the section, but about its shear centre. Only T-, L- and I-sections are explicitly referred to in BS 8110; for such sections it would seem adequate to take the shear centre as lying on the centre line of the web. For other asymmetrical sections, reference may be made to handbooks such as R. J. Roark and W. C. Young's *Formulas for Stress and Strain* (McGraw-Hill). However, it should not be forgotten that BS 8110's recommendations are essentially empirical, and it may not be wise to apply them to beams of unusual sections which are not adequately covered by the tests upon which such recommendations are based. Instead of meticulously locating the shear centre by elastic analysis (which may in any case be incompatible with the

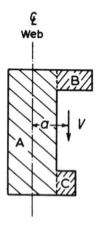

Fig. 6.11–3

space truss analogy upon which BS 8110's recommendations are based), it may be better to make some realistic simplifying assumptions. For example, with reference to the beam section in Fig. 6.11–3, the rectangle A may be designed to resist the whole of the torque Va; this would satisfy the ultimate limit state requirement. To guard against excessive cracking of the flanges B and C, these may then be reinforced to resist the respective partial torques as given by eqn (6.11–2)—that is, if the torsion shear stresses v_t corresponding to such partial torques indicate that torsion reinforcement is necessary.

Example 6.11–2
Design suitable reinforcement for the L-beam section in Fig. 6.11–4, in which the flange width b is the effective width, if for the ultimate limit state the design moment M is 215 kNm, the design shear force V is 150 kN and the design torsional moment T is 105 kNm. Given: characteristic strength of concrete $f_{cu} = 40$ N/mm^2; characteristic strength of steel $f_y = f_{yv} = 460$ N/mm^2.

SOLUTION
(a) **Bending moment**. We shall follow the I.Struct.E. Manual's procedure as explained in the steps following eqn (4.8–2).

Step 1
   Check x/d ratio:

$$K = \frac{M}{f_{cu}bd^2} = \frac{(215)\,(10^6)}{(40)\,(700)\,(750^2)} = 0.014$$

From Table 4.6–1

$$z/d = 0.94; \qquad x/d = 0.13$$

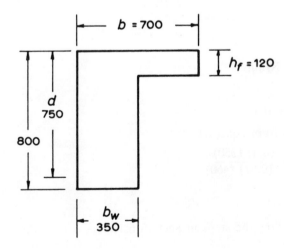

Fig. 6.11–4

Step 2

Check whether $0.9x \leq h_f$. From Step 1,

$$x = (0.13)(750) = 98 \text{ mm}$$

$$0.9x = 88 \text{ mm} < h_f$$

Hence eqn (4.8–3) is applicable:

$$A_s = \frac{M}{0.87f_y z} = \frac{(215)(10^6)}{(0.87)(460)(0.94)(750)} = \underline{762 \text{ mm}^2}$$

where 0.94 is z/d from Step 1.

Use two size 25 bars (981 mm²; $\varrho = A_s/b_w d = 0.37\%$)

(b) **Shear**. We shall follow the steps in Section 6.4.

Step 1

$$v = \frac{V}{b_v d} = \frac{150\,000}{(350)(750)} = 0.57 \text{ N/mm}^2$$

Step 2

$$0.8\sqrt{f_{cu}} = 5.1 \text{ N/mm}^2 \text{ (for } f_{cu} \text{ of 40 N/mm}^2) > 5 \text{ N/mm}^2$$

Hence the upper limit is 5 N/mm².

$$v \text{ of Step 1} < 5 \text{ N/mm}^2 \text{ OK}$$

Step 3

$$d = 750 \text{ mm and } 100A_s/b_v d = 0.37 \text{ from (a)}$$

From Table 6.4–1,

$$v_c = 0.53 \text{ N/mm}^2 \text{ by interpolation}$$

Hence

$$v > 0.5\, v_c$$

Step 4

$$v_c + 0.4 = 0.93 \text{ N/mm}^2$$

Hence

$$0.5 v_c < v < v_c + 0.4$$

Provide minimum links from eqn (6.4–2):

$$A_{sv} = \frac{0.4\, b_v s_v}{0.87 f_{yv}} = \frac{(0.4)(350)s_v}{(0.87)(460)}$$

$$\frac{A_{sv}}{s_v} = 0.35 \text{ mm}$$

(c) **Torsion**. We shall follow the steps in Section 6.11.

Step 1

Omitted.

Step 2

$$v_t = \frac{2T}{h_{min}^2[h_{max} - (h_{min}/3)]}$$

We shall first move on to Step 3 to calculate the partial torques for the web and flange component rectangles.

Step 3

$$h_{min}^3 h_{max} \text{ for web} = (350^3)(800) = (3.43)(10^{10}) \text{ mm}^4$$

$$h_{min}^3 h_{max} \text{ for flange} = (120^3)(700 - 350) = (6.04)(10^8) \text{ mm}^4$$

$$\ll (3.43)(10^{10}) \text{ mm}^4$$

Therefore, neglect the flange and design the web to resist the entire torque. From the equation in Step 2,

$$v_t = \frac{2(105)(10^6)}{(350^2)(800 - 350/3))} = 2.51 \text{ N/mm}^2$$

Step 4

Omitted.

Step 5

$$v_t \text{ (Step 3)} + v \text{ (Part (b) Step 1)} = 2.51 + 0.57 \text{ N/mm}^2$$

$$= 3.08 \text{ N/mm}^2$$

This does not exceed v_{tu} of 5 N/mm^2 from Table 6.11–1. OK

Step 6

From Table 6.11–1,

$$v_{tmin} = 0.4 \text{ N/mm}^2$$

Step 7

Since $v_t > v_{tmin}$, torsion reinforcement is required. From Table 2.5–7, use a concrete cover of, say, 30 mm. Therefore, assuming size 10 links, the link dimensions are

$$x_1 = 350 - (2)(30) - 10 = 280 \text{ mm}$$

$$y_1 = 800 - (2)(30) - 10 = 730 \text{ mm}$$

From eqns (6.11–5) and (6.11–6),

$$\frac{A_{sv}}{s_v} = \frac{(105)(10^6)}{(0.8)(280)(730)(0.87)(460)} = 1.60 \text{ mm}$$

$$A_s = (1.60)\left(\frac{460}{460}\right)(280 + 730) = 1616 \text{ mm}^2$$

<u>Use nine size 16 longitudinal bars (1809 mm^2)</u>

From Part (b), A_{sv}/s_v for shear = 0.35; hence

$$\underset{\text{(torsion)}}{\frac{A_{sv}}{s_v}} + \underset{\text{(shear)}}{\frac{A_{sv}}{s_v}} = 1.60 \text{ mm} + 0.35 \text{ mm} = 1.95 \text{ mm}$$

From Table 6.4–2,

<u>Use size 12 links at 100 mm spacing</u> (A_{sv}/s_v = 2.26)

(*Note:* The x_1 and y_1 values were based on size 10 links; these are now slightly greater than the actual values but the effect of the differences is more than offset by the higher A_{sv}/s_v value provided.)

Step 8
The reinforcement details are shown in Fig. 6.11–5. The reader should check that the requirements in Section 6.11: Step 8 are met. The size 12 transverse bars in the flange amply satisfy the transverse reinforcement requirement in Section 4.8. (These transverse bars are held in place by longitudinal bars in the flange which are not shown here.)

Example 6.11–3
Determine the maximum ultimate torsional moment that may be applied to the beam section in Example 6.11–2 if:

(a)   no torsion reinforcement is provided;
(b)   torsion reinforcement is provided.

SOLUTION
(a)   Example 6.11–2(c) shows that the web resists practically all the torsional moment. From Table 6.11–1,

$$v_{tmin} = 0.4 \text{ N/mm}^2$$

Using eqn (6.11–1),

$$T = \left(\frac{1}{2}\right) v_{tmin} h_{min}^2 \left[h_{max} - (h_{min}/3)\right]$$

$$= (0.5)\,(0.4)\,(350^2)\,[800 - 350/3]$$

$$= \underline{16.8 \text{ kNm}}$$

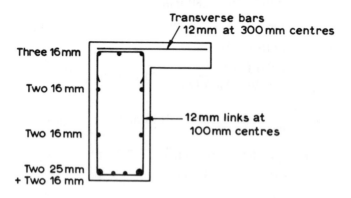

Transverse bars
12 mm at 300 mm centres

Three 16 mm

Two 16 mm

Two 16 mm

12 mm links at
100 mm centres

Two 25 mm
+ Two 16 mm

Fig. 6.11–5

(b)   From Table 6.11–1,

$$v_{tu} = 5 \text{ N/mm}^2$$

From eqn (6.11–3),

$$v_t + v = 5$$

where $v = 0.57 \text{ N/mm}^2$ from Example 6.11–2(b). Therefore

$$v_t = 5 - 0.57 = 4.43 \text{ N/mm}^2$$

Using eqn (6.11–1),

$$T = \frac{(4.43)\ (350^2)\ [800 - 350/3]}{2}$$

$$= \underline{185 \text{ kNm}}$$

6.12   Design and detailing—illustrative example

The following example is a continuation of Example 4.11–1.

Example 6.12–1
Complete Step 7 of the solution to Example 4.11–1.

SOLUTION
From Example 4.11–1 (Solution: Step 6),

$$V = 86.15 \text{ kN}$$

$$v = \frac{V}{b_v d} = \frac{(86.15)\ (10^3)}{(325)\ (320)} = 0.83 \text{ N/mm}^2$$

$$< \text{(i) } 5 \text{ N/mm}^2 \text{ and (ii) } 0.8\ \sqrt{f_{cu}}\ (= 5.1 \text{ N/mm}^2)$$

Therefore the overall dimensions are adequate for shear.
From Example 4.11–1 (Solution: Step 4),

A_s provided at interior support $= 628 \text{ mm}^2$

$$\frac{A_s}{b_v d} = \frac{628}{(325)\ (320)} = 0.60\%$$

From Table 6.4–1,

$$v_c = 0.66 \text{ N/mm}^2 \text{ by interpolation}$$

$$0.5 v_c = 0.33 \text{ N/mm}^2; \qquad v_c + 0.4 = 1.06 \text{ N/mm}^2$$

Hence

$$0.5 v_c < v\ (= 0.83 \text{ N/mm}^2) < v_c + 0.4$$

Provide minimum links, using eqn (6.4–2) (see also Example 6.4–1(b)):

$$A_{sv} = \frac{0.4 b_v s_v}{0.87 f_{yv}} = \frac{(0.4)\ (325) s_v}{(0.87)\ (250)}$$

$$\frac{A_{sv}}{s_v} = 0.60 \text{ mm}$$

$$s_{v(max)} = 0.75d = (0.75)(320) = 240 \text{ mm}$$

Provide R10 at 200 (see Fig. 4.11–2)

From Table 6.4–2,

$$\frac{A_{sv}}{s_v} \text{ (provided)} = 0.79 > 0.60 \text{ OK}$$

6.13   Computer programs

(In collaboration with **Dr H. H. A. Wong**, University of Newcastle upon Tyne)

The FORTRAN programs for this chapter are listed in Section 12.6. See also Section 12.1 for 'Notes on the computer programs'.

Problems

6.1 An acceptable rational theory for shear is not yet available, and the design methods used in practice are based on experience, laboratory tests and engineering judgement.

Practising engineers in the UK generally adopt the design procedure explained in Clause 3.4.5 of BS 8110. Comment critically on every major step of the Code's procedure; support your arguments with research results wherever possible.

Ans.   See Sections 6.2 and 6.3. Study again the comments in Section 6.4 if necessary.

6.2 Clause 2.4 of BS 8110: Part 2 explains the procedure for designing structural members for torsion. Comment critically on every major step of the Code's procedure; support your arguments with research results wherever possible.

Ans.   See Sections 6.7 to 6.10. Study again the comments in Section 6.11 if necessary.

6.3 In current design practice, the following equation is used to calculate the amount of shear reinforcement:

$$\frac{A_{sv}}{s_v} = \frac{b_v(v - v_c)}{0.87 f_{yv}}$$

Using the truss analogy, derive the equation from first principles; relate your assumptions to research results.

6.4 In current design practice, the following equation is used for calculating the torsional shear stress v_t:

$$v_t = \frac{2T}{h_{min}^2[h_{max} - (h_{min}/3)]}$$

Show that the equation may be obtained from the sand-heap analogy for plastic torsion and explain whether a plastic stress distribution exists in a concrete section at incipient failure in torsion.

6.5 BS 8110 states that the amount of torsion links should satisfy the equation:

$$\frac{A_{sv}}{s_v} \geq \frac{T}{0.8x_1 y_1(0.87f_{yv})}$$

Using the space truss analogy, derive the equation from first principles; relate your assumptions to research results.

The equation shows that the ultimate torsional resistance is independent of the concrete strength and that the torsional strength of a box beam is the same as that of a solid beam. Comment.

Briefly explain the reason for the factor 0.8 in the equation.

6.6 (a) Derive the following anchorage-bond length equation and explain your assumptions:

$$l_u = \frac{0.87f_y}{4\beta\sqrt{f_{cu}}} \phi$$

(b) Former British Codes CP 110 and CP 114 require designers to check local bond stresses, but BS 8110 does not so require. Comment. (Hint for Part (b): see Example 6.6–2.)

References

1 Evans, R. H. and Kong, F. K. Shear design and British Code CP 114. *The Structural Engineer*, **45**, No. 4, April 1967, pp. 153–8.
2 ACI–ASCE Committee 426. Shear strength of reinforced concrete members. *Proc. ASCE*, **99**, No. ST6, June 1973, pp. 1091–187. (Reaffirmed in 1980 and published by the American Concrete Institute as Publication No. 426R–74.)
3 Kotsovos, M. D. Mechanism of 'shear' failure. *Magazine of Concrete Research*, **35**, No. 123, June 1983, pp. 99–106.
4 Kotsovos, M. D. Behaviour of reinforced concrete beams with a shear span to depth ratio between 1.0 and 2.5. *Proc. ACI*, **81**, No. 3, May/June 1984, pp. 279–98.
5 Regan, P. E. Concrete Society Current Practice Sheet No. 105: Shear. *Concrete*, **19**, No. 11, Nov. 1985, pp. 25–6.
6 Hsu, T. C. C. Is the 'staggering concept' of shear design safe? *Proc. ACI*, **79**, No. 6, Nov./Dec. 1982, pp. 435–43.
7 Marti, P. Staggered shear design of simply supported concrete beams. *Proc. ACI*, **83**, No. 1, Jan./Feb. 1986, pp. 36–42.
8 Elzanaty, A. H., Nilson, A. H. and Slate, F. O. Shear capacity of reinforced concrete beams using high strength concrete. *Proc. ACI*, **83**, No. 2, March/April 1986, pp. 290–6.

246   *Shear, bond and torsion*

 9 Clark, L. A. and Thorogood, P. Shear strength of concrete beams in hogging regions. *Proc. ICE* (Part 2), **79**, June 1985, pp. 315–26.
10 Taylor, H. P. J. The fundamental behaviour of reinforced concrete beams in bending and shear. *Proceedings* ACI–ASCE Shear Symposium, Ottawa, 1973 (ACI Special Publication SP42). American Concrete Institute, Detroit, 1974, pp. 43–77.
11 Nawy, E. G. Shear, diagonal tension and torsional strength. In *Handbook of Structural Concrete*, edited by Kong, F. K., Evans, R. H., Cohen, E. and Roll, F. Pitman, London and McGraw-Hill, New York, 1983, Chapter 12, Section 3.
12 CIRIA Guide 2: *The Design of Deep Beams in Reinforced Concrete*. Ove Arup and Partners, London, and Construction Industry Research and Information Association, London, 1984.
13 Reynolds, C. E. and Steedman, J. C. *Reinforced Concrete Designer's Handbook*, 9th ed. Cement and Concrete Association, Slough, 1981 (Table 151: deep beams).
14 Kong, F. K. Reinforced concrete deep beams. In *Concrete Framed Structures*, edited by Narayanan, R. Elsevier Applied Science Publishers, Barking, 1986, Chapter 6.
15 Kong, F. K., Garcia, R. C., Paine, J. M., Wong, H. H. A., Tang, C. W. J. and Chemrouk, M. Strength and stability of slender concrete deep beams. *The Structural Engineer*, **64B**, No. 3, Sept. 1986, pp. 49–56.
16 Kong, F. K. and Sharp, G. R. Structural idealization for deep beams with web openings. *Magazine of Concrete Research*, **29**, No. 99, June 1977, pp. 81–91.
17 Kong, F. K., Sharp, G. R., Appleton, S. C., Beaumont, C. J. and Kubik, L. A. Structural idealization for deep beams with web openings—further evidence. *Magazine of Concrete Research*, **30**, No. 103, June 1978, pp. 89–95.
18 Kong, F. K., Robins, P. J. and Sharp, G. R. The design of reinforced concrete deep beams in current practice. *The Structural Engineer*, **53**, No. 4, April 1975, pp. 173–80.
19 Kong, F. K. and Singh, A. Diagonal cracking and ultimate loads of lightweight concrete deep beams. *Proc. ACI*, **69**, No. 8, Aug. 1972, pp. 513–21.
20 Cusens, A. R. and Besser, I. I. Shear strength of reinforced concrete wall-beams under combined top and bottom loads. *The Structural Engineer*, **63B**, No. 3, Sept. 1985, pp. 50–6.
21 Haque, M., Rasheeduzzafar, A. and Al-Tayyib, H. J. Stress distribution in deep beams with web openings. *ASCE Journal of Structural Engineering*, **112**, No. 5, May 1986, pp. 1147–65.
22 Rajagopolan, K. S. and Ferguson, P. M. Exploratory sheer tests emphasizing percentage of longitudinal steel. *Proc. ACI*, **65**, No. 8, Aug. 1968, pp. 634–8.
23 Kani, G. N. J. How safe are our large reinforced concrete beams? *Proc. ACI*, **64**, No. 3, March 1967, pp. 128–41.
24 Taylor, H. P. J. Shear strength of large beams. *Proc. ASCE*, **98**, No. ST11, Nov. 1972, pp. 2473–90.
25 Swamy, R. N. and Qureshi, S. A. Shear behaviour of reinforced concrete T beams with web reinforcement. *Proc. ICE*, **57** (Part 2), March 1972, pp. 35–49.
26 I.Struct.E./ICE Joint Committee. *Manual for the Design of Reinforced Concrete Building Structures*. Institution of Structural Engineers, London, 1985.
27 Kong, F. K., Robins, P. J., Singh, A. and Sharp, G. R. Shear analysis and design of reinforced concrete deep beams. *The Structural Engineer*, **50**, No. 10, Oct. 1972, pp. 405–9.
28 Kong, F. K. and Sharp, G. R. Shear strength of lightweight concrete deep

beams with web openings. *The Structural Engineer*, **51**, No. 8, Aug. 1973, pp. 267–75.

29 Kemp, E. L. Bond in reinforced concrete: behaviour and design criteria. *Proc. ACI*, **83**, No. 1, Jan. 1986, pp. 50–7.

30 Royles, R. and Morley, P. D. Further responses of the bond in reinforced concrete to high temperatures. *Magazine of Concrete Research*, **35**, No. 124, Sept. 1983, pp. 157–63.

31 Edwards, A. D. and Yannopoulos, P. J. Local bond–stress/slip relationships under repeated loading. *Magazine of Concrete Research*, **30**, No. 103, June 1978, pp. 62–72.

32 Evans, R. H. and Robinson, G. W. Bond stresses in prestressed concrete from X-ray photographs. *Proc. ICE* (Part 1), **4**, No. 2, March 1955, pp. 212–35.

33 ACI Committee 438. *Torsion in Concrete* (Bibliography No. 12). American Concrete Institute, Detroit, 1978.

34 ACI Committee 438. *Analysis of Structural Systems for Torsion* (SP–35). American Concrete Institute, Detroit, 1974.

35 Hsu, T. T. C. and Mo, Y. L. Softening of concrete in torsional members. Part 1: theory and tests; Part 2: design recommendations; Part 3: prestressed concrete. *Proc. ACI*, **82**, No. 3, May/June 1985, pp. 290–303; No. 4, July/Aug. 1985, pp. 443–52; No. 5, Sept./Oct. 1985, pp. 603–15.

36 Evans, R. H. and Sarkar, S. A method of ultimate strength design of reinforced concrete beams in combined bending and torsion. *The Structural Engineer*, **43**, No. 10, Oct. 1965, pp. 337–44.

37 Evans, R. H. and Khalil, M. G. A. The behaviour and strength of prestressed concrete rectangular beams subjected to combined bending and torsion. *The Structural Engineer*, **48**, No. 2, Feb. 1970, pp. 59–73.

38 Goode, C. D. Torsion: CP 110 and ACI 318–71 compared. *Concrete*, **8**, No. 3, March 1974, pp. 36–40.

39 Coates, R. C., Coutie, M. G. and Kong, F. K. *Structural Analysis*, 2nd edn. Van Nostrand Reinhold, 1980, pp. 397 and 412.

40 Marshall, W. T. Torsion in concrete and CP 110:1972. *The Structural Engineer*, **52**, No. 3, March 1974, pp. 83–8.

41 Evans, R. H. and Kong, F. K. The extensibility and microcracking of the in-situ concrete in composite prestressed concrete beams. *The Structural Engineer*, **42**, No. 6, June 1964, pp. 181–9.

42 Sawko, F. *Developments in Prestressed Concrete*. Applied Science, London, 1978. (Vol. 1: Section 1.4 Torsion—elastic theory; Section 1.7 Torsion—ultimate limit state.)

43 Lampert, P. and Collins, M. P. Torsion, bending and confusion—and attempt to establish the facts. *Proc. ACI*, **69**, No. 8, Aug. 1972, pp. 500–4.

44 Hsu, T. T. C. and Kemp, E. L. Background and practical application of tentative design criteria for torsion. *Proc. ACI*, **66**, No. 1, Jan. 1969, pp. 12–23

45 Hsu, T. T. C. Torsion of structural concrete—interaction surface for combined torsion, shear and bending in beams without stirrups. *Proc. ACI*, **65**, No. 1, Jan. 1968, pp. 51–60.

46 Goode, C. D. and Helmy, M. A. Bending and torsion of reinforced concrete beams. *Magazine of Concrete Research*, **20**, No. 64, Sept. 1968, pp. 156–66.

47 Goode, C. D. Reinforced concrete beams subjected to a sustained torque. *The Structural Engineer*, **53**, No. 5, May 1975, pp. 215–20.

48 Mitchell, D. and Collins, M. P. Detailing for torsion. *Proc. ACI*, **73**, No. 9, Sept. 1976, pp. 506–11.

Chapter 7
Eccentrically loaded columns and slender columns

Preliminary note: Readers interested only in structural design to BS 8110 may concentrate on the following sections:

(a) *Section 7.2: Effective column height (BS 8110).*
(b) *Section 7.3: Eccentrically loaded short columns (BS 8110).*
(c) *Section 7.5: Slender columns (BS 8110).*

7.1 Principles of column interaction diagrams

The analysis of reinforced concrete members under **combined bending and axial load** [1–4] may be based on the same assumptions as those in the general theory for ultimate flexural strengths in Chapter 4. The member is considered to be at the ultimate limit state of collapse when the concrete strain at the more highly compressed face reaches a specified value ε_{cu}, which is taken as 0.0035 in current British practice.

In the design of eccentrically loaded columns, engineers make extensive use of design charts called **column interaction diagrams**. Readers interested only in the construction of such diagrams may move direct to Example 7.1–6, omitting the rest of this section.

To obtain an insight into the properties of column interaction diagrams and the principles governing their construction, we shall first consider a plain concrete section subjected simultaneously to an axial load N and a bending moment M. The section (Fig. 7.1–1(a)) is at incipient failure, with strain and stress distributions as in Figs 7.1–1(b) and (c). The depth d_c of the concrete stress block is equal to $0.9x$ for $0.9x \leq h$. (What happens when $0.9x > h$ will be explained in Example 7.1–5.) From the equilibrium condition,

$$N(\text{concrete}) = 0.45f_{cu}bd_c = 0.405f_{cu}bx \tag{7.1–1}$$

$$M(\text{concrete}) = N\left[\frac{h}{2} - \frac{d_c}{2}\right] \quad \begin{array}{l}\text{(by taking moments about}\\ \text{the mid-depth of the section)}\end{array}$$

$$= 0.225f_{cu}bd_c(h - d_c)$$

$$= 0.203f_{cu}bx(h - 0.9x) \tag{7.1–2}$$

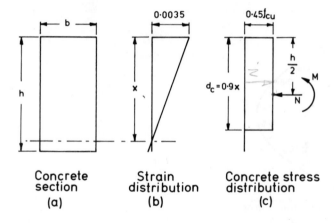

Concrete section (a) Strain distribution (b) Concrete stress distribution (c)

Fig. 7.1–1

where the word 'concrete' emphasizes that the N and M values refer to a plain concrete section.

Equations (7.1–1) and (7.1–2) can be expressed in dimensionless form:

$$\alpha_{\text{conc}} = \frac{N(\text{conc})}{f_{\text{cu}}bh} = 0.45\left[\frac{d_c}{h}\right]$$

$$= 0.405\frac{x}{h} \qquad (7.1\text{--}3)$$

$$\beta_{\text{conc}} = \frac{M(\text{conc})}{f_{\text{cu}}bh^2} = 0.225\left[\frac{d_c}{h}\right]\left[1 - \frac{d_c}{h}\right]$$

$$= 0.203\frac{x}{h}\left[1 - 0.9\frac{x}{h}\right] \qquad (7.1\text{--}4)$$

Thus for $x/h = 0.1$, $\alpha_{\text{conc}} = 0.0405$ and $\beta_{\text{conc}} = 0.0185$; for $x/h = 0.2$, $\alpha_{\text{conc}} = 0.0810$ and $\beta_{\text{conc}} = 0.0333$ and so on. Therefore by successively assigning different values to x/h, the complete $\alpha_{\text{conc}} \sim \beta_{\text{conc}}$ curve may be constructed (Fig. 7.1–2). Each point on this curve represents a state of incipient failure; points inside the curve represent safe combinations of N and M, while those outside represent unacceptable combinations. (It should be noted here that eqns 7.1–3 and 7.1–4 are based on BS 8110's simplified stress block as described in Section 4.4. The stress block is intended primarily for design so that the stress intensity of $0.45f_{\text{cu}}$ includes an allowance for the partial safety factor. Therefore at incipient failure, the N and M values will be higher than those given by these equations. However, as long as this point is understood, there is no objection to using BS 8110's stress block, since our emphasis is on principles.)

Now consider the effect of an area A_{s2} of reinforcement, as in Fig. 7.1–3(a). The contribution of A_{s2} to the axial force is

$$N(A_{s2}) = A_{s2}f_{s2} \qquad (7.1\text{--}5)$$

where f_{s2} is the compressive stress in the reinforcement which corresponds

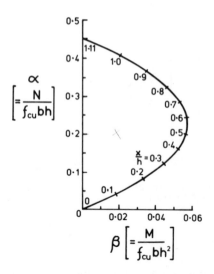

Fig. 7.1–2   Column interaction diagram—plain concrete section

to the strain ε_{s2} in Fig. 7.1–3(b). (*Note: According to current convention, reinforcement stresses in columns are designated positive when compressive.*) The contribution to the bending moment about the mid-depth of the section is

$$M(A_{s2}) = -A_{s2}f_{s2}\left(\frac{h}{2} - d_2\right) \tag{7.1-6}$$

where the negative sign is used because the compressive steel force $A_{s2}f_{s2}$ produces a moment acting in an opposite sense to that produced by the concrete stress block. In dimensionless form, these equations are

$$\alpha_{s2} = \frac{N(A_{s2})}{f_{cu}bh} = \left(\frac{A_{s2}}{bh}\right)\frac{f_{s2}}{f_{cu}} \tag{7.1-7}$$

$$\beta_{s2} = \frac{M(A_{s2})}{f_{cu}bh^2} = -\left(\frac{1}{2} - \frac{d_2}{h}\right)\alpha_{s2} \tag{7.1-8}$$

For a given column section, the quantities A_{s2}/bh, d_2/h and f_{cu} are known. Therefore α_{s2} and β_{s2} are completely defined by the steel stress f_{s2}, which for a given steel depends only on the strain ε_{s2}. From Fig. 7.1–3(b),

$$\frac{\varepsilon_{s2}}{0.0035} = \frac{d_2 - (h - x)}{x} = \frac{x/h - (1 - d_2/h)}{x/h} \tag{7.1-9}$$

That is, ε_{s2} and hence α_{s2} and β_{s2} are completely defined by the x/h ratio. For a given column section, therefore, the α_{s2} and β_{s2} ratios can be computed for a range of values assigned to x/h. When these ratios are superimposed on α_{conc} and β_{conc}, we obtain Fig. 7.1–4. The figure shows that, for any specified value of x/h, the effect of the reinforcement is represented by the vector $\alpha_{s2} + \beta_{s2}$ where

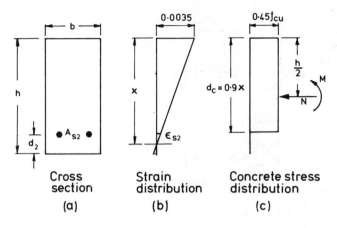

Fig. 7.1–3

$$\text{length of vector} = \sqrt{\{(\alpha_{s2})^2 + (\beta_{s2})^2\}} \tag{7.1–10}$$

$$\text{inclination of vector} = \tan^{-1}\left[\frac{\beta_{s2}(\text{eqn } 7.1–8)}{\alpha_{s2}(\text{eqn } 7.1–7)}\right]$$

$$= \tan^{-1}\left[-\left(\frac{1}{2} - \frac{d_2}{h}\right)\right] \tag{7.1–11}$$

Thus for $x/h = 0.4$, the point C_1 is moved to the point F; that is point F now represents the axial force and bending moment combination that will produce a state of incipient failure.

Figure 7.1–4 has been prepared for the case where $A_{s2}/bh = 1\%$, $f_{cu} = 40$ N/mm^2 and $d_2/h = 0.15$. The reinforcement stress/strain relation is that shown in Fig. 3.2–1(b); f_y has been taken as 460 N/mm^2, so that the stress/strain relation is linear between $f_{s2} = \pm 0.87 f_y = \pm 400$ N/mm^2, at which definite yield occurs. (*Note:* other stress/strain relations can be used if necessary—see Example 7.1–4.) E_s has been taken as 200 kN/mm^2.

The reader should verify the following statements:

(a) For $x/h = 1 - d_2/h(= 0.85$ here), the steel strain is zero; hence from eqns (7.1–7) and (7.1–8), both α_{s2} and β_{s2} are zero and the steel is inactive (point C_3).

(b) For $x/h > 1 - d_2/h$, ε_{s2} in eqn (7.1–9) is positive and the reinforcement is in compression.

(c) For $x/h < 1 - d_2/h$, ε_{s2} is negative and the reinforcement is in tension.

(d) For x/h equal to a certain critical value, which is 0.541 in this case (point C_2), the strain ε_{s2} is -0.0020 so that the reinforcement reaches its design strength in tension. The length of the vector $\alpha_{s2} + \beta_{s2}$ now reaches its maximum value (C_2 B) and remains unchanged as x/h is further reduced.

(e) As x/h moves above 0.541 the length of the vector shortens until it vanishes at $d_c/h = 0.85$ (point C_3) and then extends again in the

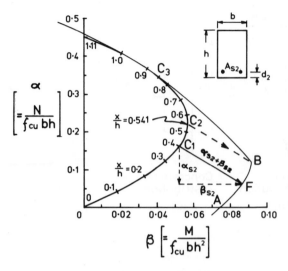

Fig. 7.1–4   Column interaction diagram ($A_{s2}/bh = 1\%$, $f_{cu} = 40$ N/mm², $f_y = 460$ N/mm², $d_2/h = 0.15$)

opposite direction for higher values of x/h. Irrespective of the value of x/h, however, the inclination of the vector is constant, being given by eqn (7.1–11).

(f)   When the axial force and bending moment acting on the column section are such that (α, β) falls on the point B, the concrete maximum compressive strain reaches 0.0035 simultaneously as the reinforcement reaches the tensile design strength; this mode of failure is referred to as a **balanced failure**. For (N,M) combinations represented by points that lie on the curve above B, the reinforcement does not reach the design strength $0.87f_y$ in tension when the concrete compression strain reaches 0.0035; similarly, for (N,M) combinations that lie on AFB, the reinforcement reaches $0.87f_y$ in tension before the concrete compressive strain reaches 0.0035. *It is thus seen that, for a given column section, whether the balanced condition is achieved at failure depends on the loading rather than (as it does in a beam) on the amount of the reinforcement.*

Equations (7.1–5) to (7.1–11) refer to an area A_{s2} of reinforcement remote from the face at which the concrete compressive strain is a maximum. If instead we have an area A'_{s1} of reinforcement near the compression face, as in Fig. 7.1–5, the reader should verify that

$$N(A'_{s1}) = A'_{s1}f'_{s1} \tag{7.1–12}$$

$$M(A'_{s1}) = A'_{s1}f'_{s1}\left(\frac{h}{2} - d'\right) \tag{7.1–13}$$

Or, in dimensionless form,

$$\alpha_{s1} = \frac{N(A'_{s1})}{f_{cu}bh} = \left(\frac{A'_{s1}}{bh}\right)\frac{f'_{s1}}{f_{cu}} \tag{7.1–14}$$

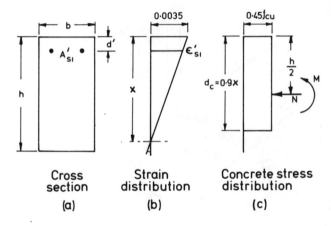

Fig. 7.1–5

$$\beta_{s1} = \frac{M(A'_{s1})}{f_{cu}bh^2} = \left(\frac{1}{2} - \frac{d'}{h}\right)\alpha_{s1} \tag{7.1–15}$$

where the compressive stress f'_{s1} in the reinforcement corresponds to the strain ε'_{s1} in Fig. 7.1–5(b):

$$\frac{\varepsilon'_{s1}}{0.0035} = \frac{x - d'}{x} = \frac{x/h - d'/h}{x/h} \tag{7.1–16}$$

For any assigned value of x/h, therefore, α_{s1} and β_{s1} can be computed, since for a given column section the quantities A'_{s1}/bh, d'/h, f_{cu} and the steel properties are known. As shown in Fig. 7.1–6, the effect of A'_{s1} is represented by the vector $\alpha_{s1} + \beta_{s1}$, which is inclined at a constant angle to the α-axis:

$$\text{inclination of vector} = \tan^{-1}\left[\frac{\beta_{s1}\ (\text{eqn 7.1–15})}{\alpha_{s1}\ (\text{eqn 7.1–14})}\right]$$

$$= \tan^{-1}\left[\frac{1}{2} - \frac{d'}{h}\right] \tag{7.1–17}$$

In Fig. 7.1–6, therefore, ALHJK is the interaction curve for the column section with reinforcement at both faces. As before, E_s has been taken as 200 kN/mm^2 and the reinforcement stress/strain curve assumed linear between $f'_{s1} = \pm 0.87f_y$ where f_y is 460 N/mm^2. (See Example 7.1–4 for other stress/strain relations.) For any point on this curve, α and β are, respectively $\alpha_{conc} + \alpha_{s2} + \alpha_{s1}$ and $\beta_{conc} + \beta_{s2} + \beta_{s1}$. In fact, adding eqns (7.1–1)(7.1–5) and (7.1–12), we have

$$N = 0.405f_{cu}bx + f'_{s1}A'_{s1} + f_{s2}A_{s2} \tag{7.1–18}$$

Similarly, adding eqns (7.1–2), (7.1–6) and (7.1–13), we have

$$M = 0.203f_{cu}bx(h - 0.9x) + f'_{s1}A'_{s1}\left[\frac{h}{2} - d'\right] - f_{s2}A_{s2}\left[\frac{h}{2} - d_2\right] \tag{7.1–19}$$

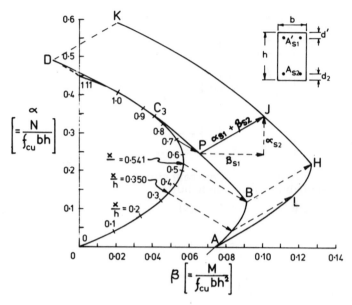

Fig. 7.1–6   Column interaction diagram $(A'_{s1}/bh = 1\%, A_{s2}/bh = 1\%, f_{cu} = 40 \text{ N/mm}^2, f_y = 460 \text{ N/mm}^2, d_2/h = 0.15, d'/h = 0.15)$

where the column reinforcement stresses f'_{s1} and f_{s2} are conventionally taken as positive if they are compressive. These two equations represent the equilibrium conditions for a column section reinforced at both faces.

Example 7.1–1
The details of a column section are: $b = 300$ mm, $h = 400$ mm, $A'_{s1}/bh = 1\%$, $A_{s2}/bh = 1\%$, $d'/h = d_2/h = 0.15$, $f_{cu} = 40 \text{ N/mm}^2$ and $f_y = 460$ N/mm^2 (notation as in Figs 7.1–5 and 7.1–6).

(a) Determine the direct load N and its eccentricity e for which a balanced failure occurs.
(b) Determine N and e for which the steel A'_{s1} reaches $0.87f_y$ in compression simultaneously as the concrete compressive strain reaches 0.0035.
(c) If the column is subjected to an eccentric load whose magnitude is gradually increasing, determine the range of eccentricities for which both the reinforcements A'_{s1} and A_{s2} would have reached $0.87f_y$ when the concrete compressive strain reaches 0.0035.

SOLUTION
(a) From statement (d) following eqn (7.1–11), the balanced condition occurs if $x/h = 0.541$. The (N,M) combination should therefore correspond to the point H in Fig. 7.1–6.
   By measurement,

$$a(\text{H}) = 0.219; \quad \beta(\text{H}) = 0.126$$

Therefore

$$N = \alpha f_{cu} bh = 0.219 \times 40 \times 300 \times 400$$

$$= \underline{1051 \text{ kN}}$$

$$M = \beta f_{cu} bh^2 = 0.216 \times 40 \times 300 \times 400^2$$

$$= \underline{242 \text{ kNm}}$$

$$e = M/N = 230 \text{ mm}$$

(b) Using the strain diagram in Fig. 7.1–5(b) and eqn (7.1–16), the reader should verify that A'_{s1} reaches the design strength $0.87 f_y$ in compression when $x/h = 0.350$. Therefore the (N, M) combination should correspond to the point L in Fig. 7.1–6.

By measurement,

$$\alpha(\text{L}) = 0.142; \qquad \beta(\text{L}) = 0.118$$

Therefore

$$N = 0.142 \times 40 \times 300 \times 400 = \underline{682 \text{ kN}}$$

$$e = M/N = \beta h/\alpha = 0.118 \times 400/0.142 = \underline{332 \text{ mm}}$$

(c) From (a) and (b) above, the range of eccentricities is

$$e \geq \underline{230 \text{ mm}}; \qquad e \leq \underline{332 \text{ mm}}$$

If the load N acts within this range of eccentricities, then at incipient failure the (α, β) values would lie on the portion LH of the curve in Fig. 7.1–6.

Example 7.1–2
The details of a column section are as in Example 7.1–1.

(a) Determine whether the following load–moment combinations are acceptable: $N = 1680$ kN, $M = 115$ kNm; $N = 480$ kN, $M = 250$ kNm.
(b) If the column section is acted on by an eccentric load of magnitude $N = 1680$ kN, determine the eccentricity e that produces incipient failure.
(c) Determine the magnitude N of the eccentric load that will produce incipient failure if the eccentricity is known to be $e = 0.317h$.
(d) For the column section in the state of incipient failure in (c) above, determine the depth d_c of the concrete stress block and the stresses f'_{s1} and f_{s2} in the reinforcement.

SOLUTION
(a) $N = 1680$ kN, $M = 115$ kNm. Therefore

$$\alpha = \frac{N}{f_{cu} bh} = \frac{1680 \times 10^3}{40 \times 300 \times 400} = 0.35$$

$$\beta = \frac{M}{f_{cu} bh^2} = \frac{115 \times 10^6}{40 \times 300 \times 400^2} = 0.06$$

These are the (α, β) values of the point P_1 in Fig. 7.1–7. Since P_1 is inside curve III, these (N,M) values are acceptable. Similarly, the reader may verify that point P_2 represents $N = 480$ kN and $M = 250$ kNm. Since P_2 lies outside curve III, these (N,M) values are not acceptable.

(b) $N = 1680$ kN. Therefore

$$\alpha = \frac{1680 \times 10^3}{40 \times 300 \times 400} = 0.35$$

In Fig. 7.1–7, the horizontal line $\alpha = 0.35$ is drawn to intersect curve III at point P_3, whose β value is, by measurement, 0.10. Therefore

$$e = \frac{M}{N} = \frac{\beta h}{\alpha} = \frac{0.10 \times 400}{0.35} = \underline{114 \text{ mm}}$$

(c) $e = 0.317h$. In Fig. 7.1–7, the straight line $e/h = 0.317$ is drawn to intersect curve III at P_4. (*Note:* From (b) above, $e/h = 0.317$ is the line whose slope $\beta/\alpha = 0.317$.) By measurement: $\alpha(P_4) = 0.33$. Therefore

$$N = \alpha f_{cu} bh = 0.33 \times 40 \times 300 \times 400 \text{ N} = \underline{1584 \text{ kN}}$$

(d) In Fig. 7.1–7, construct the vector $\mathbf{P_5P_4}//\mathrm{BH}$ and the vector $\mathbf{P_6P_5}//\mathrm{C_2B}$. (*Note:* $\mathrm{C_2B}$ and BH are the vectors for a balanced failure in which A'_{s1} yields in compression and A_{s2} yields in tension. See Example 7.1–1 and Fig. 7.1–6.) By interpolation, $x/h(P_6) = 0.67$. Therefore

$$x = (0.67)(400) = 268 \text{ mm}$$

$$d_c = 0.9x = \underline{241 \text{ mm}}$$

Also

$$f'_{s1} = \left[\frac{\text{length } P_5P_4}{\text{length BH}} \right](0.87f_y) = (1)(0.87)(460)$$

$$= \underline{400 \text{ N/mm}^2} \quad \text{(compressive)}$$

$$f_{s2} = \left[\frac{\text{length } P_6P_5}{\text{length } C_2B} \right](0.87f_y) = (0.43)(0.87)(460)$$

$$= \underline{172 \text{ N/mm}^2} \quad \text{(tensile)}$$

Therefore, within the accuracy of the measurements on Fig. 7.1–7, d_c = 241 mm, the stress f'_{s1} in the steel A'_{s1} is 400 N/mm² in compression, and that in A_{s2} is 172 N/mm² in tension.

Example 7.1–3 (contributed by **Dr C. T. Morley** of Cambridge University) A rectangular column section, of dimensions $b = 150$ mm and $h = 500$ mm, is reinforced with 1% of reinforcement near one short side (only), at $d_2/h = 0.15$. The concrete strength f_{cu} is 40 N/mm² and the reinforcement strength $f_y = 460$ N/mm². A load N acts at an eccentricity e from the column centre of 150 mm in a direction parallel to the long side, away from

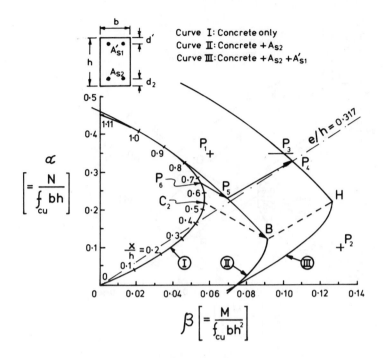

Fig. 7.1–7 **Column interaction diagram** $(A'_{s1}/bh = 1\%, A_{s2}/bh = 1\%, f_{cu} = 40 \text{ N/mm}^2, f_y = 460 \text{ N/mm}^2, d_2/h = 0.15, d'/h = 0.15)$

the reinforcement. Using a column interaction diagram, determine:

(a) the magnitude of the load, the neutral axis depth x, the depth d_c of the stress block and the stress f_{s2} in the steel at incipient failure;
(b) the additional percentage of steel required at the compression face at $d'/h = 0.15$ to increase the ultimate load to 900 kN;
(c) with this additional steel, the failure load if the eccentricity were increased to 200 mm; and
(d) the answer for the magnitude of the load in part (a) above if f_{cu} had been 30 N/mm^2 and f_y 410 N/mm^2.

SOLUTION

(a) Referring to Fig. 7.1–8, curve II is relevant. The line $e/h = 0.3$ (i.e. 150 mm/500 mm) intersects curve II at Q_1. By measurement, $\alpha(Q_1) = 0.23$. Therefore

$$N = 0.23 f_{cu} bh = (0.23)(40)(150)(500)(10^{-3}) = \underline{690 \text{ kN}}$$

Draw vector $\mathbf{Q_0 Q_1}//\mathbf{C_2 B}$ to intersect curve I at Q_0.

$\dfrac{x}{h}$ at $Q_0 = 0.68$ by interpolation

$$x = (0.68)(500) = \underline{340 \text{ mm}}$$

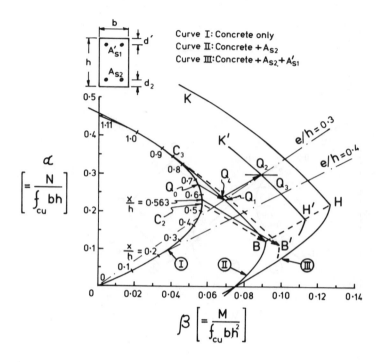

Fig. 7.1–8 Column interaction diagram $(A'_{s1}/bh = 1\%, A_{s2}/bh = 1\%, f_{cu} = 40 \text{ N/mm}^2, f_y = 460 \text{ N/mm}^2, d_2/h = 0.15, d'/h = 0.15)$

$$d_c = 0.9x = \underline{306 \text{ mm}}$$

$$f_{s2} = \left[\frac{\text{length } Q_0Q_1}{\text{length } C_2B}\right](0.87f_y)$$

$$= (0.44)(0.87)(460) = \underline{177 \text{ N/mm}^2}$$

(b) $N = 900$ kN. Therefore

$$\alpha = \frac{900 \times 10^3}{40 \times 150 \times 500} = 0.30$$

The horizontal line $\alpha = 0.30$ intersects the line $e/h = 0.3$ at Q_2. Through Q_2, draw $K'Q_2H'$ parallel to curve III. An area A'_{s1} of compression steel is required so that $K'Q_2H'$ becomes the interaction curve. The required area of A'_{s1} is determined by drawing a vector $Q_4Q_2//BH$ and comparing lengths:

$$\frac{A'_{s1}}{bh} = \frac{\text{length } Q_4Q_2}{\text{length } BH} \times 1\%$$

(since Fig. 7.1–8 has been drawn for $A'_{s1}/bh = 1\%$)

$$= \underline{0.6\%}$$

(c) $e = 200$ mm. Therefore

$e/h = 200/500 = 0.4$

Draw the line $e/h = 0.4$ (*Note: $e/h = M/Nh = \beta/\alpha$*) to intersect the interaction curve $K'Q_2H'$ at Q_3. By measurement, $\alpha(Q_3) = 0.245$. Therefore

$N = (0.245)(40)(150)(500)(10^{-3}) = \underline{735\ kN}$

(d) $f_{cu} = 30$ N/mm^2 and $f_y = 410$ N/mm^2. Equations (7.1–7) to (7.1–11) show that:
   (1)   The direction of the vector $\alpha_{s2} + \beta_{s2}$ is independent of f_{cu} and f_y.
   (2)   Corresponding to each x/h value, the length of the vector is proportional to f_{s2}/f_{cu}, i.e. $E_s\varepsilon_{s2}/f_{cu}$.
   (3)   The maximum length of that vector (which occurs at $f_{s2} = 0.87f_y$) is proportional to f_y/f_{cu}.
   (4)   The x/h value at which f_{s2} reaches $0.87f_y$ depends on f_y (see eqn 7.1–9).

Using the above information, the reader should verify that if f_{cu} were 30 N/mm^2 and f_y 410 N/mm^2, then the steel stress f_{s2} would reach the design stress of $0.87f_y$ ($= 357$ N/mm^2 in tension) at $x/h = 0.563$ and at this value of f_{s2}, $\alpha_{s2} = -0.119$ and $\beta_{s2} = 0.042$. Hence show that the curve C_3B', shown as the dotted line in Fig. 7.1–8, is the new interaction curve.

By measurement, the α value at the intersection of this dotted curve C_3B' with the line $e/h = 0.3$ is 0.24. Therefore

$N = 0.24f_{cu}bh = 0.24(30)(150)(500)(10^{-3}) = \underline{540\ kN}$

Example 7.1–4
In constructing the column interaction diagram in Fig. 7.1–6, the stress/strain relation was taken as that of BS 8110, as reproduced here in Fig. 3.2–1(b), i.e. stresses varied linearly with strains for stresses between $\pm 0.87f_y$.

(a)   How will the construction be affected if the steel stress/strain curve has an arbitrarily specified shape?
(b)   If, as stipulated in BS 8110 : Clause 3.8.2.4, an eccentricity not less than $e_{min} = 0.05h$ must be used in design, how would the column interaction diagram be affected?

SOLUTION
(a)   *Arbitrarily specified stress/strain curve.* For each value assigned to x/h the steel strain ε_{s2} is computed as usual from eqn (7.1–9). The stress f_{s2}, instead of being $E_s\varepsilon_{s2}$, is now read off the stress/strain curve. The α_{s2} and β_{s2} values are then computed as usual from eqns (7.1–7) and (7.1–8), and curve ABCD in Fig. 7.1–6 constructed. Depending on the shape of the given stress/strain curve, there may not be a distinct point (B) corresponding to the balanced failure. Similar remarks apply to the construction of curve ALHJK in Fig. 7.1–6.
(b)   *Effect of $e_{min} = 0.05h$.* The effect of stipulating a minimum

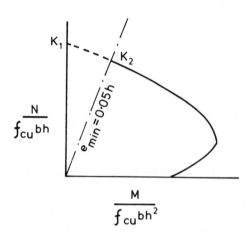

Fig. 7.1–9

eccentricity in design is to exclude part of the curve, K_1K_2, as shown in Fig. 7.1–9.

Example 7.1–5

In the paragraph preceding eqn (7.1–1), it was pointed out that, using BS 8110's simplified stress block, the depth $d_c = 0.9x$ for $d_c \leq h$ (i.e. for $x/h \leq 1/0.9$). For cases where $x/h > 1/0.9$ (i.e. >1.111) explain:

(a) how the (α, β) values are to be determined; and
(b) how the use of the column interaction diagram in Fig. 7.1–6 is affected in design.

SOLUTION
When the bending moment M is small compared with the axial load N, the neutral axis may fall sufficiently far outside the column section for x/h to exceed 1.111. When this happens, d_c is no longer equal to $0.9x$; instead $d_c/h = 1$ for all values of $x/h \geq 1.111$.

(a) **(α, β) values for $x/h > 1.111$.** α_{conc} and β_{conc} are calculated from eqns (7.1–3) and (7.1–4) but with x/h taken as 1.111 (that is, even when x/h actually exceeds 1.111). α_{s2} and β_{s2} are calculated as usual from eqns (7.1–7) to (7.1–9), none of which require any modification. Similarly, α_{s1} and β_{s1} are calculated as usual from eqns (7.1–14) to (7.1–16).
(b) **Column interaction diagram in Fig. 7.1–6.** Using the information in (a) above, the reader should verify that:
  (1)   for $x/h = 1.111$, $(\alpha, \beta) = (0.591, 0.020)$;
  (2)   for $x/h = 1.5$, $(\alpha, \beta) = (0.626, 0.008)$;
  (3)   for $x/h \geq 1.983$, $(\alpha, \beta) = (0.650, 0)$.
That is, as x/h increases beyond 1.111, the point (α, β) progressively approaches the α axis, until at $x/h = 1.983$ when both A'_{s1} and A_{s2} then reach the design strength $0.87f_y$ in compression. For $x/h =$

1.983, (α, β) falls on the α axis and remains there for all larger values of x/h. Note that the points (α, β) for $x/h > 1.111$ in fact fall within the region excluded by the BS 8110 line $e_{min}/h = 0.05$, referred to in Example 7.1–4(b). Therefore, cases where $x/h > 1.111$ are of little practical interest.

Example 7.1–6
In this section, our emphasis has been on the fundamental principles, the aim being to obtain an insight into the properties of column interaction diagrams. The worked examples have been designed with the above aim in mind. (Indeed, Example 7.1–3 has been adapted from a past examination paper of Cambridge University.) However, if our aim is merely the straightforward construction of column interaction diagrams for use as routine design charts, then we can use a very simple procedure, such as that explained below.

SOLUTION
For the typical section in Fig. 7.1–10, it is immediately seen that the direct compression is

$$N = 0.45f_{cu}bd_c + f'_{s1}A'_{s1} - f_{s2}A_{s2} \tag{7.1–20}$$

Taking moments about the mid-depth of the section,

$$M = 0.45f_{cu}bd_c\left(\frac{h}{2} - \frac{d_c}{2}\right) + f'_{s1}A'_{s1}\left(\frac{h}{2} - d'\right)$$
$$+ f_{s2}A_{s2}\left(\frac{h}{2} - d_2\right) \tag{7.1–21}$$

where $d_c = 0.9x$ and compressive stresses in the reinforcement are taken as positive.
Next proceed as follows.

Step 1
Assume a value for x/h.

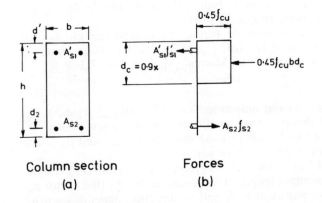

Column section
(a)

Forces
(b)

Fig. 7.1–10

Step 2
Calculate $d_c/h = (= 0.9 \, x/h \leq 1)$.

Step 3
Calculate ε'_{s1} and ε_{s2} using the strain diagram (see eqns 7.1–9 and 7.1–16 for example).
Calculate f'_{s1} and f_{s2}:

$$f'_{s1} = E_s \varepsilon'_{s1} \leq 0.87 f_y$$
$$f_{s2} = E_s \varepsilon_{s2} \leq 0.87 f_y$$

Step 4
Using the d_c/h value for Step 2 and the steel stresses in Step 3, calculate (N, M) from eqns (7.1–20) and (7.1–21).
This gives us one point on the column interaction curve for the given section.

Step 5
Assume another value for x/h and repeat Steps 2, 3 and 4 to obtain another point (N, M) on the interaction curve.

Step 6
Repeat Step 5 to complete the column interaction curve.

Step 7
Draw the BS 8110 limit $e_{min}/h = 0.05$ referred to in Example 7.1–4(b).

Example 7.1–7 (contributed by **Dr H. H. A. Wong** of the University of Newcastle upon Tyne)
BS 8110: Part 3 gives a comprehensive set of column design charts, one of which is reproduced as Fig. 7.3–1. It can be seen that the BS 8110 charts refer to symmetrically reinforced columns so that (1) $A'_{s1} = A_{s2} = A_{sc}/2$ and (2) $d' = d_2 = h - d$. It is also seen from Fig. 7.3–1 that each of BS 8110's interaction curves refers to a particular set of $(\varrho; f_{cu}; f_y; d/h)$ values, where ϱ is the steel ratio A_{sc}/bh.

(a) Explain whether it is possible to reduce the four variables $(\varrho; f_{cu}; f_y; d/h)$ to three, namely: $(\varrho/f_{cu}; f_y; d/h)$.
(b) Explain whether it is possible further to reduce the number of variables to two, namely: $(\varrho f_y/f_{cu}; d/h)$, as has been done in Appendix D of the I.Struct.E. Manual [5]. (See Fig. 7.3–2.)

SOLUTION
(a) Yes, it is possible to use only three variables: $(\varrho/f_{cu}; f_y; d/h)$. To understand why this is so, we need only consider a few points:
  (1) Equations (7.1–3) and (7.1–4) show that α_{conc} and β_{conc} are independent of any of these variables.
  (2) Equations (7.1–7) and (7.1–11) (eqns 7.1–14 and 7.1–17) show that the maximum length of the vector $\alpha_{s2} + \beta_{s2}$ (the vector $\alpha_{s1} + \beta_{s1}$) is proportional to f_y and its direction depends on d/h.
  (3) Equations (7.1–7) and (7.1–8) (eqns 7.1–14 and 7.1–15) show

that α_{s2} and β_{s2} (α_{s1} and β_{s1}) do not depend on the individual values of ϱ and f_{cu} but only on the ratio ϱ/f_{cu}.

(b) No, it is theoretically incorrect to use only the two variables ($\varrho f_{cu}/f_y$; and d/h). The reason is as follows. It is true that as long as the quantity $\varrho f_y/f_{cu}$ is kept constant, then neither the magnitude nor the direction of the vector $\alpha_{s2} + \beta_{s2}$ can be affected by the value of f_y itself—see eqns (7.1–7) and (7.1–8). However, a close examination of eqn (7.1–9) shows that the x/h value at which the vector $\alpha_{s2} + \beta_{s2}$ reaches its maximum length does depend on the value of f_y. Therefore, in Fig. 7.1–7, the position of the point C_2 will vary with f_y (even though the quantity $\varrho f_y/f_{cu}$ is kept constant). Hence curve II, and for the same reason curve III, will both vary with f_y itself.

Comments on Part (b)

We have seen that it is theoretically impossible to use the two variables ($\varrho f_y/f_{cu}$; d/h) to define a column interaction curve. However, provided we restrict the f_y values to those of the common grades of reinforcing steel, then the column curves do become practically independent of f_y as $\varrho f_y/f_{cu}$ is kept constant. Indeed, for a given value of $\varrho f_y/f_{cu}$, the curves for $f_y = 250$ N/mm^2 and $f_y = 460$ N/mm^2 are quite close to each other (see Fig. 7.1–11).

Of course it is important that the student should know the basic principles (see Section 7.1), so that he understands the limitations of the column curves in the I.Struct.E. Manual [5]. For example, Fig. 7.1–11 shows the interaction curves for a rectangular column section for which $d/h = 0.85$ and $\varrho f_y/f_{cu} = 0.80$. Specifically, the details for the curves in Fig. 7.1–11 are:

(1) $f_y = 250$ N/mm^2 (BS 4449 mild steel bars), $f_{cu} = 25$ N/mm^2, $\varrho = 8.0\%$ so that $\varrho f_y/f_{cu} = 0.80$.

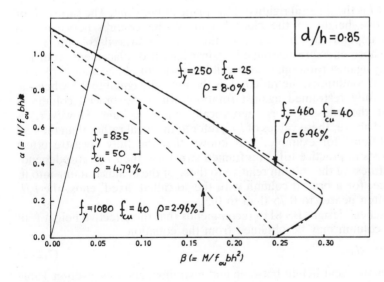

Fig. 7.1–11

(2) $f_y = 460$ N/mm^2 (BS 4449/BS 4461 high yield bars), $f_{cu} = 40$ N/mm^2, $\varrho = 6.96\%$ so that $\varrho f_y/f_{cu} = 0.80$.
(3) $f_y = 835$ N/mm^2 (BS 4486 high tensile bars), $f_{cu} = 50$ N/mm^2, $\varrho = 4.79\%$ so that $\varrho f_y/f_{cu} = 0.80$.
(4) $f_y = 1080$ N/mm^2 (BS 4486 high tensile bars), $f_{cu} = 40$ N/mm^2, $\varrho = 2.96\%$ so that $\varrho f_y/f_{cu} = 0.80$.

Comparing Fig. 7.1–11 with the column curve for $\varrho f_y/f_{cu} = 0.80$ in Fig. 7.3–2, it can be seen that the I.Struct.E. Manual's curve is essentially the average of the top two curves in Fig. 7.1–11 (i.e. for $f_y = 250$ and 460 N/mm^2). It is thus clear that the I.Struct.E. Manual's column charts are intended for the normal range of reinforcement bars. Applying the manual's design charts to very high strength bars (e.g. BS 4486 high tensile prestressing bars) can lead to structural collapse. Of course, BS 4486's high tensile bars, for use in prestressed concrete, are designated by their nominal proof stress and tensile strength. Hence the use of f_y for these bars is not really proper, though it is hoped that Fig. 7.1–11 helps to give a clear message to the student.

7.2   Effective column height (BS 8110)

The behaviour of a column is much dependent on its **effective height** [6, 7], or **effective length**, l_e where l_e is related to the elastic critical buckling load [6] N_{cr} by

$$N_{cr} = \frac{\pi^2 EI}{l_e^2} \qquad (7.2\text{–}1)$$

in which EI is the flexural rigidity in the plane of buckling. The ratio l_e/l_0 of the effective height to the clear height of the column between end restraints depends on the end conditions. For a **braced column**, i.e. a column whose ends are restrained against lateral displacements (but not necessarily against rotations), we recall from structural mechanics [6] that, under ideal conditions, the ratio l_e/l_0 is theoretically equal to 0.5 when the ends are fully restrained against rotation; when the braced column is pinned at the ends, l_e/l_0 is theoretically equal to one. However, as mentioned above, these values of l_e/l_0 refer to the ideal rather than the real end conditions; for example, it is known that the fully encastré effect hardly exists in practice [6]. In a framed structure, the l_e/l_0 ratio depends on the stiffness of the column relative to those of the beams framing into it. In practice, for a braced column with the so-called 'fixed' ends, the l_e/l_0 ratio is often nearer to 0.75 than to 0.5.

For practical design, BS 8110 recommends that the effective height l_e of a braced column may be obtained from the equation

$$l_e = \beta l_0 \qquad (7.2\text{–}2)$$

where l_0 is the clear height between end restraints, and β is given in Table 7.2–1.

Table 7.2–1 Values of $\beta(= l_e/l_0)$ for braced columns (BS 8110 : Clause 3.8.1.6)

	End condition at bottom[a]		
End condition at top[a]	(1) 'Fixed'	(2) Partially restrained	(3) 'Pinned'
(1) 'Fixed'	0.75	0.80	0.90
(2) Partially restrained	0.80	0.85	0.95
(3) 'Pinned'	0.90	0.95	1.00

[a] End conditions (1), (2) and (3) are defined as follows:

(1) The column is connected monolithically to beams on each side which are at least as deep as the overall depth of the column in the plane considered. If the column is connected to a foundation structure, this must be designed to carry moment.
(2) The column is connected to beams or slabs on each side which are shallower than the overall depth of the column in the plane considered. (The I.Struct.E. Manual [5] is more specific; it states that such beams and slabs should not be less than half the column depth, in order to satisfy this condition.)
(3) The column is connected to members which provide only nominal restraint against rotation.

7.3 Eccentrically loaded short columns (BS 8110)

7.3(a) BS 8110 design procedure

BS 8110: Clause 3.8.1.3 defines a braced **short column** as a braced column for which both the ratios l_{ex}/h and l_{ey}/b are less than 15, where l_{ex} is the effective height in respect of the major axis (Table 7.2–1); l_{ey} is the effective height in respect of the minor axis (Table 7.2–1); h is the depth in respect of the major axis; and b is the width of the column.

The design of axially loaded short columns was explained in Section 3.4. For eccentrically loaded short columns, the strength may be assessed by the following equations:

$$N = 0.405f_{cu}bx + f'_{s1}A_{s1} + f_{s2}A_{s2} \qquad (7.3–1)$$

$$M = 0.203f_{cu}bx(h - 0.9x) + f'_{s1}A'_{s1}\left(\frac{h}{2} - d'\right) - f_{s2}A_{s2}\left(\frac{h}{2} - d_2\right)$$
$$(7.3–2)$$

where N = the design ultimate axial load;
M = the design maximum moment under ultimate condition; M should not be taken as less than Ne_{min} where e_{min} is the **design minimum eccentricity**; BS 8110: Clause 3.8.2.4 stipulates that the design minimum eccentricity e_{min} should be taken as $0.05h$ or 20 mm whichever is less (see Example 7.1–4(b));
A'_{s1} = the area of compression reinforcement in the more highly compressed face (see Fig. 7.1–6);
A_{s2} = the area of reinforcement in the other face (see Fig. 7.1–6)

which may be either in compression, inactive or in tension, depending on the value of the x/h ratio (see Fig. 7.1–6);

b = the width of the column section;

d_2 = the depth from the surface to the reinforcement A_{s2} (see Fig. 7.1–6);

d' = the depth to the reinforcement A'_{s1} (see Fig. 7.1–6);

f_{cu} = the characteristic strength of the concrete;

f'_{s1} = the stress in the reinforcement A'_{s1} (compressive is +ve; tensile is −ve) which should not be taken as more than $0.87f_y$;

f_{s2} = the stress in the reinforcement A_{s2} (compressive is +ve) which should not be taken as more than $0.87f_y$;

f_y = the characteristic strength of the reinforcement;

h = the depth of the column section in the plane of bending; and

x = the neutral axis depth measured from the more highly compressed face.

Equations (7.3–1) and (7.3–2) were derived earlier as eqns (7.1–18) and (7.1–19). Using these equations in design involves a process of trial and error to see if a value of x can be found such that the right-hand sides of the equations are not less than the left-hand sides. For convenience in design, BS 8110: Part 3 provides a comprehensive set of **design column interaction diagrams** for use as design charts, for both rectangular and circular columns. One such **design chart** is reproduced in Fig. 7.3–1 and it is worth noting that:

(a)   BS 8110's design charts are based on the rectangular–parabolic stress block (Fig. 4.4–3) and not on the simplified stress block (Fig. 4.4–5).

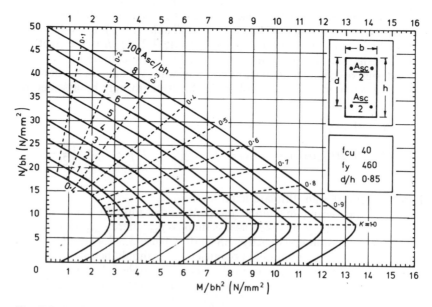

Fig. 7.3–1   Column design chart—BS 8110

(b) In using BS 8110's design charts, the bending moment M should not be taken as less than Ne_{min}, where N is the design ultimate axial load and e_{min} is the design minimum eccentricity, as explained in the definition of M under eqns (7.3–1) and (7.3–2) (see also Example 7.1–4(b)).

Figure 7.3–2 shows a column design chart reproduced from the I.Struct.E. Manual, [5]. Comparing Fig. 7.3–2 with Fig. 7.3–1 we see that:

(1) The variable $\varrho f_y/f_{cu}$ is used for the interaction curves in Fig. 7.3–2, where $\varrho = A_{sc}/bh$. Instead, Fig. 7.3–1 uses the single variable ϱ.
(2) In Fig. 7.3–2 the variables for the coordinate axes are N/bhf_{cu} and M/bh^2f_{cu}. Instead, Fig. 7.3–1 uses N/bh and M/bh^2.

The I.Struct.E. Manual's presentation of the design charts (Fig. 7.3–2) has the advantage that a relatively small number of charts will cover the designer's needs. Thus, Fig. 7.3–2 can be used for different values of f_{cu} and f_y; Fig. 7.3–1, on the other hand, can be used only for a particular set of f_{cu} and f_y values. *However, there is a danger that students who do not have a sound understanding of the properties of column interaction diagrams (see Section 7.1) may not realize that the design charts in the I.Struct.E. Manual apply only to the usual range of f_y and f_{cu} values.* It could be dangerous, for example, to use such design charts with BS 4486 high-tensile prestressing bars ($f_y = 835$ N/mm² and 1080 N/mm²). This point is discussed in some detail in Example 7.1–7(b)—see also Problem 7.8 at the end of this chapter.

Design charts such as Fig. 7.3–1 (and Fig. 7.3–2) cover bending about one axis, i.e. uniaxial bending. For **biaxial bending**, BS 8110 : Clause 3.8.4.5 gives the following recommendation for symmetrically reinforced rectangular columns. Suppose the column is subjected to (N, M_x, M_y).

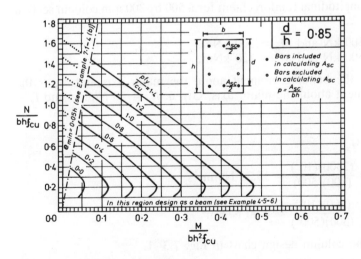

Fig. 7.3–2   Column design chart—I.Struct.E. Manual [5]

Then it can be designed either for (N, M'_x) or for (N, M'_y) depending on which of the following two conditions is valid:

(a)   For $M_x/h' \geq M_y/b'$:

$$M'_x = M_x + \beta\frac{h'}{b'}M_y \qquad\qquad (7.3\text{--}3)$$

(b)   For $M_x/h' < M_y/b'$:

$$M'_y = M_y + \beta\frac{b'}{h'}M_x \qquad\qquad (7.3\text{--}4)$$

where h' = the effective section dimension in a direction perpendicular to the major axis x–x, as shown in Fig. 7.3–3;

b' = the effective section dimension perpendicular to the minor axis y–y, as shown in Fig. 7.3–3;

β = a coefficient to be obtained from Table 7.3–1;

M_x (M_y) should not be taken as less than Ne_{min} about the x–x $(y$–$y)$ axis.

Table 7.3–1   Values of β for eqns (7.3–3) and (7.3–4) (BS 8110 : Clause 3.8.4.5)

$\dfrac{N}{f_{cu}bh}$	0	0.1	0.2	0.3	0.4	0.5	≥ 0.6
β	1.00	0.88	0.77	0.65	0.53	0.42	0.30

Examples 7.3–1 and 7.3–2 below illustrate the design procedures for uniaxial and biaxial bending. Section 7.3(b), which follows these examples, explains the behaviour of columns under biaxial bending.

Example 7.3–1 (Uniaxial bending)
Design the longitudinal reinforcement for a 500 by 300 mm column section if:

(a) $N = 2300$ kN and $M_x = 300$ kNm,
(b) $N = 2300$ kN and $M_y = 120$ kNm,

where M_x is the bending moment about the major axis and M_y is the bending moment about the minor axis. Given: $f_{cu} = 40$ N/mm^2 and $f_y = 460$ N/mm^2.

SOLUTION
(a) N and M_x.

$$\frac{N}{bh} = \frac{(2300)(10^3)}{(300)(500)} = 15.33 \text{ N/mm}^2$$

$$\frac{M_x}{bh^2} = \frac{(300)(10^6)}{(300)(500^2)} = 4.00 \text{ N/mm}^2$$

From the column design chart in Fig. 7.3–1,

$$A_{sc} = 2.3\%$$

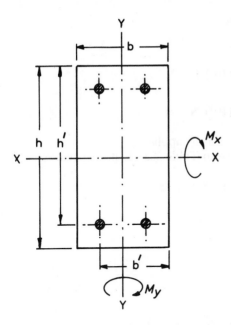

Fig. 7.3–3

2.3% of 300 mm by 500 mm = 3450 mm^2

Use 6 size 32 bars (4825 mm^2)

(b) N **and** M_y**.** (In this case, $b = 500$ mm and $h = 300$ mm.)

$$\frac{N}{bh} = \frac{(2300)(10^3)}{(500)(300)} = 15.33 \text{ N/mm}^2$$

$$\frac{M_y}{bh^2} = \frac{(120)(10^6)}{(500)(300^2)} = 2.67 \text{ N/mm}^2$$

From Fig. 7.3–1,

$A_{sc} = 1.3\%$

1.3% of 300 mm by 500 mm = 1950 mm^2

Use 4 size 25 bars (1963 mm^2)

Example 7.3–2 (Biaxial bending)

Design the longitudinal reinforcement for the 500 by 300 mm column section in Example 7.3–1 if $N = 2300$ kN, $M_x = 300$ kNm and $M_y = 120$ kNm. Given: $f_{cu} = 40$ N/mm^2 and $f_y = 460$ N/mm^2.

SOLUTION

Step 1   Calculate the enhanced moment (M'_x or M'_y)

Assuming, say, 50 mm cover to centres of bars, then (see Fig. 7.3–3)

$$h' = 500 - 50 = 450 \text{ mm}$$

$$b' = 300 - 50 = 250 \text{ mm}$$

$$\frac{M_x}{h'} = \frac{(300)(10^6)}{(450)} = (667)(10^3) \text{ N}$$

$$\frac{M_y}{b'} = \frac{(120)(10^6)}{(250)} = (480)(10^3) \text{ N}$$

Hence $M_x/h' > M_y/b'$ and eqn (7.3–3) applies.

β in the equation is obtained from Table 7.3–1, noting that

$$\frac{N}{f_{cu}bh} = \frac{(2300)(10^3)}{(40)(300)(500)} = 0.38$$

Hence, from Table 7.3–1,

$$\beta = 0.55 \text{ by interpolation}$$

From eqn (7.3–3),

$$M_x' = M_x + \beta \frac{h'}{b'} M_y$$

$$= 300 + (0.55)\frac{(450)}{(250)}(120) = 419 \text{ kNm}$$

Step 2   Design the reinforcement for (N, M_x')

$$N = 2300 \text{ kN} \quad \text{(given)}$$

$$M_x' = 419 \text{ kNm} \quad \text{(from Step 1)}$$

$$\frac{N}{bh} = \frac{(2300)(10^3)}{(300)(500)} = 15.3 \text{ N/mm}^2$$

$$\frac{M}{bh^2} = \frac{(419)(10^6)}{(300)(500^2)} = 5.59 \text{ N/mm}^2$$

From column design chart (Fig. 7.3–1),

$$A_{sc}/bh = 3.5\%$$

$$3.5\% \text{ of } 300 \text{ mm by } 500 \text{ mm} = 5250 \text{ mm}$$

Provide reinforcement as shown in Fig. 7.3–4.

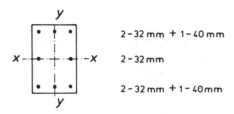

2 - 32 mm + 1 - 40 mm

2 - 32 mm

2 - 32 mm + 1 - 40mm

Fig. 7.3–4

Note that for bending about x–x axis, the two size 32 bars on that axis are not considered. Hence the effective area A_{sc} is

4 size 32: 3216 mm^2
2 size 40: 2513 mm^2
Effective A_{sc} = 5729 mm^2

7.3(b) Biaxial bending—the technical background

The solution to Example 7.3–2 is based on eqns (7.3–3) and (7.3–4), the technical background to which may be briefly explained as follows. Consider the column section in Fig. 7.3–5(a), subjected to a load N acting at eccentricities e_x and e_y. Thus,

$$M_x = Ne_y$$
$$M_y = Ne_x$$

Let

$$M = (M_x^2 + M_y^2)^{\frac{1}{2}} \tag{7.3–5}$$

If the eccentricity $e_x = 0$, then the angle α in Fig. 7.3–5(a) is zero and the column is acted on by N and M_x only; in this case, the interaction curve is A_1A_2 in Fig. 7.3–5(b). For a given load N, the magnitude of M_{ux} that causes collapse can be read off this curve (which incidentally is similar to curve III in Fig. 7.1–7). On the other hand, if $e_y = 0$ then the angle α is 90° and the column is acted on by N and M_y only; in this case the interaction curve is B_1B_2 in Fig. 7.3–5(b). Again, for a given load N, the magnitude of M_{uy} that causes collapse can be obtained from the curve (which again is similar to curve III in Fig. 7.1–7).

For the actual condition shown in Fig. 7.3–5(a), where both e_x and e_y are non-zero, then the angle α has an intermediate value between 0 and 90°. The research by Bresler and others [8, 9] has shown that for such biaxial bending, the $N \sim M$ interaction curve is C_1C_2 in Fig. 7.3–5(b). As the angle α varies from 0 to 90°, the curve C_1C_2 generates an **interaction surface** which has a shape rather like that of a quarter of a pear. For a given value of N, if we cut the interaction surface by a horizontal plane at a level ON above the base, the intersection curve gives the relation between α and the magnitude M_u of the moment M (see Fig. 7.3–5(c)) that produces collapse. That is, for a given value of N, this horizontal section shows the $M_x \sim M_y$ interaction curve for that value of N. With reference to Fig. 7.3–5(c), points falling on the shaded area represent safe combinations of M_x and M_y for that value of N; points outside represent unacceptable combinations.

The shape of the boundary curve in Fig. 7.3–5(c), i.e. the interaction curve of M_x and M_y, depends on the ratio of the actual load N to the ultimate axial-load capacity N_{uz} of the column section. If N/N_{uz} is small, then the curve may be idealized as a straight line (Fig. 7.3–5(d)):

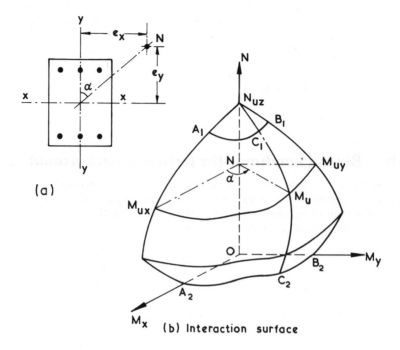

(a)

(b) Interaction surface

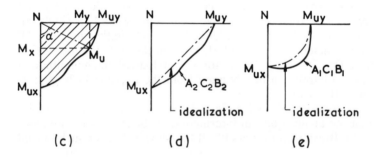

(c)   (d)   (e)

Fig. 7.3–5

$$\frac{M_x}{M_{ux}} + \frac{M_y}{M_{uy}} = 1$$

If N/N_{uz} approaches unity, idealization as an ellipse is more appropriate (Fig. 7.3–5(e)):

$$\left[\frac{M_x}{M_{ux}}\right]^2 + \left[\frac{M_y}{M_{uy}}\right]^2 = 1$$

For a general value of N/N_{uz}, we have

$$\left[\frac{M_x}{M_{ux}}\right]^{a_n} + \left[\frac{M_y}{M_{uy}}\right]^{a_n} = 1 \qquad (7.3\text{–}6)$$

where M_x = the moment about the major axis and M_y that about the minor axis, due to ultimate loads;

M_{ux} = the maximum moment capacity assuming ultimate axial load N and bending about the major axis only;

M_{uy} = the maximum capacity assuming ultimate axial load N and bending about the minor axis only;

a_n = a numerical coefficient the value of which depends on the ratio N/N_{uz} (N_{uz} being the ultimate axial-load capacity in the absence of moments). Before the publication of BS 8110, a_n in British design practice [10, 11] was obtained from Table 7.3–2 in which N_{uz} is defined by eqn (7.3–7):

$$N_{uz} = 0.45f_{cu}A_c + 0.75f_yA_s' \qquad (7.3–7)$$

where A_s' is the total area of all the longitudinal reinforcement in the column section and A_c is the nominal area of the section.

Table 7.3–2 Relationship of N/N_{uz} to a_n

N/N_{uz}	≤ 0.2	0.4	0.6	≥ 0.8
a_n	1.0	1.33	1.67	2.0

BS 8110's design procedure, using eqns (7.3–3) and (7.3–4) and Table 7.3–1, has been formulated essentially as a more convenient method of obtaining comparable results to those given by eqn (7.3–6).

We have restricted our discussions to rectangular columns. The combined axial load and biaxial bending of the general section is discussed in References 12–14.

7.4 Additional moment due to slender column effect

BS 8110 defines a **slender column** as one for which the effective height/depth ratio in respect of either the major axis or the minor axis is not less than 15. (See also the definition of a short column in Section 7.3.) The necessity to distinguish between short and slender columns arises from the fact that the strengths of slender columns are significantly reduced by the transverse deflections [15–20].

Consider a column acted on by an axial load N and end moments M_i, as shown in Fig. 7.4–1(a). The combined effect of N and M_i produces a **transverse deflection** e_{add} at mid-height. For an elastic, homogeneous column, the magnitude of e_{add} can be readily calculated using elementary techniques in structural analysis, and interested readers should do Problems 7.6 and 7.7 at the end of this chapter. Of course, reinforced concrete columns in real structures are neither elastic nor homogeneous, particularly when the ultimate limit state is approached. We shall see later how this difficulty is overcome in design. For the time being, it is sufficient to note, with reference to Fig. 7.4–1(a), that the **total moment** at the critical section is

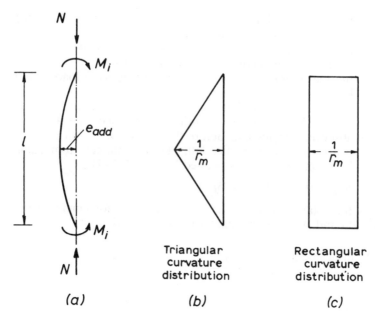

Fig. 7.4–1

$$M_t = M_i + Ne_{add} \tag{7.4–1}$$

where Ne_{add} is the **additional moment** due to the slender column effect, and the deflection e_{add} is often referred to as the **additional eccentricity**—to distinguish it from the initial eccentricity $e = M_i/N$. Suppose the applied load N is progressively increased until failure occurs; it is instructive to superimpose the $N \sim M_t$ curve on the column interaction curve, as shown in Fig. 7.4–2.

There are three cases to consider:

(a) If the column is short, the deflection e_{add} is small and hence the additional moment Ne_{add} is negligible compared with the initial moment M_i. As the load N is increased, the total moment M_t remains sensibly constant at the initial value M_i—see line AB in Fig. 7.4–2. Failure occurs when the (N, M_t) combination, as represented by the point B, falls on the column interaction curve. This type of failure is called a **material failure**; at every stage of the loading from A until B, the column is stable.

(b) If the column is slender, the deflection e_{add} is no longer small and the additional moment Ne_{add} (eqn 7.4–1) becomes significant compared with M_i; according to current design thinking, the additional moment should be considered in design if the effective height/depth ratio is not less than 15. For a slender column, $M_t = M_i + Ne_{add}$—see line AC in Fig. 7.4–2. Failure occurs when the (N, M_t) combination as represented by the point C falls on the column interaction curve. At every stage of the loading from A until C, the column is stable; the

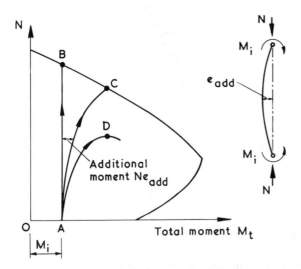

Fig. 7.4–2   Slenderness effect on column capacity—axial-load and end moments

final collapse at point C is again due to **material failure**. Note that as a result of the slenderness effect,

$$N(\text{point C}) < N(\text{point B})$$

i.e. the slenderness effect reduces the load-carrying capacity.

(c) If the column is very slender, the peak load may be reached before a material failure occurs—see line AD in Fig. 7.4–2. In this case, if the applied load N is not reduced at point D, the column will quickly collapse; such a failure is called **instability failure**.

Figure 7.4–2 refers to a column under an axial load N and constant end moments M_i. For a column under an eccentric load N, the respective $N \sim M_t$ relations up to collapse are as shown in Fig. 7.4–3.

OB: material failure of short column
OC: material failure of slender column
OD: instability failure of (very) slender column

Reinforced concrete columns in practical structures are rarely slender enough for instability failure to occur; therefore what the designer needs is a method which is sufficiently accurate for material failures and conservative (though not necessarily accurate) for instability failures [16].

For a column subjected to combined axial load and bending, it was pointed out earlier that the additional eccentricity e_{add} can be calculated easily (see Problems 7.6 and 7.7) provided the column is elastic and homogeneous. Near the ultimate condition, a reinforced concrete column is not elastic, is subject to creep and to cracking if tension occurs on the convex side of the column. Thus a conventional elastic analysis is not directly applicable (nor is a conventional plastic analysis). It turns out that, for design purpose, it is better to work from curvatures. Returning to Fig.

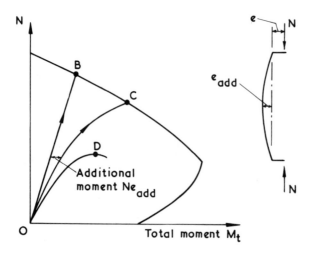

Fig. 7.4–3   Slenderness effect on column capacity—eccentric load

7.4–1(a), the additional eccentricity e_{add} depends on the curvature $1/r$ of the column and on the distribution of this curvature—see curvature–area theorem in Section 5.5. Cranston [16] of the Cement and Concrete Association has shown that, for a reinforced concrete column at the ultimate limit state, the curvature at the critical section could be assumed to depend only on the depth of the column section and the effective height/depth ratio:

$$\frac{1}{r_m} = \frac{1}{175h}\left[1 - 0.0035\frac{l_e}{h}\right] \tag{7.4–2}$$

where l_e is the effective height of the column and h is its depth in the appropriate plane of bending. The curvature distribution is not known, but may reasonably be assumed to lie somewhere between the triangular distribution which implies only one critical section (Fig. 7.4–1(b)), and the rectangular distribution which implies an infinite number of critical sections (Fig. 7.4–1(c)). Fortunately, the transverse deflection is insensitive to the curvature distribution; in fact Example 5.5–1 shows specifically that

$$e_{add} \text{ (triangular distribution)} = \frac{l^2}{12}\left(\frac{1}{r_m}\right)$$

$$e_{add} \text{ (rectangular distribution)} = \frac{l^2}{8}\left(\frac{1}{r_m}\right)$$

$$e_{add} \text{ (parabolic distribution)} = \frac{l^2}{9.6}\left(\frac{1}{r_m}\right)$$

$$e_{add} \text{ (sinusoidal distribution)} = \frac{l^2}{\pi^2}\left(\frac{1}{r_m}\right)$$

It is therefore reasonable to assume, as Cranston has suggested [16], that

$$e_{\text{add}} = \frac{1}{10}l_e^2\left(\frac{1}{r_m}\right) \tag{7.4-3}$$

Note that Cranston has replaced the actual column height l by the effective height l_e (see Table 7.2–1) to allow for the effects of the various end conditions in practice. Combining eqns (7.4–3) and (7.4–2),

$$e_{\text{add}} = \frac{h}{1750}\left[\frac{l_e}{h}\right]^2\left[1 - 0.0035\frac{l_e}{h}\right] \tag{7.4-4}$$

Before relating eqn (7.4–4) to current design practice, we shall carefully define the following symbols:

l_e = the effective column height *in the plane of bending*;
h = the depth of the column section measured *in the plane of bending*;
b = the width of the column section (i.e. the dimension of the column section perpendicular to h);
b' = the smaller dimension of the column section (i.e. $b' = h$ if $h < b$; $b' = b$ if $h > b$).

BS 8110 approximates eqn (7.4–4) to

$$a_u = \frac{1}{2000}\left[\frac{l_e}{b'}\right]^2 h$$

$$= \beta_a h \tag{7.4-5}$$

where a_u is BS 8110's symbol for the additional eccentricity e_{add}. a_u may optionally be multiplied by a reduction factor K, which will be explained in Section 7.5: Step 5. That is, a_u may be taken as $\beta_a K h$; the additional moment $N a_u$ induced by the lateral deflection is therefore

$$M_{\text{add}} = N\beta_a K h \tag{7.4-6}$$

where β_a is as defined in eqn (7.4–5). Therefore, with reference to eqn (7.4–1), the total moment is

$$M_t = M_i + M_{\text{add}} \tag{7.4-7}$$

For the simple case illustrated in Fig. 7.4–1, the initial moment M_i to be used in eqn (7.4–7) is simply the initial moment at an end of the column. For the general case where the end moments are M_1 and M_2 (M_2 being the larger), BS 8110 : Clause 3.8.3.2 recommends that M_i be taken as

$$M_i = 0.6M_2 + 0.4M_1$$

for symmetrical bending (Fig. 7.4–4(a)) and

$$M_i = 0.6M_2 - 0.4M_1 \quad \text{but} \quad \not< 0.4M_2$$

for asymmetrical bending, i.e. bending in double curvature (Fig. 7.4–4(b)). These two equations may be combined into one, namely:

$$M_i = 0.4M_1 + 0.6M_2 \geq 0.4M_2 \tag{7.4-8}$$

where M_1 = the smaller initial end moment (taken as negative if the column is bent in double curvature); and

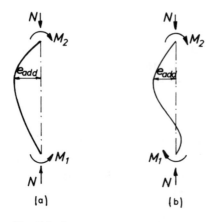

Fig. 7.4–4

M_2 = the larger initial end moment, which is always taken as positive.

Referring to eqn (7.4–7), the total moment M_t might work out to be less than M_2; this could occur when M_{add} is small, particularly for a column bent in double curvature. In such a case, the critical section is clearly where M_2 acts; hence the condition has to be imposed that

$$M_t \geq M_2 \qquad\qquad (7.4\text{–}9(a))$$

BS 8110 : Clause 3.8.3.2 imposes the further condition that

$$M_t \geq M_1 + \frac{1}{2}M_{add} \qquad\qquad (7.4\text{–}9(b))$$

Earlier, it was stated under eqn (7.3–2) that a column should be in any case designed for a moment of at least Ne_{min}. Hence we have the further condition that

$$M_t \geq Ne_{min} \qquad\qquad (7.4\text{–}9(c))$$

where e_{min} is the design minimum eccentricity, to be taken as $0.05h$ or 20 mm, whichever is less.

Summary
For a slender column, the total moment M_t to be used in design is calculated from eqn (7.4–7), subject to the further conditions imposed by eqns (7.4–9(a), (b), (c)).

7.5   Slender columns (BS 8110)

The technical background to BS 8110's design recommendations was explained in Section 7.4; BS 8110's design procedure for braced slender columns may be summarized as follows.

Step 1   The effective height l_e
    Using eqn (7.2–2) and Table 7.2–1, determine the effective column

heights for bending about the major axis and about the minor axis. If the effective height/depth ratio in respect of either axis is 15 or more, the column should be designed as a slender column.

Step 2 Bending about a minor axis

For bending about a minor axis, the column should be designed for its ultimate axial load N together with the total moment M_t which is the greatest of those given by the equations below:

$$M_t = M_i + M_{add} \tag{7.5-1}$$

$$M_t = M_2 \tag{7.5-2}$$

$$M_t = M_1 + \frac{1}{2}M_{add} \tag{7.5-3}$$

$$M_t = Ne_{min} \tag{7.5-4}$$

where
M_i = the initial moment as given by eqn (7.4–8);
M_{add} = the additional moment as given by eqn (7.4–6);
M_2 and M_1 = the initial end moments as defined under eqn (7.4–8);
e_{min} = the design minimum eccentricity, to be taken as $0.05h$ or 20 mm, whichever is the smaller, where h is the dimension in the plane of bending.

Comments on Step 2

(a) Equations (7.5–1) to (7.5–4) were derived in Section 7.4.
(b) M_{add} is calculated from eqn (7.4–6):

$$M_{add} = N\beta_a Kh$$

where β_a may be calculated from eqn (7.4–5) or obtained from Table 7.5–1; K is an optional reduction factor ($K \le 1$) which may be taken as unity in the first instance (see Step 5 below); h is the depth of the column in the plane of bending (i.e. h is the smaller dimension for minor axis bending, and is the larger dimension for major axis bending).

Table 7.5–1 Values of β_a (see eqns 7.4–5 and 7.4–6)

l_e/b' [a]	12	15	20	25	30	35	40	45	50	55	60
β_a	0.07	0.11	0.20	0.31	0.45	0.61	0.80	1.01	1.25	1.51	1.80

[a] Here, b' is the smaller dimension of the column section, as defined immediately above eqn (7.4–5).

Step 3 Bending about a major axis

For bending about a major axis there are two conditions to consider:

Condition 1: The ratio of the length of the longer side to that of the shorter side is less than 3.

Condition 2: $l_e/h \le 20$.

Case 1: If both Conditions (1) and (2) are satisfied, then the column is designed for N and M_t, using the same procedure as in Step 2. Of course, M_1 and M_2 are now the initial end moments about the major axis.

Case 2: If either Condition (1) or Condition (2) is not satisfied, then the column is designed as biaxially bent, with zero initial moment about the minor axis—see procedure in Step 4 below.

Step 4   Biaxial bending
(a)   Calculate M_{ty}. The total moment M_{ty} about the minor axis is calculated as the greatest value given by eqns (7.5–1) to (7.5–4), exactly as in Step 2.
(b)   Calculate M_{tx}. The total moment M_{tx} about the major axis is calculated as the greatest value given by eqns (7.5–1) to (7.5–4) as in Step 2, except that:
    (1) M_1 and M_2 are now the initial end moments about the major axis.
    (2) M_{add} is calculated, as usual, from eqn (7.4–6): $M_{add} = N\beta_a Kh$. However, in obtaining β_a either from eqn (7.4–5) or Table 7.5–1, b' is in this particular case to be taken as h, i.e. the dimension in the plane of bending.
(c)   Design for N, M_{tx}, M_{ty}. The column is then designed for the ultimate axial load N plus the two total moments M_{tx} and M_{ty} using eqns (7.3–3) and (7.3–4) in Section 7.3(a).

Step 5   The reduction factor K
The additional moment M_{add} in eqns (7.5–1) and (7.5–3) is given by eqn (7.4–6):

$$M_{add} = N\beta_a Kh   (K \leq 1)$$

where K is an optional **reduction factor** which can be read off Fig. 7.3–1 (and similar design charts in BS 8110 : Part 3) and is defined by

$$K = \frac{N_{uz} - N}{N_{uz} - N_{bal}} \leq 1   (7.5\text{–}5)$$

where N = the ultimate axial load;
    N_{bal} = the axial load corresponding to the balanced condition of maximum compressive strain in the concrete of 0.0035 occurring simultaneously with a maximum tensile strain in the reinforcement equal to the design yield strain, i.e. $0.87f_y/E_s$. For $f_y = 460$ N/mm^2, the design yield strain is 0.002.
    N_{uz} = the capacity of the column section under 'pure' axial load as given by eqn (7.5–6):

$$N_{uz} = 0.45f_{cu}A_c + 0.87f_y A_{sc}   (7.5\text{–}6)$$

Comments on Step 5
(a)   The load N_{bal} is that corresponding to the kinks in the column interaction diagrams (e.g. Fig. 7.3–1). Of course, the kinks in Fig. 7.3–1 correspond to the point H in Fig. 7.1–6.
(b)   Figure 7.3–1 shows, as one would expect from eqn (7.5–5), that

$K = 1$ when $N = N_{bal}$ and progressively diminishes as N increases.

(c) In fact, N_{bal} may readily be calculated from eqn (7.1–18):

$$N_{bal} = 0.405f_{cu}bx + f'_{s1}A'_{s1} + f_{s2}A_{s2}$$

where the steel stresses f'_{s1} and f_{s2} and the neutral axis depth x are to be determined from the strain diagram for the balanced condition. Consider a typical case where $f_y = 460$ N/mm^2. At the balanced condition, $\varepsilon_{s2} = 0.002$ and $f_{s2} = -0.87f_y$ (sign convention for column stresses: compressive is positive). Similarly, the reader should verify from the strain diagram (Fig. 7.1–5(b)) that $f'_{s1} = 0.87f_y$ unless d'/h exceeds 0.21 (d' assumed equal to d_2). It is also easy to verify from Fig. 7.1–5(b) that $x = 0.636 \, (h - d_2)$. Thus

$$N_{bal} = 0.258f_{cu}b(h - d_2) + 0.87f_y(A'_{s1} - A_{s2})$$

For a symmetrically reinforced column, $A'_{s1} = A_{s2}$, and we have

$$N_{bal} = 0.258f_{cu}bd \qquad (7.5-7)$$

where $d = h - d_2$. Of course, the derivation of eqn (7.5–7) has been based on the simplified rectangular stress block of Fig. 4.4–5. Using the parabolic–rectangular stress block of Fig. 4.4–4, the expression for N_{bal} can be shown to be that given in BS 8110: Clause 3.8.1.1:

$$N_{bal} = 0.25f_{cu}bd$$

(d) The reason for BS 8110's introducing the reduction factor K (eqn 7.5–5) is related to the fact that eqn (7.4–6) for M_{add} is based on eqn (7.4–5), which is in turn based on eqn (7.4–2) for the curvature at failure. Reference to the strain diagram in Fig. 7.1–3(b) will show that the curvature depends on the strain ε_{s2} of the reinforcement A_{s2} (ε_{cu} being taken always as 0.0035 at failure). The curvature expression in eqn (7.4–2) is in fact intended for the particular balanced condition of $\varepsilon_{cu} = 0.0035$ and $\varepsilon_{s2} = 0.002$ (tensile); this condition is that at the point H in the column interaction diagram in Fig. 7.1–6, where H corresponds to the steel A_{s2} reaching the tensile yield strain. Reference to Fig. 7.1–6 is helpful; as the failure load N increases, the point (α, β) moves up the curve HJK, and it can be seen that the tensile strain in A_{s2} reduces—becoming zero for $x/h = 0.85$ for the particular case illustrated. It can be visualized, therefore, that as the point (α, β) moves up the curve HJK (i.e. as N approaches the axial capacity N_{uz}) the column curvature at failure becomes progressively less, and is theoretically zero for $N = N_{uz}$. And the purpose of eqn (7.5–5) is to enable the designer to take advantage of this phenomenon: as N approaches N_{uz}, the failure curvature, and hence the additional moment Ne_{add}, are reduced and the K factor in eqn (7.5–5) enables a corresponding reduction to be made in the total moment M_t.

Example 7.5–1
Design the longitudinal reinforcement for the braced slender column in Fig. 7.5–1, **for bending about the minor axis**, if $N = 2500$ kN, $M_{1y} = 100$ kNm, $M_{2y} = 120$ kNm, $f_{cu} = 40$ N/mm^2 and $f_y = 460$ N/mm^2.

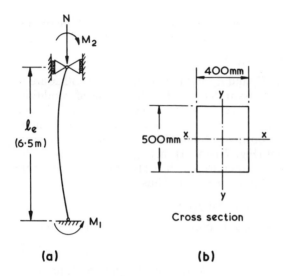

Fig. 7.5–1   Slender column ($l_e/400 = 16.25 > 15$)

SOLUTION

For bending about a minor axis, Step 2 of Section 7.5 is relevant. First calculate M_i and M_{add}. From eqn (7.4–8),

$$M_i = 0.4M_1 + 0.6M_2$$
$$= (0.4)(100) + (0.6)(120)$$
$$= 112 \text{ kNm}$$

From eqn (7.4–6),

$$M_{add} = N\beta_a Kh$$

From Table 7.5–1, for $l_e/b' = 6500/400 = 16.25$, $\beta_a = 0.13$. The optional reduction factor K is taken as unity (see Comment (c) at the end of the solution). h is the depth in the plane of bending, i.e. 400 mm. Hence

$$M_{add} = (2500)(0.13)(1)(0.4) = 130 \text{ kNm}$$

From eqn (7.5–1),

$$M_t = M_i + M_{add}$$
$$= 112 + 130 = 242 \text{ kNm}$$

From eqn (7.5–2),

$$M_t = M_2 = 120 \text{ kNm}$$

From eqn (7.5–3),

$$M_t = M_1 + \frac{1}{2}M_{add} = 100 + \frac{130}{2} = 165 \text{ kNm}$$

From eqn (7.5–4),

$$M_t = Ne_{min} \quad \text{where } e_{min} = (0.05)(400) = 20 \text{ mm}$$

$$= (2500)(0.02)$$

$$= 50 \text{ kNm}$$

Taking the greatest value given by eqns (7.5–1) to (7.5–4),

$$M_t = 242 \text{ kNm}$$

$$\frac{N}{bh} = \frac{(2500)(10^3)}{(500)(400)} = 12.5 \text{ N/mm}^2$$

$$\frac{M_t}{bh^2} = \frac{(242)(10^6)}{(500)(400^2)} = 3.03 \text{ Nmm}^2$$

From the design chart in Fig. 7.3–1 (see Comment (b) at the end of this example),

$$A_{sc} = 1.0\% \ (2000 \text{ mm}^2)$$

Provide 4 size 32 bars (3216 mm²)

Comments
(a) In Section 7.5 (and Section 7.4) the symbol h represents the depth in the plane of bending. Hence for bending about a minor axis, as in this example, $h = 400$ mm (not 500 mm).
(b) The design chart in Fig. 7.3–1 is for a particular value of the d/h ratio (i.e. $d/h = 0.85$). However, we have for convenience assumed that the chart applies to the column in this example—and also to those in Examples 7.5–2 and 7.5–3. BS 8110: Part 3, of course, contains a comprehensive set of design charts.
(c) BS 8110 allows the additional moment M_{add} to be optionally reduced by the K factor of eqn (7.5–5).

Example 7.5–2
Design the longitudinal reinforcement for the braced slender column in Fig. 7.5–1, for **bending about the major axis**, if $N = 2500$ kN, $M_{1x} = 200$ kNm, $M_{2x} = 250$ kNm, $f_{cu} = 40$ N/mm² and $f_y = 460$ N/mm².

SOLUTION
For bending about the major axis, Step 3 of Section 7.5 is relevant:

$$\text{Condition (i): } \frac{\text{longer side}}{\text{shorter side}} = \frac{500}{400} < 3$$

$$\text{Condition (ii): } \frac{l_e}{h} = \frac{6500}{500} < 20$$

Hence **Case 1** of Step 3 of Section 7.5 applies.
  From eqn (7.4–8),

$$M_i = 0.4M_1 + 0.6M_2$$

$$= (0.4)(200) + (0.6)(250)$$

$$= 230 \text{ kNm}$$

From eqn (7.4–6),

$$M_{add} = N\beta_a Kh$$

Since β_a (see Table 7.5–1) depends only on l_e/b', it has the same value as in Example 7.5–1, namely $\beta_a = 0.13$. For the time being assume $K = 1$ (with adjustment later on). Also, $h = 500$ mm. Hence

$$M_{add} = (2500)(0.13)(1) \, (0.5) = 162.5 \text{ kNm}$$

From eqn (7.5–1),

$$M_t = M_i + M_{add}$$
$$= 230 + 162.5 = 392.5 \text{ kNm}$$

From eqn (7.5–2),

$$M_t = M_2, = 250 \text{ kNm}$$

From eqn (7.5–3),

$$M_t = M_1 + \frac{1}{2}M_{add}$$
$$= 200 + (0.5)(162.5) = 281.3 \text{ kNm}$$

From Eqn (7.5–4),

$$M_t = Ne_{min} \quad \text{where } e_{min} = 0.05h$$
$$= 25 \text{ mm} > 20 \text{ mm}$$
$$\text{(hence take } e_{min} = 20)$$
$$M_t = (2500)(20)(10^{-3}) = 50 \text{ kNm}$$

Taking the greatest of the values given by eqns (7.5–1) to (7.5–4),

$$M_t = 392.5 \text{ kNm}$$

$$\frac{N}{bh} = \frac{(2500)(10^3)}{(400)(500)} = 12.5 \text{ N/mm}^2$$

$$\frac{M}{bh^2} = \frac{(392.5)(10^6)}{(400)(500^2)} = 3.93 \text{ N/mm}^2$$

From the design chart in Fig. 7.3–1,

$$K = 0.75 \text{ approximately}$$

Therefore, the revised M_t is

$$M_t = 230 + (0.75)(162.5) = 352 \text{ kNm}$$

$$\frac{N}{bh} = 12.5 \text{ N/mm}^2 \text{ as before}$$

$$\frac{M}{bh^2} = \frac{(352)(10^6)}{(400)(500^2)} = 3.52 \text{ N/mm}^2$$

From the design chart in Fig. 7.3–1,

$A_{sc} = 1.4\%$ of bh $(= 2800 \text{ mm}^2)$

Provide 4 size 32 bars (3216 mm²)

Example 7.5–3

Design the longitudinal reinforcement for the braced slender column in Fig. 7.5–1, **for biaxial bending**, if $N = 2500$ kN, $M_{1x} = 200$ kNm, $M_{2x} = 250$ kNm, $M_{1y} = 100$ kNm, $M_{2y} = 120$ kNm, $f_{cu} = 40$ N/mm² and $f_y = 460$ N/mm².

SOLUTION

For biaxial bending, Step 4 of Section 7.5 is relevant.

Step 4(a)

Calculate M_{ty}.

$$M_{ty} = 242 \text{ kNm} \quad \text{(as } M_t \text{ in Example 7.5–1)}$$

Step 4(b)

Calculate M_{tx}.

$$M_{ix} = 230 \text{ kNm} \quad \text{(as } M_i \text{ in Example 7.5–2)}$$

From eqn (7.4–6),

$$M_{add} = N\beta_a Kh$$

where $h = 500$ mm and K is for simplicity being taken as unity (see Comments at the end of the solution). To obtain β_a in Table 7.5–1, we note that, in evaluating l_e/b' for biaxial bending, b' is to be taken as the depth in the plane of bending under consideration, i.e. in this case, $b' = h = 500$ mm. Hence $l_e/b' = 6500/500 = 13$, so that $\beta_a = 0.085$ from Table 7.5–1. Hence

$$M_{add} = (2500)(0.085)(1)(500)(10^{-3}) = 106 \text{ kNm}$$

From eqn (7.5–1)

$$M_{tx} = M_{ix} + M_{add}$$
$$= 230 + 106 = 336 \text{ kNm}$$

By inspection, eqns (7.5–2) to (7.5–4) are not critical. Hence M_{tx} is taken as 336 kNm.

Step 4(c)

We shall first check which of eqns (7.3–3) and (7.3–4) is applicable. With the symbols h' and b' as defined under those equations, we have

$$\frac{M_{tx}}{h'} = \frac{(336)(10^6)}{(0.85, \text{ say})(500)} = (791)(10^3)$$

$$\frac{M_{ty}}{b'} = \frac{(242)(10^6)}{(0.85, \text{ say})(400)} = (712)(10^3)$$

Hence $M_{tx}/h' > M_{ty}/b'$ and eqn (7.3–3) is applicable:

$$M'_{tx} = M_{tx} + \beta \frac{h'}{b'} M_{ty}$$

To find β, we note that $N/(f_{cu}bh)$ works out to be 0.313; hence from Table 7.3–1, $\beta = 0.64$.

$$M'_{tx} = 336 + (0.64)\frac{(0.85 \text{ of } 500)}{(0.85 \text{ of } 400)}(242)$$

$$= 529.6 \text{ kNm}$$

$$\frac{N}{bh} = \frac{(2500)(10^3)}{(400)(500)} = 12.5 \text{ N/mm}^2$$

$$\frac{M'_{tx}}{bh^2} = \frac{(529.6)(10^6)}{(400)(500^2)} = 5.3 \text{ N/mm}^2$$

From design chart (Fig. 7.3–1),

$$A_{sc} = 2.8\% \ bh \ (5600 \text{ mm}^2)$$

Provide 4 size 40 plus 4 size 32 bars

The arrangement is shown in Fig. 7.5–2.

Note that for bending about the x–x axis, bars lying on or near that axis do not count (similarly, for bending about the y–y axis, bars lying on or near the y–y axis do not count). Hence the effective A_{sc} is in this case given by the 4 size 40 bars and the 2 size 32 bars on the y–y axis:

$$\text{Effective } A_{sc} \text{ provided} = 5026 + 1608$$

$$= 6634 \text{ mm}^2$$

Comments

We could have used the optional reduction factor K (eqn 7.5–5) to reduce M_{ty} and M_{tx}. Example 7.5–2 illustrates the use of K.

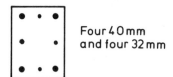

Four 40 mm
and four 32 mm

Fig. 7.5–2

7.6   Design details (BS 8110)

See Section 3.5. Note also the **slenderless limit** for columns: BS 8110: Clause 3.8.1.7 specifies that the clear distance between end restraints should not exceed 60 times the minimum thickness of the column. (For unbraced columns, the limit is further reduced; see BS 8110: Clause 3.8.1.8.)

7.7   Design and detailing—illustrative example

See Example 11.5–3.

7.8 Computer programs

(in collaboration with **Dr H. H. A. Wong**, University of Newcastle upon Tyne)

The FORTRAN programs for this chapter are listed in Section 12.7. See also Section 12.1 for 'Notes on the computer programs'.

Problems

7.1 Is the moment-carrying capacity M of a column section increased by an increase in the axial load N?

Ans. Yes and no. See Fig. 7.1–7, which shows that (a) in the region of tension failure, i.e. when α is less than that of point H, the answer is 'yes' and (b) in the region of compression failure, i.e. when α is greater than that of point H, the answer is 'no'.

7.2 Is the load-carrying capacity N of a column section increased by an increase in the bending moment M?

Ans. No. (Study Fig. 7.1–7 and think carefully.)

7.3 In Section 7.1, the principles of interaction diagrams were discussed with reference to a rectangular column section. Explain how these principles can be applied to the construction of an interaction diagram for a **circular column**.

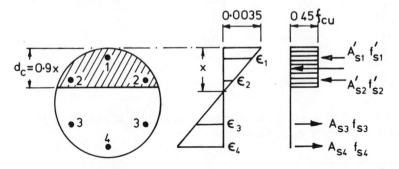

Problem 7.3

Ans. (The numerical calculations tend to be tedious, but no new principles are involved.) For each assumed value of x, determine the steel strains from compatibility and read off the corresponding steel stresses from the stress/strain curve. Then determine N and M from equilibrium consideration. Repeat the process for various x values and complete the $N \sim M$ curve.

7.4 BS 8110: Clause 3.8.4.5 gives the following equations for the design of columns under biaxial bending:

(a)   For $M_x/h' \geq M_y/b'$:

$$M'_x = M_x + \beta \frac{h'}{b'} M_y$$

(b)   For $M_x/h' < M_y/b'$:

$$M'_y = M_y + \beta \frac{b'}{h'} M_x$$

State the meanings of the symbols and briefly explain the technical background to these equations.

Ans.   See Section 7.3(b): Biaxial bending—the technical background.

7.5   Work out fully the solution to part (d) of Example 7.1–3.

7.6   A pin-ended elastic, homogeneous column of uniform flexural stiffness *EI* has an initial crookedness defined by $e_0 = e_1 \sin \pi z/l$.

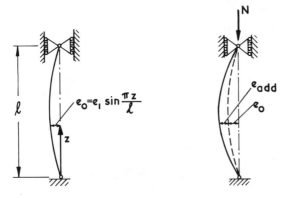

Problem 7.6

When a load *N* is applied the transverse deflection is increased, by an amount e_{add}, to $e_0 + e_{add}$. Show that

$$e_{add} = \frac{\alpha_E}{1 - \alpha_E}$$

where $\alpha_E = N/N_E$ and $N_E (= \pi^2 EI/l^2)$ is the Euler load of the column. (If necessary, read Section 9.3 of Reference 6.)

Comments

Note that the eccentricity **magnification factor** is

$$\frac{e_0 + e_{add}}{e_0} = \frac{1}{1 - \alpha_E}$$

7.7   Problem 7.6 appears at first sight to be rather restricted because it refers to a particular initial crookedness $e_0 = e_1 \sin \pi z / l$. However, any arbitrary initial crookedness may be represented by a Fourier series:

$$e_0 = e_1 \sin \frac{\pi z}{l} + e_2 \sin \frac{2\pi z}{l} + e_3 \sin \frac{3\pi z}{l} + \cdots$$

Using the same technique that you used for Problem 7.6, show that for this general case the total transverse deflection is

$$e_0 + e_{add} = \frac{e_1}{1 - \alpha_E} \sin \frac{\pi z}{l} + \frac{e_2}{1 - (\alpha_E/2^2)} \sin \frac{2\pi z}{l}$$

$$+ \frac{e_3}{1 - (\alpha_E/3^2)} \sin \frac{3\pi z}{l} + \cdots \frac{e_n}{1 - (\alpha_E/n^2)} \sin \frac{n\pi z}{l} + \cdots$$

Comments on Problem 7.7 (May be omitted during first reading)
(a) The Fourier series solution is of wide application, as illustrated by the following two examples.

First, consider an eccentrically loaded column (Fig. 7.4–3). The eccentricity e may be thought of as an initial crookedness of uniform magnitude ($= e$) along the entire column length, and represented by a Fourier series. The result of Problem 7.7 becomes directly applicable.

Next consider a column subjected to an axial load N and end moments M_i (Fig. 7.4–1(a)). The transverse deflection e_0 due to M_i acting **alone** is of course easily calculated. The total transverse deflection due to N and M_i may then be determined by considering the load N acting (in the absence of M_i) on a column having an initial crookedness e_0. By expressing e_0 as a Fourier series, the result of Problem 7.7 becomes directly applicable:

$$e_0 + e_{add} = \sum \frac{e_n}{1 - (\alpha_E/n^2)} \sin \frac{n\pi z}{l}$$

Note that $(e_0 + e_{add})$ here has the same meaning as e_{add} in Fig. 7.4–1(a).

(b) The Fourier series solution for $e_0 + e_{add}$ tends to converge rapidly. In many practical applications, a good estimate is obtained by just taking the first term:

$$e_0 + e_{add} = \frac{e_1}{1 - \alpha_E} \sin \frac{\pi z}{l}$$

$$= \frac{e_1}{1 - \alpha_E} \text{ at } z = \frac{l}{2}$$

Therefore, referring to Fig. 7.4–1(a), the total moment is, approximately:

$$M_t = M_i + \frac{Ne_1}{1 - \alpha_E}$$

Interested readers are referred to Section 9.4 of Reference 6, which shows that this simple equation forms the basis of the well-known **Perry-Robertson formula** in structural steelwork design:

$$N^2 - N[N_y + N_E(1 + \eta)] + N_E N_y = 0$$

where N is the load at material failure of the column; N_y is the squash load (= area $\times$ yield strength; N_E is the Euler load (= $\pi^2 EI/l^2$); η is a function of the slenderness ratio l/r, and values of 0.003 (l/r) and 0.3 $(l/100r)^2$ have been used for η in practical design.

It is interesting to compare the steelwork designers' Perry-Robertson formula with the reinforced concrete designers' eqn (7.4–6), and observe that both formulae are primarily concerned with material failure and that, in each formula, the load capacity N is regarded as being significantly dependent on the column slenderness.

7.8 BS 8110 : Part 3 gives a comprehensive set of column design charts, one of which is reproduced as Fig. 7.3–1. It can be seen that each of BS 8110's interaction curves refers to a particular set of $(\varrho; f_{cu}; f_y; d/h)$ values, where ϱ is the steel ratio A_{sc}/bh.

The I.Struct.E. Manual [5], on the other hand, uses only two variables $(\varrho f_y/f_{cu}; d/h)$, with the great advantage that only a very small number of charts are then necessary for design. Explain whether the I.Struct.E. Manual [5] is justified in using only two variables $\varrho f_y/f_{cu}$ and d/h.

Ans.   In Section 7.1, the authors use three variables: $(\varrho/f_{cu}; f_y$ and $d/h)$. Theoretically, it is impossible to use only two variables $(\varrho f_y/f_{cu}; d/h)$. For a detailed explanation, see Example 7.1–7.

References

1   ACI Committee 318. *Commentary on Building Code Requirements for Reinforced Concrete* (ACI 318R–83). American Concrete Institute, Detroit, 1983.
2   Nawy, E. G. Columns in uniaxial and biaxial bending. In *Handbook of Structural Concrete*, edited by Kong, F. K., Evans, R. H., Cohen, E. and Roll, F. Pitman, London, and McGraw-Hill, New York, 1983, Chapter 12, Sections 2.5 and 2.6.
3   ACI–ASCE Committee 441. *Reinforced Concrete Columns* (ACI Publication SP–50). American Concrete Institute, Detroit, 1975.
4   Evans, R. H. and Lawson, K. T. Tests on eccentrically loaded columns with square twisted steel reinforcement. *The Structural Engineer*, **35**, No. 9, Sept. 1957, pp. 340–8.
5   I.Struct.E./ICE Joint Committee. *Manual for the Design of Reinforced Concrete Building Structures*. Institution of Structural Engineers, London, 1985.
6   Coates, R. C., Coutie, M. G. and Kong, F. K. *Structural Analysis*, 2nd edn. Van Nostrand Reinhold, Wokingham, 1980, pp. 286–93 and 306.
7   Smith, I. A. Column design to CP110. *Concrete*, **8**, Oct. 1974, pp. 38–40.
8   Bresler, B. Design criteria for reinforced concrete columns under axial load and biaxial bending. *Proc. ACI*, **57**, No. 5, Nov. 1960, pp. 481–90.
9   Furlong, R. W. Ultimate strength of square columns under biaxially eccentric loads. *Proc. ACI*, **57**, No. 9, March 1961, pp. 1129–40.
10   CP 110 : 1972. *Code of Practice for the Structural Use of Concrete*. British Standards Institution, London, 1972.

11   Beeby, A. W. *The Design of Sections for Flexure and Axial Load according to CP 110* (Development Report No. 2). Cement and Concrete Association, Slough, 1978.

12   Kawakami, M., Tokuda, H., Kagaga, M. and Hirata, M. Limit states of cracking and ultimate strength of arbitrary concrete sections under biaxial bending. *Proc. ACI*, **82**, No. 2, March/April 1985, pp. 203–12.

13   Davister, M. D. Analysis of reinforced concrete columns of arbitrary geometry subjected to axial load and biaxial bending—a computer program for exact analysis. *Concrete International*, **8**, No. 7, July 1986, pp. 56–61.

14   Kwan, K. H. and Liauw, T. C. Computer aided design of reinforced concrete members subjected to axial compression and biaxial bending. *The Structural Engineer*, **63B**, No. 2, June 1985, pp. 34–40.

15   MacGregor, J. G., Breen, J. E. and Pfrang, E. O. Design of slender concrete columns. *Proc. ACI*, **67**, No. 1, Jan. 1970, pp. 6–28.

16   Cranston, W. B. *Analysis and Design of Reinforced Concrete Columns*. (Research Report No. 20). Cement and Concrete Association, Slough, 1972.

17   Beal, A. N. The design of slender columns. *Proc. ICE* (Part 2), **81**, Sept. 1986, pp. 397–414.

18   Kong, F. K., Garcia, R. C., Paine, J. M., Wong, H. H. A., Tang, C. W. J. and Chemrouk, M. Strength and stability of slender concrete deep beams. *The Structural Engineer*, **64B**, No. 3, Sept. 1986, pp. 49–56.

19   Kong, F. K., Paine, J. M. and Wong, H. H. A. Computer aided analysis and design of slender concrete columns. *Proceedings* of the First International Conference on Computer Applications in Concrete. Singapore Concrete Institute/Nanyang Technological Institute, Singapore, March 1986, pp. C68–C98.

20   Kong, F. K. and Wong, H. H. A. Buckling failure of slender concrete columns—a computer method. *Proceedings* of the International Conference on Structural Failure. Singapore Concrete Institute/Nanyang Technological Institute, Singapore, March 1987.

Chapter 8
Reinforced concrete slabs and yield-line analysis

Preliminary note: Readers interested only in structural design to BS 8110 may concentrate on the following sections:
(a) *Section 8.1: Flexural strength (BS 8110).*
(b) *Section 8.7: Shear strength (BS 8110).*
(c) *Section 8.8: Design of slabs (BS 8110).*

8.1   Flexural strength of slabs (BS 8110)

For practical purposes, the ultimate moment of resistance [1] of reinforced concrete slabs may be determined by the methods explained in Chapter 4 for beams. The beam design chart in Fig. 4.5–2 may of course be used for slab design, but since practical slabs are almost always under-reinforced, only the initial portion of the lowest curve in the chart is really relevant. Designers generally prefer to use the formulae given by the I.Struct.E. Manual [1]. These formulae are of course the same as those explained in Section 4.6(c) for beams:

$$A_s = \frac{M}{0.87 f_y z} \qquad\qquad (8.1\text{--}1)$$

$$M_u = K' f_{cu} b d^2 \qquad\qquad (8.1\text{--}2)$$

where the lever-arm distance z is obtained from Table 4.6–1 and K' is taken as 0.156. (*Note:* Where there is moment redistribution, use Table 4.7–2 for z and Table 4.7–1 for K'.) Of course, the symbols in eqns (8.1–1) and (8.1–2) have the same meanings as in the corresponding equations in Section 4.6(c), except that A_s, M and M_u here all refer to a width b of the slab (b is normally taken as 1 m).

One-way slabs, which as the name implies, span in one direction, are in principle analysed and designed as beams and present no special problems. The design of **two-way slabs** presents varying degrees of difficulty depending on the boundary conditions. The simpler cases of rectangular slabs may be designed by using the moment coefficients in BS 8110: Clause 3.5.3. For irregular cases, the yield-line theory provides a powerful design tool as we shall see in the following sections of this chapter.

8.2   Yield-line analysis

The **yield-line theory** pioneered by Johansen [2, 3] is an ultimate-load theory for slab design and is based on assumed collapse mechanisms and plastic properties of under-reinforced slabs. The assumed **collapse mechanism** is defined by a pattern of **yield lines**, along which the reinforcement has yielded and the location of which depends on the loading and boundary conditions. For the yield-line theory to be valid, shear failures, bond failures and primary compression failures in flexure must all be prevented. The moment/curvature relationship must resemble that of Fig. 4.9–1, having a long horizontal portion when the yield capacity of the slab is reached; in practice, this restriction presents no difficulties because slabs are usually very much under-reinforced.

Figure 8.2–1 shows some typical **yield-line patterns** for slabs in the collapse state under uniformly distributed loads. A full line represents a **positive yield line** caused by a sagging yield moment, so that the concrete cracks in tension on the bottom face of the slab; a broken line represents a **negative yield line** caused by a hogging yield moment so that tensile cracking occurs on the top face. The **convention for support conditions** is as follows: single hatching represents a simply supported edge, double hatching represents a built-in edge, while a line by itself represents a free edge. The following comments [3] should be noted:

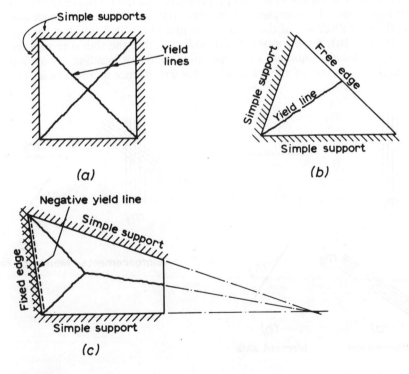

Fig. 8.2–1

(a) The yield lines divide the slab into several regions, called **rigid regions**, which are assumed to remain plane, so that all rotations take place in the yield lines.
(b) Yield lines are straight and they end at a slab boundary.
(c) A yield line between two rigid regions must pass through the intersection of the axes of rotation of the two regions (Fig. 8.2–1(c): the supports form the axes of rotation).
(d) An axis of rotation usually lies along a line of support and passes over columns.

A yield-line pattern indicates how a slab collapses, just as a plastic-hinge mechanism indicates how a framework collapses.

8.3 Johansen's stepped yield criterion

This is the yield criterion in common use. It is based on the assumptions that all reinforcement crossing the yield line has yielded and that all reinforcement stays in its original direction, i.e. there is no 'kinking' [4] of the reinforcement bars in crossing the yield line.

Before describing the stepped yield criterion, we shall first explain the **moment-axis notation**. Suppose a slab is reinforced with a band of reinforcement such that the yield moment of resistance per unit width of slab is m for bending about an axis perpendicular to the reinforcement; this information may be represented in abbreviated form by a line drawn normal to the direction of the reinforcement and labelled m. Thus, in Fig. 8.3–1(a) and (b), the m_1 moment axis indicates that the slab is reinforced to give a resistance moment of m_1 per unit width for bending about that axis; similarly the m_2 moment axis indicates a resistance moment of m_2 per

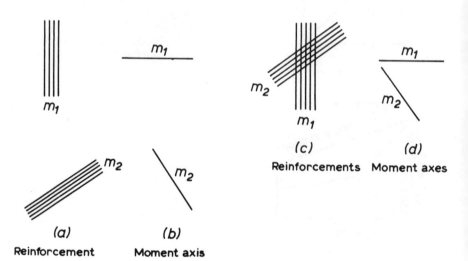

(a)
Reinforcement

(b)
Moment axis

(c)
Reinforcements

(d)
Moment axes

Fig. 8.3–1

unit width. Where two moment axes are shown for a slab, it means that there are two bands of reinforcement. Thus, in Fig. 8.3–1(c) and (d), the presence of both m_1 and m_2 moment axes shows that there is a band of reinforcement m_1 normal to the m_1 axis, and another band m_2 normal to the m_2 axis. Note in particular that m_1 refers to the resistance moment due to the band of reinforcement m_1 taken on its own; that is, neglecting any interaction effect of the other band. Similarly, m_2 refers to that of the reinforcement band labelled m_2 taken on its own. In accordance with our convention, a moment axis shown as a full line indicates positive flexural strength (i.e. reinforcement near bottom face of slab); a moment axis shown as a broken line indicates negative flexural strength (i.e. reinforcement near top face of slab).

Figure 8.3–2 shows part of a yield line in a slab in which the reinforcement bar spacing is s, measured in the direction of the moment axis. The quantity $A_s f_y$ is the yield force in the reinforcement bar, having components of $A_s f_y \cos \phi_1$ and $A_s f_y \sin \phi_1$ respectively in the directions normal and tangential to the yield line, where ϕ_1 is the angle between the yield line and the moment axis. The label m_1 on the moment axis shows that the slab has a yield moment of m_1 per unit width, i.e.

$$m_1 s = A_s f_y z$$

where z is the lever arm.

If we let m_n denote the normal yield moment (or the **normal moment** as it is usually called) per unit length of slab along the yield line, and let m_{ns} denote the **twisting moment** per unit length of slab along the yield line, then

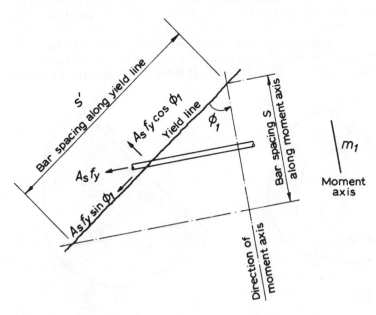

Fig. 8.3–2

$$m_n s' = (A_s f_y \cos \phi_1)z$$

$$= m_1 s \cos \phi_1 \quad \text{(from the expression for } m_1 s \text{ above)}$$

$$m_{ns} s' = (A_s f_y \sin \phi_1)z$$

$$= m_1 s \sin \phi_1$$

Since from Fig. 8.3–2, $s/s' = \cos \phi_1$, these equations may be written as

$$m_n = m_1 \cos^2 \phi_1 \tag{8.3-1}$$

$$m_{ns} = m_1 \sin \phi_1 \cos \phi_1 \tag{8.3-2}$$

Note that in these equations ϕ_1 is the acute angle measured *anticlockwise* from the yield line to the moment axis. In Fig. 8.3–3(a), $\mathbf{m_n}$ and $\mathbf{m_{ns}}$ are represented by the conventional double-headed arrow **moment vectors**, the direction of the arrow being that of the advance of a right-handed screw turned in the same sense as the moment.

The reader should now verify that if the yield line makes an acute angle ϕ_1 measured *clockwise* from yield line to moment axis as in Fig. 8.3–3(b), then eqns (8.3–1) and (8.3–2) become

$$m_n = m_1 \cos^2 \phi_1 \quad \text{(as in eqn 8.3–1)} \tag{8.3-3}$$

$$m_{ns} = -m_1 \sin \phi_1 \cos \phi_1 \tag{8.3-4}$$

We shall now return to **Johansen's stepped yield criterion**, which states that:

(a)   The yield line may be considered to be divided into small steps with sides respectively parallel and perpendicular to the reinforcement. On the sides parallel to the reinforcement (that is, on those sides perpendicular to the moment axis, such as a'b, b'c ... e'f in Fig. 8.3–4) there is neither normal nor twisting moment. On the sides perpendicular to the reinforcement (i.e. parallel to the moment axis, such as aa', bb' ... ff' in Fig. 8.3–4) there is only a normal moment

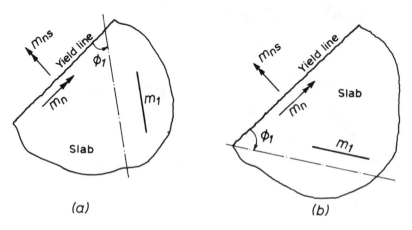

(a)   (b)

Fig. 8.3–3

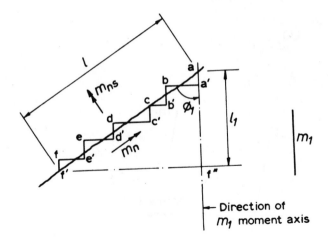

Direction of
m_1 moment axis

Fig. 8.3–4

m_1, where m_1 is the yield moment per unit length for bending about the moment axis. The values of the normal moment m_n and the twisting moment m_{ns} on the yield line are such as to be in equilibrium with the yield moments m_1 on the steps parallel to the moment axis.

(b)   When there are several bands of reinforcement, the moments m_n and m_{ns} on a yield line may be obtained by considering each band of reinforcement in turn, the resultant values of m_n and m_{ns} being the algebraic sum of the values separately calculated for each band taken on its own.

   With reference to Fig. 8.3–4, it follows from Johansen's statement (a) that

$$(\mathbf{m}_n + \mathbf{m}_{ns})l = \mathbf{m}_1(aa' + bb' + cc' + \ldots + ff')$$

i.e.

$$(\mathbf{m}_n + \mathbf{m}_{ns})l = \mathbf{m}_1 l_1 \tag{8.3–5}$$

where l is the length of the yield line being considered, l_1 the projection of l on the direction of the m_1 moment axis, and the sense of $\mathbf{m}_1$ is that of the component of $\mathbf{m}_n$ in the direction of the m_1 moment axis. That is, referring to Fig. 8.3–4, $\mathbf{m}_1$ is directed from f" to a.

   If there are two bands of reinforcement, as indicated by the two moment axes in Fig. 8.3–5, then from Johansen's statement (b)

$$(\mathbf{m}_n + \mathbf{m}_{ns})l = \mathbf{m}_1 l_1 \quad \text{if } m_1 \text{ alone exists}$$

$$(\mathbf{m}_n + \mathbf{m}_{ns})l = \mathbf{m}_2 l_2 \quad \text{if } m_2 \text{ alone exists}$$

Adding,

$$(\mathbf{m}_n + \mathbf{m}_{ns})l = \mathbf{m}_1 l_1 + \mathbf{m}_2 l_2 \tag{8.3–6}$$

where, from eqns (8.3–1) to (8.3–4),

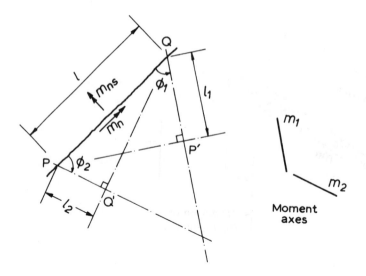

Fig. 8.3–5

$$m_n = m_1 \cos^2 \phi_1 + m_2 \cos^2 \phi_2 \qquad (8.3-7)$$

$$m_{ns} = m_1 \sin \phi_1 \cos \phi_1 - m_2 \sin \phi_2 \cos \phi_2 \qquad (8.3-8)$$

If the two bands of reinforcement are perpendicular to each other, the slab is said to be **orthotropically reinforced**; in this case, the m_1 and m_2 moment axes are mutually perpendicular and the angles ϕ_1 and ϕ_2 are complementary. Hence eqns (8.3–7) and (8.3–8) reduce to

$$m_n = m_1 \cos^2 \phi_1 + m_2 \sin^2 \phi_1 \qquad (8.3-9)$$

$$m_{ns} = (m_1 - m_2) \sin \phi_1 \cos \phi_1 \qquad (8.3-10)$$

If, in an orthotropically reinforced slab, the yield moments m_1 and m_2 are equal (say $m_1 = m_2 = m$), the slab is said to be **isotropically reinforced**; eqns (8.3–9) and (8.3–10) further reduce to

$$m_n = m \qquad (8.3-11)$$

$$m_{ns} = 0 \qquad (8.3-12)$$

That is, in an isotropically reinforced slab the normal moment m_n on a yield line has the same value whatever the orientation of the yield line; and the twisting moment m_{ns} is always zero. It should be noted that the definition of an isotropically reinforced slab is based on the yield moments m_1 and m_2 being equal. If a slab is reinforced in two mutually perpendicular directions with two identical bands of reinforcement laid one on top of the other, the yield moments m_1 and m_2 are not strictly equal—because of the difference in the effective depths for the two bands of reinforcement. However, in yield-line analysis, such a small difference is ignored and the slab is regarded as isotropically reinforced; the yield moment m is calculated on the basis of the average effective depth for the two layers

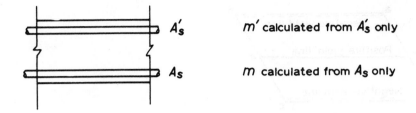

Fig. 8.3–6   **Slab with top and bottom reinforcement**

of reinforcement. Similarly, where a slab is reinforced with both top and bottom reinforcement, as in Fig. 8.3–6, the positive yield moment m is calculated from the bottom steel A_s by ignoring the existence of the top steel A_s'; the negative yield moment m' is calculated from the top steel A_s' by ignoring the existence of the bottom steel A_s. (See Example 8.4–4.)

Example 8.3–1
A skew slab is rigidly supported along two opposite edges and unsupported along the other two edges (Fig. 8.3–7(a)). The top and bottom reinforcement each consists of two unequal bands of bars respectively parallel to the supported and unsupported edges. Determine the normal moments on all the yield lines for the assumed patterns in Fig. 8.3–7(b) and (c).

SOLUTION

For yield-line pattern in Fig. 8.3–7 (b)

$$(m_n)_{ab} = (m_n)_{fe} = \underline{m_2 \cos^2 \alpha} \quad \text{(from Fig. 8.3–8(a) and eqn 8.3–7)}$$

For yield-line pattern in Fig. 8.3–7 (c)

$$(m_n)_{ef} = \underline{m_2 \cos^2 \alpha} \quad \text{(as for } (m_n)_{ab} \text{ above)}$$

$$(m_n)_{eh} = \underline{m_1 \cos^2 \alpha} \quad \text{(from Fig. 8.3–8(b) and eqn 8.3–7)}$$

$$(m_n)_{ej} = m_1 \cos^2 \left(\frac{\pi}{2} - \beta\right) + m_2 \cos^2 (\beta - \alpha)$$

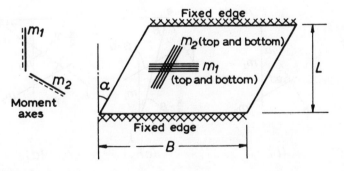

Fig. 8.3–7(a)

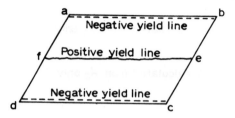

Fig. 8.3–7(b)

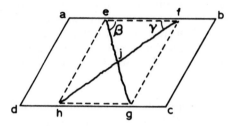

Fig. 8.3–7(c)

(from Fig. 8.3–8(c) and eqn 8.3–7)

$$= m_1 \sin^2 \beta + m_2 \cos^2 (\beta - \alpha)$$

$$(m_\mathrm{n})_\mathrm{fj} = m_1 \cos^2 \left(\frac{\pi}{2} - \gamma\right) + m_2 \cos^2 (\gamma + \alpha)$$

(from Fig. 8.3–8(d) and eqn 8.3–7)

$$= m_1 \sin^2 \gamma + m_2 \cos^2 (\gamma + \alpha)$$

This worked example demonstrates the usefulness of the concept of moment axes.

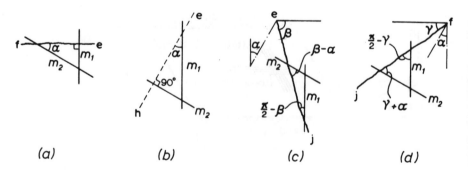

Fig. 8.3–8   Angles between yield line and moment axes

8.4 Energy dissipation in a yield line

Figure 8.4–1(a) shows a positive yield line ab of length l and making angles α_A and α_B respectively with the two axes eg and df, about which the rigid regions A and B rotate through the small angles θ_A and θ_B. Figure 8.4–1(b) shows a cross-section taken perpendicular to the yield line; the angle between the rigid regions is $\theta_{nA} + \theta_{nB}$, where θ_{nA} is the component, in the direction of the yield line, of the actual rotation θ_A of the rigid region A, and θ_{nB} is the component of θ_B, as shown in the vector diagrams in Fig. 8.4–1(a). The sign convention used for the **rotation vectors** needs some explanation. Since the work done on any yield line is always positive and since Fig. 8.4–1(a) shows clearly that positive work results if the sense of rotation of a rigid region is opposite to that of the normal moment m_n acting on that region, the usual practice is to adopt a sign convention for rotation vectors which is opposite to that for moment vectors. In other words, while the right-handed screw rule is used for moment vectors, the left-handed screw rule will be used for rotation vectors—as in Fig. 8.4–1(a).

From Fig. 8.4–1,

energy dissipation per unit length of yield line

$$= m_n(\theta_{nA} + \theta_{nB})$$
$$= m_n\theta_A \cos \alpha_A + m_n\theta_B \cos \alpha_B$$

Therefore

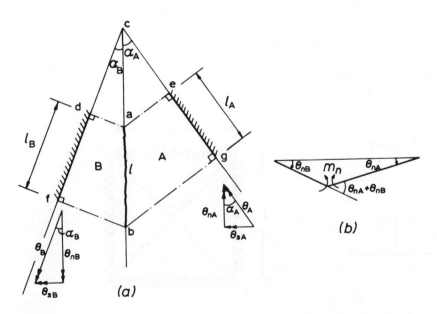

(a)

(b)

Fig. 8.4–1

energy dissipation for the length l

$= m_n \theta_A l \cos \alpha_A + m_n \theta_B l \cos \alpha_B$

$= m_n \theta_A l_A + m_n \theta_B l_B$

where l_A and l_B are respectively the projections of l on the axes of rotation for the rigid regions A and B. That is,

energy dissipation for length l of yield line

$$= m_n \sum \begin{bmatrix} \text{projection of } l \\ \text{on an axis} \end{bmatrix} \begin{bmatrix} \text{rotation of rigid region} \\ \text{about that axis} \end{bmatrix} \qquad \textbf{(8.4–1)}$$

where m_n is the normal moment per unit length on the yield line.

Example 8.4–1

A square slab with built-in edges is isotropically reinforced with top and bottom steel (Fig. 8.4–2). Determine the intensity q of the uniformly distributed load that will cause collapse of the slab.

SOLUTION

A reasonable pattern of positive and negative yield lines is that shown in Fig. 8.4–2. Consider a unit virtual deflection at point e:

external work done by the loading $= \frac{1}{3} q L^2$

(where $\frac{1}{3}$ unit is the average deflection of the load). From eqn (8.4–1),

energy dissipation on the positive yield line ae

$$= m_n \left[(\text{ae}) \cos 45° \times \frac{1}{L/2} + (\text{ae}) \cos 45° \times \frac{1}{L/2} \right]$$

$$= 2m \text{ since } (\text{ae}) = \frac{L}{2} / \cos 45°$$

and from eqn (8.3–11),

$m_n = m$

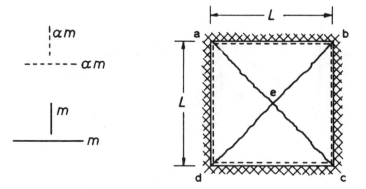

Fig. 8.4–2

energy dissipation on the negative yield line ab

$$= m_n L \frac{1}{L/2} = 2m_n$$

$$= 2am \text{ since, from eqn (8.3–11)}, m_n = am$$

Therefore

total energy dissipation on all the positive and negative yield lines

$$= 4[2m + 2am] = 8m(1 + \alpha)$$

Equating this to the work done by the loading,

$$\tfrac{1}{3}qL^2 = 8m(1 + \alpha)$$

Therefore

$$q = \frac{24(1 + \alpha)m}{L^2}$$

Comments

(a) Example 8.4–1 illustrates the so-called **work method** of solution: a yield-line pattern is assumed, and the work done by the loading during a small motion of the rigid regions is equated to the energy dissipated in the yield lines. A similarity between the yield-line analysis of slabs and the plastic analysis of frames [5, 6] is therefore obvious: the yield-line pattern assumed for the former corresponds to the collapse mechanism assumed for the latter. The **upper-bound theorem** [6] for plastic analysis states that for a given frame subjected to a given loading, the magnitude of the loading which is found to correspond to any assumed collapse mechanism must be either greater than or equal to the true collapse loading. Thus the collapse load of $q = 24(1 + \alpha)m/L^2$ as determined in Example 8.4–1 may be regarded as an upper-bound solution, and other reasonable yield-line patterns may be investigated to see whether these would give lower values for the collapse load. However, because of the membrane action [7] in the slab and because of the effect of strain hardening of the reinforcement after yielding, the so-called upper-bound collapse load obtained by yield-line analysis tends in practice to be much lower than the actual value. Thus, the search for the worst yield-line pattern need not be carried out exhaustively. For design purposes, trying a few simple and obvious patterns is usually sufficient.

(b) Example 8.4–1 shows that, provided the equation $m = qL^2/24(1 + \alpha)$ is satisfied, the slab will have the required load-carrying capacity. Thus, if the top and bottom reinforcements are made equal ($\alpha = 1$), $m = qL^2/48$; if no top steel is used ($\alpha = 0$), $m = qL^2/24$. In either case, the strength requirement is satisfied, but in the slab without top steel severe cracking on the top face is likely to occur under service loading. In using yield-line analysis, the designer must remember that it gives no information on cracking or deflections under service load.

Example 8.4–2
The simply supported rectangular slab in Fig. 8.4–3 is isotropically reinforced with bottom steel, such that the yield moment of resistance per unit width of slab is m for bending about any axis. Determine the required value of m if the slab is to carry a uniformly distributed load of intensity q.

SOLUTION

A reasonable yield-line pattern, defined by the parameter α, is shown in Fig. 8.4–3. Consider unit virtual deflection along EF.

External work done by the load

$$= \int qz\,dA \quad \text{(where } z \text{ is the deflection of the elementary area } dA\text{)}$$

$$= q \times \text{(volume swept by the slab as EF undergoes unit deflection)}$$

$$= q \left[\underbrace{2 \times \tfrac{1}{3}b\alpha l}_{\substack{\text{end} \\ \text{pyramids}}} + \underbrace{\tfrac{1}{2}b(1 - 2\alpha)l}_{\substack{\text{central} \\ \text{prism}}} \right]$$

$$= \frac{qbl}{6} (3 - 2\alpha) \tag{8.4-2}$$

From eqn (8.4–1),

energy dissipation for yield line AE

$$= m \left[\frac{\alpha l}{b/2} + \frac{b/2}{\alpha l} \right]$$

energy dissipation for yield line EF

$$= m \left[2 \times \frac{(1 - 2\alpha)l}{b/2} \right]$$

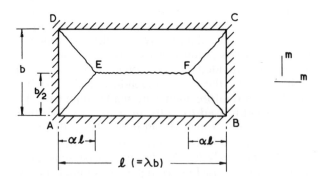

Fig. 8.4–3

Therefore, total energy dissipation for yield lines AE, DE, BF, CF and EF

$$= 4m \left[\frac{al}{b/2} + \frac{b/2}{al} \right] + 2m \frac{(1 - 2a)l}{b/2}$$

$$= \frac{2m(1 + 2\lambda^2 a)}{\lambda a} \quad \left(\text{where } \lambda = \frac{l}{b} \right)$$

The work equation is therefore

$$\frac{2m(1 + 2\lambda^2 a)}{\lambda a} = \frac{qbl}{6} (3 - 2a)$$

$$m = \frac{ql^2}{12} \cdot \frac{(3 - 2a)a}{1 + 2\lambda^2 a} \quad \left(\text{where } \lambda = \frac{l}{b} \right) \qquad (8.4\text{--}3)$$

The minimum required value of m is the maximum as given by the above work equation.

$$\frac{dm}{da} = 0$$

gives

$$a = \frac{1}{2\lambda^2} (+ \sqrt{(1 + 3\lambda^2)} - 1) \qquad (8.4\text{--}4)$$

On substitution of this value of a into the work equation,

$$m = \frac{ql^2}{24} \cdot \frac{(\sqrt{(1 + 3\lambda^2)} - 1)^2}{\lambda^4} \quad Ans.$$

Recognizing that $a^2 = (\sqrt{(1 + 3\lambda^2)} - 1)^2/4\lambda^4$, the answer may also be expressed as

$$m = \frac{ql^2}{6} a^2$$

where a has the value determined by eqn (8.4–4).

Comments
In this example, a full algebraic procedure was followed, in which the value of a corresponding to $dm/da = 0$ was determined and substituted back into the work equation to obtain the value of $m(\max)$. In practice, $m(\max)$ is frequently obtained directly from the work equation, i.e. eqn (8.4–3), by calculating the values of m for various trial values of a. Usually, this arithmetic method of trial and error yields a close approximation to the correct answer in a comparatively short time; another advantage of the arithmetic method over the algebraic method is that gross mistakes are easily noticed.

Example 8.4–3
A rectangular slab is of length 10 m and width 5 m, and is isotropically reinforced with top and bottom reinforcements such that the yield moment of resistance is m per metre width of slab, both for positive and for negative

bending about any axis. The slab is simply supported along three edges and fully fixed along the fourth edge. By considering a reasonable yield-line pattern, such as the one shown in Fig. 8.4–4, determine the intensity q of the uniformly distributed load that will cause collapse.

SOLUTION
The yield-line pattern in Fig. 8.4–4 is defined by the parameters x and y. Consider a unit virtual deflection along EF.

External work done by the load (see eqn 8.4–2 of Example 8.4–2)

$= q \times$ volume swept

$= q[(2)(\frac{1}{3})(5x) + (\frac{1}{2})(5)(10 - 2x)]$

$= (25 - 1.667x)q$

From eqn (8.4–1),

energy dissipation for yield lines AE and BF

$= 2m\left(\frac{x}{y} + \frac{y}{x}\right)$

energy dissipation for yield lines DE and CF

$= 2m\left(\frac{x}{5 - y} + \frac{5 - y}{x}\right)$

energy dissipation for yield line EF

$= m\left(\frac{10 - 2x}{y} + \frac{10 - 2x}{5 - y}\right)$

energy dissipation for negative yield line AB

$= m\left(\frac{10}{y}\right)$

Total energy dissipation

$= 10m\left(\frac{1}{x} + \frac{2}{y} + \frac{1}{5 - y}\right)$

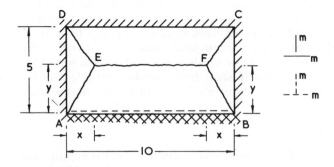

Fig. 8.4–4

Therefore the work equation is

$$10m\left(\frac{1}{x} + \frac{2}{y} + \frac{1}{5-y}\right). = (25 - 1.667x)q$$

$$\frac{q}{m} = \frac{10\left(\frac{1}{x} + \frac{2}{y} + \frac{1}{5-y}\right)}{25 - 1.667x} \tag{8.4-5}$$

The worst layout of the yield line pattern satisfies the conditions

$$\frac{\partial(q/m)}{\partial y} = 0 \quad \text{and} \quad \frac{\partial(q/m)}{\partial x} = 0$$

Now

$$\frac{\partial(q/m)}{\partial y} = 0 \quad \text{when} \quad \frac{\partial}{\partial y}\left(\frac{1}{x} + \frac{2}{y} + \frac{1}{5-y}\right) = 0$$

i.e.

$$\frac{2}{y^2} + \frac{(-1)}{(5-y)^2} = 0 \quad \text{giving } y = 2.93$$

That is q/m is a maximum when $y = 2.93$, irrespective of the value of x. (As expected, $y > 2.5$ because of the hogging moment of resistance along the edge AB.)

For $y = 2.93$, eqn (8.4–5) reduces to

$$\frac{q}{m} = \frac{\dfrac{10}{x} + 11.657}{25 - 1.667x} \tag{8.4-6}$$

$$\frac{\partial(q/m)}{\partial x} = 0 \quad \text{gives } x = 2.82$$

Substituting $x = 2.82$ into eqn (8.4–6), or $x = 2.82$ and $y = 2.93$ into eqn (8.4–5),

$$q = \underline{0.75m}$$

Example 8.4–4
In yield-line analysis, where a slab is reinforced with both top and bottom reinforcements, the assumption is made that the sagging yield moment m may be calculated from the bottom steel A_s by ignoring the existence of the top steel; similarly, it is assumed that the hogging yield moment may be calculated from the top steel A'_s by ignoring the existence of the bottom steel. Comment critically on the error introduced by this assumption.

SOLUTION
The assumption amounts to completely ignoring any **interaction between the top and bottom steels**. To investigate the error thus introduced, let us study again Fig. 4.5–2, which shows M_u/bd^2 curves for a range of tension and compression steel ratios ϱ and ϱ'. Figure 4.5–2 clearly shows that the interaction between the compression and tension reinforcements in a

slab (or beam) depends strongly on the tension steel ratio A_s/bd. For example, if the tension steel ratio $\varrho(= A_s/bd)$ is 3%, the effect of a 3% compression steel is to increase the ultimate moment of resistance from $7.4bd^2$ to $10.3bd^2$—an increase of about 39%. However, if $\varrho = 0.5\%$, the effect of a 0.5% compression steel (or even 3 or 4%) is practically zero! In yield-line analysis, the tension (or compression) steel ratio rarely exceeds 1%, values between 0.5% and 0.8% being quite common; hence the interaction between compression and tension reinforcements is usually negligible.

8.5 Energy dissipation for a rigid region

The use of eqn (8.4–1) requires the calculation of the normal moment m_n for each yield line and can be cumbersome where the slab is reinforced with skew bands of reinforcement. In this section another method, called the **component vector method** (after Jones and Wood [8]), is explained; in this the total internal work is calculated as the sum of the work due to the separate rotations of each rigid region.

Let us return briefly to Fig. 8.3–5 and eqn (8.3–6), which states that for a yield line of length l

$$(m_n + m_{ns})l = m_1 l_1 + m_2 l_2$$

This equation may be written in the equivalent form

$$(\mathbf{m_n} + \mathbf{m_{ns}})l = m_1 \mathbf{l_1} + m_2 \mathbf{l_2} \tag{8.5–1}$$

where, with reference to Fig. 8.3–5, the magnitude of the vector $\mathbf{l_1}$ is given by the projection of l on the m_1 moment axis and its sense is that of the component of $\mathbf{m_n}$ in the direction of the m_1 moment axis; that is, in Fig. 8.3–5, l_1 is directed from P' to Q where PQ is a positive yield line—if PQ had been a negative yield line, l_1 would have been directed from Q to P'. The vector l_2 is similarly defined with reference to the m_2 moment axis.

Consider a typical rigid region A bounded by the yield lines ab, bc and the simple support ac (Fig. 8.5–1). Applying eqn (8.5–1) to the yield line ab,

$$(\mathbf{m_n}l + \mathbf{m_{ns}}l)_{ab} = (m_1 \mathbf{l_1} + m_2 \mathbf{l_2})_{ab} \tag{8.5–2}$$

where l_1 is now the vector $\mathbf{ab_1}$ and l_2 is now the vector $\mathbf{ab_2}$.
For the yield line bc, we have

$$(\mathbf{m_n}l + \mathbf{m_{ns}}l)_{bc} = (m_1 \mathbf{l_1} + m_2 \mathbf{l_2})_{bc} \tag{8.5–3}$$

where

$$l_1 = \mathbf{b_1 c_1} \quad \text{and} \quad l_2 = \mathbf{b_2 c_2}$$

Adding eqns (8.5–2) and (8.5–3), we have

$$(\mathbf{m_n}l + \mathbf{m_{ns}}l)_{ab \text{ and } bc} = m_1[(l_1)_{ab} + (l_1)_{bc}]$$
$$+ m_2[(l_2)_{ab} + (l_2)_{bc}]$$
$$= m_1 \mathbf{ac_1} + m_2 \mathbf{ac_2}$$

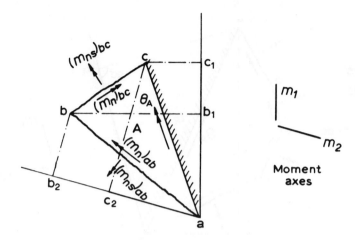

Fig. 8.5–1

That is

$$\sum (\mathbf{m}_n l + \mathbf{m}_{ns} l) = m_1 \times \text{projection of } l_R \text{ on } m_1 \text{ moment axis}$$
$$+ m_2 \times \text{projection of } l_R \text{ on } m_2 \text{ moment axis}$$
$$(8.5–4)$$

where

$$l_R = l_{ab} + l_{bc} \text{ (i.e. } l_R = ac\text{)}$$

Equation (8.5–4) can obviously be extended to a rigid region bounded by many yield lines, ab, bc, ... de, etc., as in Fig. 8.5–2; referring to that figure,

$$\sum (\mathbf{m}_n l + \mathbf{m}_{ns} l) = m_1(\mathbf{ae}_1) + m_2(\mathbf{ae}_2) \qquad (8.5–5)$$

Now if the rigid region A in Fig. 8.5–2 undergoes a rotation θ_A about its axis of rotation ae, then the work done by the normal moment $\mathbf{m}_n$ and the twisting moment $\mathbf{m}_{ns}$ on the various yield lines is given by the scalar product sum $\sum (\mathbf{m}_n l + \mathbf{m}_{ns} l) \cdot \theta_A$. From eqn (8.5–5),

$$\sum (\mathbf{m}_n l + \mathbf{m}_{ns} l) \cdot \theta_A = m_1(\mathbf{ae}_1 \cdot \theta_A) + m_2(\mathbf{ae}_2 \cdot \theta_A) \qquad (8.5–6)$$

That is (using the rule for scalar products of vectors),

$$\sum \begin{matrix} \text{energy dissipation} \\ \text{for region A} \end{matrix} = m_1[\text{projection of } l_R \text{ on } m_1 \text{ moment axis}]$$
$$\times [\text{projection of } \theta_A \text{ on } m_1 \text{ moment axis}]$$
$$+ m_2[\text{projection of } l_R \text{ on } m_2 \text{ moment axis}]$$
$$\times[\text{projection of } \theta_A \text{ on } m_2 \text{ moment axis}]$$
$$(8.5–7)$$

where (see Fig. 8.5–2) the sign of each term on the right-hand side is positive if the projection of l_R on the respective moment axis is of the same

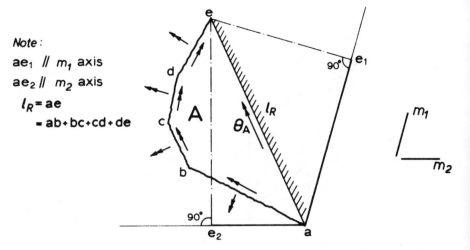

Fig. 8.5-2

sense as that of θ_A. From statement (b) of Johansen's stepped yield criterion, eqn (8.5-7) can be extended to slabs with several bands of reinforcement, though in practice designers rarely use more than two bands of reinforcement near the bottom face and two bands near the top face.

Example 8.5-1
With reference to eqn (8.5-7), explain how the projection of θ_A on a moment axis may be found.

SOLUTION
Consider the typical rigid region A in Fig. 8.5-3, bounded by the positive yield lines ab, bc, cd, de and by the axis of rotation ae. Consider a unit virtual deflection at, say, point d. Then

$$\theta_A = \frac{1}{dd'}$$

where dd' is measured in a direction normal to the axis of rotation. The magnitude of the projection of θ_A on the m_1 moment axis is $\theta_A \cos \alpha$ where α is the angle between the m_1 axis and the axis of rotation. Therefore

$$\theta_A \cos \alpha = \left(\frac{1}{dd'}\right)\left(\frac{dd'}{dd_1}\right) = \frac{1}{dd_1}$$

That is, considering a unit deflection at any typical point d on the rigid region A, the magnitude of the projection of θ_A on the m_1 moment axis is $1/dd_1$ where dd_1 is the distance from d to the axis of rotation measured in a direction perpendicular to the m_1 moment axis.

Similarly, for the same unit deflection at d, the magnitude of the projection of θ_A on the m_2 moment axis is $1/dd_2$, where the point d_2 is as shown in Fig. 8.5-3. (*Note:* for clarity, the m_2 moment axis in Fig. 8.5-3 has been shown different from that in Fig. 8.5-2.)

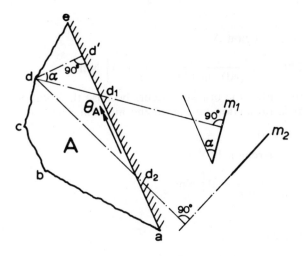

Fig. 8.5–3

Example 8.5–2
An isotropically reinforced triangular slab is simply supported along two edges and carries a uniformly distributed load q N/m^2 (Fig. 8.5–4(a)). Assuming a collapse mode consisting of a single yield line extending from the vertex to the free edge, determine the collapse load.

SOLUTION
Since the slab is isotropically reinforced, we are entitled (see eqn 8.3–11) to assume for convenience that the reinforcement arrangements on the two sides of the yield line are as shown in Fig. 8.5–4(b).

Consider unit deflection at d:

$$\text{external work} = (\tfrac{1}{3})(q)(\tfrac{1}{2} \times 8 \times 6 \sin 70°) = 8q \sin 70° \text{ Nm}$$

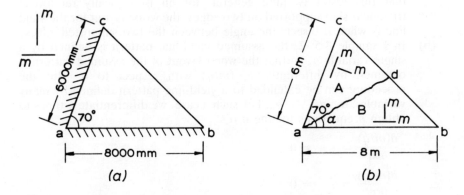

Fig. 8.5–4

From eqn (8.5–7),

energy dissipation for region A

$$= m \left[(ad) \cos (70° - a) \frac{1}{(ad) \sin (70° - a)} + 0 \right] \text{Nm}$$

(where the zero within the brackets occurs because the rotation vector has a zero projection on one of the moment axes)

$$= m \cot (70° - a) \text{ Nm}$$

energy dissipation for region B

$$= m \left[(ad) \cos a \frac{1}{(ad) \sin a} + 0 \right] \text{Nm}$$

$$= m \cot a \text{ Nm}$$

The work equation is therefore

$$m \cot (70° - a) + m \cot a = 8q \sin 70°$$

which simplifies to

$$\frac{m}{q} = 8 \sin a \sin (70° - a)$$

The worst layout for the assumed yield-line pattern is when

$$\frac{d(m/q)}{da} = 0 = -\sin a \cos (70° - a) + \sin (70° - a) \cos a$$

or

$$\sin(70° - 2a) = 0 \qquad \text{therefore } a = 35°$$

The work equation then becomes

$$\frac{m}{q} = 8 \sin^2 35° \quad \text{or} \quad \underline{q = 0.38m \text{ N/m}^2}$$

Comments

(a) Example 8.5–2 shows that $a = 35° = \frac{1}{2}(70°)$. The reader should prove that this result is quite general: for an isotropically reinforced triangular slab supported on two edges, the worst layout for the yield line is when it bisects the angle between the two supported edges.

(b) In Example 8.5–2, the assumed yield-line pattern is defined by a single variable a, so that the **worst layout** of the assumed pattern is obtained by differentiating (m/q) with respect to a. But the procedure can be extended to a yield-line pattern defined by **many variables**, $a_1, a_2, \ldots a_n$. For such a case, we differentiate n times to obtain n equations for the n a's:

$$\frac{\partial(m/q)}{\partial a_1} = \ldots = 0$$

$$\frac{\partial(m/q)}{\partial a_2} = \ldots = 0$$

$$\frac{\partial(m/q)}{\partial\alpha_n} = \ldots = 0$$

...

Example 8.5–3

If the triangular slab in Example 8.5–2 is orthotropically reinforced, as indicated by the values for moment axes in Fig. 8.5–5(a), calculate the energy dissipation per unit deflection at point d, for $\alpha = 25°, 30°, 35°, 40°$ and $45°$. Plot a graph of α against energy dissipation.

(a) Determine the worst position of the yield line and the corresponding collapse load q.

(b) Can your solution be used as evidence that your minimum collapse load is the lowest possible load for any collapse mode?

(c) Does your answer in (a) above provide a good basis for design purpose?

SOLUTION

(a) For a general value of α, if we consider unit deflection at d, then the energy dissipations (U) are given by eqn (8.5–7):

$$U(\text{Region A}) = m[\text{ad}]_m[\theta_A]_m + 0.75m[\text{ad}]_{0.75m}[\theta_A]_{0.75m}$$

$$= m(\text{de})\left(\frac{1}{\text{df}}\right) + 0.75m(\text{ae})\left(\frac{1}{\text{dg}}\right)$$

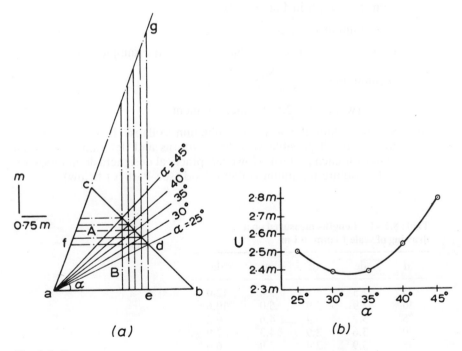

(a)   *(b)*

Fig. 8.5–5

(See Example 8.5–1 for proof that the projection of θ_A on the m moment axis is 1/df and that on the 0.75m moment axis is 1/dg.)

$$U(\text{Region B}) = m[\text{ad}]_m[\theta_B]_m + 0.75m[\text{ad}]_{0.75m}[\theta_B]_{0.75m}$$

$$= m(\text{de})(0) + 0.75m(\text{ae})\left(\frac{1}{\text{de}}\right)$$

$$= 0.75(\text{ae})/(\text{de})$$

Therefore

$$U(\text{A and B}) = m\left[\frac{\text{de}}{\text{df}} + 0.75\frac{\text{ae}}{\text{dg}} + 0.75\frac{\text{ae}}{\text{de}}\right]$$

The lengths de, df, ae, dg, and de for the various α values may conveniently be measured from a drawing drawn to scale. The reader should verify that these are as recorded in Table 8.5–1.

Therefore

$$U(\alpha = 25°) = m\left[\frac{2.5}{4.5} + 0.75 \times \frac{5.4}{12.0} + 0.75 \times \frac{5.4}{2.5}\right] = 2.51m$$

Similarly

$$U = 2.39m(\alpha = 30°) \qquad 2.40m(\alpha = 35°)$$

$$2.55m(\alpha = 40°) \quad \text{and} \quad 2.81m(\alpha = 45°)$$

From the graph in Fig. 8.5–5(b),

$$U(\text{minimum}) = 2.38m$$

Since the external work is 8q sin 70° as in Example 8.5–2,

$$q(\text{minimum}) = \frac{2.38m}{8 \sin 70°} = \underline{0.317m}$$

$$\alpha(\text{worst}) = \underline{32.5°} \text{ by measurement}$$

(b)   No, to obtain the absolute minimum collapse load, we need to investigate all possible yield-line patterns and determine the worst layout for each pattern. However, practical designers do not look for such absolute minimum collapse load (see Part (c) below).

Table 8.5–1   Lengths measured on a drawing of scale 10 mm to 1 m

α	de	df	ae	dg
25°	2.5	4.5	5.4	12.0
30°	2.9	3.9	5.0	10.6
35°	3.2	3.4	4.6	9.2
40°	3.6	2.9	4.3	7.9
45°	3.9	2.4	3.9	6.6

(c) Yes (see Comment (a) following Example 8.4–1: that comment emphasizes the conservative nature of the yield-line analysis).

Comments

(a) The algebraic method was used in Example 8.5–2 whereas the arithmetic method was used in Example 8.5–3. Much can be said in favour of the arithmetic method. Admittedly it is sometimes argued that the algebraic method has the advantage that an expression is obtained which may be used repeatedly in future. However, in practice the likelihood of repeated future use applies only to slabs of standard shapes with common loading condition; and for such standard cases, solutions are already available in handbooks [2]. For the 'one-off' type slab of unusual shape or loading, there is little point in obtaining an answer in the form of an algebraic expression; in any case, the algebra may become too complex for the solution to be economical for practical design purposes.

(b) It should also be pointed out that sometimes the quantities (m/q) may not have a stationary maximum value. Consider the rectangular slab in Fig. 8.5–6(a), with three simply supported edges and a free edge. The slab carries a uniformly distributed load of intensity q. The assumed yield-line pattern is defined by the variable x. The reader should verify that:

total energy dissipation for the three rigid regions

$$= 0.08xm + 10m/x$$

and that:

total external work done $= (25 - 1.67x)q$

Therefore

$$\frac{m}{q} = \frac{25x - 1.67x^2}{0.08x^2 + 10}$$

whence

$$\frac{\mathrm{d}(m/q)}{\mathrm{d}x} = 0 \quad \text{gives} \quad x = 5.6$$

But $x = 5.6$ exceeds half the length of the slab, so that (m/q) does not have a stationary maximum for the assumed yield-line pattern.

We have to consider other patterns, such as those in Fig. 8.5–6(b) and (c). By inspection the pattern in (b) is not critical because the external work done by q in the shaded region is less than that in the same area in (c). Therefore, of the two patterns, only that in (c) need be considered. Using z as the variable this time, the reader should verify that the work equation is

$$2m\left(1 + \frac{1}{z}\right) = (25 - 1.67z)q$$

$$\frac{m}{q} = \frac{25 - 1.67z}{2(1 + 1/z)}$$

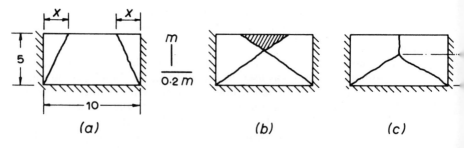

Fig. 8.5–6

$$\frac{d(m/q)}{dz} = 0 \quad \text{gives} \quad z = 3$$

which corresponds to a stationary maximum for (m/q) because, for $z = 3$, the pattern is still valid. Substituting $z = 3$ into the work equation gives $q = 0.133m$ as the stationary minimum collapse load.

Example 8.5–4
The skew slab in Fig. 8.3–7(a) is subjected to a central point load Q and a uniformly distributed load q. Considering a yield-line pattern as shown in Fig. 8.3–7(c), where the length hg is l, determine the algebraic relation between Q, q and l.

SOLUTION
For clarity, Fig. 8.3–7(c) is redrawn in Fig. 8.5–7.

Consider a unit deflection at the central point j:

external work done $= \frac{1}{3}qlL + Q$

energy dissipation in region A (using eqn 8.5–7):

(1)   Positive yield lines ej and jh:

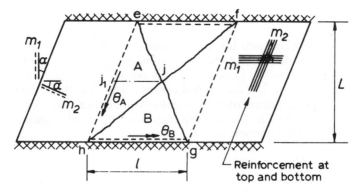

Fig. 8.5–7

m_1 [projection of eh on m_1 moment axis]

× [projection of θ_A on m_1 moment axis]

+ m_2 [projection of eh on m_2 moment axis]

× [projection of θ_A on m_2 moment axis]

$$= m_1 L \frac{1}{\text{jj}_1} + 0 \quad (\text{since } \theta_A \text{ has zero projection on } m_2 \text{ moment axis})$$

$$= 2m_1 L/l$$

(2) Negative yield line eh:

$2m_1 L/l$ as for ej plus jh

Therefore the total energy dissipation for region A = $4m_1 L/l$. Similarly, the reader should verify that the total energy dissipation for region B = $4m_2 l \cos^2 \alpha/L$. Therefore the total energy dissipation for the four rigid regions is

$$2[4m_1 L/l + 4m_2 l \cos^2 \alpha/L]$$

The required algebraic relation between (Q, q) and l is given by the work equation

$$\tfrac{1}{3} qlL + Q = 8[m_1 L/l + m_2 l \cos^2 \alpha/L]$$

Comments

(a) The main purpose of Example 8.5–4 is to demonstrate the power-fulness of eqn (8.5–7) in dealing with a fairly complicated problem. The reader should attempt a solution using eqn (8.4–1); he will immediately realize how cumbersome the solution becomes.

(b) In practice, the designer would first investigate some simpler patterns, such as that in Fig. 8.3–7(b). Also, where a heavy concentrated load is present, a slab may fail by the so-called **fan mechanism**. Figure 8.5–8(a) shows part of a slab which is isotro-pically reinforced with top and bottom steel. Suppose the slab supports a uniformly distributed load q and a rather heavy point load Q. Considering the circular fan mechanism with a unit deflection at the centre,

external work done = $\tfrac{1}{3} q\pi r^2 + Q$

where r is the radius of the fan mechanism. To determine the internal energy dissipation, consider the typical rigid region A in Fig. 8.5–8(b). Since the slab is isotropically reinforced, we are entitled to assume that, locally, the moment axes are as shown.

Using eqn (8.5–7),

energy dissipation for region A

$$= m[ed][1/r] + m'[ed][1/r]$$

$$= \frac{(m + m')}{r} [ed]$$

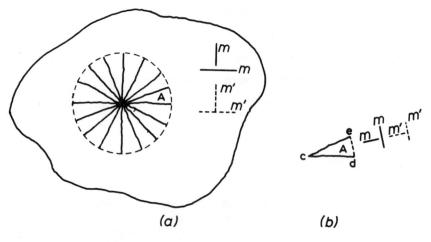

(a)   (b)

Fig. 8.5–8   Fan mechanism

Therefore, the total energy dissipation for entire fan mechanism

$$= \frac{m + m'}{r} \sum(ed) = \frac{m + m'}{r} 2\pi r$$

$$= 2\pi(m + m')$$

The work equation is therefore

$$\frac{\pi}{3} qr^2 + Q = 2\pi(m + m')$$

Therefore, if the distributed load is absent,

$$Q = 2\pi(m + m')$$

and is independent of the radius of the fan. Conversely, if the point load is absent,

$$\frac{\pi}{3}qr^2 = 2\pi(m + m')$$

or

$$q = \frac{6(m + m')}{r^2}$$

which reduces as r increases. In other words, where a slab supports a distributed load, a fan mechanism always extends to a slab boundary; where a slab supports a point load only, the fan mechanism gives a constant value for the collapse load which is independent of the radius of the mechanism.

(c)   In all of the Examples in Sections 8.4 and 8.5 the work method has been used, in which the solution is obtained by equating the external work done by the applied loads to the internal energy dissipation. All yield-line patterns which the designer is likely to encounter in practice can be dealt with by the work method. However, in many instances, an alternative technique in yield-line theory, namely, the so-called **equilibrium method** or **nodal-force method**, would give a solution more quickly. Readers interested in this alternative approach should consult the works of Jones and Wood, Morley and others [7–10].

8.6 Hillerborg's strip method

In comment (a) following Example 8.4–1 on yield-line analysis, the upper-bound theorem was mentioned. In plastic theory [5, 6, 11] there is another theorem called the **lower-bound theorem** [5, 6, 11], which states that, for a structure under a system of external loads, if a stress distribution throughout the structure can be found such that (a) all the conditions of equilibrium are satisfied and (b) the yield condition is nowhere violated, then the structure is safe under that system of external loads. It helps to consider the application of this theorem to a simple case: say a frame structure—in which only bending moments need be considered. The lower-bound theorem then states that, if a distribution of bending moments can be found such that the structure is in equilibrium under the external loading, and such that nowhere is the yield moment of resistance of any structural member exceeded, then the structure will not collapse under that loading, however 'unlikely' that distribution of moments may appear [12].

To illustrate the application of the lower-bound theorem to design [12], consider the span AB of a continuous reinforced concrete beam in Fig. 8.6–1(a), in which the load for the ultimate limit is 20 kN/m. According to the theorem, any of the bending moment diagrams in Figs 8.6–1(b), (c) and (d) may be used as a basis for design, provided we are concerned only with strength capacity and provided the beam is sufficiently under-reinforced to exhibit plastic behaviour at collapse. Adopting Fig. 8.6–1(b) as the design bending moment diagram will

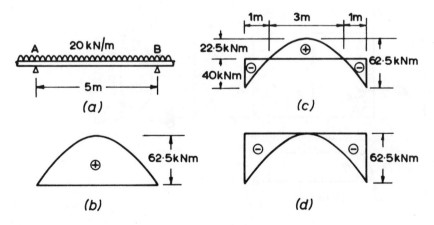

Fig. 8.6–1

result in wide top cracks at A and B under service loading; adopting Fig. 8.6–1 (d) will similarly result in wide bottom cracks at midspan; but either way the structure is safe though it may not be serviceable. If the points of zero moment in Fig. 8.6–1(c) correspond nearly to the actual points of contraflexure under service loading, then using that moment diagram for design will lead to a safe and serviceable structure. In practical design, therefore, the need is for skill and judgement in choosing a suitable bending moment distribution—but the message is clear: the engineer may design his structure on the basis of any equilibrium distribution of bending moments chosen by his structural sense (that is, if he does have structural sense). An analysis of the structure is not an essential part of the design process [12].

Hillerborg's strip method of slab design is based on the lower-bound concept and on the designer's intuitive 'feel' of the way the structure transmits the load to the supports. The method was propounded by Hillerborg of Sweden in 1956, but remained relatively obscure until a decade later, when the far-reaching research of Wood and Armer [13, 14] drew attention to its potential and possibilities. The description of the method as given below is essentially the work of Armer [13] (reproduced from the Building Research Establishment Current Paper CP 81/68, by permission of the Director, BRE):

Hillerborg's strip method is quite general [15], but we shall only deal with the simple part of the method which covers uniformly loaded slabs supported continuously; but the slab may be of any shape. The method assumes that at failure the load is carried either by bending in the x-direction or by separate bending in the y-direction, but no load is carried by the twisting strength of the slab; that is, the load is carried by pure **strip action**—hence the title 'strip method'.

Consider the simply supported rectangular slab in Fig. 8.6–2. The dotted lines on the slab are sometimes called **lines of discontinuity**; they

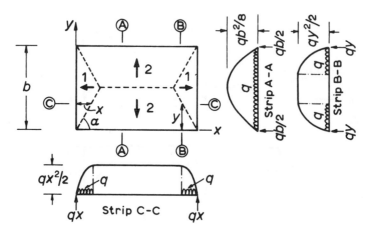

Fig. 8.6–2

indicate that the designer, using his intuition, has decided to carry all the load in the areas 1 by x-strips (i.e. strips spanning in the x-direction) and to carry all the load in the areas 2 by y-strips. Suppose the uniformly distributed load on the slab is of intensity q. Then a y-strip, such as A–A, will be loaded along its entire length, as shown in Fig. 8.6–2, so that the bending moment diagram is of parabolic shape with a maximum ordinate of $qb^2/8$. The y-strip B–B will be loaded only for a length y at each end and unloaded at the centre, because in the central length $(b - 2y)$ the load is carried by x-strips. Similarly, the x-strip C–C is loaded as shown. Thus, once the decision is made regarding the lines of discontinuity, the designer immediately has (a) all the bending moment values with which to calculate the required reinforcement and (b) the support reactions required for designing the supporting beams. But this is where design judgement comes in. Referring to Fig. 8.6–2, the angle α defining the line of discontinuity between areas 1 and 2 is up to the designer to decide. If, for example, α is made equal to 90°, then the result is a slab reinforced for one-way bending. By the lower-bound theorem, it is safe; but it is not serviceable, because excessive cracking will occur near the edges in the y-direction. Hillerborg has suggested that for such a simply supported slab, the angle α should be 45°.

In Fig. 8.6–2, the bending moment diagrams for typical strips are as shown. It is obviously impracticable to reinforce a slab to match these moments exactly. Hillerborg chose to reinforce the full length of each strip to withstand the maximum moment acting on it. However, even these maxima themselves vary with the position of the strip. For example, the maximum moment for a y-strip A–A is $qb^2/8$ while that for B–B is $qy^2/2$. Hillerborg decided to have strips of uniform reinforcement giving a slab yield moment equal to the average of the maximum moments found in that strip. Thus the slab in Fig. 8.6–3(a) might be divided into three x-strips of widths b_1, b_2 and b_1 respectively, and three y-strips of widths l_1, l_2 and l_1

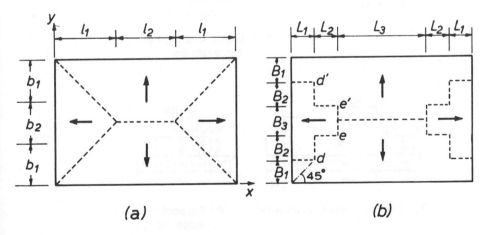

(a)   (b)

Fig. 8.6–3

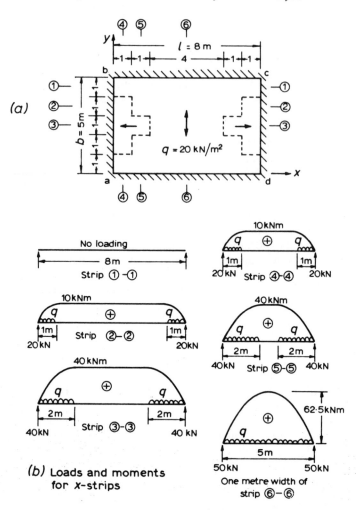

(a)

(b) Loads and moments
for x-strips

(c) Loads and moments
for y-strips

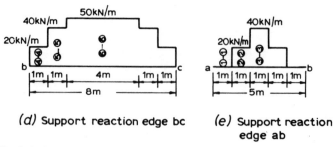

(d) Support reaction edge bc

(e) Support reaction
edge ab

Fig. 8.6–4

respectively. Within each strip the reinforcement would be uniformly arranged.

Wood and Armer have pointed out that the discontinuity pattern in Fig. 8.6–3(a) need not have been chosen, and have suggested the alternative pattern in Fig. 8.6–3(b) in which the points d, e, d′ and e′ lie on the 45° diagonals. The alternative pattern avoids the troubles arising from oddly shaped loaded portions on the strips and the consequent averaging of moments.

Example 8.6–1

Figure 8.6–4(a) shows a simply supported rectangular slab abcd. If the design load for the ultimate limit state is 20 kN/m² determine:

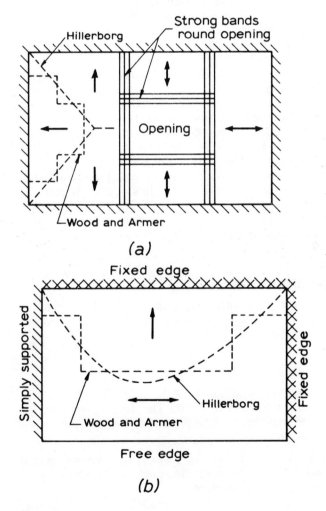

(a)

(b)

Fig. 8.6–5

(a) the design bending moment diagrams for typical strips;
(b) the design load diagrams for the edge beams.

Use lines of discontinuity as proposed by Wood and Armer.

SOLUTION
(a) The design moment diagrams for the *x*-strips are shown in Fig. 8.6–4(b) and those for the *y*-strips in Fig. 8.6–4(c).
(b) The reactions required for the design of the supporting beams are shown in Fig. 8.6–4(d) and (e).

Comments
(a) Figure 8.6–4(b) shows that there is no load and hence no moments on the *x*-strip 1–1. These strips must, however, be reinforced to the minimum level required by BS 8110 (see Section 8.8).
(b) The slab in Example 8.6–1 is simply supported. If it had been continuous over the supports, then the moment diagrams in Fig. 8.6–4(b) and (c) are still acceptable from the strength point of view, but unacceptable from the serviceability point of view. Wood and Armer have suggested that for such a **continuous slab** the designer may assume that in each strip the points of contraflexure are located from the supports at 0.2 times the span for that strip. Consider, for example, the strip 6–6; the points of contraflexure would be assumed at $0.2 \times 5 = 1$ m from each end support, so that the bending moment diagram for that strip becomes as shown in Fig. 8.6–1(c). The moment diagram for each of the other strips is similarly obtained by lifting the base line by the appropriate amount so that the positions of zero moment coincide with the assumed points of contraflexure.
(c) Wood and Armer's suggestions for stepped lines of discontinuity can be applied to slabs with openings or slabs with different support conditions, as illustrated in Fig. 8.6–5.

8.7   Shear strength of slabs (BS 8110)

The shear strength of slabs is governed by the same general principles as for beams (see Chapter 6). In design, the nominal design shear stress *v* is calculated from eqn (6.4–1):

$$v = \frac{V}{bd} \tag{8.7–1}$$

where *V* is the shear force due to ultimate loads, *b* the width of the slab under consideration and *d* the effective depth. The design procedure for slabs is essentially the same as that for beams (see Section 6.4), and only a few comments are necessary:

(a) The design shear stress *v* from eqn (8.7–1) should not exceed $0.8\sqrt{f_{cu}}$ or 5 N/mm², whichever is less. Increase the slab thickness, if necessary, to satisfy this requirement.
(b) If *v* is less than v_c in Table 6.4–1, no shear reinforcement is required.
(c) If $v_c \leq v \leq (v_c + 0.4)$, provide minimum links in accordance with eqn (6.4–2):

$$A_{sv} \text{ (minimum links)} \geq \frac{0.4bs_v}{0.87f_{yv}} \tag{8.7-2}$$

where A_{sv} is the area of the shear links within the width b (b is normally taken as 1 m), and s_v and f_{yv} have the same meanings as in eqn (6.4–2).

(d) If $v > (v_c + 0.4)$, provide shear links in accordance with eqn (6.4–3):

$$A_{sv} \geq \frac{(v - v_c)bs_v}{0.87f_{yv}} \tag{8.7-3}$$

where the symbols have the same meanings as in eqn (8.7–2). Bent-up bars might be used as shear reinforcement, either on their own or in combination with links—see BS 8110 : Clause 3.5.5.3 for details.

(e) The use of shear reinforcement in slabs less than 200 mm thick is considered impractical; hence, for such shallow slabs, v should not exceed v_c.

8.8 Design of slabs (BS 8110)

Moment and shear forces (BS 8110 : Clause 3.5)
In design, the flexural and shear strengths are calculated from the methods of Sections 8.1 and 8.7. In current British design practice, the moments and shear forces on slabs are usually determined by elastic analysis. For a continuous **one-way slab**, the moments and shear forces may conveniently be obtained from Table 11.4–2, extracted from BS 8110. For a **two-way rectangular slab**, the moment and shear force coefficients in BS 8110 : Clause 3.5.3 may be used.

For more complicated cases, an **elastic analysis** may be carried out using the **finite difference** or the **finite element** methods, as described in Reference 11. BS 8110 permits the use of the powerful yield-line method and Hillerborg's strip method. However, these two methods provide no information on deflection and cracking; BS 8110 : Clause 3.5.2.1 requires that, where these methods are used, the ratios between support and span moments should be similar to those obtained by elastic analysis.

Limit state of deflection (BS 8110 : Clause 3.5.7)
In design, deflections are usually controlled by limiting the ratio of the span to the effective depth. The procedure is essentially the same as that explained in Section 5.3 for beams.

Step 1
Select the basic span/depth ratio from Table 5.3–1; namely, 7 for cantilever slabs; 20 for simply supported slabs and 26 for continuous slabs.

Step 2
The basic span/depth ratio is now multiplied by a modification factor, obtained from Table 5.3–2, to allow for the effect of the tension reinforcement.

Comments
(a)   If the actual span/depth ratio does not exceed that obtained in Step 2 above, then BS 8110's deflection requirements are met.
(b)   The modification factors in Table 5.3–2 are based on the M/bd^2 values. In using Table 5.3–2 for slabs, M is the design ultimate moment at the centre of the span; for a cantilever, M is that at the support [1].

Limit state of cracking (BS 8110: Clause 3.5.8)
In design, crack widths are usually controlled by limiting the spacing of the reinforcement. BS 8110's requirements are as follows:

(a)   The clear spacing between bars should not exceed $3d$ or 750 mm, whichever is less, where d is the effective depth of the slab. This requirement applies to both main bars and distribution bars. (*Note:* The I.Struct.E. Manual [1] gives separate, but more restrictive, recommendations for main bars and distribution bars: main bar spacings should not exceed $3d$ or 300 mm; distribution bar spacings should not exceed $3d$ or 400 mm.)
(b)   Having checked that the above requirement is met, no further check on bar spacing is necessary, provided that at least one of the three conditions below is met:
   (1)   for $f_y = 250$ N/mm^2: $h \leq 250$ mm (where h is the overall slab thickness);
   (2)   for $f_y = 460$ N/mm^2: $h \leq 200$ mm;
   (3)   the steel ratio $\varrho \ (= A_s/bd) < 0.3\%$.
(c)   If none of the three conditions in (b) is met, then the main-bar spacing should be limited as stipulated below:

Case (i) $\varrho \ (= A_s/bd) \geq 1\%$

The clear spacing between the main bars should not exceed the appropriate value given in Table 5.4–1.

Case (ii) $\varrho \ (= A_s/bd) < 1\%$

The clear spacing between the main bars should not exceed the appropriate value in Table 5.4–1, divided by the steel ratio ϱ (where ϱ is expressed as a percentage).

Comments on Requirement (b)(3)
For the condition $\varrho \ (= A_s/bd) < 0.3\%$, BS 8110: Clause 3.12.11.2.7 defines A_s as the 'minimum recommended area', which might confuse some readers. The authors believe the Code's intention is that A_s is the area of steel actually provided.

Comments on requirement (c)
In using Table 5.4–1, the moment redistribution percentage may be taken as (-15%) for support moments and zero for span moments.

Minimum area of reinforcement (BS 8110: Clauses 3.12.5.3 and 3.12.11.2.9)
The area of reinforcement in each direction should not be less than:

(a) 0.13% of *bh* for $f_y = 460$ N/mm^2;

(b) 0.24% of *bh* for $f_y = 250$ N/mm^2.

Where the control of shrinkage and temperature cracking is important, the minimum areas should be increased to:

(a) 0.25% of *bh* for $f_y = 460$ N/mm^2;

(b) 0.3% of *bh* for $f_y = 250$ N/mm^2.

Comments
(a) The I.Struct.E. Manual [1] recommends in addition that main bars should not be less than size 10 and that the percentage of distribution bars should be at least one-quarter that of the main bars.
(b) With reference to Fig. 8.8–1, suppose the bottom distribution bars 'mark 2' already satisfy the 0.13% requirement (or 0.24%, etc. as the case may be). Then, according to a strict interpretation of BS 8110:Cl 3.12.5.3, the top distribution bars 'mark 1' need only satisfy the functional requirement of keeping the main top bars in position. However, the authors share the view that the area of the distribution reinforcement provided to the main tension bars should be at least 0.13% of *bh*. For the slab shown in Fig. 8.8–1, the top bars are the main reinforcement; hence the top distribution bars 'mark 1' should satisfy the minimum requirement of 0.13%, irrespective of the area of the bars 'mark 2'.

Concrete cover for durability (BS 8110:Clause 3.3)
Table 2.5–7 in Section 2.5(e) gives the nominal covers to meet the durability requirements for slabs and other structural members.

Note that in Table 2.5–7 the meaning of **nominal cover** is as defined in BS 8110:Clause 3.3.1.1, namely: the nominal cover is the design depth of concrete to all reinforcement, including links. It is the dimension used in design and indicated on the drawings.

Fire resistance (BS 8110:Clause 3.3.6)
The fire resistance of a slab depends on its overall thickness and the concrete cover to the reinforcement, as shown in Table 8.8–1.

Simplified rules for curtailment of bars (BS 8110:Clause 3.12.10.3.1)
The following simplified rules may be used if (i) the loading is predominatly uniformly distributed, and (ii) for continuous slabs, the spans do

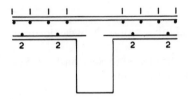

Fig. 8.8–1

Table 8.8–1 Fire resistance requirements for slabs (BS 8110 : Part2 : Clause 4.3.1)

Fire rating (hours)	Minimum thickness (mm)		Concrete cover to MAIN reinforcement (mm)	
	Simply supported	Continuous	Simply supported	Continuous
1	95	95	20	20
2	125	125	35	25
3	150	150	45[a]	35
4	170	170	55[a]	45[a]

[a] See BS 8110 : Part 2 : Clause 4 for protection against spalling.

not differ by more than 15% of the longest span [1]. In these rules, l refers to the effective span length and ϕ the bar size.

(a) **Simply supported slabs:** All the tension bars should extend to within $0.1l$ of the centres of the supports. At least 40% of these bars should further extend for at least 12ϕ (or its equivalent in hooks or bends) beyond the centres of the supports.

(b) **Cantilever slabs:** All the tension bars at the support should extend a distance of $l/2$ or 45ϕ, whichever is greater. At least 50% of these bars should extend to the end of the cantilever.

(c) **Continuous slabs:**
  (1) All the tension bars at the support should extend $0.15l$ or 45ϕ from the face of the support, whichever is greater. At least 50% of these bars should extend $0.3l$ from the face of the support.
  (2) All the tension bars at midspan should extend to within $0.2l$ of the centres of supports. At least 40% of these bars should extend to the centres of supports.
  (3) At a simply supported end, the detailing should be as given in (a) for a simply supported slab.

8.9 Design and detailing—illustrative example

See Example 11.5–1.

8.10 Computer programs

(in collaboration with **Dr H. H. A. Wong**, University of Newcastle upon Tyne)

The FORTRAN programs for this chapter are listed in Section 12.8. See also Section 12.1 for 'Notes on the computer programs'.

Problems

8.1 In yield-line analysis, where a slab is reinforced with both top and bottom reinforcements, the assumption is made that the sagging yield moment *m* may be calculated from the bottom steel A_s by ignoring the existence of the top steel; similarly, it is assumed that the hogging yield moment may be calculated from the top steel A_s' by ignoring the existence of the bottom steel. Comment critically on the error introduced by this assumption, which amounts to ignoring completely any **interaction between the top and bottom reinforcement bars**.

Ans. See Example 8.4–4 if necessary.

8.2 An isotropically reinforced slab is simply supported along two edges and carries a uniformly distributed load. For a collapse mode consisting of a single yield line from the vertex to the free edge, verify that the worst layout is when the yield line bisects the angle between the two supported edges, i.e. when $\alpha = \phi/2$. (See Example 8.5–2: Comment (a) at the end of the solution.)

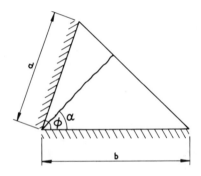

Problem 8.2

8.3 A rectangular slab, simply supported along all four edges, is isotropicaly reinforced to give a yield moment of 27 kNm per metre width

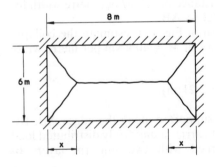

Problem 8.3

of slab. The slab measures 8 by 6 m; by considering a reasonable collapse mode, such as that shown in the figure, calculate the value of the uniformly distributed load q that would just cause collapse. Check your answer against the general algebraic solution of Example 8.4–2.

Ans.   13.91 kN/m².

8.4* Figure (a) shows a uniform rectangular slab simply supported along the edges BC, CD and DA and fully fixed along AB; it carries a uniformly distributed load of intensity q. The slab is reinforced at the bottom face with two uniform layers of reinforcement, giving a moment of resistance M per metre width for bending about an axis parallel to AB and $0.5M$ per metre width for bending about an axis parallel to AD. It is also reinforced

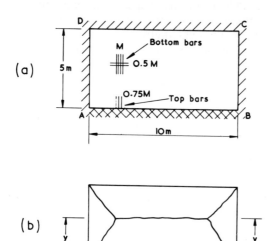

Problem 8.4

at the top face to give a moment of resistance of $0.75M$ per metre width for hogging bending about an axis parallel to AB.

By considering the yield-line pattern in Fig. (b), estimate the collapse value of q according to yield-line theory, if $M = 40$ kNm per metre width of slab.

Ans.   24.48 kN/m²($y = 2.85$ m; $x = 2.21$ m).

8.5* Figure (a) shows a slab simply supported along the edges AB, BC and CD and unsupported along AD; it carries a uniformly distributed load of intensity q. The slab is reinforced with two uniform layers of

* Cambridge University Engineering Tripos: Part II (past examination question).

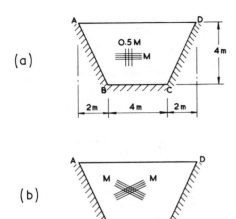

(a)

(b)

Problem 8.5

reinforcement at the bottom face, giving a moment of resistance of M per metre width for bending about an axis perpendicular to BC, and $0.5M$ per metre width for bending about an axis parallel to BC. Using the yield-line method, determine the collapse value of q.

If the slab had been reinforced at the bottom face with two uniform skew layers of reinforcement, in directions perpendicular respectively to AB and CD, as shown in Fig. (b), explain briefly how the collapse load could be determined.

Ans. 1st Part: $q = 0.28M$, for a collapse mode with two yield lines, extending from B and C to the unsuported edge. (Worst layout of yield-line pattern: each yield line has a projection of 1.80 m on BC.) 2nd Part: Use graphical method: see Example 8.5–3.

References

1 I.Struct.E/ICE Joint Committee. *Manual for the Design of Reinforced Concrete Building Structures*. Institution of Structural Engineers, London, 1985.
2 Johansen, K. W. *Yield-line Formulae for Slabs*. Cement and Concrete Associaton, London, 1972.
3 Hognestad, E. Yield-line theory for the ultimate flexural strength of reinforced concrete slabs. *Pro. ACI*, **49**, No. 7, March 1953, pp. 637–56.
4 Morley, C. T. Experiments on the yield criterion of isotropic reinforced concrete slabs. *Proc. ACI*, **64**, No. 1, Jan. 1967, pp. 40–5.
5 Baker, Sir John and Heyman, J. *Plastic Design of Frames. 1: Fundamentals*. Cambridge University Press, 1968.
6 Kong, F. K. and Charlton, T. M. the fundamental theorems of the plastic theory of structures. *Proceedings* of the M. R. Horne Conference on Instability and Plastic Collapse of Steel Structures (Editor: L. J. Morris), Manchester 1983. Granada Publishing, London, 1983, pp. 9–15.

7   Wood, R. H. Slab design: past, present and future. ACI Publication SP-39. American Concrete Institute, Detroit, 1971, pp.203-21.

8   Jones, L. L. and Wood, R. H. *Yield-line Analysis of Slabs*. Thames and Hundson, Chatto and Windus, London, 1967.

9   Morley, C. T. The ultimate bending strength of reinforced concrete slabs. University of Cambridge, Ph.D. Thesis, 1965.

10 Morley, C. T. Equilibrium methods for exact upper bounds of rigid plastic plates. *Magazine of Concrete Research*: special publication on recent developments in yield-line theory, 1965.

11 Coates, R. C., Coutie, M. G. and Kong, F. K. *Structural Analysis*, 2nd edn. Van Nostrand Reinhold UK, 1980, pp. 418, 452 and 497.

12 Kong, F. K. Discussion of: 'Why not WL/8?' by A. W. Beeby. *The Structural Engineer*, **64A**, No. 7, July 1986, pp. 184-5.

13 Armer, G. S. T. *The Strip Method: A New Approach to the Design of Slabs* (Current·Paper CP 81/68). Watford, Building Research Establishment, 1968.

14 Wood, R. H. and Armer, G. S. T. The theory of the strip method for design of slabs. *Proc. ICE*, **41**, Oct. 1968, pp. 285-311.

15 Hillerborg, A. The advanced strip method—a simple design tool. *Magazine of Concrete Research*, **34**, No. 121, Dec. 1982, pp. 175-81.

Chapter 9
Prestressed concrete simple beams

Preliminary note: For class teaching, it has been found very effective to use the following C and CA booklet as a supplementary text: A. H. Allen, 'An Introduction to Prestressed Concrete', supplied free of charge by the C and CA to universities and colleges. Students should be asked to read, in their own time, A. H. Allen's booklet, particularly the excellent descriptive chapters on prestressing methods, materials and equipments. The class lectures would then follow Chapters 9 and 10 of this book as the main text.

9.1   Prestressing and the prestressed section

Prestressed concrete construction has now developed [1–4] to such a stage that a general understanding of its principles and of the design procedures is necessary for the structural engineer. Prestressed concrete may be defined as concrete in which internal stresses of such magnitude and distribution have been introduced that the stresses resulting from the given applied loading are counteracted to a desired degree.

BS 8110 divides prestressed concrete members into three* classes: no tensile stress is permitted in **Class 1 members**; in **Class 2 members** the permissible tensile stresses are kept sufficiently low so that no visible cracking occurs; in **Class 3 members**, the tensile stresses are restricted such that crack widths do not exceed 0.1 mm for very severe environments and 0.2 mm for other conditions. However, we shall consider only Class 1 and Class 2 members; hence, except when considering the ultimate limit state of collapse, the members are analysed and designed as uncracked members, i.e. the ordinary elastic beam theory applies.

Consider a beam subjected to a prestressing force P, as in Fig. 9.1–1, which also shows the stresses obtained by the elastic theory. The **sign convention** in the figure, which conforms to the principles adopted in this book, is that used in current British design practice:

(a)   Moment due to applied load: sagging is positive and hogging is negative.

---

* Class 1 members are conventionally referred to as fully prestressed members, while Class 2 and (particularly) Class 3 members are referred to as being partially prestressed. Partial prestressing offers decided economic advantages over full prestressing and may be expected to achieve growing importance in the near future.

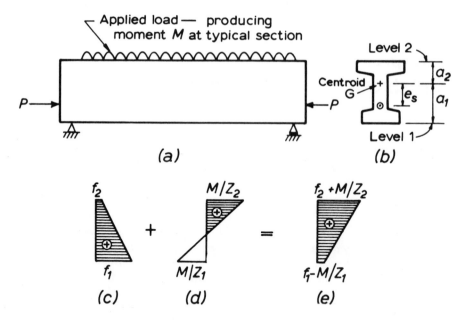

Fig. 9.1–1   **Prestressed beam and stresses at a typical section**

(b)   Concrete stress: compressive is positive and tensile is negative.
(c)   Tendon stress: tensile is positive (the term 'tendon' is defined later).
(d)   Tendon eccentricity: downwards is positive, and upwards is negative.

Also, the subscripts 1 and 2 in the symbols refer respectively to the bottom and the top faces of the beam. Thus, in Fig. 9.1–1, f_1 is the prestress at the bottom fibres, and f_2 that at the top fibres. Similarly, Z_1 and Z_2 are the elastic section moduli referred to the bottom and the top fibres respectively; i.e. $Z_1 = I/a_1$ and $Z_2 = I/a_2$, where I is the second moment of area of the cross-section about a horizontal axis through the centroid G.

In practice, the prestressing force (P in Fig. 9.1–1(a)) is usually applied by means of **tendons**, which may be:

(a)   7-wire strands of typical characteristic strength f_{pu} of 1770 N/mm^2 (BS 5896);
(b)   cold-drawn wires of typical f_{pu} 1570 N/mm^2 (BS 5896);
(c)   high-tensile alloy bars of typical f_{pu} 1030 N/mm^2 (BS 4486).

Note that both BS 5896 and BS 4486 use the term **characteristic breaking load**, the breaking load being the tensile strength times the cross-sectional area. However, in this book, we follow BS 8110 and use the term characteristic strength (f_{pu}).

The tendons are usually tensioned to an **initial prestress** of about 70%, and occasionally up to 80%, of the characteristic strength. Two methods of prestressing are in general use. In **pre-tensioning**, the tendons pass through the mould, or moulds for a number of similar members arranged end to end, and are tensioned between external end anchorages, by which the tension is maintained while the concrete is placed. When the concrete has hardened sufficiently the ends of the tendons are slowly released from the anchorages. During this operation, which is known as **transfer**, the force in the tendons is transferred to the concrete by bond stress [5–7]. The length required at each end of a member to transmit the full tendon force to the concrete is called the **transmission length** and is roughly about 65 diameters for crimped wires and 25 diameters for strands (see BS 8110 : Clause 4.10 for values to use in design). In **post-tensioning**, the concrete member is cast incorporating ducts for the tendons. When the concrete has hardened sufficiently, the tendons are tensioned by jacking against one or both ends of the member, and are then anchored by means of **anchorages** which bear against the member or are embedded in it.

Pre-tensioning is more suitable for mass production of standard members in a factory; usually straight tendons only are used. Post-tensioning is generally used on site for members cast in their final place; within limits, tendon profiles of any shape can be used.

Prestressed concrete has several important advantages over reinforced concrete. First, reference to the stress diagram in Fig. 9.1–1(e) shows that the entire concrete section is effective in resisting the applied moment M, whereas only the portion of the section above the neutral axis is fully effective in reinforced concrete; this leads to greatly reduced deflections under service conditions. Second, the use of curved tendon profiles (in post-tensioning) enables part of the shear force to be carried by the tendons. Also, as we shall see in Section 9.6, the precompression in the concrete tends to reduce the diagonal tension. In general, the same applied load can be carried by a lighter section in prestressed concrete; this yields more clearance where it is required and enables longer spans to be used. The absence or near absence of cracks under service loading is another advantage.

9.2   Stresses in service: elastic theory

In contrast to the design of reinforced concrete members, the design of Class 1 and Class 2 prestressed concrete members is generally governed by the stress criteria in service or at transfer, rather than by their ultimate strengths, though the latter must be checked. Hence the elastic theory is very relevant in prestressed concrete design. Designs are normally based on the conditions in service, but the stresses at transfer (see Section 9.3) must be checked.

Consider again the simply supported beam in Fig. 9.1–1. Suppose the prestressing force in the tendon is P at transfer. When the beam is in service, the prestressing force will be less than P, because of loss of prestress, which topic will be discussed in Section 9.4. In the meantime it is sufficient to note that under service condition the **effective prestressing force** will be

$$P_e = P\alpha \qquad\qquad (9.2–1)$$

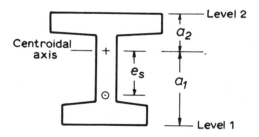

Fig. 9.2–1

where P is the prestressing force at transfer, and α is the **loss ratio**, $\alpha = 0.8$ being a typical value.

For clarity, Fig. 9.1–1(b) is redrawn in Fig. 9.2–1. Under service conditions. we have for the typical section:

$$f_1 = \frac{P_e}{A} + \frac{P_e e_s}{Z_1} \tag{9.2–2}$$

$$f_2 = \frac{P_e}{A} - \frac{P_e e_s}{Z_2} \tag{9.2–3}$$

where (see notation and sign convention in Section 9.1)

f = the compressive prestress in the concrete;
A = the concrete section area (usually taken as the nominal area of the cross-section);
Z = the elastic section modulus;
e_s = the eccentricity of the prestressing force P_e;
    subscript 1 refers to the bottom fibre and
    subscript 2 refers to the top fibre.

Note that where only one tendon is used, the eccentricity e_s of the prestressing force is that of the tendon; where more than one tendon is used, e_s is the eccentricity of the centroid of the tendons. For simplicity, e_s will be referred to as the **tendon eccentricity**; also, where several tendons are used they are often collectively referred to as the tendon. In post-tensioned beams, e_s usually varies along the beam; however, the inclination of the tendon to the beam axis is sufficiently small in practice for the horizontal component of the tendon force to be taken as equal to the tendon force itself. Hence in eqns (9.2–2) and (9.2–3), and all equations to follow, no distinction need be made between the tendon force and its horizontal component.

Still considering the typical section in Fig. 9.2–1, let us introduce symbols as follows:

M_d = sagging moment due to dead load

$M_{imax}(M_{imin})$ = maximum (minimum) sagging moment due to imposed load

M_r = the moment range $M_{imax} - M_{imin}$

The prestress values at a typical section must be such that the following stress criteria are satisfied under service conditions:

$$f_1 - \frac{M_{\mathrm{imax}} + M_{\mathrm{d}}}{Z_1} \geq f_{\mathrm{amin}} \tag{9.2-4}$$

$$f_1 - \frac{M_{\mathrm{imin}} + M_{\mathrm{d}}}{Z_1} \leq f_{\mathrm{amax}} \tag{9.2-5}$$

$$f_2 + \frac{M_{\mathrm{imax}} + M_{\mathrm{d}}}{Z_2} \leq f_{\mathrm{amax}} \tag{9.2-6}$$

$$f_2 + \frac{M_{\mathrm{imin}} + M_{\mathrm{d}}}{Z_2} \geq f_{\mathrm{amin}} \tag{9.2-7}$$

where f_{amax} is the **maximum allowable stress** in the concrete and f_{amin} the **minimum allowable stress**, the sign convention being, as usual, positive for compression; BS 8110 gives the values in Tables 9.2–1 and 9.2–2, which include allowances for the partial safety factor.

Comments on eqns (9.2–4) to (9.2–7)
With some loss in generality, it is possible to use only two of these four equations. See the **simplified design procedure in Problem 9.1 at the end of this chapter.**

Table 9.2–1   **Compressive stresses in concrete for the serviceability limit states** (BS 8110 : Clause 4.3.4.2)

Nature of loading	Allowable compressive stresses (f_{amax})
Design load in bending	$0.33f_{\mathrm{cu}}$ (in continuous beams this may be increased to $0.4f_{\mathrm{cu}}$ within the range of support moments)
Design load in direct compression	$0.25f_{\mathrm{cu}}$

Table 9.2–2   **Flexural tensile stresses for Class 2 members: serviceability limit state of cracking** (BS 8110 : Clause 4.3.4.3)

	Allowable tensile stress $(-f_{\mathrm{amin}})$ (N/mm^2)			
Characteristic strength f_{cu} (N/mm^2)	30	40	50	60
Pre-tensioned members	—	2.9	3.2	3.5
Post-tensioned members	2.1	2.3	2.6	2.8

Note:
(a)   For Class 1 members, $f_{\mathrm{amin}} = 0$.
(b)   Table 9.2–2 gives the allowable stresses in tension and hence a negative sign should be used when assigning these stress values to f_{amin}.
(c)   Designers usually limit the tensile stresses under service conditions to less than the limiting values in Table 9.2–2. For example, for a post-tensioned member of $f_{\mathrm{cu}} = 50$ N/mm^2, f_{amin} may well be taken as, say, -2 N/mm^2, instead of -2.6 N/mm^2 as permitted by BS 8110.

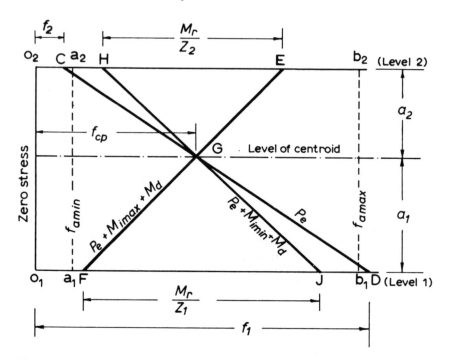

Fig. 9.2–2   Stresses in a prestressed section

Figure 9.2–2 shows the stresses in a prestressed beam section. The line $O_1a_1b_1$ represents level 1, i.e. the beam-soffit; the line $O_2a_2b_2$ represents level 2, i.e. the beam top. Line CGD is the stress distribution due to P_e; therefore O_1D is the prestress f_1 and O_2C is f_2. Line HGJ is the stress distribution due to $P_e + M_{imin} + M_d$ and EGF that due to $P_e + M_{imax} + M_d$. The application of a sagging moment rotates the stress-distribution line clockwise about G. O_1O_2 is the ordinate for zero stress; similarly a_1a_2 and b_1b_2 are the ordinates for the stresses f_{amin} and f_{amax} respectively. The reader should note that:

(a)   Equation (9.2–4) represents the condition that point F must not pass beyond the line a_1a_2.
(b)   Equation (9.2–5) represents the condition that point J must not pass beyond the line b_1b_2.
(c)   Similarly, eqns (9.2–6) and (9.2–7) represent the conditions that E and H must not lie outside the region a_2b_2.
(d)   Under service condition, the maximum change of stress at the bottom fibres is M_r/Z_1, where M_r is the range of imposed moments $M_{imax} - M_{imin}$.
(e)   Similarly, the maximum change of service stress at the top fibres is M_r/Z_2.
(f)   Therefore the minimum Z's to be provided must satisfy the conditions:

$$\left.\begin{array}{c}Z_1 \\ Z_2\end{array}\right\} \geq \frac{M_{\text{imax}} - M_{\text{imin}}}{f_{\text{amax}} - f_{\text{amin}}} \qquad (9.2\text{-}8)$$

that is, the minimum Z's are independent of M_d, f_1 and f_2.

(g) The **critical section** of the beam is where the imposed-load moment range $M_r(= M_{\text{imax}} - M_{\text{imin}})$ is a maximum. In particular, M_{imax} at the critical section is not necessarily larger than that at another section, and M_{imin} at the critical section is not necessarily smaller than that at another section.

(h) If, at the critical section, the actual Z_1 (or Z_2) is exactly equal to the minimum value of eqn (9.2-8), then the prestress f_1 (or f_2) must have that unique value which makes eqns (9.2-4) and (9.2-5) (or eqns 9.2-6 and 9.2-7) identities. Referring to Fig. 9.2-2, for such a case the points F and J (or H and E) fall on a_1 and b_1 (or a_2 and b_2) respectively.

(i) In practice it is rare for the Z's actually provided to be exactly the minima required. Therefore, f_1 and f_2 may vary within limits. The minimum f_1 required is that at which point F coincides with point a_1; the minimum f_2 required is that at which H coincides with a_2. Similarly, the maximum permissible f_1 is that which makes J coincide with b_1, and the maximum permissible f_2 makes E coincide with b_2. In other words, the minimum required prestresses are those that make eqns (9.2-4) and (9.2-7) identities:

$$\text{min. reqd } f_1 = f_{\text{amin}} + \frac{M_{\text{imax}} + M_d}{Z_1} \qquad (9.2\text{-}9)$$

$$\text{min. reqd } f_2 = f_{\text{amin}} - \frac{M_{\text{imin}} + M_d}{Z_2} \qquad (9.2\text{-}10)$$

Similarly, the maximum permissible prestresses are those that make eqns (9.2-5) and (9.2-6) identities:

$$\text{max. perm. } f_1 = f_{\text{amax}} + \frac{M_{\text{imin}} + M_d}{Z_1} \qquad (9.2\text{-}11)$$

$$\text{max. perm. } f_2 = f_{\text{amax}} - \frac{M_{\text{imax}} + M_d}{Z_2} \qquad (9.2\text{-}12)$$

(j) Referring to Fig. 9.2-2, the prestress at the centroid of the section is

$$f_{\text{cp}} = \frac{P_e}{A} \qquad (9.2\text{-}13)$$

where A is the cross-sectional area. Minimum f_{cp} is compatible with minimum f_1 and f_2; hence, from eqn (9.2-13), the required **minimum prestressing force**, P_{emin}, is that which gives the minimum f_1 and f_2. From eqns (9.2-2) and (9.2-3),

$$P_e = \frac{(f_1 Z_1 + f_2 Z_2)A}{Z_1 + Z_2} \qquad (9.2\text{-}14)$$

$$e_s = \frac{(f_1 - f_2)Z_1 Z_2}{(f_1 Z_1 + f_2 Z_2)A} \qquad (9.2\text{-}15)$$

To obtain P_{emin}, and the e_s to be used with P_{emin}, it is only necessary to insert in these equations the minimum f_1 and f_2. Substituting eqns (9.2–9) and (9.2–10) into eqns (9.2–14) and (9.2–15),

$$P_{emin} = \frac{[f_{amin}(Z_1 + Z_2) + M_r]A}{Z_1 + Z_2} \tag{9.2–16}$$

$$\frac{e_s}{(\text{for } P_{emin})} = \frac{Z_2 M_{imax} + Z_1 M_{imin} + (Z_1 + Z_2)M_d}{[f_{amin}(Z_1 + Z_2) + M_r]A} \tag{9.2–17}$$

Example 9.2–1
A Class 1 pre-tensioned concrete beam is simply supported over a 10 m span. The characteristic imposed load Q_k is a 100 kN force at midspan. The concrete characteristic strength is 50 N/mm² and the unit weight of concrete is 23 kN/m³.

(a)   Determine the minimum required sectional moduli for the service condition.
(b)   If the section adopted is of area 120 000 mm² and exactly the minimum required moduli, determine the effective prestressing force P_e required under service condition and the tendon eccentricity e_s at midspan.

SOLUTION
For Class 1 members, $f_{amin} = 0$. From Table 9.2–1,

$$f_{amax} = 0.33 \times 50 = 16.5 \text{ N/mm}^2$$

design imposed load for the service condition $= 1.0 \, Q_k = 100$ kN

Therefore

$$M_{imax} = \tfrac{1}{4} \times 100 \times 10 = 250 \text{ kNm}; \qquad M_{imin} = 0$$

(a)   From eqn (9.2–8),

$$\text{min. reqd } Z = \frac{250 \times 10^6}{16.5 - 0} = \underline{15.15 \times 10^6 \text{ mm}^3}$$

(b)   Adopted section: $A = 120\,000$ mm²; $Z_1 = Z_2 = 15.15 \times 10^6$ mm³.

$$\text{design dead load} = \frac{120 \times 10^3}{10^6} \times 23 = 2.76 \text{ kN/m}$$

$$M_d = \tfrac{1}{8} \times 2.76 \times 10^2 = 34.5 \text{ kNm}$$

Since exactly the minimum required Z's have been used, eqns (9.2–4) to (9.2–7) become identities, as explained in statement (h) above. Equations (9.2–4) and (9.2–5) will give the same value for f_1; similarly eqns (9.2–6) and (9.2–7) will give the same f_2. Use, say, eqns (9.2–4) and (9.2–6):

$$f_1 - \frac{250 \times 10^6 + 34.5 \times 10^6}{15.15 \times 10^6} = 0 \qquad \text{therefore } f_1 = 18.78 \text{ N/mm}^2$$

$$f_2 + \frac{250 \times 10^6 + 34.5 \times 10^6}{15.15 \times 10^6} = 16.5 \quad \text{therefore } f_2 = -2.28 \text{ N/mm}^2$$

Substituting into eqns (9.2–14) and (9.2–15) (or eqns 9.2–2 and 9.2–3),

$$P_e = 990 \text{ kN} \qquad e_s = 161 \text{ mm}$$

Example 9.2–2

A Class 2 post-tensioned concrete beam is simply supported over a 10 m span. The characteristic imposed load consists of a single 100 kN force at midspan. The characteristic concrete strength is 50 N/mm^2 and the unit weight of concrete is 23 kN/m^3. The beam is of uniform section having the following properties: area $A = 120\,000$ mm^2, Z_1 (bottom) $= 19.0 \times 10^6$ mm^3, Z_2 (top) $= 21.7 \times 10^6$ mm^3. Determine for the service condition:

(a) the minimum effective prestressing force required (P_{emin}) and the corresponding midspan tendon eccentricity (e_s);
(b) the maximum effective prestressing force (P_{emax}) that may safely be used, and the midspan tendon eccentricity (e_s) for this force.

SOLUTION
From Table 9.2–1,

$$f_{amax} = 0.33 \times 50 = 16.5 \text{ N/mm}^2$$

From Table 9.2–2,

$$f_{amin} = -2.55 \text{ N/mm}^2$$

(But see note (c) following Table 9.2–2.)

By inspection, the critical section is at midspan, where

$$M_{imax} = 250 \text{ kNm} \qquad M_{imin} = 0 \quad \text{and} \quad M_d = 34.5 \text{ kNm}$$

(all as in Example 9.2–1).

(a) P_{emin} is associated with the minimum required values of f_1 and f_2. From eqns (9.2–9) and (9.2–10),

$$\text{min. reqd } f_1 = -2.55 + \frac{250 \times 10^6 + 34.5 \times 10^6}{19.0 \times 10^6} = 12.42 \text{ N/mm}^2$$

$$\text{min. reqd } f_2 = -2.55 - \frac{0 + 34.5 \times 10^6}{21.7 \times 10^6} = -4.14 \text{ N/mm}^2$$

Substituting into eqns (9.2–14) and (9.2–15) (or eqns 9.2–2 and 9.2–3):

$$P_{emin} = 431 \text{ kN} \qquad e_s = 390 \text{ mm}$$

Alternatively, P_{emin} and e_s can be obtained directly from eqns (9.2–16) and (9.2–17) without first calculating f_1 and f_2.

(b) P_{emax} is associated with the maximum permissible values of f_1 and f_2. From eqns (9.2–11) and (9.2–12),

$$\text{max. perm. } f_1 = 16.5 + \frac{0 + 34.5 \times 10^6}{19.0 \times 10^6} = 18.32 \text{ N/mm}^2$$

$$\text{max. perm. } f_2 = 16.5 - \frac{250 \times 10^6 + 34.5 \times 10^6}{21.7 \times 10^6} = 3.39 \text{ N/mm}^2$$

Substituting into eqns (9.2–14) and (9.2–15) (or eqns 9.2–2 and 9.2–3),

$$\underline{P_{\text{emax}} = 1243 \text{ kN}} \qquad \underline{e_s = 122 \text{ mm}}$$

Example 9.2–3

Equation (9.2–16) shows that the minimum required effective prestressing force is independent of the dead load. Explain how you could have arrived at this conclusion by common-sense reasoning.

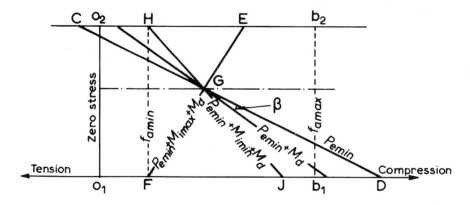

Fig. 9.2–3   Stress distributions when P_{emin} is used

SOLUTION

Figure 9.2–3 shows the stress distributions when the effective prestressing force has the minimum required value P_{emin}.

Line EGF represents the condition $(P_{\text{emin}} + M_{\text{imax}} + M_d)$

Line HGJ represents the condition $(P_{\text{emin}} + M_{\text{imin}} + M_d)$

The limiting stress conditions, namely that points F and H fall on the f_{amin} stress ordinate, are completely defined by the $(P_{\text{emin}} + M_{\text{imax}} + M_d)$ line and the $(P_{\text{emin}} + M_{\text{imin}} + M_d)$ line. For specified values of M_{imax} and M_{imin}, these two lines are fixed relative to the $(P_{\text{emin}} + M_d)$ line. Irrespective of the value of the dead load moment M_d, the $(P_{\text{emin}} + M_d)$ line can be held in a prescribed absolute position by rotating the P_{emin} line, i.e. by changing the angle β in the figure. Since changing β necessitates changing only the tendon eccentricity, it means that the prestressing force P_{emin} cannot be affected by the dead load.

The permissible tendon zone

In practice the designer usually tries to keep the prestressing force constant along the beam. If the effective prestressing force has the value P_{emin} as calculated from eqn (9.2–16) for the critical section, then at this section the tendon eccentricity must have the particular value given by eqn (9.2–17). At any other section where $(M_{imax} - M_{imin})/(Z_1$ or $Z_2)$ is less than that at the critical section, this value of the prestressing force will be larger than the absolute minimum required for that section, and the tendon eccentricity e_s may be allowed to vary within a zone called the **permissible tendon zone**. In general if the effective prestressing force used is within the limits P_{emin} and P_{emax}, then throughout the beam the centroid of the tendons may lie anywhere within a permissible tendon zone.

The limits of the permissible tendon zone are in fact given by eqns (9.2–4) to (9.2–7) if f_1 and f_2 in these equations are expressed in terms of P_e and e_s (using eqns 9.2–2 and 9.2–3). The reader should verify that:

$$e_s \geq \frac{M_{imax} + M_d}{P_e} - \frac{Z_1}{A} + \frac{Z_1 f_{amin}}{P_e} \quad \text{(from eqn 9.2–4)} \quad (9.2–18)$$

$$e_s \geq \frac{M_{imax} + M_d}{P_e} + \frac{Z_2}{A} - \frac{Z_2 f_{amax}}{P_e} \quad \text{(from eqn 9.2–6)} \quad (9.2–19)$$

$$e_s \leq \frac{M_{imin} + M_d}{P_e} - \frac{Z_1}{A} + \frac{Z_1 f_{amax}}{P_e} \quad \text{(from eqn 9.2–5)} \quad (9.2–20)$$

$$e_s \leq \frac{M_{imin} + M_d}{P_e} + \frac{Z_2}{A} - \frac{Z_2 f_{amin}}{P_e} \quad \text{(from eqn 9.2–7)} \quad (9.2–21)$$

If the beam is of uniform cross-section and if the prestressing force P_e is constant along the beam, then the curves of eqns (9.2–18) and (9.2–19) are parallel, and so are those of eqns (9.2–20) and (9.2–21). Therefore, for such a beam, it is only necessary to use all the four equations at one section to determine which two are the governing equations. Where the cross-section or the prestressing force varies along the beam, then all four equations must be applied to a number of sections to draw the limits of the tendon zone.

Example 9.2–4*

If in Example 9.2–2 the effective prestressing force adopted in the design is the mean of the minimum required force and the maximum permissible force, plot the permissible tendon zone.

SOLUTION
From Example 9.2–2,

$$P_{emax} = 1243 \text{ kN} \qquad P_{emin} = 431 \text{ kN}$$

$$f_{amax} = 16.5 \text{ N/mm}^2 \qquad f_{amin} = -2.6 \text{ N/mm}^2$$

$$A = 120\,000 \text{ mm}^2 \qquad Z_1 = 19.0 \times 10^6 \text{ mm}^3 \qquad Z_2 = 21.7 \times 10^6 \text{ mm}^3$$

Therefore

$$P_e = \tfrac{1}{2}(P_{emax} + P_{emin}) = 837 \text{ kN}$$

---

* Thanks are due to Mr J. P. Withers for the solution to this example.

Consider midspan section

$$M_{\text{imax}} = 250 \text{ kNm} \qquad M_{\text{imin}} = 0 \qquad M_{\text{d}} = 34.5 \text{ kNm}$$

(all from Example 9.2–2).
Therefore, from eqn (9.2–18),

$$e_{\text{s}} \geq \frac{250 \times 10^6 + 34.5 \times 10^6}{837 \times 10^3} - \frac{19.0 \times 10^6}{120,000}$$
$$+ \frac{19.0 \times 10^6 \times (-2.6)}{837 \times 10^3}$$
$$\geq \underline{124 \text{ mm}}$$

Similarly,

$$e_{\text{s}} \geq 93 \text{ mm} \quad \text{(from eqn 9.2–19)}$$
$$e_{\text{s}} \leq 257 \text{ mm} \quad \text{(from eqn 9.2–20)}$$
$$e_{\text{s}} \leq 288 \text{ mm} \quad \text{(from eqn 9.2–21)}$$

Therefore, eqns (9.2–18) and (9.2–20) govern (at this section and at all other sections).

Consider section at 3 m from support
$M_{\text{imax}} = 150 \text{ kNm}$, $M_{\text{imin}} = 0$, $M_{\text{d}} = 29.1 \text{ kNm}$ (and the reader should verify these values).

From eqn (9.2–18): $e_{\text{s}} \geq \underline{-2 \text{ mm}}$

From eqn (9.2–20): $e_{\text{s}} \leq \underline{251 \text{ mm}}$

Consider section from 1 m from support
$M_{\text{imax}} = 50 \text{ kNm}$, $M_{\text{imin}} = 0$, $M_{\text{d}} = 12.4 \text{ kNm}$ (again the reader should verify these).

From eqn (9.2–18): $e_{\text{s}} \geq \underline{-142 \text{ mm}}$

From eqn (9.2–20): $e_{\text{s}} \leq \underline{231 \text{ mm}}$

Consider support section
$$M_{\text{imax}} = M_{\text{imin}} = M_{\text{d}} = 0$$

From eqn (9.2–18): $e_{\text{s}} \geq \underline{-216 \text{ mm}}$

From eqn (9.2–20): $e_{\text{s}} \leq \underline{216 \text{ mm}}$

The permissible tendon zone is as plotted in Fig. 9.2–4.

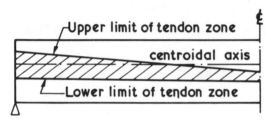

Fig. 9.2–4   Permissible tendon zone

Shear in prestressed concrete beams

Figure 9.2–5(a) shows a beam with a curved tendon profile; such a profile is commonly used in post-tensioned beams. At a typical section, the vertical component of the tendon force P_e produces a shear force acting on the beam:

$$V_p = -P_e \sin \beta = -P_e \left[\frac{de_s}{dx} \right] \tag{9.2–22}$$

where the negative sign is consistent with the sign convention in Fig. 9.2–5(b). Therefore the net shear force acting on the concrete section is

$$V_c = V + V_p \tag{9.2–23}$$

where V is the shear force due to the imposed load and the dead load. Specifically.

$$V_{cmax} = V_{imax} + V_d + V_p \tag{9.2–24}$$

$$V_{cmin} = V_{imin} + V_d + V_p \tag{9.2–25}$$

where V_{cmax} is the maximum net shear force acting on the concrete at that section, V_{imax} is the maximum shear force due to the imposed load, and so on. Ideally, the tendon profile should be such as to result in net shear forces of the least magnitude; this occurs if $V_{cmax} = -V_{cmin}$, i.e. as V_{imin} changes to V_{imax}, the shear force V_c is exactly reversed. Putting $V_{cmax} = -V_{cmin}$ in eqn (9.2–24) gives

$$V_p = -\tfrac{1}{2}[V_{imax} + V_{imin} + 2V_d] \tag{9.2–26}$$

Using eqn (9.2–22),

$$\frac{de_s}{dx} = \frac{1}{2P_e}[V_{imax} + V_{imin} + 2V_d] \tag{9.2–27}$$

Thus, the **ideal tendon profile** for shear is one having a slope at any point given by eqn (9.2–27). For simply supported beams, the load distribution producing V_{imax} (V_{imin}) usually also produces the moments M_{imax} (M_{imin}), and a further simplification of eqn (9.2–27) is possible, by integrating with respect to x:

$$\int de_s = \frac{1}{2P_e} \int \left[\frac{dM_{imax}}{dx} + \frac{dM_{imin}}{dx} + 2\frac{dM_d}{dx} \right] dx$$

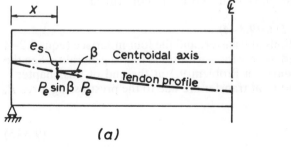

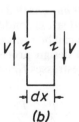

(a)   (b)

Fig. 9.2–5

(since $V = dM/dx$), or

$$e_s = \frac{1}{2P_e}[M_{imax} + M_{imin} + 2M_d] + C \tag{9.2-28}$$

where C is a constant of integration. Since the shear force V_p depends only on the shape of the tendon profile e_s but not on its absolute value, we are free to choose C such that the profile lies within the permissible tendon zone. If a profile of the ideal shape cannot be accommodated within the permissible zone, then a shape close to the ideal but lying within the zone may be used.

For the ideal profile of eqn (9.2–28), we have, from eqns (9.2–24) to (9.2–26),

$$\left.\begin{array}{r} V_{cmax} \\ V_{cmin} \end{array}\right\} = \pm\tfrac{1}{2}[V_{imax} - V_{imin}] \tag{9.2-29}$$

Consider an example: Suppose $V_{imax} = 1500$ kN, $V_{imin} = 300$ kN, $V_d = 400$ kN. Then without prestressing, $V_{cmax} = 1500 + 400 = 1900$ kN. With prestress and an ideal profile, $V_{cmax} = \tfrac{1}{2}[1500 - 300] = 600$ kN only.

9.3   Stresses at transfer

Service stresses were considered in Section 9.2. In design it is necessary to consider also the stresses at transfer. The stress criteria at transfer are expressed by eqns (9.2–4) to (9.2–7); if the subscript t is used to denote stresses at transfer and if the terms M_{imax} and M_{imin} are neglected (since they are negligible at transfer), then these equations become

$$f_{1t} - \frac{M_d}{Z_1} \geq f_{amint} \tag{9.3-1}$$

$$f_{1t} - \frac{M_d}{Z_1} \leq f_{amaxt} \tag{9.3-2}$$

$$f_{2t} + \frac{M_d}{Z_2} \leq f_{amaxt} \tag{9.3-3}$$

$$f_{2t} + \frac{M_d}{Z_2} \geq f_{amint} \tag{9.3-4}$$

where eqns (9.3–1) and (9.3–3) are normally not critical.

Comments on eqns (9.3–1) to (9.3–4)
These four stress conditions at transfer, and the four in service (eqns 9.2–4 to 9.2–7) can be reduced to only four, with some loss of generality. See the **simplified design procedure** in Problem 9.1 at the end of this chapter.

The prestresses f_{1t} and f_{2t} at transfer are due to the prestressing force P, namely

$$f_{1t} = \frac{P}{A} + \frac{Pe_s}{Z_1} \tag{9.3-5}$$

$$f_{2t} = \frac{P}{A} - \frac{Pe_s}{Z_2} \qquad (9.3\text{--}6)$$

where, as in eqn (9.2–1), P is related to the effective prestressing force P_e by the loss ratio

$$P = P_e/\alpha \qquad (9.3\text{--}7)$$

Table 9.3–1   Allowable compressive stresses at transfer (BS 8110 : Clause 4.3.5)

Nature of stress distribution	f_{amaxt}
Extreme fibre	$0.5f_{ci}$[a]
Uniform or near uniform	$0.4f_{ci}$

[a] f_{ci} = cube strength at transfer.

In practice, the stress criteria in eqns (9.3–1) and (9.3–3) are almost invariably met; but eqns (9.3–2) and (9.3–4) must be checked. BS 8110 limits f_{amaxt} to the values in Table 9.3–1. f_{amint} for Class 1 members should not be less than -1 N/mm^2 (i.e. tension not to exceed 1 N/mm^2); f_{amint} values for Class 2 members are as in Table 9.2–2, where f_{cu} is now to be interpreted as the cube strength f_{ci} at transfer. As explained in note (c) following Table 9.2–2, designers usually do not set f_{amin} for the service conditions to the limiting stresses in that table; however, stresses at transfer are of a temporary nature, and designers are more willing to set f_{amint} to the limiting values.

Example 9.3–1

Given that for the Class 2 post-tensioned beam in Example 9.2–2, the prestress loss ratio α is 0.8 and the cube strength f_{ci} at transfer is 40 N/mm^2, determine the permissible tendon zone for an effective prestressing force P_e of 837 kN, considering both the service stresses and the stresses at transfer. Allowable stresses are as in BS 8110.

SOLUTION

(For the service condition, the permissible tendon zone has been worked out in Example 9.2–4 and plotted in Fig. 9.2–4.) At transfer, $P = P_e/\alpha = 837/0.8 = 1046$ kN. Consider midspan section: from eqns (9.3–2) and (9.3–4),

$$f_{1t} \leq f_{amaxt} + \frac{M_d}{Z_1}$$

$$\leq 20 \ (\text{Table 9.3--1}) + \frac{34.5 \times 10^6}{19.0 \times 10^6} \leq 21.82 \ \text{N/mm}^2$$

$$f_{2t} \geq f_{amint} - \frac{M_d}{Z_2}$$

$$\geq -2.3 \ (\text{Table 9.2--2}) - \frac{34.5 \times 10^6}{21.7 \times 10^6} \geq -3.89 \ \text{N/mm}^2$$

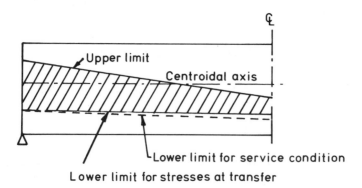

Fig. 9.3–1

Hence, from eqns (9.3–5) and (9.3–6),

$$\frac{P}{A} + \frac{Pe_s}{Z_1} \leq 21.82$$

$$\frac{P}{A} - \frac{Pe_s}{Z_2} \geq -3.89$$

With $P = 1046$ kN and A, Z_1 and Z_2 as in Example 9.2–2, we have $e_s \leq$ 238 mm from the first condition (and, less critically, $e_s \leq 262$ mm from the second).

Similarly, the reader should verify that, at transfer,

$e_s \leq 233$ mm at 3 m from support

$e_s \leq 217$ mm at 1 m from support

$e_s \leq 205$ mm at support

Figure 9.3–1 shows the permissible tendon zone which satisfies the stress conditions both in service and at transfer. To increase the depth of the zone at midspan, a lower effective prestressing force than $P_e = 837$ kN has to be used.

9.4   Loss of prestress

Design calculations for the loss of prestress are usually straightforward; simple practical procedures are given in BS 8110: Clauses 4.8 and 4.9.

In general, allowance should be made for losses of prestress resulting from:

(a) Relaxation of the steel comprising the tendons: Typically this produces a loss of about 5%.

(b) Elastic deformation of the concrete: In a pre-tensioned beam there is an immediate loss of prestress at transfer, resulting from the elastic shortening of the beam. In post-tensioning, the elastic shortening of the concrete occurs when the tendons are actually being tensioned;

therefore where there is only one tendon, there is no loss due to elastic deformation. But if there are several tendons, then the tensioning of each tendon will cause a loss in those tendons that have already been tensioned. Very roughly, the loss of pretress due to elastic deformation is about 5–10% for pretensioned beams and 2 or 3% for post-tensioned beams.

(c) Shrinkage and creep of the concrete: Typically, these produce a 10–20% loss.

(d) Slip of the tendons during anchoring: This loss is important where the tendons are short, e.g. in some post-tensioned beams.

(e) Friction between tendon and duct in post-tensioned beams: Varies from 1 to 2% in simple beams, with a fairly flat tendon profile, to over 10% in continuous beams.

Example 9.4–1
For the prestressed concrete section in Fig. 9.4–1, determine the loss of prestress due to a **steel relaxation** f_r (N/mm^2).

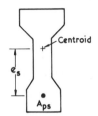

Section data:

Area = A

2nd moment of area = I

Young's moduli : E_s, E_c

Tendon fully bonded.

Fig. 9.4–1

SOLUTION
Let δf_s = loss of prestress in the tendon due to steel relaxation;
δf_c = the corresponding reduction of the concrete compressive stress at tendon level.

In the hypothetical case of zero change in the length of the tendon, the loss of prestress in the tendon will by definition be f_r. Therefore if the actual loss of prestress is δf_s, then there must have been a unit extension of the tendon of $(f_r - \delta f_s)/E_s$.
    Hence the compatibility condition is

$$\frac{f_r - \delta f_s}{E_s} = \frac{\delta f_c}{E_c}$$

The equilibrium condition is, by statics,

$$\delta f_c = \frac{\delta f_s A_{ps}}{A}\left[1 + \frac{e_s^2}{k^2}\right]$$

where $k = \sqrt{(I/A)}$ is the radius of gyration.
    Eliminating δf_c from the compatibility and equilibrium conditions,

$$\delta f_s = \frac{f_r}{1 + \frac{E_s A_{ps}}{E_c A}\left[1 + \frac{e_s^2}{k^2}\right]}$$

(9.4–1)

For practical prestressed concrete sections, the quantity $E_s A_{ps}(1 + e_s^2/k^2)/E_c A$ is small compared with unity; a numerical example, on p. 110 of Reference 2, gives 0.05 as a typical value. Therefore, in design the **relaxation loss** is usually taken simply as

$$\delta f_s \text{ (relaxation)} = f_r$$

(9.4–2)

Example 9.4–2
For the pretensioned concrete section shown in Fig. 9.4–1, determine the loss of prestress due to the **elastic deformation** of the concrete, if the initial prestress in the tendon is f_s immediately before transfer.

SOLUTION
Let δf_s = loss of prestress in the tendon due to elastic deformation;
$\quad\quad f_c$ = concrete compressive stress at tendon level after transfer.
The equilibrium condition is, from statics,

$$f_c = \frac{(f_s - \delta f_s)A_{ps}}{A}\left[1 + \frac{e_s^2}{k^2}\right]$$

where $k = \sqrt{(I/A)}$ is the radius of gyration. The compatibility condition is

$$\frac{f_c}{E_c} = \frac{\delta f_s}{E_s}$$

Eliminating f_c from the equilibrium and compatibility conditions,

$$\delta f_s = \frac{f_s \alpha_e \varrho\left[1 + \frac{e_s^2}{k^2}\right]}{1 + \alpha_e \varrho\left[1 + \frac{e_s^2}{k^2}\right]}$$

(9.4–3)

where $\alpha_e = E_s/E_c$ and $\varrho = A_{ps}/A$. As explained in the paragraph following eqn (9.4–1), the denominator on the right-hand side of eqn (9.4–3) is nearly equal to unity. Therefore, in design, the **elasticity loss** is usually taken as

$$\delta f_s \text{ (elasticity)} = f_s \alpha_e \varrho\left[1 + \frac{e_s^2}{k^2}\right]$$

(9.4–4)

Example 9.4–3
For the prestressed concrete section in Fig. 9.4–1, determine the loss of prestress due to a **concrete shrinkage** ε_{cs}.

SOLUTION
Let δf_s = loss of prestress in the tendon due to concrete shrinkage;
$\quad\quad \delta f_c$ = the corresponding reduction in the concrete compressive stress at tendon level.

The equilibrium condition is

$$\delta f_c = \frac{\delta f_s A_{ps}}{A} \left[1 + \frac{e_s^2}{k^2} \right] \quad \left(\text{where } k = \sqrt{\frac{I}{A}} \right)$$

The compatibility condition is

$$\varepsilon_{cs} - \frac{\delta f_c}{E_c} = \frac{\delta f_s}{E_s}$$

Eliminating δf_c from the two equations,

$$\delta f_s = \frac{E_s \varepsilon_{cs}}{1 + \dfrac{E_s A_{ps}}{E_c A} \left[1 + \dfrac{e_s^2}{k^2} \right]} \tag{9.4-5}$$

As explained in the paragraph following eqn (9.4–1), the quantity $E_s A_{ps}(1 + e_s^2/k^2)/E_c A$ is small compared with unity. Therefore in design the **skinkage loss** is usually taken as

$$\delta f_s \text{ (shrinkage)} = E_s \varepsilon_{cs} \tag{9.4-6}$$

Example 9.4–4

For the prestressed concrete section in Fig. 9.4–1, determine the loss of prestress due to the **creep of concrete**, if the creep coefficient is ϕ.

SOLUTION

Let δf_s = loss of prestress in the tendon due to creep of concrete;
$\quad f_s$ = prestress in the tendon immediately after transfer;
$\quad f_c$ = concrete compressive stress at tendon level immediately after transfer;
$\quad f_{c,f}$ = final concrete compressive stress at tendon level, after loss of prestress δf_s.

In the hypothetical case of the concrete stress remaining constant at f_c,

$$\text{creep strain} = \phi \frac{f_c}{E_c}$$

$$\delta f_s = E_s \times \text{creep strain}$$

$$= \alpha_e \phi f_c \quad (\text{where } \alpha_e = E_s/E_c)$$

Similarly, in the hypothetical case of the concrete stress being kept at a constant value equal to $f_{c,f}$, we have

$$\delta f_s = \alpha_e \phi f_{c,f}$$

In fact, as the loss of prestress occurs, the concrete stress decreases continually from f_c to $f_{c,f}$. Therefore, we know that

$$\delta f_s < \alpha_e \phi f_c \tag{9.4-7}$$

and

$$\delta f_s > \alpha_e \phi f_{c,f} \tag{9.4-8}$$

Taking a mean value for design, say,

$$\delta f_s = \alpha_e \phi \frac{f_c + f_{c,f}}{2} \tag{9.4–9}$$

From statics,

$$\frac{f_{c,f}}{f_c} = \frac{f_s - \delta f_s}{f_s}$$

i.e.

$$f_{c,f} = f_c \left(1 - \frac{\delta f_s}{f_s}\right)$$

Substituting into eqn (9.4–9),

$$\delta f_s = \alpha_e \phi f_c \left[1 - \frac{\delta f_s}{2f_s}\right]$$

or

$$\delta f_s = \alpha_e \phi f_c \left[1 + \alpha_e \phi \frac{f_c}{2f_s}\right]^{-1}$$

$$= \alpha_e \phi f_c \left[1 - \alpha_e \phi \frac{f_c}{2f_s} + \left(\alpha_e \phi \frac{f_c}{2f_s}\right)^2 \right.$$

$$\left. - (\quad)^3 + (\quad)^4 - \cdots \right]$$

In practice, the quantity $\alpha_e \phi f_c / 2f_s$ is usually not small enough compared with unity to be neglected; however, the second- and higher-order terms certainly are negligible. Hence, the **creep loss** is

$$\delta f_s \text{ (creep)} = \alpha_e \phi f_c \left[1 - \frac{\alpha_e \phi f_c}{2f_s}\right] \tag{9.4–10}$$

$$\doteqdot \alpha_e \phi f_c$$

where $\alpha_e = E_s/E_c$;
ϕ = creep coefficient;
f_c = concrete compressive stress at tendon level before loss occurs;
f_s = prestress in tendon before loss.

Comments
(a) For design purpose, creep may be assumed to be proportional to stress both in tension and compression. Therefore, the initial concrete stresses, linearly distributed across the beam section, will cause creep strains which are also linearly distributed with zero strain at the neutral axis, which remains stationary. Any change in the tendon force will produce stress changes which are also linearly distributed. Therefore, as the creep loss occurs, plane sections will continue to remain plane, with zero stress at the original level of the neutral axis (but see Comment (b) below).
(b) With reference to Comment (a) above, it is appropriate to quote

p. 113 of Reference 2: 'strictly speaking, as the concrete creeps the neutral axis in fact shifts slightly towards the tendon, since the tendon becomes relatively stiffer. However, such slight shifting of the neutral axis is of little significance in practice, and it makes sense to assume that the position of the neutral axis remains unchanged.'

Example 9.4–5
The cross-section of a post-tensioned concrete beam is as shown in Fig. 9.4–1, with $A = 5 \times 10^4$ mm^2, $I = 4.5 \times 10^8$ mm^4, $A_{ps} = 350$ mm^2, $E_s = 200$ kN/mm^2, $E_c = 34$ kN/mm^2. The beam is simply supported over a 10 m span and the tendon profile is as shown in Fig. 9.4–2; the effective prestress f_s immediately after transfer is 1290 N/mm^2. Calculate the loss of prestress δf_s for the middle third of the span, if the steel relaxation is 129 N/mm^2, the concrete shrinkage ε_{cs} is 450×10^{-6} and the creep coefficient ϕ is 2.

SOLUTION

$$P = A_{ps}f_s = 350 \times 1290$$

$$= 451.5 \text{ kN immediately after transfer}$$

Concrete prestress at tendon level is

$$f_c = \frac{P}{A}\left(1 + \frac{e_s^2}{k^2}\right) \quad (\text{where} \quad k^2 = I/A)$$

Within the middle third of span:

$$f_c = \frac{451.5}{5 \times 10^4}\left(1 + \frac{100^2}{4.5 \times 10^8/5 \times 10^4}\right) \text{ kN/mm}^2$$

$$= 19.06 \text{ N/mm}^2$$

From Examples 9.4–1, 9.4–3 and 9.4–4 the total loss of prestress due to relaxation, shrinkage and creep is

$$\delta f_s = f_r + E_s\varepsilon_{cs} + \alpha_e\phi f_c\left(1 - \frac{\alpha_e\phi f_c}{2f_s}\right)$$

$$= 129 + 200 \times 10^3 \times 450 \times 10^{-6}$$

$$+ \frac{200 \times 10^3}{34 \times 10^3} \times 2 \times 19.06\left(1 - \frac{200 \times 10^3}{34 \times 10^3} \times \frac{2 \times 19.06}{2 \times 1290}\right)$$

$$= 424 \text{ N/mm}^2$$

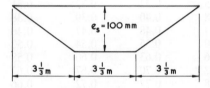

$e_s = 100$ mm

$3\frac{1}{3}$ m $\quad$ $3\frac{1}{3}$ m $\quad$ $3\frac{1}{3}$ m

Fig. 9.4–2

Comments
Examples 9.4–1 to 9.4–5 all deal with prestressed beams with a single tendon. The loss of prestress in members with **multi-level tendons** is dealt with in Problems 9.5 and 9.6.

9.5   The ultimate limit state: flexure (BS 8110)

Having designed the member for the service conditions and checked the stresses at transfer, it is still necessary to check that the ultimate limit state requirements are satisfied; for this latter purpose, the partial safety factors should be those for the ultimate limit state (see Tables 1.5–1 and 1.5–2). First consider the flexural strength. Using BS 8110's rectangular stress block (see Fig. 4.4–5) the resistance moment of a rectangular beam is immediately seen to be

$$M_u = f_{pb} A_{ps}(d - 0.45x) \tag{9.5–1(a)}$$

where f_{pb} = the tensile stress in the tendons at beam failure;
A_{ps} = the area of the prestressing tendons in the tension zone. (Prestressing tendons in the compression zone should be ignored in using this equation.);
d = the effective depth to the centroid of A_{ps}; and
x = the neutral axis depth.

This then is BS 8110's equation for the ultimate flexural strength; it applies to rectangular beams and to flanged beams in which the neutral axis lies within the flange. Values of f_{pb} and x for such beams may be taken from Table 9.5–1 for bonded tendons.

Table 9.5–1   Conditions at the ultimate limit state for pre-tensioned beams or bonded post-tensioned beams (BS 8110 : Clause 4.3.7.3)

$\dfrac{f_{pu}A_{ps}}{f_{cu}bd}$	Design stress in tendons as a proportion of the design strength, $f_{pb}/0.87f_{pu}$			Ratio of depth of neutral axis to that of the centroid of the tendons in the tension zone, x/d		
	$f_{pe}/f_{pu} =$			f_{pe}/f_{pu}		
	0.6	0.5	0.4	0.6	0.5	0.4
0.05	1.0	1.0	1.0	0.11	0.11	0.11
0.10	1.0	1.0	1.0	0.22	0.22	0.22
0.15	0.99	0.97	0.95	0.32	0.32	0.31
0.20	0.92	0.90	0.88	0.40	0.39	0.38
0.25	0.88	0.86	0.84	0.48	0.47	0.46
0.30	0.85	0.83	0.80	0.55	0.54	0.52
0.35	0.83	0.80	0.76	0.63	0.60	0.58
0.40	0.81	0.77	0.72	0.70	0.67	0.62
0.45	0.79	0.74	0.68	0.77	0.72	0.66
0.50	0.77	0.71	0.64	0.83	0.77	0.69

For **unbonded tendons**, values of f_{pb} and x for use in eqn (9.5–1(a)) may be taken as

$$f_{pb} = f_{pe} + \frac{7000}{l/d}\left[1 - 1.7\frac{f_{pu}A_{ps}}{f_{cu}bd}\right]$$

$$\leq 0.7f_{pu} \qquad\qquad (9.5\text{–}1(b))$$

$$x = 2.47\left[\frac{f_{pu}A_{ps}}{f_{cu}bd}\right]\left[\frac{f_{pb}}{f_{pu}}\right]d \qquad\qquad (9.5\text{–}1(c))$$

where f_{pe} = design effective prestress in the tendons after all losses;
$\quad\ f_{pu}$ = characteristic strength of the tendons;
$\qquad l$ = normally taken as the length of the tendons between the end anchorages;
$\qquad b$ = width of the rectangular beam or the effective width of a flanged beam;

and the other symbols have their usual meanings.

With reference to the use of Table 9.5–1 and eqns (9.5–1), the term effectively **bonded post-tensioned beam** refers to a beam in which the space between the tendon and the duct is grouted after tensioning; similarly, an **unbonded post-tensioned beam** is one where the duct or ducts are not grouted. Where the neutral axis of a flanged beam lies outside the flange, eqn (9.5–1(a)) is not applicable, and a more general approach is required (see Fig. 9.5–4 and Comment (b) at the end of Example 9.5–1).

A general flexural theory

The general flexural theory in Section 4.2 may be modified for application to prestressed concrete. Consider the beam section in Fig. 9.5–1(a). The tendon strain ε_{pb} at the ultimate condition may be considered to be made up of two parts: (a) the strain ε_{pe} due to the effective tendon prestress after losses and (b) the additional strain ε_{pa} produced by the applied loading. Thus

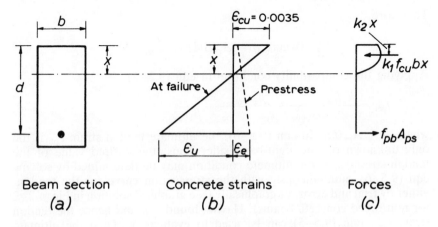

Beam section   Concrete strains   Forces
(a)   (b)   (c)

Fig. 9.5–1

$$\varepsilon_{pb} = \varepsilon_{pe} + \varepsilon_{pa} \tag{9.5-2}$$

The prestress strain ε_{pe} is f_{pe}/E_s if the stress is within the elastic limit, but may in any case be determined from f_{pe} and the stress/strain curve. The additional strain ε_{pa} can be evaluated by considering the change in concrete strain at the level of the tendon. In Fig. 9.5–1(b), the broken line represents the strain distribution in the concrete produced by the effective prestressing force. Thus ε_e is $(1/E_c)$ times the concrete prestress at the tendon level. The strain ε_u is the average concrete strain at that level at the ultimate condition. Where effective bond exists, the additional strain in the tendon is $\varepsilon_{pa} = \varepsilon_e + \varepsilon_u$. In an unbonded post-tensioned beam ε_{pa} will be less than $\varepsilon_e + \varepsilon_u$. In general, we can write

$$\varepsilon_{pa} = \beta_1\varepsilon_e + \beta_2\varepsilon_u \tag{9.5-3}$$

where the **bond factors** β_1 and β_2 may be taken as unity for pre-tensioned beams or bonded post-tensioned beams; for unbonded post-tensioned beams, β_1 is often taken as 0.5 and β_2 as between 0.1 and 0.25. From the geometry of Fig. 9.5–1(b), ε_u may be expressed in terms of ε_{cu}, so that

$$\varepsilon_{pa} = \beta_1\varepsilon_e + \beta_2\frac{d-x}{x}\varepsilon_{cu} \tag{9.5-4}$$

Now

$$\varepsilon_{pb} = \varepsilon_{pe} + \varepsilon_{pa}$$

$$= \varepsilon_{pe} + \beta_1\varepsilon_e + \beta_2\varepsilon_{cu}\frac{d-x}{x}$$

whence

$$\frac{x}{d} = \frac{\beta_2\varepsilon_{cu}}{\beta_2\varepsilon_{cu} + \varepsilon_{pb} - \varepsilon_{pe} - \beta_1\varepsilon_e} \tag{9.5-5}$$

Applying the equilibrium condition to Fig. 9.5–1(c),

$$f_{pb}A_{ps} = k_1f_{cu}bx$$

Therefore

$$f_{pb} = \frac{k_1f_{cu}}{\varrho}\frac{x}{d} \quad \text{(where } \varrho = A_{ps}/bd\text{)}$$

Applying the compatibility condition in eqn (9.5–5),

$$f_{pb} = \frac{k_1f_{cu}}{\varrho}\frac{\beta_2\varepsilon_{cu}}{\beta_2\varepsilon_{cu} + \varepsilon_{pb} - \varepsilon_{pe} - \beta_1\varepsilon_e} \tag{9.5-6}$$

where $\varrho = A_{ps}/bd$. In eqn (9.5–6), the ultimate tendon strain ε_{pb} is the only unknown on the right-hand side. Thus, the desired value of the tendon stress f_{pb} at the ultimate condition may be determined by solving eqn (9.5–6) simultaneously with the stress/strain curve for the tendon, either by trial and error, or graphically (see graphical solution in Fig. 4.2–2 for reinforced concrete beams). Having found f_{pb}, and hence the tendon strain ε_{pb}, eqn (9.5–5) can be used to evaluate x. Then the ultimate moment of resistance is immediately obtained:

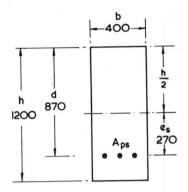

Fig. 9.5–2   Beam section

$$M_u = f_{pb}A_{ps}(d - k_2 x) \qquad (9.5-7)$$

The coefficient k_2 in eqn (9.5–7), and k_1 in eqn (9.5–6), may be read off from Fig. 4.4–4 or Fig. 4.4–5.

If BS 8110's rectangular stress block (Fig. 4.4–5) is used, then the above general approach may easily be modified for application to, say, flanged beams in which the neutral axis lies outside the flange, or to other non-rectangular beams.

Example 9.5–1

A bonded prestressed concrete beam is of rectangular section 400 mm by 1200 mm, as shown in Fig. 9.5–2. The tendon consists of 3300 mm^2 of standard strands, of characteristic strength 1700 N/mm^2, stressed to an effective prestress of 910 N/mm^2, the strands being located 870 mm from the top face of the beam. The concrete characteristic strength is 60 N/mm^2 and its modulus of elasticity 36 kN/mm^2. The stress/strain curve of the tendon is as shown in Fig. 9.5–3, with Young's modulus equal to 200 kN/mm^2 for stresses up to 1220 N/mm^2.

(a) Working from first principles, calculate the ultimate moment of resistance of the beam section.

(b) Suppose, as a result of a site error, the strands have not been tensioned, i.e. the effective tendon prestress is zero. Calculate the ultimate moment of resistance of the beam section.

Use BS 8110's rectangular stress block in Fig. 4.4–5.

SOLUTION

(a) Refer to the strain and force diagrams in Fig. 9.5–1, and reason from first principles.

$$A = 400 \times 1200 = 4.8 \times 10^5 \text{ mm}^2 \qquad A_{ps} = 3300 \text{ mm}^2$$

$$I = \tfrac{1}{12} \times 400 \times 1200^3 = 5.76 \times 10^{10} \text{ mm}^4$$

$$P_e = 3300 \times 910 = 3.003 \times 10^6 \text{ N}$$

$$e_s = 870 - 1200/2 = 270 \text{ mm}$$

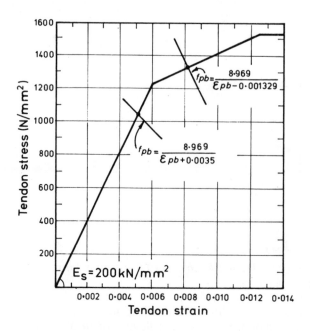

Fig. 9.5–3   Tendon stress/strain curve

Concrete stress at tendon level

$$= \frac{3.003 \times 10^6}{4.8 \times 10^5} + \frac{3.003 \times 10^6 \times 270^2}{5.76 \times 10^{10}} = 10.06 \text{ N/mm}^2$$

Concrete prestrain at tendon level, ε_e,

$$= \frac{10.06}{3.6 \times 10^3} = 0.000279$$

Concrete strain ε_u at tendon level at collapse

$$= \frac{870 - x}{x} \times 0.0035 = \left(\frac{3.045}{x} - 0.0035\right)$$

Change in tendon strain due to ultimate moment, ε_{pa},

$$= \text{change in concrete strain, } \varepsilon_e + \varepsilon_u$$

$$= 0.00279 + \frac{3.045}{x} - 0.0035 = \frac{3.045}{x} - 0.003221$$

Tendon prestrain $= \dfrac{910}{200 \times 10^3} = 0.00455$

Therefore, tendon strain ε_{pb} at collapse is given by

$$\varepsilon_{pb} = 0.00455 + \frac{3.045}{x} - 0.003221$$

$$= 0.001329 + \frac{3.045}{x}$$

or

$$x = \frac{3.045}{\varepsilon_{pb} - 0.001329} \qquad (9.5-8)$$

Concrete compression force

$$= k_1 f_{cu} bx \quad \text{(where } k_1 = 0.405 \text{ from Fig. 4.4–5)}$$
$$= 0.405 f_{cu} bx$$

Tendon force $= A_{ps} f_{pb}$
$$= 3300 f_{pb}$$

where f_{pb} is the as yet unknown tendon stress at beam collapse.
   Equating the concrete and tendon forces,

$$3300 f_{pb} = (0.405)(60)(400x)$$

i.e.

$$f_{pb} = 2.945x \qquad (9.5-9)$$

Eliminating x from the compatibility and equilibrium equations (eqns 9.5–8 and 9.5–9),

$$f_{pb} = \frac{8.969}{\varepsilon_{pb} - 0.001329} \qquad (9.5-10)$$

This equation is now solved with the stress/strain curve, as shown in Fig. 9.5–3, giving

$$f_{pb} = 1327 \text{ N/mm}^2; \qquad \varepsilon_{pb} = 0.0081$$

Using eqn (9.5–9) (or eqn 9.5–8),

$$x = 450.6 \text{ mm}$$

Obtain the ultimate resistance moment by taking moments about the centroid of the concrete compression block:

$$M_u = f_{pb} A_{ps}[d - k_2 x] \quad \text{(where } k_2 = 0.45, \text{ see Fig. 4.4–5)}$$
$$= 1327(3300)[870 - (0.45)(450.6)] \text{ Nmm}$$
$$= \underline{2922 \text{ kNm}}$$

(b) Strands not tensioned.
   concrete strain ε_u at tendon level at beam collapse

$$= \left[\frac{870 - x}{x}\right](0.0035) = \frac{3.045}{x} - 0.0035 \text{ as before}$$

Change in tendon strain due to M_u:

$$\varepsilon_{pa} = \varepsilon_u + \varepsilon_e (= 0)$$
$$= \varepsilon_u = \frac{3.045}{x} - 0.0035$$

Tendon strain at collapse is (see eqn 9.5–2)

$$\varepsilon_{pb} = \varepsilon_{pa} + \varepsilon_{pe} \, (= 0) = \varepsilon_{pa}$$

$$= \frac{3.045}{x} - 0.0035$$

i.e.

$$x = \frac{3.045}{\varepsilon_{pb} + 0.0035} \tag{9.5–11}$$

$$f_{pb} = 2.945x \text{ as before} \tag{9.5–12}$$

Eliminating x from eqns (9.5–11) and (9.5–12),

$$f_{pb} = \frac{8.969}{\varepsilon_{pb} + 0.0035}$$

which is solved with the stress/strain curve in Fig. 9.5–3, giving

$$f_{pb} = 1046 \text{ N/mm}^2; \qquad \varepsilon_{pb} = 0.0051$$

Using eqn (9.5–12), $x = 355.2$ mm,

$$M_u = (1046)(3300)[870 - (0.45)(355.2)] \text{ Nmm}$$

$$= 2451 \text{ kNm}$$

Comments

(a) The site error in part (b) only leads to a small reduction in the M_u value, from 2922 kNm to 2451 kNm. However, the reduction in the cracking moment may be serious, and so may the increase in the working load deflection.

(b) The general principles as illustrated in this example can of course be applied to **flanged beams**. With reference to Fig. 9.5–4, if at collapse the neutral axis is within the flange thickness, then the method of solution is as for a rectangular beam; in particular, eqns (9.5–5) to (9.5–7) can be applied without modification. If the neutral axis depth x exceeds the flange thickness h_f, the compatibility condition is still represented by eqn (9.5–5), but the equilibrium condition has to be worked out from Fig. 9.5–4. Using the rectangular stress block of Fig. 4.4–5, the compression in the shaded area (1) in Fig. 9.5–4 is $0.405f_{cu}b_wx$ and that in the areas (2) is $0.45f_{cu}(b - b_w)h_f$. Therefore, the equilibrium condition is

$$f_{pb}A_{ps} = 0.405f_{cu}b_wx + 0.45f_{cu}(b - b_w)h_f \tag{9.5–13}$$

Eliminating x from eqns (9.5–13) and (9.5–5) gives the following equation which relates the unknown tendon stress f_{pb} to the unknown tendon strain ε_{pb}:

$$f_{pb} = \frac{0.45f_{cu}}{\varrho} \cdot \left[\frac{0.9\beta_2\varepsilon_{cu}}{\varepsilon_{pb} - \varepsilon_{pe} - \beta_1\varepsilon_e + \beta_2\varepsilon_{cu}} \cdot \frac{b_w}{b} \right.$$

$$\left. + \left(1 - \frac{b_w}{b}\right)\frac{h_f}{d} \right] \tag{9.5–14}$$

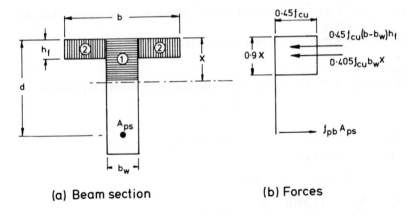

(a) Beam section   **(b) Forces**

Fig. 9.5–4

where $\varrho = A_{ps}/bd$ and for a pretensioned beam or a bonded post-tensioned beam, the bond factors $\beta_1 = \beta_2 = 1$.

Equation (9.5–14) may now be solved with the tendon stress/strain curve to obtain f_{pb} and ε_{pb}. The value of x may then be obtained from eqn (9.5–5), and the ultimate moment of resistance evaluated by taking moments, say, about the tendon:

$$M_u = 0.405 f_{cu} b_w x (d - 0.45x) + 0.45 f_{cu} (b - b_w) h_f (d - 0.5h_f)$$
$$(9.5-15)$$

(c) If the tendon consists of several **cables at different levels**, then for the ith cable the strain ε_{pbi} is related to the neutral axis depth x by the compatibility condition

$$\frac{x}{d_i} = \frac{\beta_2 \varepsilon_{cu}}{\varepsilon_{pbi} - \varepsilon_{pei} - \beta_1 \varepsilon_{ei} + \beta_2 \varepsilon_{cu}} \qquad (9.5-16)$$

which is similar to eqn (9.5–5), except that subscripts i have been added to indicate reference to the ith cable. Therefore, for any assumed value of x, the strain ε_{pbi} may be calculated for each cable in turn, since on the right-hand side of eqn (9.5–16) the only unknown is ε_{pbi}. Using the tendon stress/strain curve, the corresponding stress f_{pbi} is found for each cable. Next calculate the compression force in the concrete, using Fig. 9.5–1(c) or Fig. 9.5–4(b) as the case may be. If the total force in the cables does not balance that in the concrete, adjust the value of x by inspection and repeat the process until a reasonble balance is achieved. Then calculate M_u by taking moments, say, about the neutral axis.

(d) In Fig. 9.5–3, the **modulus of elasticity** E_s of the prestressing tendon is shown as 200 kN/mm². More specifically, BS 8110 : Clause 2.5.3 gives the following values:
(1)   205 kN/mm² for wire to section two of BS 5896;
(2)   195 kN/mm² for strand to section three of BS 5896;

(3)   206 kN/mm^2 for rolled or rolled, stretched and tempered bars to BS 4486)

(4)   165 kN/mm^2 for rolled and stretched bars to BS 4486.

Example 9.5–2
Repeat Example 9.5–1 using BS 8110's **design table** as here reproduced in Table 9.5–1.

SOLUTION

$$f_{pu} = 1700 \text{ N/mm}^2 \qquad f_{cu} = 60 \text{ N/mm}^2$$

$$A_{ps} = 3300 \text{ mm}^2$$

$$b = 400 \text{ mm} \qquad d = 870 \text{ mm}$$

$$\frac{f_{pu}}{f_{cu}} \cdot \frac{A_{ps}}{bd} = \frac{(1700)(3300)}{(60)(400)(870)} = 0.27$$

$$\frac{f_{pe}}{f_{pu}} = \frac{910}{1700} = 0.54$$

From Table 9.5–1,

$$\frac{f_{pb}}{0.87f_{pu}} = 0.86 \text{ by interpolation}$$

$$\frac{x}{d} = 0.50 \text{ by interpolation}$$

Hence

$$f_{pb} = (0.86)(0.87)(1700) = 1272 \text{ N/mm}^2$$

$$x = (0.50)(870) = 435 \text{ mm}$$

From Eqn (9.5–1(a)),

$$M_u = f_{pb}A_{ps}(d - 0.45x)$$

$$= (1272)(3300)[870 - (0.45)(435)] \text{ Nmm}$$

$$= \underline{2830 \text{ kNm}}$$

Comments
In Example 9.5–1(a), M_u was worked out from first principles to be 2922 kNm. By comparison, BS 8110's value here is more conservative, being 3% lower.

9.6   The ultimate limit state: shear (BS 8110)

There are two cases to consider.

Case 1   Sections uncracked in flexure
Consider a concrete element at the centroidal axis of the beam, subjected to a longitudinal compressive stress f_c and a shear stress v_{c0} (Fig. 9.6–1).

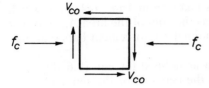

Fig. 9.6–1

Then a Mohr circle analysis will quickly show that the magnitude of the principal tensile stress is

$$f_t = \tfrac{1}{2}\sqrt{(f_c^2 + 4v_{c0}^2)} - \tfrac{1}{2}f_c$$

or

$$v_{c0} = \sqrt{(f_t^2 + f_c f_t)}$$

For a rectangular section of width b_v and depth h, it is well known from elementary mechanics of materials that

$$v_{c0} = \frac{3}{2}\frac{V_{c0}}{b_v h}$$

where V_{c0} is the shear force acting on the concrete section. Therefore the above equation becomes

$$V_{c0} = 0.67 b_v h \sqrt{(f_t^2 + f_c f_t)}$$

For design purposes, BS 8110 states that f_c should be taken as $0.8 f_{cp}$ (defined below). Therefore

$$V_{c0} = 0.67 b_v h \sqrt{(f_t^2 + 0.8 f_{cp} f_t)} \tag{9.6–1}$$

V_{c0} is then BS 8110's ultimate shear resistance of the concrete, if the principal tensile stress f_t is assigned the numerical value of $0.24\sqrt{f_{cu}}$. In eqn (9.6–1), f_{cp} is the concrete compressive stress at the centroidal axis due to the effective prestress, h is the beam depth and b_v the beam width.

For flanged members, the width b_v should be interpreted as the web width b_w. In flanged members where the centroidal axis is within the flange thickness, eqn (9.6–1) should be applied to the junction of the flange and web; f_t is taken equal to $0.24\sqrt{f_{cu}}$ as before, but f_{cp} is to be interpreted as the concrete prestress at the flange/web junction.

Case 2 Sections cracked in flexure
BS 8110 gives the following empirical equation for the ultimate shear resistance of a section cracked in flexure:

$$v_{cr} = \left(1 - 0.55\frac{f_{pe}}{f_{pu}}\right)v_c b_v d + \frac{M_0}{M}V \tag{9.6–2}$$

$$\not< 0.1 b_v d \sqrt{f_{cu}}$$

where f_{pe}/f_{pu} = the ratio of the effective tendon prestress to the characteristic strength of the tendons;

v_c = the design shear stress taken from Table 6.4–1, in which A_s is now interpreted as the sum of the area A_{ps} and that of any ordinary longitudinal reinforcement bars that may be present;

b_v = the width of the beam as defined for eqn (9.6–1);

d = the effective depth to the centroid of the tendons.

M_0 = the moment necessary to produce zero stress in the concrete at the extreme tension fibre. For the purpose of this equation, M_0 is to be calculated as $0.8[f_{pt}I/y]$ where f_{pt} is the concrete compressive stress at the extreme tension fibre due to the effective prestressing force, I the second moment of area of the beam section, and y is the distance of the extreme tension fibre from the centroid of the beam section (see Step 3 of Example 9.6–1);

V and M = the shear force and bending moment respectively at the section considered, due to ultimate loads (ignoring the vertical component of the tendon force if any).

Comments

(a) The derivation of eqn (9.6–2) is given on pp. 42–47 of Reference 2, which also explains the pre-cracking and post-cracking behaviour of prestressed concrete beams. Until recently, it was thought adequate to use elastic theory for shear design; that is, to calculate the principal tensile stresses under service condition and limit them to a specified value. On p. 20 of Reference 2, four major reasons are given to explain why the elastic theory is not adequate.

(b) BS 8110 states that, in using eqn (9.6–1), f_t is to be taken as $0.24\sqrt{f_{cu}}$. The tensile strength of concrete is usually between

$$0.3\sqrt{f_{cu}} \quad \text{and} \quad 0.4\sqrt{f_{cu}}$$

Therefore, for design purposes it is reasonable to take f_t as

$$0.3\sqrt{(f_{cu}/\gamma_m)} = 0.24\sqrt{f_{cu}}$$

when 1.5 is substituted for the partial safety factor γ_m. Note also that in eqn (9.6–1), BS 8110 applies a factor of 0.8 to f_{cp}. This is because the value $0.24\sqrt{f_{cu}}$ (for f_t) includes a factor of $1/\sqrt{1.5} \doteq 0.8$.

(c) Equation (9.6–1) might at first sight appear to be applicable to rectangular sections only, since its derivation is based on the equation

$$v_{c0} = \frac{3V_{c0}}{2b_v h}$$

For a general section, the shear stress is of course given by the well-known formula

$$v_{c0} = \frac{V_{c0}A\bar{y}}{b_v I}$$

where the product $A\bar{y}$ is the statical moment (taken about the centroidal axis of the entire cross-section) of the area above the level at which v_{c0} occurs, b_v is the beam width at that level, and I is the second moment of area of the entire cross section taken about the centroidal axis. Therefore, for a general section, eqn (9.6–1) should take the modified form

$$V_{c0} = \frac{bvI}{A\bar{y}}\sqrt{(f_t^2 + 0.8f_{cp}f_t)}$$

where, for practical I sections, the quantity $b_v I/A\bar{y}$ usually works out to be about $0.8b_v h$ so that eqn (9.6–1) errs on the safe side. However, for such sections, the maximum principal tensile stress does not necessarily occur at the centroidal axis (though the maximum shear stress exists there) but frequently at the junction of the web and the tensile flange; eqn (9.6–1) refers only to the condition at the centroidal axis and in this respect errs on the unsafe side. The two effects tend to cancel out [8], so that in practice eqn (9.6–1) is applied to both rectangular and I-sections. For similar reasons, eqn (9.6–1) is judged suitable for use with L- and T-beams also.

Design procedure for shear (BS 8110)

Step 1
Calculate the shear force V_L and the bending moment M due to the design ultimate loads. The shear force V_L, which is due to the external loading, is then adjusted as explained below.

Case 1   (section uncracked in flexure):

$$V = V_L - P_e\sin\beta \qquad\qquad (9.6–3)$$

where the term $P_e\sin\beta$ (see Fig. 9.2–5(a) and eqn 9.2–22) allows for the effect of the tendon force.

Case 2   (section cracked in flexure):

$$V = V_L - P_e\sin\beta \quad\text{or}\quad V_L \qquad\qquad (9.6–4)$$

whichever is greater.

Step 2
Calculate V_{c0} from eqn (9.6–1).

Step 3
Calculate V_{cr} from eqn (9.6–2).

Step 4
The design ultimate shear resistance V_c is taken as follows:

(a)   **uncracked sections:** where M from Step 1 is less than M_0 as defined for eqn (9.6–2), the section is considered uncracked in flexure, in which case $V_c = V_{c0}$ of Step 2

(b)   **cracked sections:** where M from Step 1 is not less than M_0 as defined for eqn (9.6–2), the section is considered cracked in flexure, in which case

$$V_c = V_{c0} \text{ of Step 2} \quad\text{or}\quad V_{cr} \text{ of Step 3}$$

whichever is the lesser.

Step 5
Check that in no case should V/b_vd exceed $0.8\sqrt{f_{cu}}$ or 5 N/mm² which-ever is the lesser. (These stress limits include an allowance for γ_m.)

Step 6
If $V < 0.5V_c$, no shear reinforcement is required.

Step 7
If $V \geq 0.5V_c$ but $\leq (V_c + 0.4b_vd)$, provide shear links as follows:

$$\frac{A_{sv}}{s_v} = \frac{0.4b_v}{0.87f_{yv}} \tag{9.6-5}$$

where symbols have their usual meanings (see eqn 6.4–2 if necessary).

Step 8
If $V > (V_c + 0.4b_vd)$, provide shear links as follows:

$$\frac{A_{sv}}{s_v} = \frac{V - V_c}{0.87f_{yv}d_t} \tag{9.6-6}$$

where A_{sv}, s_v ($s_v \leq 0.75d_t$) and f_{yv} have their usual meanings (see eqn 6.4–3 if necessary), and d_t is the depth from the extreme compression fibre to the centroid of the tendons or to the longitudinal corner bars around which the links pass, whichever is the greater.

Comments
BS 8110's shear design procedure is essentially similar to that of the previous Code CP 110. Smith [9] has written a useful article on the subject.

Example 9.6–1
Design the shear reinforcement for a symmetrical prestressed I-section, given that:

$$V_L = 400 \text{ kN} \qquad M = 800 \text{ kNm}$$

rib width $b_w = 200$ mm   overall depth $h = 1000$ mm
area $A = 310 \times 10^3$ mm² $I = 36 \times 10^9$ mm⁴

$$A_{ps} = 1803 \text{ mm}^2 \qquad f_{pu} = 1750 \text{ N/mm}^2 \qquad f_{pe} = 0.6f_{pu}$$

$e_s = 290$ mm (and tendon is inclined at $\beta = 3°$ at the section considered)

$$f_{cu} = 50 \text{ N/mm}^2 \qquad f_{yv} = 250 \text{ N/mm}^2$$

SOLUTION

$$P_e = A_{ps}f_{pe} = 1803 \times 0.6 \times 1750 = 1893 \text{ kN}$$

$$f_{cp} = P_e/A = 1893 \times 10^3/310\,000 = 6.10 \text{ N/mm}^2$$

$$f_t = 0.24\sqrt{f_{cu}} = 0.24\sqrt{50} = 1.7 \text{ N/mm}^2$$

Step 1
  Case 1:   From eqn (9.6–3),

$$V = V_L - P_e \sin\beta$$

$$= 400 - 1893 \sin 3° = 301 \text{ kN}$$

Case 2: From eqn (9.6-4),

$$V = V_L = 400 \text{ kN}$$

Step 2
From eqn (9.6-1),

$$V_{c0} = (0.67)(200)(1000)\sqrt{[(1.7)^2 + (0.8)(6.1)(1.7)]}$$
$$= 448.2 \text{ kN}$$

Step 3
The preliminary calculations for eqn (9.6-2) are as follows:

$$f_{pe}/f_{pu} = 0.6$$

$$\frac{A_{ps}}{b_w d} = \frac{1803}{(200)(790)} = 1.14\%$$

(*Note:* $d = h/2 + e_s = 790$ mm)
Hence, from Table 6.4-1,

$$v_c = 0.77 \text{ N/mm}^2 \text{ by interpolation}$$

$$f_{pt} = \frac{P_e}{A} + \frac{P_e e_s y}{I}$$

$$= \frac{(1893)(10^3)}{310000} + \frac{(1893)(10^3)(290)(500)}{(36)(10^9)}$$

$$= 13.73 \text{ N/mm}^2 \text{ compressive}$$

$$M_0 = 0.8 f_{pt} \frac{I}{y} = (0.8)(13.73)\left[\frac{(36)(10^9)}{(500)}\right]$$

$$= 791 \text{ kNm}$$

Substituting into eqn (9.6-2),

$$V_{cr} = [1 - (0.55)(0.6)](0.77)(200)(790) + \frac{(791)}{(800)}(400)(10^3)$$

$$= 477 \text{ kN}$$

$$0.1 b_w d \sqrt{f_{cu}} = (0.1)(200)(790)\sqrt{50} = 112 \text{ kN}$$

$$< 477 \text{ kN}$$

Therefore

$$V_{cr} = 477 \text{ kN}$$

Step 4

$$M = 800 \text{ kNm} \quad \text{(given)}$$

$$M_0 = 791 \text{ kNm} \quad \text{(from Step 3)}$$

Hence the section is uncracked.

$$V_c = V_{c0} \text{ of Step 2} \quad (\text{since } V_{c0} < V_{cr})$$
$$= \underline{448 \text{ kN}}$$

Step 5

$$0.8\sqrt{f_{cu}} = 0.8\sqrt{50} = 5.66 \text{ N/mm}^2 > 5 \text{ N/mm}^2$$

Hence the upper limit on $V/b_v d$ is 5 N/mm^2.
Actual $V/b_v d$ (see Step 1: Case 2)

$$= \frac{(400)(10^3)}{(200)(790)} = 2.53 \text{ N/mm}^2 \text{ OK}$$

Step 6

$$V = 400 \text{ kN} \quad (\text{Step 1: Case 2})$$
$$V_c = 448 \text{ kN}$$

Hence V is not less than $0.5V_c$. Move to Step 7.

Step 7

By inspection,

$$V > 0.5V_c \quad \text{but} \quad < (V_c + 0.4b_v d)$$

From eqn (9.6–5),

$$\frac{A_{sv}}{s_v} = \frac{(0.4)(200)}{(0.87)(250)}$$
$$= 0.37 \text{ mm}$$

From Table 6.4–2,
Provide size 10 links at 300 mm centres ($A_{sv}/s_v = 0.52$)

For further reading on shear in prestressed concrete, the reader is referred to References 1–3 and 8–13.

9.7   The ultimate limit state: torsion (BS 8110)

The torsional resistance of a prestressed concrete member is significantly higher than that of the corresponding reinforced concrete member [1–3]. However, BS 8110 is cautious; it recommends that the same procedure as explained here in Section 6.11 for ordinary reinforced concrete members should be used for the torsion design of prestressed members.

The effect of prestressing on the torsional behaviour of concrete members, both under service condition and ultimate-load condition, is explained on pp. 57–65 of Reference 2, which also includes comments on design procedure.

9.8   Short-term and long-term deflections

In Chapter 5 it was pointed out that, in assessing the deflections of reinforced concrete beams, an efficient and general method was to work

Curvature diagram	Midspan deflection
(a) 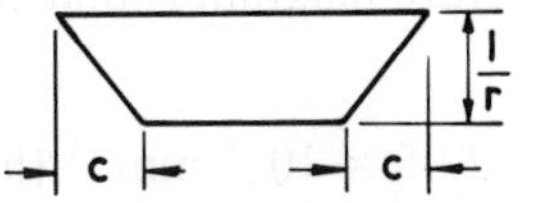	$\dfrac{\ell^2}{8}\left[1-\dfrac{4}{3}\left(\dfrac{c}{\ell}\right)^2\right]\left(\dfrac{1}{r}\right)$
(b) 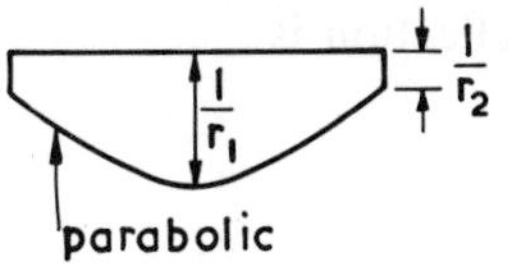 parabolic	$\dfrac{\ell^2}{9.6}\left[\left(\dfrac{1}{r_1}\right)+\dfrac{1}{5}\left(\dfrac{1}{r_2}\right)\right]$
(c) 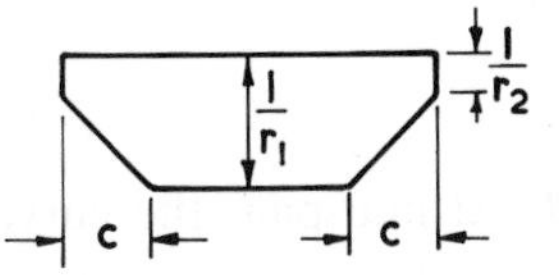	$\dfrac{\ell^2}{8}\left[\left\{1-\dfrac{4}{3}\left(\dfrac{c}{\ell}\right)^2\right\}\left(\dfrac{1}{r_1}\right)+\dfrac{4}{3}\left(\dfrac{c}{\ell}\right)^2\left(\dfrac{1}{r_2}\right)\right]$

Fig. 9.8–1 Deflection–curvature relations for simply supported beams

out the curvatures and then apply the curvature–area theorem, as illustrated by Example 5.5–1. The same principles can be applied to prestressed concrete beams. In this connection, it will prove convenient to augment the curvature diagrams in Fig. 5.5–1 with those in Fig. 9.8–1, as the latter curvatures correspond to tendon profiles commonly used in practice.

Short-term deflections
To calculate the short-term deflections, it is only necessary to apply the curvature–area theorem (see Section 5.5) using the EI value of the *uncracked* section; creep and shrinkage effects do not come in.

Example 9.8–1
A prestressed concrete beam is simply supported over a 10 m span and carries a uniformly distributed imposed load of 3 kN/m, half of which is non-permanent. The tendon follows a trapezoidal profile: the eccentricity e_s is 100 mm within the middle third of the span and varies linearly from the third-span points to zero at the supports. The tendon area A_{ps} is 350 mm^2 and the effective prestress f_s immediately after transfer is 1290 N/mm^2. The beam is of uniform cross-section, the pertinent properties of which are

$$\text{Area} \quad A = 5 \times 10^4 \text{ mm}^2$$
$$I \text{ (uncracked)} = 4.5 \times 10^8 \text{ mm}^4$$
$$E_c = 34 \text{ kN/mm}^2$$
$$E_s = 200 \text{ kN/mm}^2$$

Calculate the **short-term deflections**. (Assume unit weight of concrete = 23.6 kN/m^3.)

SOLUTION
(a) ***Deflection due to prestressing.*** The curvature diagram is that shown in Fig. 9.8–1(a); in this example

$$\frac{1}{r} = \frac{P_{es}}{E_c I} = \frac{350 \times 1290 \times 100}{34 \times 10^3 \times 4.5 \times 10^8} = 2.95 \times 10^{-6} \text{ mm}^{-1} \text{ (hogging)}$$

From Fig. 9.8–1(a), the midspan deflection is

$$a = \frac{l^2}{8}\left[1 - \frac{4}{3}\left(\frac{c}{l}\right)^2\right]\frac{1}{r}$$

$$= \frac{(10 \times 10^3)^2}{8}[1 - \tfrac{4}{3}(\tfrac{1}{3})^2]2.95 \times 10^{-6}$$

$$= 31 \text{ mm (upwards)}$$

(b) ***Deflection due to non-permanent load.*** At midspan, the curvature is

$$\frac{1}{r} = \frac{M}{E_c I}$$

$$= \frac{1}{8} \times \frac{1.5 \text{ (N/mm)} \times (10 \times 10^3)^2}{34 \times 10^3 \times 4.5 \times 10^8}$$

$$= 1.23 \times 10^{-6} \text{ mm}^{-1} \text{ (sagging)}$$

The curvature distribution is parabolic, and Fig. 5.5–1(c) applies. The midspan deflection is then as given in Example 5.5–1(c), namely:

$$a = \frac{l^2}{9.6}\frac{1}{r}$$

$$= \frac{(10 \times 10^3)^2}{9.6} \times 1.23 \times 10^{-6}$$

$$= 12.8 \text{ mm (downwards)}$$

(c) ***Deflection due to permanent load***

$$\text{Self-weight} = \frac{5 \times 10^4}{10^6} \times 23.6 \text{ kN/m} = 1.18 \text{ kN/m}$$

Permanent load = 1.18 + 1.5 = 2.68 kN/m

From the result of (b) above, the midspan deflection is

$$a = \frac{2.68}{1.5} \times 12.8 = 22.8 \text{ mm (downwards)}$$

(d) ***Short-term deflections.*** The short-term deflection when the non-permanent load is acting is

$$a = -31 \text{ mm} + 12.8 \text{ mm} + 22.8 \text{ mm}$$

$$= 5 \text{ mm (say)}$$

The short-term deflection when the non-permanent load is not acting is

$$a = -31 \text{ mm} + 22.8 \text{ mm}$$

$$= -8 \text{ mm (say)} \quad \text{i.e. 8 mm upwards}$$

Comments
In the simple example here considered, the applied loads are uniformly distributed. Hence the corresponding deflections could have been written down straight away using the formula

$$a = \frac{5}{384} \frac{ql^4}{E_c I}$$

However, it is thought that readers should become familiar with the curvature–area theorem, as it represents a powerful tool for the more complicated loadings and, especially, for dealing with shrinkage and creep effects. (See Problem 5.3 at the end of Chapter 5.)

Long-term deflections
As explained at the beginning of this section, an efficient and general procedure is to work out the curvatures (using the *EI* value of the *uncracked* section) and then apply the curvature–area theorem. We saw in Example 9.8–1 that curvatures are produced by (a) the prestress and (b) the applied load. These are considered in more detail below.

To explain the general principles, it is sufficient to consider a simply supported beam. The **curvature due to prestress** is made up of three parts:

(a) $\quad \dfrac{1}{r} = \dfrac{Pe_s}{E_c I}$ (hogging) $\hfill (9.8-1)$

This is the instantaneous curvature at transfer.

(b) $\quad \dfrac{1}{r} = \dfrac{(\delta P)e_s}{E_c I}$ (sagging) $\hfill (9.8-2)$

This is the change in curvature corresponding to the loss of prestress δP due to relaxation, shrinkage and creep.

(c) $\quad \dfrac{1}{r} = \phi \dfrac{[P + (P - \delta P)]e_s}{2E_c I}$ (hogging) $\hfill (9.8-3)$

This is the increase in curvature due to the creep of concrete. Note that $\frac{1}{2}[P + (P - \delta P)]$ represents the average value of the prestressing force; ϕ is the creep coefficient.

Adding eqns (9.8–1) to (9.8–3) together, the total long-term curvature due to the prestress is

$$\frac{1}{r} \text{ (prestress, long term)} = \frac{Pe_s}{E_c I} - \frac{(\delta P)e_s}{E_c I} + \phi \frac{(P + P - \delta P)e_s}{2E_c I}$$

$$= \frac{Pe_s}{E_c I} \left[\frac{P - \delta P}{P} + \frac{1 + \dfrac{P - \delta P}{P}}{2} \phi \right]$$

Noting that $(P - \delta P)/P$ is the prestress loss ratio α, we have

$$\frac{1}{r} \text{ (prestess, long term)} = \frac{Pe_s}{E_c I}\left(\alpha + \frac{1 + \alpha}{2}\phi\right) \text{ (hogging)} \quad \textbf{(9.8--4)}$$

where P = the prestressing force immediately after transfer;
$\quad e_s$ = the tendon eccentricity at the section considered;
$\quad \phi$ = the creep coefficient for the time interval;
$\quad E_c$ = the modulus of elasticity of the concrete at transfer (Table 2.5--6);
$\quad I$ = the second moment of area of the uncracked section;
$\quad \alpha$ = the prestress loss ratio, i.e. $\alpha = (P - \delta P)/P = (f_s - \delta f_s)/f_s$ (see Examples 9.4--1 to 9.4--4 for δf_s).

Of course, the **curvatures due to the applied load** are simply

$$\frac{1}{r} \text{ (load, short term)} = \frac{M}{E_c I} \text{ (sagging)} \quad (9.8-5)$$

$$\frac{1}{r} \text{ (load, creep)} = \phi\frac{M}{E_c I} \text{ (sagging)} \quad (9.8-6)$$

Adding together,

$$\frac{1}{r} \text{ (load, long term)} = (1 + \phi)\frac{M}{E_c I} \text{ (sagging)} \quad \textbf{(9.8-7)}$$

The right-hand side of eqn (9.8--7) is sometimes written as

$$\frac{M}{(\text{long-term } E_c)I}$$

where the long-term or **effective modulus** E_c is $E_c/(1 + \phi)$, as in eqn (5.5--3).

Comments
(a) See Problem 9.3 for the legitimacy of eqn (9.8--6) (which is occasionally questioned by the brighter students!).
(b) When calculating the long-term deflections of ordinary reinforced concrete beams, it helps to use the concept of an effective modulus, as in eqn (5.5--3). However, for prestressed concrete beams, it is necessary to consider the effects of the loss of prestress (see eqns (9.8--2) and (9.8--3)) and the use of an effective modulus can load to confusion. Indeed, the important eqn (9.8--4) cannot be conveniently expressed in terms of an effective modulus of elasticity.

Example 9.8--2
Calculate the **long-term deflections** of the prestressed concrete beam in Example 9.8--1 if:

$\qquad$ concrete creep coefficient $\phi = 2.0$

$\qquad$ concrete shrinkage $\varepsilon_{cs} \qquad = 450 \times 10^{-6}$

$\qquad\qquad$ tendon relaxation $f_r = 10\%$ of $f_s = 129$ N/mm^2

SOLUTION

Step 1   Loss of prestress
From Example 9.4–5,

$$\text{loss of prestress } \delta f_s = 424 \text{ N/mm}^2, \text{ within middle third of span}$$

$$\text{prestress loss ratio } \alpha = \frac{f_s - \delta f_s}{f_s} = \frac{1290 - 424}{1290}$$

$$= 0.67, \text{ within middle third of span}$$

Comments on Step 1
The 424 N/mm² loss of prestress is in a sense fictitious, because the effect of the applied load has been ignored. The applied-load bending moment is greatest at midspan and decreases to zeo at the support. For practical design purposes, it can safely be said that no serious consequences will result from neglecting the applied-load stresses this way.

Step 2   Long-term curvature due to prestress
From eqn (9.8–4),

$$\frac{1}{r} \text{ (prestress, long term)} = \frac{Pe_s}{E_c I}\left(\alpha + \frac{1 + \alpha}{2}\phi \right)$$

Within middle third of span:

$$\frac{1}{r} \text{ (prestress, long term)}$$

$$= \frac{451.5 \times 10^3 \times 100}{34 \times 10^3 \times 4.5 \times 10^8}\left(0.67 + \frac{1 + 0.67}{2} \times 2 \right)$$

$$= 6.905 \times 10^{-6} \text{ mm}^{-1} \quad \text{(hogging)}$$

At the supports, $e_s = 0$; therefore

$$\frac{1}{r} \text{ (prestress)} = 0$$

Of course, if e_s had not been zero at the supports, it would have been necessary also to calculate the loss ratio α for the support sections in Step 1.

Step 3   Long-term deflection due to prestress
It is sufficiently accurate to assume that the curvature distribution is similar to the tendon profile. Using the curvature–area theorem and referring to Fig. 9.8–1(a).

$$\text{midspan deflection} = \frac{l^2}{8}\left[1 - \frac{4}{3}\left(\frac{c}{l}\right)^2 \right]\frac{1}{r}$$

$$= \frac{(10 \times 10^3)^2}{8}[1 - \tfrac{4}{3}(\tfrac{1}{3})^2] \times 6.905 \times 10^{-6}$$

$$= 73.5 \text{ mm} \quad \text{(upwards)}$$

Comments on Steps 2 and 3

For preliminary calculations of deflections, the long-term curvature due to prestress is often taken simply as

$$\frac{1}{r} = (1 + \phi)\frac{Pe_s}{E_c I}$$

This simplified equation leads to a deflection of 94.3 mm (see Problem 9.4) as against the 73.5 mm calculated in Step 3—an error of about 28%. An error of this magnitude is not serious in preliminary calculations, but the reader must not let the simplified equation obscure his understanding of structural behaviour.

Step 4   Long-term deflection due to premanent load

Long term deflection $= (1 + \phi) \times$ (short-term deflection)

$$= (1 + 2) \times 22.8 \text{ mm}$$

(from Example 9.8–1)

$$= 68.4 \text{ mm}   \text{(downwards)}$$

Comments on Step 4

If the short-term deflection of 22.8 mm had not been available, it would have been necessary to calculate the curvature from eqn (9.8–7) and then use the result of Example 5.5–1(c).

Step 5   Short-term deflection due to non-permanent load

From Example 9.8–1,

short-term deflection $= 12.8 \text{ mm}$ (downwards)

Step 6   Total long-term deflections

(a)   When the non-permanent load is acting:

midspan deflection $=$ Step 3 $+$ Step 4 $+$ Step 5

$$= -73.5 \text{ mm} + 68.4 \text{ mm} + 12.8 \text{ mm}$$

$$= 8 \text{ mm (say)}$$

(b)   When the non-permanent load is not acting:

midspan deflection $=$ Step 3 $+$ Step 4

$$= 73.5 \text{ mm} + 68.4 \text{ mm}$$

$$= -5 \text{ mm (say)   i.e. (5 mm upwards)}$$

9.9   Summary of design procedure

Step 1

Select a suitale cross-section (eqn 9.2–8).

Step 2

Select a suitable effective prestressing force P_e and the tendon profile e_s at the critical section (eqns 9.2–9 to 9.2–12).

Step 3
Determine the permissible tendon zone (eqns 9.2–18 to 9.2–21).

Step 4
Compute the initial prestressing force P on the basis of an estimated loss ratio α (eqn 9.3–7).

Step 5
Determine the number and arrangement of tendons [10,14–16]. Select a suitable prestressing system (see specialist texts) [10,14–16).

Step 6
Calculate the loss ratio α (see BS 8110: Clauses 4.8 and 4.9). If α thus calculated differs too much from the estimated value in Step 4, revise Steps 4 and 5.

Step 7
Check stresses at transfer and determine the permissible tendon zone for conditions at transfer.

Step 8
Select the tendon profile. Use ideal shear profile (eqn 9.2–27) if possible.

Step 9
Check ultimate flexural strength as explained in Section 9.5.

Step 10
Check ultimate shear resistance and design shear reinforcement if necessary, as explained in Section 9.6.

Step 11
Check ultimate torsional resistance and design torsion reinforcement if necessary; see Section 9.7.

Step 12
Check short-term and long-term deflections; see Section 9.8.

9.10 Computer programs

(in collaboration with **Dr H. H. A. Wong**, University of Newcastle upon Tyne)

The FORTRAN programs for this Chapter are listed in Section 12.9. See also Section 12.1 for 'Notes on the computer programs'.

Problems

9.1 The design procedure of Section 9.9 first considers the stress conditions in service (eqns 9.2–4 to 9.2–7) and then those at transfer (eqns 9.3–1 to 9.3–4). Of the eight stress conditions considered, the following four are often not critical:

     eqns (9.2–5) and (9.2–7)   eqns (9.3–1) and (9.3–3)

Show that, by totally disregarding these four equations, it is possible to draw up a **simplified design procedure** which does not necessitate the separate checking of the stress conditions in service and at transfer. Specifically, show that in the simplified procedure the following equations give the minimum required Z values that will simultaneously satisfy both the stress conditions in service and those at transfer:

$$Z_1 \geq \frac{M_{\text{imax}} + (1 - \alpha)M_{\text{d}}}{\alpha f_{\text{amaxt}} - f_{\text{amin}}}$$

$$Z_2 \geq \frac{M_{\text{imax}} + (1 - \alpha)M_{\text{d}}}{f_{\text{amax}} - \alpha f_{\text{amint}}}$$

(*Hint:* Write $f_1 = \alpha f_{1t}$ and $f_2 = \alpha f_{2t}$ in eqns 9.2–4 and 9.2–6; then eliminate f_{1t} and f_{2t} using eqns 9.3–2 and 9.3–4. If necessary, see pp. 7–9 of Reference 2.)

9.2 A pretensioned concrete beam is of rectangular section 150 mm × 1100 mm. The tendon consists of 1130 mm^2 of standard strands, of characteristic strengths 1700 N/mm^2, stressed to an effective prestress of 910 N/mm^2, the tendon eccentricity being 250 mm below the centroid of the section. The tendon stress/strain curve is as in Fig. 9.5–3. The concrete characteristic strength f_{cu} is 60 N/mm^2 and its modulus of elasticity may be taken as 36 kN/mm^2 for stresses up to $0.4f_{\text{cu}}$. Determine the ultimate moment of resistance.

Ans. (For method of solution, see Example 9.5–1. For complete solution, see Reference 2: pp. 35–41.)

9.3 According to BS 8110 : 1985 and the CEB–FIP Model Code (1978), the creep coefficient ϕ is defined, with reference to concrete under a constant stress, by the equation

creep strain $= \phi \times$ elastic strain

Show that eqn (9.8–6), namely

creep curvature $= \phi \times$ elastic curvature

is compatible with the BS 8110/CEB–FIP definition of ϕ.
(*Hint:* If necessary, see the comments following Example 9.4–4. Satisfy yourself that each fibre of the beam section creeps under a sustained bending stress which does not change with time.)

9.4 In preliminary calculations for long-term deflections, the following simplified formula is often used for estimating the curvature due to the prestress:

$$\frac{1}{r} = (1 + \phi)\frac{Pe_{\text{s}}}{E_{\text{c}}I}$$

Using this formula, calculate the long-term deflections of the beam in

Example 9.8–2 and compare your answers with those in Step 6 of the solution to that example.

Ans. (a) When non-permanent load is acting, deflection $= -94.3$ mm $+ 68.4$ mm $+ 12.8$ mm $= -13$ mm (compared with $+8$ mm in the solution).

(b) When non-permanent load is not acting, deflection $= -94.3$ mm $+ 68.4$ mm $= -26$ mm (compared with -5 mm in the solution).

Comments on Problem 9.4
The simplified equation neglects the effect of the prestress loss, as represented by eqns (9.8–2) and (9.8–3). Indeed, in eqn (9.8–4), i.e. in

$$\frac{1}{r} = \frac{Pe_\mathrm{s}}{E_\mathrm{c}I}\left(\alpha + \frac{1 + \alpha}{2}\phi\right)$$

if the prestress loss ratio α is taken as unity, the simplfied equation results.

9.5 A pretensioned concrete beam is of uniform cross-section, the area of which is A and the radius of gyration k. It has n layers of steel strands at eccentricities $e_{s1}, e_{s2} \ldots e_{sn}$ respectively from the centroidal axis of the section; these are tensioned to an initial stress of $f_{s1}, f_{s2} \ldots f_{sn}$ respectively. The areas of the n layers of strands are respectively $A_{ps1}, A_{ps2} \ldots A_{psn}$. The modulus of elasticity of the concrete is E_c and that of the steel E_s. Write down a system of n simultaneous equations, which can be solved for the prestress losses $\delta f_{s1}, \delta f_{s2} \ldots \delta f_{sn}$ that occur in the n layers of strands, as a result of the elastic deformation of the concrete at transfer.

Ans. $\dfrac{\delta f_{s1}}{E_s} = \dfrac{1}{E_c A}\displaystyle\sum_{i=1}^{i=n}(f_{si} - \delta f_{si})A_{psi}\left(1 + \dfrac{e_{si}}{k^2}e_{s1}\right)$

$$\cdots \quad \cdots \quad \cdots$$

$$\cdots \quad \cdots \quad \cdots$$

$\dfrac{\delta f_{sn}}{E_s} = \dfrac{1}{E_c A}\displaystyle\sum_{i=1}^{i=n}(f_{si} - \delta f_{si})A_{psi}\left(1 + \dfrac{e_{si}}{k^2}e_{sn}\right)$

Hint: If $n = 1$, the n equations reduce to that of Example 9.4–2, namely:

$$\frac{\delta f_s}{E_s} = \frac{(f_s - \delta f_s)A_{ps}}{E_c A}\left(1 + \frac{e_s^2}{k^2}\right)$$

9.6 For the prestressed concrete beam of Problem 9.5, write down a system of n simultaneous equations which can be solved for the prestress losses due to a concrete shrinkage ε_{cs}.

Ans. $\dfrac{\delta f_{s1}}{E_s} = \varepsilon_{cs} - \dfrac{1}{E_c A}\displaystyle\sum_{i=1}^{i=n}\delta f_{si}A_{psi}\left(1 + \dfrac{e_{si}}{k^2}e_{s1}\right)$

$$\cdots \quad \cdots \quad \cdots$$

$$\cdots \quad \cdots \quad \cdots$$

$$\frac{\delta f_{sn}}{E_s} = \varepsilon_{cs} - \frac{1}{E_cA} \sum_{i=1}^{i=n} \delta f_{si} A_{psi}\left(1 + \frac{e_{si}}{k^2}e_{sn}\right)$$

Hint: If $n = 1$, the above n equations reduce to that of Example 9.4–3, namely

$$\frac{\delta f_s}{E_s} = \varepsilon_{cs} - \frac{1}{E_c} \cdot \frac{\delta f_s A_{ps}}{A}\left(1 + \frac{e_s^2}{k^2}\right)$$

References

1 Lin, T. Y. and Chow, P. Looking into the future. In *Handbook of Structural Concrete*, edited by Kong, F. K., Evans, R. H., Cohen, E. and Roll, F. Pitman, London and McGraw-Hill, New York, 1983, Chapter 1. Warner, R. F. Prestressed concrete and partially prestressed concrete. In ibid., Chapter 19.

2 Kong, F. K. Bending, shear and torison, In *Developments in Prestressed Concrete*, edited by Sawko, F. Applied Science, London, 1978, Vol. 1, Chapter 1. Evans, R. H. and Kong, F. K. Creep of prestressed concrete. In ibid., Chapter 3.

3 T. Y. Lin Symposium on Prestressed Concrete: Past, Present, Future. University of California, 5 June 1976. *Journal of Prestressed Concrete Institute*, Special Commemorative Issue, Sept. 1976.

4 Evans, R. H. Institution of Civil Engineers' Unwin Memorial Lecture: Research and developments in prestressing. *Journal ICE*, **35**, Feb. 1951, pp. 231–61.

5 Evans, R. H. and Robinson, G. W. Bond stresses in prestressed concrete from X-ray photographs. *Proc. ICE*, Part I, **4**, March 1955, pp. 212–35.

6 Evans, R. H. and Williams, A. The use of X-rays in measuring bond stresses in prestressed concrete. *Proceedings*, World Conference on Prestressed Concrete, University of California, 1957, pp. A32–1 to A32–8.

7 Mayfield, B., Davies, G. and Kong, F. K. Some tests of the transmission length and ultimate strength of pre-tensioned concrete beams incorporating Dyform strand. *Magazine of Concrete Research*, **22**, No. 73, Dec. 1970, pp. 219–26.

8 Reynolds, G. C., Clarke, J. L. and Taylor, H. P. J. *Shear Provisions for Prestressed Concrete in the Unified Code, CP 110:1972* (Technical Report 42.500). Cement and Concrete Association, Slough, 1974.

9 Smith, I. A. Shear in prestressed concrete to CP 110. *Concrete*, **8**, No. 7, July 1974, pp. 39–41.

10 Abeles, P. W. and Bardhan-Roy, B. K. *Prestressed Concrete Designer's Handbook*. Cement and Concrete Association, Viewpoint Publication, Slough, 1981.

11 Evans, R. H. and Hosny, A. H. H. Shear strength of post-tensioned prestressed concrete beams. *Proceedings*, FIP 3rd Congress, Berlin, 1958, pp. 112–32.

12 Evans, R. H. and Schumacher, E. G. Shear strength of prestressed beams without web reinforcement. *Proc. ACI*, **60**, Nov. 1963, pp. 1621–42.

13 MacGregor, J. G., Sozen, M. A. and Siess, C. P. Effect of draped reinforcement on behaviour of prestressed concrete beams. *Proc. ACI*, **57**, Dec. 1960, pp. 649–78.

14 Guyon, Y. *Limit-state Design of Prestressed Concrete*. Applied Science, London, Vol. 1, 1972, 485pp.; Vol. 2, 1974.
15 Green, J. K. *Detailing for Standard Prestressed Concrete Bridge Beams*. Cement and Concrete Association, London, 1973.
16 Lin, T. Y. and Burns, N. H. *Design of Prestressed Concrete Structures*, 3rd edn. John Wiley, New York, 1981.

Chapter 10
Prestressed concrete continuous beams

10.1 Primary and secondary moments

In prestressed concrete, one important difference between simple beams and continuous beams is that, in the latter, prestressing generally induces support reactions. Consider the simple beam in Fig. 10.1–1(a). If for the time being, we do not consider the effects of the dead and imposed loads, then whatever the magnitude of the prestressing force and the tendon profile, there will be no reactions at the supports A and B.* Of course, the prestressing force produces a moment $-P_e e_s$, where the negative sign is used because $P_e e_s$ is a hogging moment for positive values of e_s (Fig. 10.1–1(b)). This moment is called the **primary moment**, M_1. For the tendon profile shown here, the primary moment causes the beam to deflect upwards (Fig. 10.1–1(c)). Suppose the upward deflection at a point C is a_C. If the beam is restrained against deflection at C by an additional support (Fig. 10.1–1(d)), then the support C must exert a reaction R_C on the beam. R_C also induces reactions R_A and R_B, so that the three reactions form an equilibrium set of forces; these support reactions cause a **secondary moment**, M_2, to act on the beam (Fig. 10.1–1(e)) such that the downward deflection at C due to M_2 is numerically equal to a_C. The algebraic sum of the primary moment and the secondary moment is called the **resulting moment** (Fig. 10.1–1(e)):

$$
\begin{array}{ccccc}
M_3 & = & M_1 & + & M_2 \\
\text{resulting} & & \text{primary} & & \text{secondary} \\
\text{moment} & & \text{moment} & & \text{moment}
\end{array}
\qquad (10.1–1)
$$

where, it should be pointed out, the so-called secondary moment may sometimes be of larger magnitude than the primary moment.

At any section of the continuous beam, the effect of the prestressing force P_e and the resulting moment M_3 is equivalent to that of a force P_e acting at an eccentricity e_p, where

$$
e_p = -M_3/P_e
\qquad (10.1–2)
$$

(If for any reason, such as friction, P_e cannot be regarded as constant along

---

* It is assumed that provisions are made to permit horizontal displacement at supports.

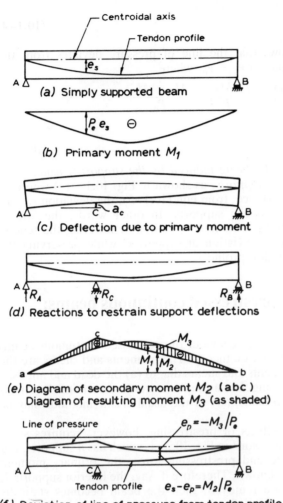

(a) Simply supported beam

(b) Primary moment M_1

(c) Deflection due to primary moment

(d) Reactions to restrain support deflections

(e) Diagram of secondary moment M_2 (abc)
    Diagram of resulting moment M_3 (as shaded)

(f) Deviation of line of pressure from tendon profile

Fig. 10.1–1

the beam, then the local value of P_e should be used in this equation.) The line having the eccentricity e_p is called the **line of pressure** or the **line of thrust** (Fig. 10.1–1(f)); it represents the locus of the centre of compression in the concrete section. In a simple beam, of course, no secondary moments exist and the tendon profile is the line of pressure.

In applying eqns (9.2–2) and (9.2–3) to continuous beams, the tendon eccentricity e_s must be replaced by the eccentricity e_p of the line of pressure:

$$f_1 = \frac{P_e}{A} + \frac{P_e e_p}{Z_1} \qquad (10.1–3)$$

$$f_2 = \frac{P_e}{A} - \frac{P_e e_p}{Z_2} \tag{10.1-4}$$

Figure 10.1–1(f) shows that the line of pressure deviates from the tendon profile by an amount $e_s - e_p$, where

$$e_s - e_p = \left(-\frac{M_1}{P_e}\right) - \left(-\frac{M_3}{P_e}\right) = \frac{M_3 - M_1}{P_e}$$

i.e.

$$e_s - e_p = M_2/P_e \tag{10.1-5}$$

The secondary moment M_2, being induced by the support reactions, can vary only linearly between supports, as shown in Fig. 10.1–1(e) where ac and cb are straight lines. Hence, from eqn (10.1–5), the deviation $e_s - e_p$ can only vary linearly between supports. In other words, the line of pressure in Fig. 10.1–1(f) is obtained by raising or lowering the tendon profile by the appropriate deviation at support C while preserving the intrinsic shape of the tendon profile.

10.2   Analysis of prestressed continuous beams: elastic theory

Under service conditions, the behaviour of prestressed continuous beams may be studied using the elastic theory. The moments and shears are the algebraic sum of the moments and shears due to (a) the dead and imposed loads and (b) the prestressing. It is a simple matter to determine the effects of the dead and imposed loads; any of the usual elastic methods may be used.

The analysis for the effects of prestressing, however, requires some explanation. Consider the beam* in Fig. 10.2–1, which has n supports. For any given tendon profile and prestressing force, the primary moment diagram ($M_1 = -P_e e_s$) is known. Therefore if, say, the interior supports 2, 3, ..., i, ..., $n - 1$ are removed, then the beam becomes statically determinate and the upward deflections at these support positions caused by the primary moment are readily determined. Let these upward deflections be

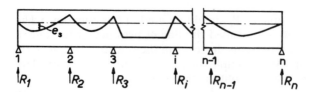

Fig. 10.2–1

---

* In Fig. 10.2–1, the tendon profile is shown to have sharp bends at various locations. In practice, these bends are smoothed out locally.

$a_{2,\,P};\ a_{3,\,P};\ \ldots\ a_{i,\,P}\ \ldots\ a_{n-1,\,P}$ respectively

Also, for the beam simply supported at 1 and n, let a_{ij} be the upward deflection at support position i due to a unit upward force at position j. Since in fact there is no deflection at the typical support i, the following compatibility condition is satisfied:

$$a_{i2}R_2 + a_{i3}R_3 + \ldots + a_{ii}R_i + \ldots\ a_{i,\,n-1}R_{n-1} + a_{i,\,P} = 0$$
$$(10.2\text{--}1)$$

where R_i is the upward reaction at support i of the continuous beam caused by the primary moment; the coefficients a_{ij} and $a_{i,\,P}$ are deflection coefficients of the simple beam and are hence readily determined. Equation (10.2–1) is satisfied for all values of i from $i = 2$ to $i = n - 1$. Therefore, we have $n - 2$ simultaneous equations to solve for the $n - 2$ support reactions, using standard computer routines if necessary. This method of analysis is of general applicability. However, for practical purposes, a more convenient method is available. Before describing this method, brief reference will be made to some well-known relations in elementary structural mechanics. Referring to the short length of beam in Fig. 10.2–2, the shear force V, the bending moment M and the load q are related by

$$V = dM/dx \qquad\qquad (10.2\text{--}2)$$

$$q = -dV/dx = -d^2M/dx^2 \qquad\qquad (10.2\text{--}3)$$

$$\int_A^B q\ dx = \int_A^B - dV = V_A - V_B$$

$$= \left[\frac{dM}{dx}\right]_A - \left[\frac{dM}{dx}\right]_B \qquad\qquad (10.2\text{--}4)$$

The following general statements may be made:

(a) The load between two sections of a beam is equal to the change in slope of the bending moment diagram between the points (see eqn 10.2–4).

(b) If the curvature of the bending moment diagram is constant between the two sections, then the load is uniformly distributed (see eqn 10.2–3). In practice where the tendon profile is circular or parabolic, or is approximately circular or parabolic, the load due to prestressing is taken as uniformly distributed.

(c) If the bending moment diagram for a certain length of the beam

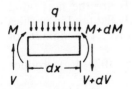

Fig. 10.2–2

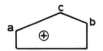

(a) Bending moment diagram

(b) Shear force diagram

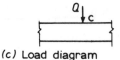

(c) Load diagram

Fig. 10.2–3

consists of two straight lines, such as ac and cb in Fig. 10.2–3(a), then from eqn (10.2–2) the shear force is constant from a to c and also from c to b (Fig. 10.2–3(b)); therefore the load must be a concentrated one at section c, its magnitude Q being equal to the change in shear force from one side of c to the other (Fig. 10.2–3(c)). That is, Q is equal to the change in slope of the bending moment diagram at section c.

In a prestressed concrete beam, the tendon profile represents to scale the primary moment diagram; hence the transverse load due to the prestressing can be worked out directly from the tendon profile, as explained in Example 10.2–1.

Example 10.2–1
Figure 10.2–4(a) shows the tendon profile in a continuous beam of uniform cross-section. If the prestressing force is 5000 kN, determine the line of pressure. Hence determine the support reactions induced by the prestressing.

SOLUTION
The loads produced by the prestressing are as shown in Fig. 10.2–4(b). Thus, at A there is a concentrated load of 5000 kN × 0.082 radians = 410 kN (see statement (c) above). Between A and C there is a uniformly distributed load of (5000 kN × 0.18 radians)/25 m = 36 kN/m (see statement (b)) and so on. The axial forces at A and B are the horizontal components of the tendon force and, for the relatively flat tendon profiles used in practice, are taken as equal to the tendon force. The moment of 500 kNm at B is obtained as 5000 kN × 0.1 m.

Of the forces in Fig. 10.2–4(b), only those in Fig. 10.2–4(c) produce bending moments acting on the beam.

The beam is next analysed for the continuity moments at the supports. Hardy Cross's moment distribution method is used here (Fig. 10.2–4(d)),

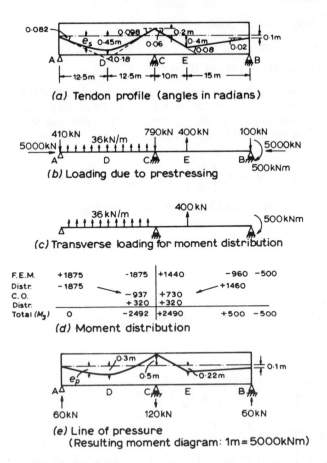

(a) Tendon profile (angles in radians)

(b) Loading due to prestressing

(c) Transverse loading for moment distribution

(d) Moment distribution

(e) Line of pressure
   (Resulting moment diagram: 1m = 5000kNm)

Fig. 10.2-4

but where there are five or more spans a solution using a standard computer program may be quicker. The final moments in Fig. 10.2–4(d) are the resultant moments due to the effect of the prestressing on the continuous beam. Hence the eccentricity e_p of the line of pressure at C must be $-M_3/P_e = -2491$ kNm/5000 kN = -0.5 m, as shown in Fig. 10.2–4(e). The complete line of pressure is then obtained by a process called linear transformation, which is described in Section 10.3.

The support reactions induced by the prestressing may be determined from the secondary moments M_2. Equation (10.1–1) states that the secondary moment diagram is the difference between the resulting moment diagram (which is Fig. 10.2–4(e)) and the primary moment diagram (which is Fig. 10.2–4(a)); that is the secondary moment diagram is a triangle (Rule 3 in Section 10.3 will make this point clear) in which

$$M_2 \text{ at A} = (M_3 \text{ at A}) - (M_1 \text{ at A}) = 0$$

$$M_2 \text{ at C} = (M_3 \text{ at C}) - (M_1 \text{ at C})$$

$$= 5000 \times 0.5 - 5000 \times 0.2$$

$$= 1500 \text{ kNm}$$

$$M_2 \text{ at } B = (M_3 \text{ at } B) - (M_1 \text{ at } B) = 0$$

The support reactions induced by the prestressing are now calculated from these secondary moments M_2, and are as shown in Fig. 10.2–4(e). These support reactions and the tendon profile e_s completely define the shear force V_p at any section produced by the prestressing. For example at a section between C and E,

$$V_p = 60 \text{ kN} - 120 \text{ kN} - P_e \left[\frac{de_s}{dx} \right] \quad \text{(see eqn 9.2–22)}$$

$$= 60 \text{ kN} - 120 \text{ kN} - 500 \text{ kN} \times 0.06 \text{ rad} \quad \text{(see Fig. 10.2–4(a))}$$

$$= - 360 \text{ kN}$$

(A more convenient method of determining V_p is given in Example 10.4–1.)

Example 10.2–2

With reference to the beam in Example 10.2–1, determine the prestress at the top and bottom fibres at section E in terms of the sectional properties of the beam.

SOLUTION
From eqns (10.1–3) and (10.1–4),

$$\text{bottom fibre prestress } f_1 = \frac{P_e}{A} + \frac{P_e e_p}{Z_1}$$

$$\text{top fibre prestress } f_2 = \frac{P_e}{A} - \frac{P_e e_p}{Z_2}$$

From Fig. 10.2–4(e), $e_p = 0.22$ m at E; also $P_e = 5000$ kN as given. Therefore

$$f_1 = 5000 \times 10^3/A + 5000 \times 10^3 \times 0.22 \times 10^3/Z_1 \text{ N/mm}^2$$

$$f_2 = 5000 \times 10^3/A - 5000 \times 10^3 \times 0.22 \times 10^3/Z_2 \text{ N/mm}^2$$

where A and Z are respectively in mm^2 and mm^3 units.

Example 10.2–3

If the prestressed beam in Example 10.2–1 is acted on by a uniformly distributed load of 50 kN/m, determine the resulting line of pressure due to the combined action of the prestressing and the imposed load.

SOLUTION
The solution is summarized in Fig. 10.2–5. The resulting line of pressure is obtained by superposition, i.e.

Fig. 10.2–5(e) = Fig. 10.2–5(b) + Fig. 10.2–5(d)

The reader should go through the solution properly; study the solution to Example 10.2–1 again if necessary.

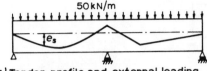

(a) Tendon profile and external loading

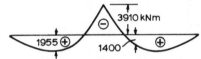

(b) Line of pressure due to prestressing

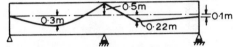

(c) Bending moment diagram for external loading

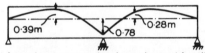

(d) Line of pressure due to external loading

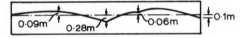

(e) Resulting line of pressure

Fig. 10.2–5

10.3   Linear transformation and tendon concordancy

In the design of prestressed concrete continuous beams, it is advantageous to be familiar with several terms to be defined below. A **transformation profile** is a tendon profile which consists of straight lines between the supports and in which the eccentricity is zero at simple end supports. Examples of transformation profiles are shown in Fig. 10.3–1(a) and (b); the profile in Fig. (c) is not a transformation profile because the eccentricity is not zero at the simple end support D.

By **linear transformation** is meant the raising or lowering of a tendon at the *internal* supports while maintaining the intrinsic shape of the tendon profile between supports. Thus, each of the profiles (a) and (b) in Fig. 10.3–2 is obtained from the original tendon profile (shown in full line) by linear transformation, and the new profiles (a) and (b) are referred to as linearly transformed profiles of the original, or often simply as the linear transformations of the original profile. Linear transformation can be considered as the superposition of a transformation profile on a tendon profile. For example, the profile in Fig. 10.2–4(e) is the sum of the

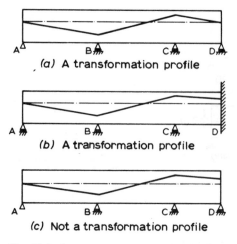

(a) A transformation profile

(b) A transformation profile

(c) Not a transformation profile

Fig. 10.3–1

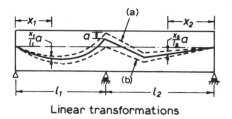

Linear transformations

Fig. 10.3–2

profile in Fig. 10.2–4(a) and a transformation profile having an eccentricity of −0.3 m at the internal support.

A tendon in a continuous beam is said to be a **concordant tendon** if it produces a line of pressure coincident with the tendon itself; the profile of a concordant tendon is called a **concordant profile**. By definition, therefore, a concordant tendon induces no support reactions and no secondary moments. (In this sense, all tendons in simply supported beams are concordant tendons.) Consequently, if (and only if) concordant tendons are used then eqns (9.2–2) to (9.2–7) inclusive become directly applicable to continuous beams.

We now list below several simple rules which will be found useful in design; studying these rules is incidentally also highly valuable for developing a sound understanding of the mechanics of prestressed concrete continuous beams:

Rule 1

The line of pressure is a linear transformation of the tendon profile, i.e. it can be obtained by superimposing a transformation profile on to the tendon profile.

Proof: The proof of this rule follows directly from eqn (10.1–5), noting that the secondary moment M_2 varies only linearly between supports.

Rule 2
Linear transformations of the tendon profile have no effect on the position of the line of pressure. In other words, any transformation profile may be superimposed on the tendon profile without affecting the line of pressure.

Proof: The intrinsic shape of the primary moment diagram within each individual span depends only on the intrinsic shape of the tendon profile. In linear transformation, the tendon between two supports is moved as a rigid body without changing its intrinsic shape. Hence (see statements (a)–(c) under eqn 10.2–4) linear transformation does not affect the transverse load exerted by the tendon on the span, and hence the resulting moment M_3 is not affected; therefore (see eqn 10.1–2), the line of pressure is not affected.

Rule 3
The line of pressure obtained from any tendon profile is itself a concordant profile.

Proof: From Rule 1, the line of pressure is a linear transformation of the tendon profile. From Rule 2, if the tendon is moved to follow the original line of pressure, that line of pressure remains unchanged, so that the new tendon profile is coincident with the line of pressure.

Rule 4
Any bending moment diagram for a continuous beam with non-yielding supports, produced by any set of transverse forces or moments, is a concordant profile.

Proof: For a given beam, the deflections (including support deflections) are completely defined by the bending moment diagram. For a continuous beam with rigid supports, every bending moment diagram due to external loads is computed on the basis of no support deflection. Since any tendon profile following such a bending moment diagram will produce a primary moment diagram with ordinates everywhere proportional to that bending moment diagram, such a tendon profile will produce no deflections at support positions and hence will induce no support reactions; therefore it is a concordant profile. (Where the prestressing force is not constant along the beam, the eccentricity of the concordant profile at any section will be that of the bending moment diagram divided by the prestressing force at that section.)

Rule 5
Superposition of any number of concordant profiles will result in a concordant profile. Therefore any number of concordant tendons acting together will form a concordant tendon, in the sense that the locus of their centre of gravity coincides with the resultant line of pressure due to these tendons.

Proof: The proof follows from Rule 4. Any concordant profile represents to some scale a bending moment diagram due to a certain set of external loads. Hence the superposition of any number of concordant profiles will result in a profile which represents, to some scale, a bending moment diagram due to some combination of external loads and which is therefore concordant.

Example 10.3–1
(a) Determine a concordant profile for a 5000 kN tendon force to neutralize the bending moments due to the loading in Fig. 10.3–3(a).
(b) If for some reason the eccentricity of the profile in (a) has to be raised to 0.4 m above the beam axis at B, by how much should the eccentricites be changed elsewhere to maintain its concordancy?
(c) Does the tendon, in its new position in part (b), still fulfil the purpose specified in part (a)?
(Given: the beam is of uniform cross-section.)

SOLUTION
(a) The bending moment diagram due to the applied loads is (and the reader should verify this) as shown in Fig. 10.3–3(b). The profile in Fig. 10.3–3(c) is obtained by dividing the ordinates of the above moment diagram by the tendon force. From Rule 4, this is a concordant profile. Hence the resulting moments due to the pretressing are equal in magnitude but opposite in sense to the moments produced by the applied loads.
(b) The profile in Fig. 10.3–3(d) is (the reader should verify this) similar to the bending moment diagram due to a unit moment applied at the end B of the beam, and is hence a concordant profile (Rule 4). If the tendon in Fig. 10.3–3(c) is to be raised to 0.4 m above the centroidal axis of the beam at support B, it is only necessary (Rule 5) to

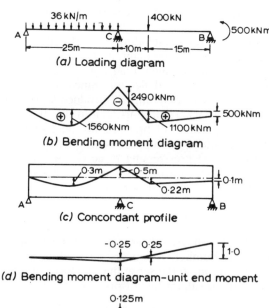

(a) Loading diagram

(b) Bending moment diagram

(c) Concordant profile

(d) Bending moment diagram–unit end moment

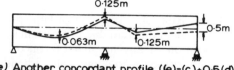

(e) Another concordant profile ((e)=(c)+0·5(d))

Fig. 10.3–3

superimpose 0.5 times Fig. 10.3–3(d) to the profile in Fig. 10.3–3(c), i.e.

Fig. 10.3–3(e) = Fig. 10.3–3(c) + 0.5 × Fig. 10.3–3(d)

Specifically, the tendon must be lowered by 0.125 m at the interior support C. Within each of the spans AC and CB, the tendon is moved as a rigid body without changing its shape.

(c)   In its new position, the tendon does not fulfil its original purpose, because the new line of pressure is no longer that shown in Fig. 10.3–3(c) but is as shown by the full line in Fig. 10.3–3(e).

10.4   Applying the concept of the line of pressure

Equations (9.2–2) to (9.2–7) become applicable to continuous beams if the eccentricity e_p of the line of pressure is substituted for the tendon eccentricity e_s. Thus, for a continuous. beam, the prestressing force P_e and the profile of the line of pressure must satisfy the conditions:

$$\frac{P_e}{A} + \frac{P_e e_p}{Z_1} - \frac{M_{imax} + M_d}{Z_1} \geq f_{amin} \qquad (10.4\text{--}1)$$

$$\frac{P_e}{A} + \frac{P_e e_p}{Z_1} - \frac{M_{imin} + M_d}{Z_1} \leq f_{amax} \qquad (10.4\text{--}2)$$

$$\frac{P_e}{A} - \frac{P_e e_p}{Z_2} + \frac{M_{imax} + M_d}{Z_2} \leq f_{amax} \qquad (10.4\text{--}3)$$

$$\frac{P_e}{A} - \frac{P_e e_p}{Z_2} + \frac{M_{imin} + M_d}{Z_2} \geq f_{amin} \qquad (10.4\text{--}4)$$

where e_p is the eccentricity of the line of pressure at the section considered, and the meanings of the other symbols are as in eqns (9.2–1) to (9.2–7). Rearrangement of these equations gives the limits of the **permissible pressure zone**, i.e. the permissible zone for the line of pressure:

$$e_p \geq \frac{M_{imax} + M_d}{P_e} - \frac{Z_1}{A} + \frac{Z_1 f_{amin}}{P_e} \qquad (10.4\text{--}5)$$

$$e_p \geq \frac{M_{imax} + M_d}{P_e} + \frac{Z_2}{A} - \frac{Z_2 f_{amax}}{P_e} \qquad (10.4\text{--}6)$$

$$e_p \leq \frac{M_{imin} + M_d}{P_e} - \frac{Z_1}{A} + \frac{Z_1 f_{amax}}{P_e} \qquad (10.4\text{--}7)$$

$$e_p \leq \frac{M_{imin} + M_d}{P_e} + \frac{Z_2}{A} - \frac{Z_2 f_{amin}}{P_e} \qquad (10.4\text{--}8)$$

If a concordant profile is used then the line of pressure is coincident with the tendon profile and e_s and e_p are synonymous; the above equations then become identical to eqns (9.2–18) to (9.2–21).

Equation (9.2–8) is applicable to continuous beams, irrespective of whether the tendon is concordant or not:

$$\left.\begin{array}{l} Z_1 \text{ (bottom)} \\ Z_2 \text{ (top)} \end{array}\right\} \geq \frac{M_{imax} - M_{imin}}{f_{amax} - f_{amin}} \qquad (10.4\text{--}9)$$

Similarly, the reader should verify that, irrespective of whether the tendon is concordant or not, the minimum required prestressing force is still given by eqn (9.2–16), namely

$$P_{emin} = \frac{[f_{amin}(Z_1 + Z_2) + M_r]A}{Z_1 + Z_2} \qquad (10.4-10)$$

However, eqn (9.2–17) no longer refers to the tendon eccentricity; it now specifies the required position of the line of pressure at the critical section:

$$e_p = \frac{Z_2 M_{imax} + Z_1 M_{imin} + (Z_1 + Z_2)M_d}{[f_{amin}(Z_1 + Z_2) + M_r]A} \qquad (10.4-11)$$

Shear in prestressed concrete continuous beams

The concept of the line of pressure can also be applied to the determination of the shear forces in prestressed concrete continuous beams. From Rule 1, the line of pressure is the sum of the tendon profile and a transformation profile, and hence we are entitled to consider the tendon profile as being made up of the sum of the profile of the line of pressure and a transformation profile:

$$e_s = e_p + e_t \qquad (10.4-12)$$

where the transformation profile e_t is actually given by M_2/P_e (see eqn 10.1–5).

Applying Rule 3 of Section 10.3, the actual tendon profile may be considered to be made up of two profiles: a concordant profile e_p and a transformation profile e_t. By definition, a transformation profile consists of straight lines between supports and is associated with point loads at support positions only. Therefore the profile e_t has nothing to do with the bending moments and shear forces in the beam; these are associated only with the profile e_p. Therefore eqn (9.2–22) becomes

$$V_p = -P_e\left[\frac{de_p}{dx}\right] \qquad (10.4-13)$$

where V_p is the shear force due to the prestressing, and de_p/dx is the slope of the line of pressure at the section considered. Note that the actual tendon force P_e may have induced support reactions, but these do not enter into the shear expression in eqn (10.4–13); this is because we have split the actual tendon profile e_s into two fictitious profiles e_p and e_t. Profile e_t produces no shear while profile e_p produces no support reactions!

Example 10.4–1

Referring to the beam in Example 10.2–1 and Fig. 10.2–4, determine the shear force at a section between C and E.

SOLUTION

(See also final part of solution to Example 10.2–1).

$$V_p = -P_e\left[\frac{de_p}{dx}\right]$$

From Fig. 10.2–4(e),

$$\frac{de_p}{dx} = \frac{0.5 \text{ m} + 0.22 \text{ m}}{10 \text{ m}} = 0.072$$

Therefore

$$V_p = -5000 \times 0.072 \text{ kN} = \underline{-360 \text{ kN}}$$

This shear force of -360 kN includes the effects of the support reactions induced by the prestressing.

10.5   Summary of design procedure

The design of prestressed concrete continuous beams may be summarized in the steps below; it is suggested that, before studying these steps, the reader should first review Section 9.9 on the design procedure for simply supported beams.

Step 1
Using the known values of M_{imax} and M_{imin} determine the minimum required Z's from eqn (10.4–9). Using these minimum required Z's as a guide, assume a member cross-section, remembering that if a larger section than the absolute minimum required is provided, the depth of the permissible zone for the line of pressure will be increased, and hence the design of the tendon profile will be easier.

Step 2
M_d is now known. P_{emin} is then computed from eqn (10.4–10). Using this value as a guide, select a suitable force P_e, noting that if P_e is somewhat larger than P_{emin} there will be more freedom in choosing a tendon profile.

Step 3
Plot the permissible zone for the line of pressure, using eqns (10.4–5) to (10.4–8). Too wide a zone indicates that an excess of prestressing force or of concrete section has been provided. If the limits of the zone should cross at a particular location, then the prestressing force or the cross-section needs modifying.

Step 4
Choose a trial tendon profile within the permissible zone obtained in Step 3. If the tendon profile is concordant, then the line of pressure is within the permissible zone and the tendon profile is satisfactory in regard to stress conditions in eqns (10.4–1) to (10.4–4). If the chosen profile is non-concordant, then the line of pressure has to be determined using the procedure of Example 10.2–1. If the line of pressure is found to lie wholly or partly outside the permissible zone, a new tendon profile has to be tried. If the line of pressure lies within the zone, the trial non-concordant profile may be accepted; or it may optionally be linearly transformed to follow the line of pressure (see Rules 1, 2, 3 in Section 10.3).
In this trial and error process, much labour and boredom may be

saved by trying concordant profiles; in this respect, Rule 4 is helpful. Rule 5 will be found useful if the trial profile has to be modified while maintaining its concordancy (see Example 10.3–1(b)).

Step 5

If the profile selected in Step 4 is such that the tendon is too near the beam top or the beam soffit at a section, increased concrete cover may be achieved by linear transformation—which renders the tendon non-concordant but which does not affect the position of the line of pressure. Similarly, if the 'kink' or change of slope of the tendon is too sharp over a support (consequence: too much loss of prestress due to friction), this may be eased by linear transformation; for example, in Fig. 10.3–2, the 'kink' over the interior support is less in the profile (b) than in the profile (a).

Step 6

Check stresses at transfer and calculate loss of prestress; revise design as necessary.

Step 7

Check ultimate flexural strength and ultimate shear resistance at critical sections. See Sections 9.5 and 9.6.

Step 8

Design end blocks if necessary [1, 2].

Example 10.5–1

A post-tensioned beam of uniform cross-section is continuous over three equal spans of 10 m, and carries imposed loads of 100 kN acting simultaneously at each of the third-span points. The allowable stresses in service are $f_{amax} = 14$ N/mm^2, $f_{amin} = 0$ and those at transfer are $f_{amaxt} = 16$ N/mm^2 and $f_{amint} = -1$ N/mm^2. The prestress loss ratio (ratio of the prestressing force in service to that at transfer) is $\alpha = 0.85$. Design:

(a) the concrete section;
(b) the prestressing force; and
(c) the tendon profile.

SOLUTION

Step 1

The reader should verify that the imposed-load moments M_i and the dead-load moments M_d are as shown in Fig. 10.5–1.

By inspection, section B is critical. From eqn (10.4–9),

$$Z(\text{min}) = \frac{M_{imax} - M_{imin}}{f_{amax} - f_{amin}} = \frac{0 - (-266)}{14 - 0} = 19.0 \times 10^6 \text{ mm}^3$$

To obtain a reasonable depth of the permissible zone for the line of pressure, choose a section with, say, the following properties:

$$Z_1(\text{bottom}) = 22 \times 10^6 \text{ mm}^3 \qquad Z_2(\text{top}) = 23 \times 10^6 \text{ mm}^3$$

$$\text{area } A = 160,000 \text{ mm}^2 \qquad \text{overall depth } h = 770 \text{ mm}$$

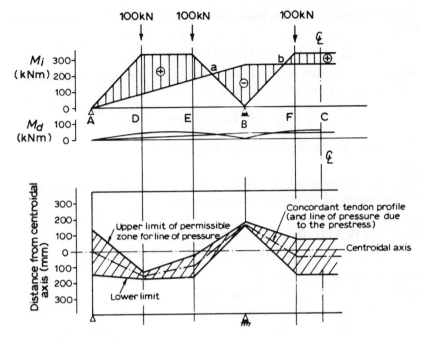

Fig. 10.5–1

centroidal distance from soffit $a_1 = 394$ mm

(In practice, sections may be chosen from tables, such as those in Appendix B of Reference 1.)

Step 2
From eqn (10.4–10),

$$P_{emin} = \frac{[f_{amin}(Z_1 + Z_2) + M_r]A}{Z_1 + Z_2}$$

$$= \frac{(0 + 266 \times 10^6) \times 160\,000}{22 \times 10^6 + 23 \times 10^6} = 946 \text{ kN}$$

Use

$$P_e = 105\% P_{emin}(\text{say}) = 1000 \text{ kN(say)}$$

Step 3
The permissible zone for the line of pressure is given by eqns (10.4–5) to (10.4–8). Substituting in numerical values it will be found that eqns (10.4–5) and (10.4–8) are the two critical ones. The reader should verify that the limiting values for e_p at sections A, B, C, D, E and F are as plotted in Fig. 10.5–1. (If in doubt see Example 9.2–4 for method of calculation.)

Step 4
Try a concordant profile; say, one following the imposed-load bending

moment diagram. Let the tendon eccentricity e_s at the interior support B be

$$e_s = \tfrac{1}{2}(-174.3 - 159.1) = -167 \text{ mm}$$

where -174.3 mm and -159.1 mm are the limiting values of the eccentricity e_p (not e_s) at B. Then

$$e_s \text{ at E} = \frac{M_i \text{ at E}}{M_i \text{ at B}} \times (-167) \text{ mm etc.}$$

The tendon profile is as shown in Fig. 10.5–1.

Step 5
Since the tendon profile in Step 4 is concordant (Rule 4) and since it lies within the permissible zone, it is acceptable.

Step 6
From eqn (9.3–7), the prestressing force at transfer is

$$P = P_e/\alpha = 1000/0.85 = 1175\text{kN}$$

Consider Section B (Fig. 10.5–1)
From eqn (10.4–1), bottom fibre stress is

$$\frac{P}{A} + \frac{Pe_p}{Z_1} - \frac{M_{\text{imax}} + M_d}{Z_1}$$

$$= \frac{1175 \times 10^3}{160 \times 10^3} + \frac{1175 \times 10^3(-167)}{22 \times 10^6} - \frac{0 - 36.8 \times 10^6}{22 \times 10^6}$$

$$= 0.097 \text{ N/mm}^2 > f_{\text{amint}} \text{ of } -1 \text{ N/mm}^2$$

Therefore this is acceptable.

From eqn (10.4–3), top fibre stress is

$$\frac{P}{A} - \frac{Pe_p}{Z_2} + \frac{M_{\text{imax}} + M_d}{Z_2}$$

$$= \frac{1175 \times 10^3}{160 \times 10^3} - \frac{1175 \times 10^3 \times (-167)}{23 \times 10^6} + \frac{0 - 36.8 \times 10^6}{23 \times 10^6}$$

$$= 14.3 \text{ N/mm}^2 < f_{\text{amaxt}} \text{ of } 16 \text{ N/mm}^2$$

Therefore this is acceptable.

Equations (10.4–2) and (10.4–4) are not critical. Similarly, it can be shown that stresses are everywhere within the allowable values at transfer. (*Note:* If the 'kink' of the tendon at B is considered too sharp, this can be reduced by linear transformation. After such linear transformation, the tendon may be outside the shaded zone in Fig. 10.5–1, but this does not matter, because (Rule 2) the line of pressure remains within the zone.)

Example 10.5–2
Explain whether the following statements are true or false:

(a) The line of pressure (line of thrust) in a prestressed concrete continuous beam is a linear transformation of the tendon profile.

(b) Provided the intrinsic shapes of the tendon profile within the individual spans are maintained, raising or lowering the tendon over any support will not change the position of the line of pressure.

(c) A tendon, which follows the line of pressure obtained from another non-concordant tendon, is a non-concordant tendon.

(d) Any bending moment diagram for a continuous beam with non-yielding supports, produced by any system of transverse loads, represents a concordant profile irrespective of whether the tendon force is constant along the beam.

(e) If a concordant tendon is raised or lowered at a simple end support, its concordancy may be restored by suitable adjustments of its eccentricities at the interior supports. These adjustments will also restore the line of pressure to its original position.

(f) Since linear transformation does not change the position of the line of pressure, it follows that linear transformation does not change the secondary moments.

SOLUTION

(a) True. See Rule 1 in Section 10.3.

(b) True for movements over internal supports only. See definition of linear transformation in Section 10.3.

(c) False. See Rule 3.

(d) False. True only if the tendon froce is constant along the beam. For varying tendon force, the ordinate of the moment diagram should be divided by the local value of the tendon force.

(e) First part true; see Example 10.3–1(b). Second part false; since the tendon remains concordant, then by definition the line of pressure must have moved to follow the new tendon profile. See Example 10.3–1(c).

(f) False. Linear transformation produces transverse forces, which are directly transferred to the supports; therefore it induces changes in support reactions and hence induces changes in secondary moments. However, the changes in secondary moments are exactly equal and opposite to the changes it produces in the primary moments. Hence the resulting moments are unchanged (and hence the line of pressure is unchanged).

Example 10.5–3

In Fig. 10.5–2, the force in each of the tendons A and B is P, and the profile of tendon B is the line of pressure of tendon A. Explain whether the following statements are true or false:

(a) Since profile B is the line of pressure of tendon A, it follows that profile A must be the line of pressure of tendon B.

(b) The hogging bending moment acting on the beam is

$$P(e_s \text{ of B} + e_s \text{ of A})$$

where e_s is the tendon eccentricity at the section considered, so that at a section such as X–X, where e_s of B and e_s of A are numerically equal but are on opposite sides of the centroidal axis of the beam, the beam will experience no bending moment.

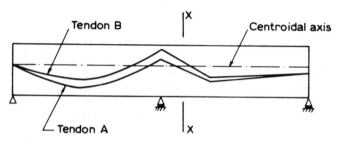

Fig. 10.5–2

(c) If the general shapes of profiles A and B are as shown, then the prestressing force in tendon B would produce an upward reaction at the middle support.

(d) Since tendon B follows the line of pressure of tendon A, it (tendon B) cannot produce any shear force on the beam.

SOLUTION

(a) False. In fact profile B is concordant; see Rule 3 in Section 10.3.

(b) False. The hogging bending moment at a typical section is

$$P(e_p \text{ of } B + e_p \text{ of } A)$$

That is, the eccentricity e_p of the line of pressure should be used, and not the tendon eccentricity e_s. However, in this case e_p of A = e_s of B (why?) = e_p of B (why?). Therefore hogging bending moment = $2P \times (e_p \text{ of } B)$. This applies everywhere, even at section X–X!

(c) False. Tendon B is concordant (why?); therefore no reaction is induced.

(d) False. From eqn (10.4–13), the shear force due to tendon B is

$$V_p(B) = -P\frac{d(e_p \text{ of } B)}{dx}$$

Note that $V_p(A) = V_p(B)$ so that the shear force experienced by the beam is twice that given above.

Comments

The authors' experience is that students tend to have difficulties with the analysis and design of prestressed concrete continuous beams. *These difficulties are due to the lack of a clear understanding of the principles of mechanics as applied to prestressing. This chapter has been written with the aim of helping the student develop such an understanding and Examples 10.5–2 and 10.5–3 provide a test of his mastery of the fundamental principles.*

References 3–8 provide further insight into the fundamental behaviour of prestressed concrete beams.

Problems

10.1 The figure shows a continuous prestressed concrete beam of uniform cross-section. The cross-sectional area is 200×10^3 mm^2 and the sectional modulus is 78×10^6 mm³. The tendon profile consists of straight lines within the end spans and an approximately parabolic curve within the interior span. The tendon force, which may be taken as uniform throughout the beam, is 2000 kN.

Working from first principles and neglecting the effects of the self-weight of the beam, determine:

(a) the prestress at the mid-span section of the interior span;
(b) the shear force at a section 18 m from an end support; and
(c) the reaction at an end support.

Ans. (a) $f_2 = 5.1$ N/mm², $f_1 = 14.9$ N/mm² (both compressive).
     (b) 146.5 kN.
     (c) 13.2 kN (upwards).

10.2 With reference to the beam in Problem 10.1 suppose, in addition to the existing tendon, there is a second tendon which carries a uniform 500 kN force and which follows the line of pressure of the existing tendon. Determine the new values of the quantities (a), (b) and (c) in Problem 10.1. *Work from first principles.*

Ans. (a) $f_2 = \dfrac{2000 + 500}{2000} \times 5.1 = 6.4$ N/mm² (compressive)

$f_1 = \dfrac{2000 + 500}{2000} \times 14.9 = 18.6$ N/mm² (compressive)

(b) 183.1 kN.
(c) 13.2 kN, upwards (as in Problem 10.1).

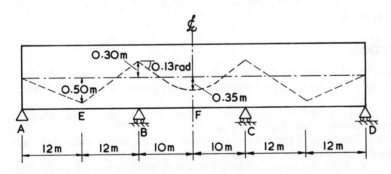

Problem 10.1

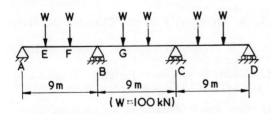

Problem 10.3

10.3*   A post-tensioned concrete beam ABCD is of uniform cross-section and is continuous over three equal spans of 9 m each; it carries point loads of 100kN at each of the third-span points, as shown in the figure. The tendon force is 1000 kN, and the self-weight of the concrete is negligible.

Working from first principles, obtain solutions to the following problems:

(a)   Design a tendon profile which will simultaneously satisfy two conditions: first, the stressing of the tendon should not induce any change in the support reactions and, second, when the tendon is fully stressed the beam should carry the given loads without the concrete section experiencing a bending moment anywhere.

(b)   With the tendon profile designed as in (a) above, and the loading applied, determine the shear force carried by the concrete at a section at the midspan of AB.

(c)   Suppose a site error is made in setting out the tendon profile, so that the eccentricity is 50 mm too high at B, but is correct at A, C, D and within the span CD; within the spans AB and CB, the eccentricity error is proportional to the distance from A and C respectively. With the tendon profile incorrectly set out this way, calculate the shear force carried by the concrete at the midspan of AB. What is now the support reaction at A? If the concrete is to remain free of bending stress everywhere, what changes must be made to the applied loading?

Ans.   (a) $e_A = 0$, $e_E = 0.22$ m, $e_F = 0.14$ m, $e_B = -0.24$ m,
$e_G = 0.06$ m (+ ve downwards);
(b)   Zero shear; note that $V = dM/dx = 0$, everywhere;
(c)   (i) Zero shear; (ii) $R_A = 67.7$ kN (up); (iii) no changes required.

References

1   Abeles, P. W. and Bardhan-Roy, B. K. *Prestressed Concrete Designer's Handbook*. Viewpoint Publication, Cement and Concrete Association, Slough, 1981.
2   Rowe, R. E. End-block stresses in post-tensioned concrete beams. *The Structural Engineer*, **41**, Feb. 1963, pp. 54–68.
3   Lin, T. Y. Strengths of continuous prestressed concrete beams under static and repeated loads. *Proc. ACI*, **51**, June 1955, pp. 1037–59.
4   Lin, T. Y. Load balancing method for design and analysis of prestressed concrete structures. *Proc. ACI*, **60**, June 1963, pp. 719–42.
5   Lin, T. Y. and Burns, N. H. *Design of Prestressed Concrete Structures*, 3rd edn. Wiley, New York, 1981.
6   Evans, R. H. and Bennett, E. W. *Prestressed Concrete*. Chapman and Hall, London, 1958.
7   Sawko, F. (ed.). *Developments in Prestressed Concrete*. Applied Science, London, 1978, Vols 1 and 2.
8   Kong, F. K., Evans, R. H., Cohen, E. and Roll, F. (eds). *Handbook of Structural Concrete*. Pitman, London and McGraw-Hill, New York, 1983.

* Cambridge University Engineering Tripos: Part II (Past examination question).

Chapter 11
Practical design and detailing

In collaboration with *Dr B. Mayfield*, University of Nottingham

11.1   Introduction

All professional engineers are concerned with *design* in some form or other. The word *design* has different meanings to different professions, but is here taken to mean the formulation, in the mind, of some scheme or plan. This normally entails a proposition followed by some form of analysis which either proves or disproves the original proposition. The formulation of the original idea is usually based on experience and, therefore, the structural engineering student, meeting structural design for the first time, tends to find the discipline confusing since he is, to a large extent, forced into an 'opinion' as opposed to a 'fact' situation. Frustration follows with the subsequent need for the normally necessary refinement or amendment of the original proposition. The only method of overcoming the initial confusion and associated frustration is by observation and practice, i.e. by increasing the personal experience. The use of Codes of Practice in the design process has been succinctly stated by the main authors in Section 1.1.

The student must also realize the importance of the *presentation* of his design calculations and the consequent drawings. Design ideas normally need to be communicated for construction, and confusion *must* be minimized so as to avoid costly or dangerous errors due to misunderstanding or misinterpretation. Various publications [1–3] have been produced in recent years to aid this necessary transference and the worked examples in this chapter will use these texts as appropriate. Higgins and Rogers's book [3] on design and detailing can, in particular, be recommended in this connection.

11.2   Loads—including that due to self-mass

Chapter 1 introduced the concepts of **characteristic loads** and the **dead**, **imposed** and **wind** components. The appropriate definitions and values to be taken can be found in BS 6399 [4] and further help is given in Reference 6. This chapter will initially confine attention to dead and imposed loading.

Dead loads
Examples of dead loads are those due to finishes, linings, waterproofing asphalt, isulation, partitions, brickwork and, most importantly, self-mass. Typical values can be found in References 4 and 6 and some examples are given in Table 11.2–2 (see p. 000). The unit mass of reinforced concrete is considerable and a typical value is 2400 kg/m^3 (normally taken as 24 kN/m^3 in design calculations) and it is here that the student meets his first problem. To design even a simply supported beam, he needs to guess the beam size before he can include its self-weight in the analysis.

The *Manual for the Design of Reinforced Concrete Structures* (Institution of Structural Engineers, 1985) gives some help in this respect for practising engineers. Undergraduates, however, are mainly concerned with simple structural members such as slabs, beams and columns; for them the following simplified procedure may be adequate for preliminary design and **member sizing** purposes.

Step 1   Fire resistance
Fire resistance requirements [7, 8] may at times dictate the minimum size of a structural member, even though the loading may be comparatively low. Hence a useful starting point would be the fire resistance tables:
(a)   Slabs: Table 8.8–1;
(b)   Beams: Table 4.10–3;
(c)   Columns: Table 3.5–1.

Step 2   Concrete cover
The concrete cover to be used depends both on the fire resistance requirement and the durability requirement (Table 2.5–7). The concrete cover to be used should be the larger of that required by Table 2.5–7 and that by the relevant fire resistance table in Step 2.

Step 3   Span/depth ratio of beams and slabs
Table 5.3–1 gives the span/effective depth ratios for beams and slabs. These are the *minimum* values which should not normally be exceeded. In particular, in the preliminary sizing of beams, it is advisable to assume a span/depth ratio of about 12 for simply supported beams, 6 for cantilevers and, say, 15 for continuous beams.

Step 4   Resistance moment of beams and slabs
The adequacy of the assumed member size can be checked with Fig. 11.2–1, where $\varrho = A_s/bd$ and $\varrho' = A_s'/bd$. The figure has been prepared using the beam design chart in Fig. 4.5–2. 'Rules of thumb' suggest that the width b of a rectangular beam should be between 1/3 and 2/3 of the effective depth d, and Fig. 11.2–1 can be used as a guide in selecting b.

Step 5   Shear resistance of beams
Guidance on the shear resistance of beams is given in Fig. 11.2–2, where $\varrho = A_s/b_v d$—see Step 4 of the shear-design procedure in Section 6.4. Shear is not normally a problem in slabs supported on beams.

Step 6   Effective height of columns
The ratio of the effective height to the smaller lateral dimension should

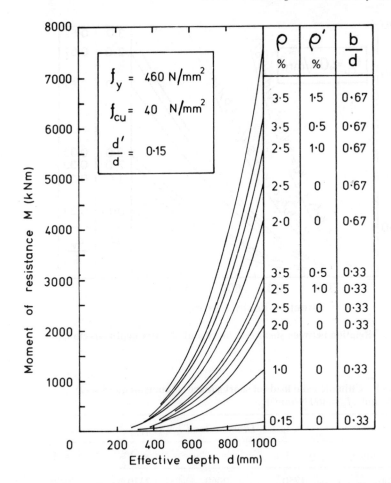

Fig. 11.2–1   Relation between resistance moment and effective depth—rectangular beams

not exceed 15, so that the column satisfies BS 8110's definition of a 'short column'. The effective height is defined by eqn (7.2–2), but it is simplest at this stage to take it as the clear height from floor to ceiling.

Step 7   Ultimate load of columns
   Further assistance in the selection of the size of a column is given in Table 11.2–1, where the ultimate loads have been calculated from eqn (3.4–2) for short columns.

Imposed loads
These include those due to stored solid materials and liquids, people, natural phenomena in addition to those due to moving vehicles and equipment [4, 6, 9]. Some examples are given in Table 11.2–2. As was discussed in Section 1.5, the student must recognize that the most severe

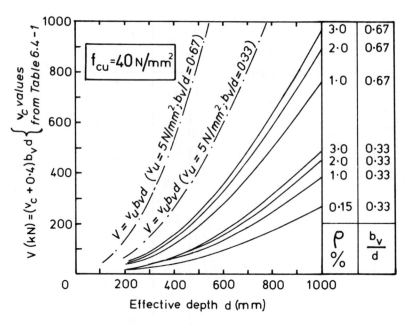

Fig. 11.2–2   Relation between shear resistance and effective depth—rectangular beams

Table 11.2–1   Ultimate axial loads of short square columns—eqn (3.4–2)
$(f_{cu} = 40 \text{ N/mm}^2; f_y = 460 \text{ N/mm}^2)$

Column size (square, mm)	Ultimate load kN (eqn 3.4–2)			
	$\varrho = 1\%$	$\varrho = 2\%$	$\varrho = 3\%$	$\varrho = 4\%$
300	1750	2060	2370	2680
350	2380	2800	3220	3650
400	3110	3660	4210	4760
450	3930	4630	5330	6030
500	4860	5720	6580	7450

stresses may occur in a particular part of a structure when the imposed load is completely withdrawn from some other part—see Section 1.5 and Fig. 4.9–6.

BS 6399 : Part 1 [4] gives the design imposed loading for many types of structure from art galleries to workshops. Two alternatives are given, uniformly distributed and concentrated and as an example the values given for reading rooms in libraries are 4.0 kN/m^2 and 4.5 kN respectively. The floor slabs have to be designed to carry whichever of these produces the greater stresses in the part under consideration. Since it is unlikely that at one particular time all floors will simultaneously be carrying the

Table 11.2–2   Some typical values of dead and imposed loads

Dead loads		Imposed loads	
	kN/m^2		kN/m^3
Brickwork		**Liquids**	
120 mm (4.5 in)		Water (1000 kg/m³)	9.81
nominal	2.6	Bottled beer (in cases)	4.5
		Cement slurry	14.0
Plaster		**Solids**	
Two-coat 12 mm thick	0.22	Bricks (stacked)	17.5
Solid cored		Hops (in sacks)	1.7
plasterboard		Basic slag	17.0
(12.7 mm thick)	0.11	Sugar (loose)	7.8
		Coarse aggregates	16.0
Asphalt		Fine aggregates	17.0
(Roof) 2 layers,		Cement	14.2
20 mm thick	0.42		
(Floors) 25 mm thick	0.53	Steel (7850 kg/m³)	77.01

maximum loadings, BS 6399 : Part 1 permits some reductions in the design of columns, foundation and other supporting members (see Table 11.2–3).

Since, at the design stage, the *actual* loading on a structure is a somewhat nebulous number, and recourse in the design is usually made to past experience or some inexact information, it is usually sufficient to assume that the weight of a 1 kg mass is 10 N rather than 9.81 N. A conversion factor of 10 is after all much more convenient in practice.

Notwithstanding the apparent accuracy of the electronic computer, it should be evident that with the accuracy of the basic data being what it is, final results quoted to more than three significant figures cannot normally be justified.

Table 11.2–3   Reduction in imposed floor loads (BS 6399 : Part 1)

Number of floors (including roof) supported by the member	1	2	3	4	5–10	Over 10
Reduction of imposed load on all floors	0	10%	20%	30%	40%	50%

11.3   Materials and practical considerations

Steel reinforcement is commercially available with characteristic strengths f_y of 250 and 460 N/mm². It is sold by weight with a basic price for that of size 16 mm and above, and smaller sizes commanding a higher unit price. The normally available sizes, their mass/metre length and their normal maximum lengths are given in Table 11.3–1. The basic price usually applies to bars up to standard lengths of 12 m.

Lengths longer than those shown in Table 11.3–1 are usually provided by lapping as explained in Section 4.10 or by proprietary connectors or welding. It is good practice to use the maximum diameter consistent with the design requirements concerning bond and crack widths. The number of different diameters to be used on a particular job should be minimized to simplify ordering, stocking and sorting. There is an oft-quoted 'natural' law, with many variations, which states that if something *can* go wrong it *will*—hence it is unwise to use small changes (e.g. 2 mm) in bar diameter since they are not readily separately distinguishable on site.

Academic courses, in this subject, normally limit their attention to the provision of the necessary reinforcement at a particular section and thereby ignore the problems associated with **detailing** [7]. The Institution of Structural Engineers [10] has pointed out that 'bad detailing can lead to disaster just as surely as a defective overall scheme—in fact it is much the more frequent cause of trouble'.

It is surprising that comparatively little attention has been paid to this part of the design and construction process, and to the tests [11–15] which have shown the shortcomings of some 'standard' reinforcement details. The student must be aware of the difficulties involved in curtailment, lapping beam-column joints, short cantilever brackets, deep beams, etc.

Concrete mixes can be designed depending upon the available basic materials (see Section 2.7) to give a large variety of commercial characteristic strengths up to a current practical limit of, say, 70 N/mm^2. It is obviously desirable to limit this variety on any particular job, and preferable to have a limitation between jobs also. For this reason it is recommended that for normal dense aggregate reinforced concrete, **grades** of 30, 35 and 40 are used (f_{cu} = 30, 35 and 40 N/mm^2 respectively). Having decided on the grade of concrete to be used, Table 2.5–6 will give estimates of the moduli of elasticity, if required.

Table 11.3–1   Steel Reinforcement data (see also Tables A2–1 and A2–2)

Size (mm)	8	10	12	16	20	25	32	40
Normal max. length (m)	10	10	10	12	18	18	20	20
Mass/metre (kg/m)	0.395	0.617	0.888	1.58	2.47	3.85	6.31	9.86

11.4   The analysis of framed structure (BS 8110)

11.4(a)   General comments

Although the now normally accepted method of section design is based on ultimate conditions, the stage of development of corresponding methods of structural analysis for concrete structures is such that further work is needed before they can be accepted (see Section 4.9 and References 17, 18). In current design practice, elastic analysis is normally used to obtain the member forces and bending moments in structural frames. A redistribution of these moments as explained in Section 4.9 is permitted, to make allowance for what may happen under ultimate conditions.

In the analysis of the structure to determine the member forces and bending moments, the properties of materials (e.g. the modulus of elasticity—see Table 2.5–6) should be those associated with their characteristic strengths, irrespective of which limit state is being considered. According to BS 8110 : Clause 2.5.2, the relative stiffness of the members may be based on the **second moment of area** I, calculated on any of the following sections (but a consistent approach should be used for all elements of the structure):

(1)   The **concrete section:** The entire concrete cross-section, ignoring the reinforcement.
(2)   The **gross section:** the entire concrete cross-section, including the reinforcement on the basis of the modular ratio α_e, which may be taken as 15.
(3)   The **transformed section:** the compression area of the concrete cross-section combined with the reinforcement on the basis of the modular ratio α_e, which may be taken as 15.

The I.Struct.E. Manual [20] additionally gives the following recommendation for calculating the **stiffness of flanged beams**: the flange width of T-beams may be taken as the actual flange width, or 0.14 times the effective span plus the web width, whichever is the less; the flange width of L-beams may be taken as the actual flange width, or 0.07 times the effective span plus the web width, whichever is the less.

BS 8110 allows a structure to be analysed by partitioning it into sub-frames. The sub-frames that can be used depend on the type of structure being analysed, namely braced or unbraced, since a rigid frame's reaction is different for the two cases. A **braced frame** is designed to resist vertical loads only; therefore the building must incorporate, in some other way, the resistance to lateral loading and sidesway. Such resistance can be provided by bearing walls, shear walls or cores, truss or tubular systems [16]. It is obviously better to provide a regular and symmetrical system of stiffening walls [16], as otherwise lateral loading may also induce undesirable torsional effects in the frames that are being assumed to resist vertical loads only. The **unbraced frame**, where the building incorporates none of these stiffening systems, has to be designed to resist both vertical and lateral loads.

BS 8110 allows the moments, loads and shear forces in the individual

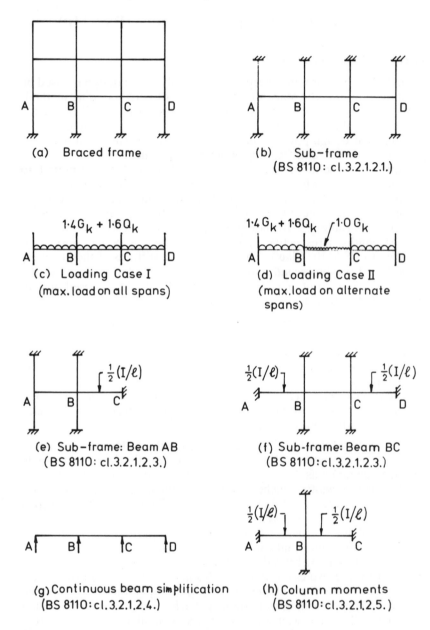

(a)   Braced frame

(b)   Sub-frame
(BS 8110 : cl.3.2.1.2.1.)

(c) Loading Case I
(max. load on all spans)

(d) Loading Case II
(max. load on alternate
spans)

(e) Sub-frame: Beam AB
(BS 8110: cl.3.2.1.2.3.)

(f) Sub-frame: Beam BC
(BS 8110: cl.3.2.1.2.3.)

(g) Continuous beam simplification
(BS 8110: cl.3.2.1.2.4.)

(h) Column moments
(BS 8110: cl.3.2.1.2.5.)

Fig. 11.4–1   Sub-frames for braced frame analysis (BS 8110 : Clause 3.2.1.2)

columns and beams to be derived from an elastic analysis of a series of sub-frames. For a *braced* frame, three methods of simplification may be used:

(a) **Sub-frames type I** (BS 8110 : Clause 3.2.1.2.1). Each sub-frame is taken to consist of the beams at one level together with the columns above and below. The ends of the columns may be assumed to be fixed unless the assumption of a pinned end is clearly more appropriate.

Thus, for the braced frame in Fig. 11.4–1(a), the sub-frame in Fig. 11.4–1(b) may be used for the beams AB, BC, CD and for the column moments at that level. According to BS 8110: Clause 3.2.1.2.2, it is normally necessary to consider only two **loading arrangements** for a braced frame (Fig. 11.4–1(a)) or its associated sub-frames (Fig. 11.4–1(b)):

(1) All spans loaded with the maximum design ultimate load $(1.4G_k + 1.6Q_k)$. See Fig. 11.4–1(c).

(2) Alternate spans loaded with $(1.4G_k + 1.6Q_k)$ and all other spans with the minimum design ultimate load $1.0G_k$. Thus, for the span moment in AB, the loading would be as shown in Fig. 11.4–1(d). Similarly, for the span moment in BC, the loading for that span would be $(1.4G_k + 1.6Q_k)$ while those for AB and CD would be $1.0G_k$.

(b) **Sub-frames type II** (BS 8110: Clause 3.2.1.2.3). The moments and forces in each individual beam may be found by considering a sub-frame consisting only of that beam, the columns attached to the ends of that beam, and the beams on each side. The column and beam ends remote from the beam under consideration may be assumed to be fixed, unless the assumption of pinned ends is clearly more reasonable. The stiffness of the beams on each side of the beam under consideration should be taken as *half* their actual values if they are taken as fixed at their outer ends. Thus, the sub-frame in Fig. 11.4–1(e) would be used to analyse the beam AB; similarly, that in Fig. 11.4(f) would be used for the beam BC. *The loading arrangements for this type of sub-frames are the same as those explained above for sub-frames type I.*

The moments in an individual column may be found from this type of sub-frame, provided that the sub-frame has its central beam the longer of the two spans framing into the column under consideration. If, in Fig. 11.4–1, beam AB is longer than beam BC, then the sub-frame in Fig. 11.4–1(e) should be used for the column at B (and also for that at A, of course). On the other hand, if beam BC is longer than AB, then the sub-frame in Fig. 11.4–1(f) should be used for the column at B.

(c) **'Continuous beam' simplification** (BS 8110: Clause 3.2.1.2.4.). As a more conservative alternative to the sub-frames described above, the moments and shear forces in the beams at one level may be obtained by considering the beams as a continuous beam over supports providing no restraint to rotation. Thus, the beams at the level ABCD in the frame in Fig. 11.4–1(a) may be analysed as a continuous beam on simple supports, as shown in Fig. 11.4–1(g). *The loading arrangements to be considered are the same as for the sub-frames described above—see illustration in Figs 11.4–1(c) and (d).*

Where the continuous beam simplification (Fig. 11.4–1(g)) is used, the column moments may be calculated by simple moment distribution procedure, on the assumption that the column and beam ends remote from the junction under consideration are fixed and that the beams possess *half* their actual stiffness, as shown in Fig. 11.4–1(h). The loading arrangement should be such as to cause the maximum moment in the column; thus, referring to Fig. 11.4–1(h), the longer of the beams AB and BC would carry the load $(1.4G_k + 1.6Q_k)$ and the shorter of the two beams would carry the load $1.0G_k$.

An even greater simplification may be used in the continuous beam analysis if:

(1)   the characteristic imposed load Q_k does not exceed the characteristic dead load G_k;
(2)   the load is fairly uniformly distributed over three or more spans;
(3)   the variation in the spans does not exceed 15% of the largest.

If these conditions are met, BS 8110 : Clause 3.4.3 states that the ultimate bending moments and shear forces are to be obtained from Table 11.4–1.

It is convenient here to explain the **loading arrangement for slabs**. BS 8110 : Clause 3.5.2.3 states that slabs may be designed for a single loading case of maximum design ultimate load $(1.4G_k + 1.6Q_k)$ on all spans provided that the following conditions are met:

(1)   In a one-way slab, the area of each bay exceeds 30 m². In this context, a bay means a strip across the full width of a structure bounded on the two other sides by lines of supports. Thus, referring to the typical floor plan in Fig. 11.5–1, each bay is 5.5 by 16 m = 88 m².
(2)   The ratio $Q_k/G_k \leq 1.25$, where Q_k is the characteristic imposed load and G_k the characteristic dead load.
(3) Q_k does not exceed 5 kN/m², excluding partitions.
(4)   The variation in the spans does not exceed 15% of the longest. (BS 8110 : Clause 3.5.2.4 uses the phrase 'approximately equal span'; the specific reference to 15% has been taken from the I.Struct.E. Manual [20]).

If these conditions are met, BS 8110 : Clause 3.5.2.4 states that the moments and shear forces in continuous one-way slabs may be obtained from Table 11.4–2.

Table 11.4–1   Beams—Ultimate bending moments and shear forces (BS 8110 : Clause 3.4.3)

	At outer support	Near middle of end span	At first interior support	At middle of interior span	At interior supports
Moment[a]	0	0.09Fl[b]	−0.11Fl	0.07Fl	−0.08Fl
Shear	0.45F	—	0.60F	—	0.55F

[a] No further moment redistribution is allowed if the moment values of this table are used.
[b] F is the design load $(1.4G_k + 1.6Q_k)$ and l is the effective span of the beam.

Table 11.4–2 **One-way slabs—ultimate bending moments and shear forces** (BS 8110: Clause 3.5.2.4)

	End support	End span	First interior support	Interior spans	Interior supports
Moment[a]	0	$0.086Fl$[b]	$-0.086Fl$	$0.063Fl$	$-0.063Fl$
Shear	$0.4F$	—	$0.6F$	—	$0.5F$

[a] No further moment redistribution is allowed if the moment values of this table are used.
[b] F is the design ultimate load $(1.46G_k + 1.6Q_k)$ and l is the span of the slab.

If the structure is *unbraced*, the frame will have to resist both vertical and lateral forces. BS 8110: Clause 3.2.1.3.2 then states that the unbraced frame design may be based on the moments, loads and shears resulting from *either* a braced-frame vertical load analysis as described above (see Fig. 11.4–1) *or*, if more severe, the *sum of the effects* from (a) and (b) below:

(a) An elastic analysis of the type of sub-frame shown in Fig. 11.4–1(b) loaded only vertically, with a uniform $1.2(G_k + Q_k)$ throughout.
(b) An elastic analysis of the complete frame under lateral load only, of magnitude 1.2 times the characteristic wind load W_k, and assuming points of contraflexure at the centres of all beams and columns, as shown in Fig. 11.4–2.

This is readily justified if the column feet are 'fixed' and if the beams and columns are of similar stiffness, but the assumption will be incorrect if either of these conditions does not apply. For instance, if a column were pin-footed then the point of zero moment would be at the bottom of the column, and the mid-point assumption would underestimate the moment at the top of the column.

The final loading pattern to be considered concerns the overall stability of the building. Even though instability due to soil failure, e.g. by sliding or

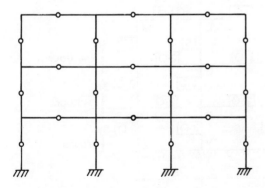

Fig. 11.4–2 **Assumed points of contraflexure for lateral-loading analysis** (BS 8110: Clause 3.2.1.3.2)

exceeding ultimate ground pressure values, may have been taken care of by proper consideration of the foundation material, it is still possible for a tall, narrow building to fail by overturning. Such a mode of failure should be investigated with a load application of $1.0G_k + 1.4W_k$ (BS 8110: Clause 3.2.1.3.2).

Some examples of the application of these loading arrangements to the given sub-frames will be given in Sections 11.4(b) and (c).

We have explained the analysis for braced and unbraced frames. It is necessary to bear in mind that the designer should aim for safe, robust and durable structures; unbraced frames should be avoided if possible. Indeed the I.Struct.E. Manual [20] specifically recommends that *'lateral stability in two orthogonal directions should be provided by a system of strongpoints within the structure so as to produce a braced structure, i.e. one in which the columns will not be subject to sway moments'*. Examples of such strongpoints are the shear walls and core walls referred to earlier in this section.

Robustness (BS 8110: Clause 3.1.4)
All members of the structure should be effectively held together with **ties** in the longitudinal, transverse and vertical directions [20]. Detailed provisions are given in BS 8110: Clause 3.1.4 and in the I.Struct.E. Manual [20]: Clause 4.11. The applications of the BS 8110 provisions are illustrated in Section 11.5: in Step 9 of Example 11.5–1 and Step 9 of Example 11.5–2.

11.4(b)   Braced frame analysis

Example 11.4–1(a)
Figure 11.4–3 shows a braced structural frame, such that the lateral loading is resisted by suitable shear walls or other means. Calculate the moments in the beams AB, BC and CD if the characteristic dead load g_k is

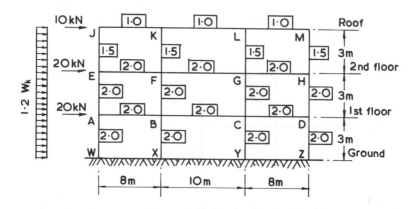

Fig. 11.4–3   Structural frame—relative *EI* values as shown in boxes

36 kN/m and the characteristic imposed load q_k is 45 kN/m; *use the sub-frame of Fig. 11.4–1(g).*

SOLUTION
As previously mentioned, the first difficulty encountered is the assessment of the relative stiffness values of the various members, which are assumed to be rigidly connected. Some guesses have to be made in order to make a start and the ones indicated in the figure are based on the assumption that the first- and second-floor loadings will be the same and greater than that on the roof, and that the second-floor columns are of smaller section than those of the other two floors.

In all the following examples the moments and shears calculated are those obtained direct from elastic analyses, i.e. *prior to* the redistribution of moments. The actual process and implications of redistribution have already been covered in Section 4.9. The discerning student will have noted that the horizontal dimensions are as those in Example 4.9–1. Thus that example can be taken here as a demonstration of the sub-frame shown in Fig. 11.4–1(g).

Strictly speaking, the relative span dimensions and also the relative magnitudes of the dead ($g_k = 36$ kN/m) and imposed loads ($q_k = 45$ kN/m) preclude (see BS 8110: Clause 3.4.3) the use of the coefficients given in Table 11.4–1, but their quick calculation is helpful in giving a comparison with those values shown in Fig. 4.9–7(b). It is important to remember that in those circumstances where the use of the coefficients is allowed and they are used, *no* redistribution of the moments so obtained is permitted. A more extensive list of similar coefficients for two-, three-, four- and five-span continuous beams and also incorporating point load effects is given in Reference 6.

The maximum distributed load

$$= 1.4g_k + 1.6q_k = (1.4 \times 36) + (1.6 \times 45) = 122 \text{ kN/m}$$
$$\text{(to 3 significant figures)}$$

$$F = 122 \times \text{span}$$

(1)   Near the middle of the end span (sagging)

$0.09Fl = 0.09(122 \times 8)(8) = 702$ kNm; (730 kNm; −3.8%)

Note: The bracketed figures are the equivalent values from Fig. 4.9–7(b) and the corresponding percentage difference.

(2)   At the first interior support (hogging)

$0.11Fl = 0.11(122 \times 10)(10) = 1342$ kNm (1097 kNm; +22%)

Note: The greater of the two spans meeting at the support is used.

(3)   At the middle of the interior span (sagging)

$0.07Fl = 0.07(122 \times 10)(10) = 854$ kNm (769 kNm; +11%)

It can be seen, notwithstanding their strictly speaking inapplicability here, that *the coefficients of Table 11.4–1 do give a rapid and useful estimate of the bending moments.* The student is advised to compare their accuracy when applied to relevant continuous beams. This example does show, however, the main danger in the use of such coefficients in that they give no indication even of the possibility of a hogging moment at the centre of this three-span continuous beam (which was clearly shown in Fig. 4.9–7(b)). This possibility should always be considered when such coefficients are used.

This continuous beam sub-frame (Fig. 11.4–1(g)) is an idealization which ignores any end fixity that may be imparted by the columns. If this sub-frame is used in a rigid frame analysis, as here, it is obviously prudent to provide some reinforcement in the top of the beam at the ends A and D even though the sub-frame analysis does not indicate any such need. In Example 11.4–1(b) that follows, it will be seen that the value at the end (110 kNm) is some 57% of the initial fixed end moment (192 kNm). This is because of the high stiffnesses of the columns relative to that of the beam, and the normal figure may well be somewhat less. Hence, a value of some 30–40% of the initial fixed end moment may be more appropriate for use in design.

Example 11.4–1(b)
With reference to the braced frame in Example 11.4–1(a) and Fig. 11.4–3, calculate the maximum sagging moment in the span BC; *use the sub-frame of Fig. 11.4–1(b)*. Compare the moment so calculated with that shown in Fig. 4.9–6: Case 2.

SOLUTION
For clarity, Fig. 11.4–1(b) is redrawn as Fig. 11.4–4, which also shows the relevant dimensions and loadings.

For the dead and imposed loading as above, and using the method of moment distribution with relative *EI* values as given in Fig. 11.4–3:

Distribution factors (DF) (Table 11.4–3):

$$\text{At A, D:} \quad \frac{\Sigma\text{cols}}{AB} = \frac{(2 \times 2/3)}{(2/8)} = \frac{0.667/0.792}{0.125/0.792} = \frac{84\%}{16\%}$$

$$\text{At B, C:} \quad \text{BA: } (\Sigma\text{cols}): \text{BC} = 2/8: (2 - 2/3): 2/10$$

$$= 14\%: 75\%: 11\%$$

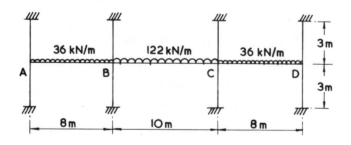

Fig. 11.4–4   Storey sub-frame

Table 11.4-3 Moment distribution

	A			B			C			D
	Σcols	AB	BA	Σcols	BC	CB	Σcols	CD	DC	Σcols
DF(%)	84	16	14	75	11	11	75	14	16	84
FEM (kNm)	0	−192	+192	0	−1020	+1020	0	−192	+192	0
Balance	+161	+31	+116	+621	+91	−91	−[a]	−[a]	−[a]	−[a]
Carry-over	−49	+58	+16	+23	−46	+46	−[a]	−[a]	−[a]	−[a]
Balance		−9	+4		+3	−3	−[a]	−[a]	−[a]	−[a]
CO	−2	+2	−5	+4	−1	+1	−[a]	−[a]	−[a]	−[a]
Balance		0	+1		+1	−1	−[a]	−[a]	−[a]	−[a]
Σ(kNm)	+110	−110	+324	+648	−972	+972	−648	−324	+110	−110

[a] The symmetry of these and the later values at C, D, with those at A, B, permits non-repetition.

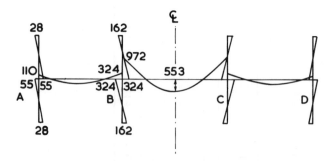

Fig. 11.4–5   Storey sub-frame bending moments (kNm)—loading as in Fig. 11.4–4

Fixed end moments (FEM) (Table 11.4–3):

Spans AB, CD: $wl^2/12 = 36 - 8^2/12 = 192$ kNm

Span BC: $wl^2/12 = 122 \times 10^2/12 = 1020$ kNm

With the 'free' bending moment in span BC equal to 1525 kNm (i.e. $122 \times 10^2/8$), the maximum sagging moment at the centre of span BC is 533 kNm (i.e. 1525–972). This is to be compared with the value of 769 kNm, as given in Fig. 4.9–6, Case 2, and the difference is due to the introduction of column restraint. Figure 11.4–5 shows the corresponding bending moment diagram.

Example 11.4–1(c)
With reference to the braced frame in Example 11.4–1(a) and Fig. 11.4–3, calculate the bending moments in the beam AB; *use the sub-frame of Fig. 11.4–1(e).*

SOLUTION
For clarity, Fig. 11.4–1(e) is redrawn as Fig. 11.4–6 with appropriate dimensions and loadings included.

Distribution factors (DF) (Table 11.4–4):

At A: (Σcols): AB = 84%: 16%   (as in Example 11.4–1(b))

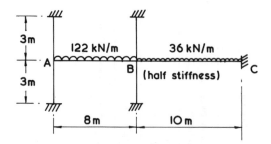

Fig. 11.4–6   Sub-frame for beam AB

Table 11.4–4 Moment distribution

	A			B		C
	Σcols	AB	BA	Σcols	BC	CB
DF(%)	84	16	15	79	6	—
FEM (kNm)	0	−651	+651	0	−300	+300
Balance	+547	+104	−53	−277	−21	
CO		−26	+52			−10
Balance	+22	+4	−8	−41	−3	
CO		−4	+2			−2
Balance	+3	+1		−2		
Σ(kNm)	+572	−572	+644	−320	−324	+288

At B: BA: (Σcol): BC = 15%: 79%: 6%

Fixed end moments (FEM) (Table 11.4–4):

Span AB: $\dfrac{122 \times 8^2}{12} = 651$ kNm

Span BC: $\dfrac{36 \times 10^2}{12} = 300$ kNm

Since the moments 572 and 644 kNm at the ends of the beam AB are of comparable magnitude, the sagging moment at the centre of the beam is very nearly the maximum and has a value of 368 kNm, i.e. $(122 \times 8^2/8)$ − $(572 + 644)/2$. This is to be compared with the value of 730 kNm given in Fig. 4.9–6: Case 3, which assumes zero moment at the support A.

An analysis using the complete storey sub-frame, as in Example 11.4–1(b) with the full imposed loading on spans AB and CD, gives values of 573, 642 and 318 kNm for the moments designated AB, BA and BC respectively. This indicates the relative accuracy of this two-span sub-frame.

Example 11.4–1(d)
With reference to the braced frame in Example 11.4–1(a) and Fig. 11.4–3, calculate the maximum sagging moment in the span BC; *use the sub-frame of Fig. 11.4–1(f)*.

SOLUTION
Since the span BC is greater than either of the two adjacent spans, the solution for BC using the sub-frame in Fig. 11.4–1(f) can also be taken to give the column design moments (BS 8110: Clause 3.2.1.2.3). This sub-frame is obviously of much more use when the structure has more spans than the one being here analysed. Since the analysis of the complete

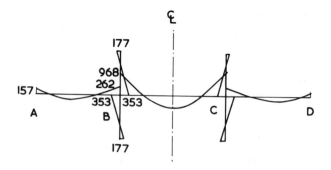

Fig. 11.4–7   Bending moment diagram (kNm)—Example 11.4–1(d)

storey as given in Example 11.4–1(b) is very similar, this is now left as an exercise for the student. The bending moment diagram is as given in Fig. 11.4–7.

Example 11.4–1(e)
With reference to the braced frame in Example 11.4–1(a) and Fig. 11.4–3, explain the *use of the sub-frame of Fig. 11.4–1(h)* for finding the column moments.

SOLUTION
Note that the sub-frame of Fig. 11.4–1(h) can be used to find column moments only. The loading applied must be that which produces the worst out-of-balance moment at the beam-column joint, i.e. in this case BC is to be loaded with the full design load $1.4g_k + 1.6q_k = 122$ kN/m and CD with the minimum design dead load $1.0g_k = 36$ kN/m.

Figure 11.4–8 gives the resulting bending moment diagram which should be checked. Comparison of the column moment values obtained with those given in Examples 11.4–1(b) and (d) shows that even this minimum sub-frame gives a good approximation in a satisfyingly rapid manner.

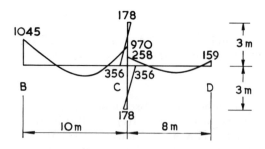

Fig. 11.4–8   Bending moment diagram (kNm)—Example 11.4–1(e)

11.4(c)   Unbraced frame analysis

The task remaining is the consideration of the frame, shown in Fig. 11.4–3, as unbraced and therefore to be assessed under the action of lateral loads. It will be remembered that BS 8110: Clause 3.2.1.3.2 states that, in such a frame, the design moments for the individual members may be obtained from (a) or (b) below, whichever gives the larger values:

(a)   those obtained by simplified analyses of the types given in Examples 11.4–1(a) to 1(e) above; or

(b)   the sum of the effects of:
  (1)   single storey analyses, idealized as in Fig. 11.4–4 and loaded throughout with $1.2 \times (g_k + q_k)$; and
  (2)   the analysis of the complete frame loaded with $1.2w_k$ *only* (Fig. 11.4–3), and assuming points of contraflexure at the centres of all beams and columns (Fig. 11.4–2).

The single-storey analysis under a uniform loading of $1.2 \times (g_k + q_k)$, i.e. $1.2(36 + 45) = 97$ kN/m, is similar to that given in Example 11.4–1(b) and results in the bending moment diagram as shown in Fig. 11.4–9.

If the **wind loading** (w_k) is not given, then recourse has to be made to CP 3: Chapter V: Part 2 [5], which gives two methods for the estimation of the forces to be taken as acting on the structure.

The procedure for the first method is as follows:

(1)   Determine the **basic wind speed** V (m/s) for the appropriate UK area from a given table of values (e.g. Aberdeen 49 m/s; Nottingham 43 m/s; London 37 m/s).

(2)   Determine values for three factors, from tables:
  S1—Topography factor. Usually 1.0.
  S2—Ground roughness, building size and height above ground factor. This varies with the various parameter values (e.g. open country, town or city) and has a range from 0.47 to 1.27.
  S3—Statistical factor. This depends upon the degree of security required and the number of years that the structure is expected to be exposed to the wind. The normal value is given as 1.0.

(3)   Calculate the **design wind speed** V_s (m/s) where

$$V_s = V \times S1 \times S2 \times S3 \text{ m/s}$$

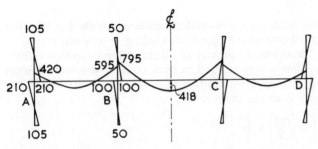

Fig. 11.4–9   Bending moment diagram (kNm) for unbraced frame with vertical load $1.2(G_k + Q_k)$

and convert to dynamic pressure q (N/m^2) using

$$q = 0.613V_s^2 \text{ N/m}^2$$

Values of q, therefore, vary between about 61 N/m^2 for a V_s value of 10 m/s and 3000 N/m^2 when $V_s = 70$ m/s.

(4)   Using given external (C_{pe}) and internal (C_{pi}) **pressure coefficients**, and taking their signs (i.e. negative if suction) into account, the force F can be taken as

$$F = (C_{pe} - C_{pi})q \cdot A$$

where A is the surface area of the element or structure being considered.

C_{pe} values vary considerably depending upon the building's aspect ratios (height : width; length : width), the wind direction and the surface being considered. When the wind is normal to the windward face the value of C_{pe} for that face is generally + 0.7, whereas the leeward face then has a value varying between -0.2 and -0.4 depending upon the plan dimensions.

C_{pi} is also very variable, but it has a value of -0.3 if all four faces are equally permeable, i.e. for the windward face mentioned the value of ($C_{pe} - C_{pi}$) is equal to 1.0.

The second method, which is only applicable to a limited range of rectangular (in plan) building shapes, gives the force directly as

$$F = C_f \times q \times A_e$$

where C_f = the **force coefficient** obtained from tabulated values, which range from 0.7 to 1.6 depending upon aspect ratios;
q = obtained as above;
A_e = the effective frontal area of the structure.

Figure 11.4–3 gives the equivalent forces due to the wind (1.2w_k) for the example here, applied at roof and floor levels, as is normally assumed.

A further assumption is now made with regard to the distribution of these forces across the frame. A choice has to be made between the two methods usually used:

(1)   The cantilever method; or
(2)   The portal method.

(1)   The assumption made in the so-called **cantilever method** is that the axial force in a column, due to the wind loading, is proportional to its distance from the centre of gravity of all the columns in that frame.

When the system is symmetrical in both column size and position the calculation is straightforward, e.g. in Fig. 11.4–10 where the centre of gravity is at the centre:

$$\left|\frac{N1}{13}\right| = \left|\frac{N2}{5}\right| = \left|\frac{N3}{5}\right| = \left|\frac{N4}{13}\right|$$

i.e.

$$|N1| = |N4| = |2.6N2| = |2.6N3|$$

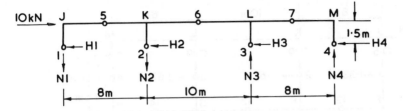

Fig. 11.4–10 Lateral loading for roof (kN)—cantilever method

It is evident that $N1$ and $N2$ are of opposite sign to $N3$ and $N4$, and that the analysis of this assumed statically determinate system in now follow. Taking moments about point 4, for the whole structure,

$$26N1 + 18N2 = 10 \times 1.5 + 8N3$$

Since $N1 = 2.6N2$ and $N3 = N2$,

$$N2 = 15/(26 \times 2.6 + 18 - 8) = 0.19 \text{ kN}$$

Also $N3 = 0.19$ kN,

$$N1 = N4 = 2.6 \times 0.19 = 0.5 \text{ kN}$$

And taking moments about point 5, for all the forces to the left,

$$H1 = \frac{4N1}{1.5} = 1.33 \text{ kN}$$

In the symmetrical case the other horizontal forces can now be inferred; alternatively they are obtained by taking moments about point 6:

$$1.5H1 + 1.5H2 - 13N1 - 5N2 = 0$$

and

$$H2 = H3 = 3.64 \text{ kN}$$

Since $H4 = H1 = 1.33$ kN, a round-off error of 0.06 kN is apparent. The resulting bending moment diagram for the roof is given in Fig. 11.4–11.

Moving down the structure to the second floor, Fig. 11.4–12 gives the forces acting.

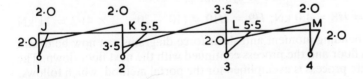

Fig. 11.4–11 Roof bending moments (kNm) due to lateral loading—cantilever method

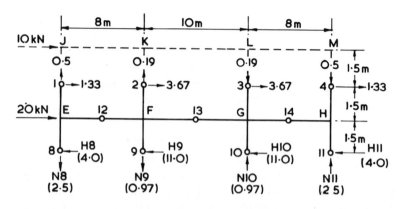

Fig. 11.4–12   Lateral loading for second floor (kN)—cantilever method

The dotted roof portion is intended to indicate that the effect of this on the second-floor sub-frame can be replaced by the forces as given acting at points 1, 2, 3 and 4 (from the above calculation). It will be noted that the horizontal forces at 2, 3 have been amended from 3.64 to 3.67 kN to make the horizontal sum at the 1–4 level equal to 10 kN.

Dividing the vertical loading as above,

$$N8 = N11 = 2.6N9 = 2.6N10$$

and taking moments about point 11 for the whole structure above the level of point 8–11,

$$26N8 + 18N9 = (20 \times 1.5 + 10 \times 4.5) + 8N10$$

Since $N8 = 2.6N9$ and $N10 = N9$, we have

$$N9 = 75/(26 \times 2.6 + 18{-}8) = 0.97 \text{ kN}$$

Also, $N10 = N9 = 0.97$ kN and

$$N8 = N11 = 2.6N9 = 2.6 \times 0.97 = 2.5 \text{ kN}$$

Then taking moments about point 12, for the left-hand portion,

$$(H8 + 1.33)1.5 = (2.5 - 0.5)4$$

$$H8 = 4.0 \text{ kN}$$

and using the argument of symmetry,

$$H11 = H8 = 4.0 \text{ kN}; \quad H9 = H10 = (10 + 20 - 2 \times 4)/2 = 11 \text{ kN}$$

The bending moment and shear force diagrams can now be drawn for this floor and the process continued with the next floor down. The complete process is exemplified for the portal method, which follows.

(2)   The alternative **portal method** assumes that the members at a given level can be split into a series of portal frames, and that the lateral

loading on each of these is in proportion to its horizontal span. The lateral loads given in Fig. 11.4–13 are so calculated.

The total lateral loading H is 10 kN and the sum of the spans is 26 m; thus, in Fig. 11.4–13

$$H1 = H3 = \frac{10 \times 8}{26} = 3.1 \text{ kN}$$

$$H2 = \frac{10 \times 10}{26} = 3.8 \text{ kN}$$

(Check: $H1 + H2 + H3$ = total lateral force H, i.e. 10 kN.)

The essence of the portal method is therefore

$$H1 = H\frac{l1}{\Sigma l} \qquad H2 = H\frac{l2}{\Sigma l} \qquad H3\frac{l3}{\Sigma l}$$

and

$$H1 + H2 + H3 = H$$

By symmetry (Fig. 11.4–13),

$$X1 = \tfrac{1}{2}H1 \qquad X2 = \tfrac{1}{2}H2$$

By taking moments about points 5 and 6,

$$|V1| = |V2| = \frac{(X1)h}{(l1)/2} = \frac{(H1)h}{l1} = H\frac{h}{\Sigma l}$$

$$|V2'| = |V3| = \frac{(X2)h}{(l2)/2} = \frac{(H2)h}{l2} = H\frac{h}{\Sigma l}$$

Therefore $|V2| - |V2'| = 0$.

This means that the portal method implies that *all* the vertical reaction due to the lateral loading is carried by the outside columns only.

Taking moments for the complete roof frame about point 4,

$$V1 = 10 \times 1.5/26 = 0.58 \text{ kN}$$

and from above

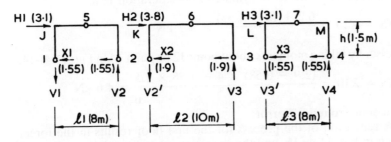

Fig. 11.4–13 Lateral loading for roof (kN)—portal method

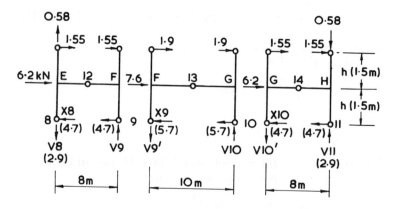

Fig. 11.4–14   Lateral loading for second floor (kN)—portal method

$$V4 = V1 = 0.58 \text{ kN}$$

The horizontal reactions at the portal feet are readily deduced either by symmetry or by taking moments about the assumed point of contraflexure in the beam:

$$X1(= X3) = \frac{(V1) \times 4}{1.5} = \frac{0.58 \times 4}{1.5} = 1.55 \text{ kN}$$

$$X3 = 1.55 \text{ kN}$$

Figure 11.4–14 gives the ensuing situation for the second storey. In the figure, the lateral loads 6.2, 7.6 and 6.2 kN at the level E–F–G–H of course sum to the total load of 20 kN; similarly the lateral loads 1.55, 1.9 kN, etc. at the higher level sum to 10 kN. As demonstrated earlier, the portal method implies that the vertical loading is carried by the outside columns only. Hence

$$V8 = \frac{20 \times 1.5 + 10 \times 4.5}{26} = 2.9 \text{ kN}$$

$$V11 = V8 = 2.9 \text{ kN}$$

We can write down from symmetry consideration that

$$X8 = X10 = \frac{6.2 + (2 \times 1.55)}{2} = 4.7 \text{ kN}$$

or else, by taking moments about point 12 or point 14 in Fig. 11.4–14,

$$X8 = X10 = \frac{(2.9 - 0.58) \times 4 - 1.55 \times 1.5}{1.5} = 4.6 \text{ kN}$$

with some round-off error.

A repetition of the process for the first floor results in the forces given in Fig. 11.4–15, and the consequent bending moment diagram of Fig. 11.4–16.

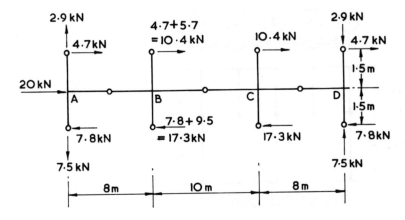

Fig. 11.4–15 Lateral loading for first floor (kN)—portal method

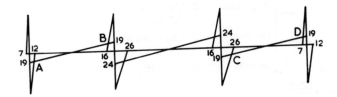

Fig. 11.4–16 First-floor bending moments (kNm) due to lateral loading—portal method

Thus, for the maximum elastic moments in the span BC, and its adjacent columns in this instance, the choice has to be made between those given in Fig. 11.4–5 and the summation of those in Figs 11.4–9 and 16. In this example it can be easily seen that the latter combination produces the least bending moments, and those in Fig. 11.4–5 are the ones to use in constructing an elastic moment envelope for design.

11.5 Design and detailing—illustrative examples

The examples in this section follow the style of Higgins and Rogers's book [3], which can be strongly recommended for further reading. These examples illustrate the design and detailing of the main structural members of the multistorey building shown in Fig. 11.5–1. The structure incorporates suitable shear walls and bracings to resist lateral loads. The dimensions listed in Fig. 11.5–1 are based on preliminary member-sizing calculations, using design aids such as Table 11.2–1 and Figs 11.2–1 and 11.2–2. The design information is given below:

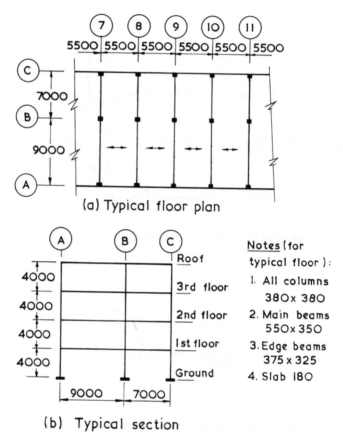

(a) Typical floor plan

(b) Typical section

Notes (for typical floor):

1. All columns 380 x 380
2. Main beams 550 x 350
3. Edge beams 375 x 325
4. Slab 180

Fig. 11.5–1   **Typical floor plan and cross-section of multistorey building [3]**

Exposure condition—internal	mild
external	moderate
Fire resistance	1 hour
Dead loads—partitions and finishes	1.5 kN/m^2
external cladding	5 kN/m
Imposed loads—roof	1.5 kN/m^2
floors	3 kN/m^2
Allowable soil-bearing pressure	200 kN/m^2
Characteristic strengths	
Concrete: f_{cu}	40 N/mm^2
Reinforcement: f_y (main bars)	460 N/mm^2
f_{yv} (links)	250 N/mm^2

Example 11.5–1

Design and detail the reinforcement for one panel of the typical floor shown in Fig. 11.5–1(a).

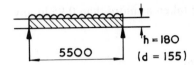

Fig. 11.5–2

SOLUTION
(See Fig. 11.5–2).

Step 1   Durability and fire resistance
From Table 2.5–7,
nominal cover for mild exposure condition = 20 mm

From table 8.8–1,
fire resistance of 180 mm slab with 20 mm cover to main bars is not less
than 1 hour

$$\underline{\text{Nominal cover} = 20 \text{ mm}}$$

$$\underline{\text{Fire resistance}} \quad \text{OK}$$

Step 2   Loading—per metre width of slab
Self-weight = (0.180 m) (24 kN/m³) (5.5 m) = 23.8

Patitions and finishes = (1.5) (5.5)   = $\underline{8.3}$

Characteristic dead load G_k = 32.1 kN/m width

Characteristic imposed load Q_k = (3) (5.5)   = 16.5 kN/m width

Design load $F = 1.4\ G_k + 1.6 Q_k$

$$= 44.9 + 26.4$$

$$= 71.3 \text{ kN/m width}$$

$$G_k = 32.1 \text{ kN/m} \qquad Q_k = 16.5 \text{ kN/m} \qquad F = 71.3 \text{ kN/m}$$

Step 3   Ultimate moments
From Table 11.4–2,

M at supports = $0.063Fl$ = (0.063) (71.3) (5.5) = 24.7 kNm/m

M at midspan = $0.063Fl$ = 24.7 kNm/m

Step 4   Main reinforcement

Effective depth $d = 180 - 20 - \frac{1}{2}$ bar $\phi = 154$ mm, say

Supports: $\dfrac{M}{f_{cu}bd^2} = \dfrac{(24.7)\ (10^6)}{(40)\ (1000)\ (154^2)} = 0.026$

From Table 4.6–1,

$$z/d = 0.94 \qquad x/d = 0.13$$

(*Note:* As explained at the end of Section 4.5,

BS 8110 does not allow z/d to be taken as more than 0.95 in any case.)

From eqn (4.6–12),

$$A_s = \frac{M}{(0.87)f_y z} = \frac{(24.7)\ (10^6)}{(0.87)\ (460)\ (0.94)\ (154)} = 426 \text{ mm}^2/\text{m}$$

From Section 8.8,

minimum tension steel = 0.13% bh = 234 mm²/m

< 426 mm²/m

Hence

A_s = 426 mm²/m OK

From Table A2–2,

Top: T10 at 150 (523 mm²/m)

Midspan: $\dfrac{M}{f_{cu}bd^2}$ = 0.026 (as at supports)

Bottom: T10 at 150 (523 mm²/m)

Step 5   Shear *(see Section 8.7)*
From Table 11.4–2,

$V = 0.5F = (0.5)\ (71.3 \text{ from Step 2}) = 35.7 \text{ kN/m}$

$$v = \frac{V}{b_v d} = \frac{(35.7)\ (10^3)}{(1000)\ (154)} = 0.23 \text{ N/mm}^2$$

$< 0.8\sqrt{f_{cu}}\ (= 5.1 \text{ N/mm}^2)$

From Table 6.4–1,

For $A_s/b_v d$ = 523/(1000) (154) = 0.34%

v_c = 0.65 N/mm² > v

From Section 8.7,

Shear resistance OK

Step 6   Deflection *(see Section 8.8)*
From Table 5.3–1,

basic span/depth ratio = 26

$$\frac{M}{bd^2} = \frac{(24.7)\ (10^6)}{(1000)\ (154^2)} = 1.04$$

From Table 5.3–2,

modification factor = 1.38

Hence

allowable span/depth ratio = (26) (1.38) = 35.9

$$\text{actual span/depth ratio} = \frac{5500}{154} = 35.7 \text{ OK}$$

Deflection OK

Comments on Step 6

If the actual span/depth ratio had slightly exceeded 35.9, we could use eqn (5.3–1(b)) to calculate the service stress f_s and then obtain an enhanced modification factor either from eqn (5.3–1(a)), or from Table 5.3–2 by interpolation.

Step 7   Cracking *(see Section 8.8)*

$$3d = (3)(154) = 462 \text{ mm}$$

clear spacing between bars $= 150 - \text{bar } \phi$

$$< 462 \text{ mm OK}$$

$$h = 180 \text{ mm} < 200 \text{ mm}$$

From Section 8.8, no further checks are required.

Cracking OK

Step 8   Secondary reinforcement *(see Section 8.8)*

From Section 8.8,

minimum secondary reinforcement $= 0.13\% \ bh$

$$= (0.0013)(1000)(180) = 234 \text{ mm}^2/\text{m}$$

T10 at 300 (262 mm²/m)

Step 9   Robustness *(see Section 11.4(a) and BS 8110 : Clauses 3.1.4.3 and 3.12.3.4)*

Longitudinal tie force $F_t = 20 + 4$ times   (no. of storeys)

$$= 20 + (4)(4) = 36 \text{ kN/m}$$

Check:

$$\frac{g_k + q_k}{7.5}\left[\frac{l_r}{5}\right]F_t = \frac{(32.1/5.5) + (16.5/5.5)}{7.5}\left[\frac{5.5}{5}\right](36)$$

$$= 46.7 \text{ kN/m} > F_t$$

$$\text{Minimum continuous internal tie} = \frac{(46.7)(10^3)}{(0.87)^*(460)} = 116.7 \text{ mm}^2/\text{m}$$

Bottom T10 at 400 (196 mm²/m)

From Table 4.10–2,

full anchorage lap length $= 32\phi$

---

* γ_m could have been taken as 1.0. See BS 8110 : Clause 3.12.3.2.

From eqns (6.6–3(a), (b)),

$$\text{the required lap length} = (32\phi)\left[\frac{f_s}{0.87f_y}\right]$$

$$= (32\phi)\left[\frac{A_{s,req}(=116.7)}{A_{s,prov}(=196)}\right]$$

$$= 19\phi = 190 \text{ mm}$$

From Section 4.10,

minimum lap length = 300 mm > 19ϕ

Tie lap length = 300 mm

Step 10   Reinforcement details

The reinforcement details are shown in Fig. 11.5–3, which conforms to the Standard Method of Detailing of Structural Concrete [1], as explained in Example 3.6–3. *Thus the label 104T10-1-150B1 tells us that* there are 104 deformed bars ($f_y = 460 \text{ N/mm}^2$) of size 10, that these bars have the 'bar mark 1', and that the bar spacing is 150 mm; B1 tells us that these bars are at the bottom outer layer. (Similarly, B2 denotes bottom upper layer; T1 denotes top outer layer and T2 top lower layer.) The bar marks indicate the probable sequence of fixing; thus bars 'mark 1' are fixed first, then 'mark 2', then 'mark 3' and so on. Note the following comments regarding BS 8110's detailing requirements:

(a) **Main bars at midspan** (mark 1)
   Curtailment: As explained in Section 8.8, in a continuous slab the main tension bars at midspan should extend to within 0.2*l* of the supports, and at least 40% should extend into the support.
   In Fig. 11.5–3, the bars 'mark 1' are staggered, 50% being curtailed at 925 + 175 = 1100 mm (= 0.2*l*) from the supports.
   Bar spacing: As explained in Section 8.8, the clear spacing should not exceed 3*d* (462 mm) or 750 mm whichever is the less.

(b) **Secondary bars** (marks 2, 3, 4 and 6)
   Bar spacing: As explained in Section 8.8, the clear spacing should not exceed 3*d* (462 mm) or 750 mm whichever is the less.
   Minimum area: All secondary bars exceed the minimum area requirement of 0.13%*bh* (see Section 8.8).
   Minimum lap length: As explained in Section 4.10, the minimum lap length should not be less than 15ϕ or 300 mm whichever is the less.

(c) **Main bars at supports** (mark 5)
   Curtailment: As explained in Section 8.8, the main bars at supports should extend a distance of at least 0.15*l* or 45ϕ, from the face of support whichever is greater, and at least 50% should extend 0.3*l*.
   In Fig. 11.5–3, all bars 'mark 5' extend a distance of at least 850 mm from the face of the support; this exceeds 0.15*l* or 45ϕ. The bars are staggered, 50% being extended 850 + 850 = 1700 mm (>0.3*l*) from the face of the support.

(d) *Transverse reinforcement across main beam* (mark 5)

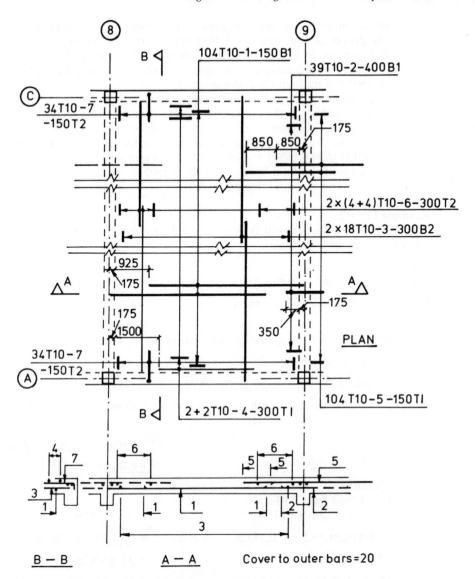

Fig. 11.5–3 **Typical floor slab [3]**

The slab and the main beam act integrally as a flanged beam. Therefore (see Section 4.8 or Section 4.10) transverse reinforcement of area not less than $0.15\% lh_f$ should be provided over the top surface and across the full effective width of the T-beam flange. From Section 4.8, the effective width is $b_w + 0.2l_z = 350 + (0.2)(0.7$ of $9000) = 1610$ mm; $1610/2 = 805$ mm. That is, the transverse reinforcement must extend at least 805 mm into the span on each side of the main beam, and the area of this transverse reinforcement should not be less than (0.0015) (5500) $(180) = 1485$ mm^2/5.5 m = 270 mm^2/m. In Fig. 11.5–3, the bars 'mark 5' extend $850 + 175 = 1025$ mm (> 805 mm) into the span, and they have an area of (T10 at 150) 523 mm^2/m (> 270 mm^2/m).

(e) **Transverse reinforcement across edge beam** (mark 7)
From Section 4.8, the effective flange width of the edge beam is $b_w + 0.1l_z = 710$ mm (see also Step 5 of Example 4.11–1); the minimum area of the transverse reinforcement is $0.15\% lh_f = (0.0015)$ (5500) $(180) = 1485$ mm$^2 = 270$ mm^2/m. In Fig. 11.5–3, the area of the bars 'mark 7' (T10 at 150) is 523 mm^2/m and the bars extend the full effective width.

Example 11.5–2

With reference to the typical floor of the building in Fig. 11.5–1, design and detail a main floor beam.

SOLUTION
See Fig. 11.5–4 and also the comments at the end of the solution.

Step 1   Durability and fire resistance
From Table 2.5–7,

nominal cover for mild exposure condition = 20 mm

cover to main bars = 20 + link ϕ = 35 mm, say

From Table 4.10–3,

for 350 mm beam width and 35 mm cover to main bars, the fire resistance exceeds 1 hour

Nominal cover = 20 mm

Fire resistance OK

Step 2   Loading
Dead load from 180 slab:
(from Example 11.5–1)   = 31.7 kN/m
Self-weight $(0.55 - 0.18) \times 0.35 \times 24$ = 3.1 kN/m
Characteristic dead load g_k = 34.8 kN/m
Characteristic imposed load $q_k = 3 \times 5.5 = 16.5$ kN/m

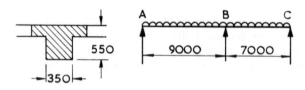

Fig. 11.5–4

Design load $F = 1.4g_k + 1.6q_k$

$$= 48.7 + 26.4 = 75.1 \text{ kN/m}$$

$$g_k = 34.8 \text{ kN/m}; \quad q_k = 16.5 \text{ kN/m}; \quad F = 75.1 \text{ kN/m}$$

Step 3   Ultimate moments

BS 8110: Clause 3.2.1.2.4 allows continuous beams in a framed structure to be analysed by *assuming that the columns provide vertical restraint but no rotational restraint.* (See Comments on Step 3 at the end of the solution.) Figure 11.5–5 shows the free bending moments and the bending moments at support B for the three loading cases. (See Step 5 for the design bending moments at the end supports A and C.)

Figure 11.5–6 shows the **bending moment envelope**; for a length of the

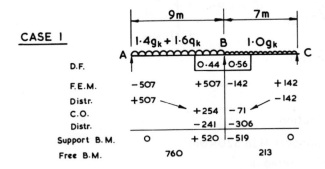

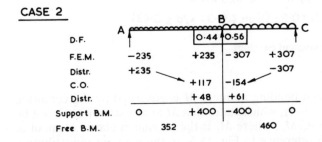

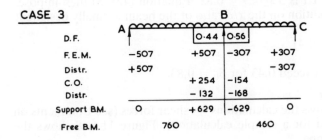

Fig. 11.5–5

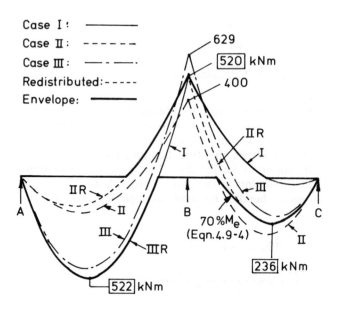

Case I :   ⎯⎯⎯
Case II :   ⎯ ⎯ ⎯ ⎯
Case III : ⎯ · ⎯ ·
Redistributed: - - - - -
Envelope: ⎯⎯⎯

629
520 kNm
400

IIR
I
I
III
IIR
A
II
III
IIIR
B
70%Me
(Eqn.4.9-4)
II
236 kNm
C
522 kNm

Redistribution

Case I : No redistribution

Case II : Increase support M to 520 kNm

i.e. 30% increase

Case III : Reduce support M to 520 kNm

i.e. βb = 0·83 (Eqn. 4.9-3 : x/d ≤ 0·43)

Fig. 11.5-6   Bending moment envelope

span BC, the design bending moment M is governed by the condition imposed by eqn (4.9-4), which states that M at any section must not be taken as less than $0.7M_e$, where M_e is the maximum elastic moment at that section. With reference to Fig. 11.5-6, the moment redistribution ratio β_b for Case III is 520/629 = 0.83. Equation (4.9-3) then imposes the following condition on the x/d ratio of the beam as finally designed:

$$\frac{x}{d} \le (\beta_b - 0.4)$$

i.e. x/d must not exceed 0.43 for $\beta_b = 0.83$.

Step 4   Shear forces

Figure 11.5-7 shows the calculations for shear forces (see Comments on Step 4 at the end for a sample calculation). Figure 11.5-8 shows the **shear force envelope**.

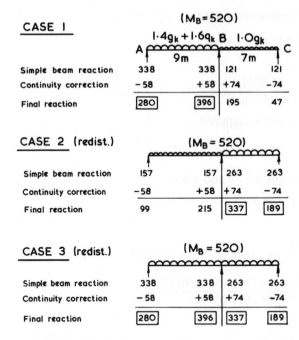

CASE 1

(M_B=520)

$1.4g_k + 1.6q_k$ B $1.0g_k$

A ⌒⌒⌒⌒⌒⌒⌒⌒⌒⌒ ⌒⌒⌒⌒⌒⌒⌒⌒ C
         9m   7m

Simple beam reaction	338	338	121	121
Continuity correction	−58	+58	+74	−74
Final reaction	280	396	195	47

CASE 2 (redist.)

(M_B = 520)

Simple beam reaction	157	157	263	263
Continuity correction	−58	+58	+74	−74
Final reaction	99	215	337	189

CASE 3 (redist.)

(M_B = 520)

Simple beam reaction	338	338	263	263
Continuity correction	−58	+58	+74	−74
Final reaction	280	396	337	189

Fig. 11.5–7

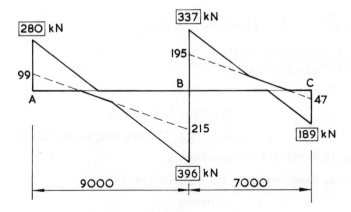

Fig. 11.5–8 Shear force envelope

Step 5 Longitudinal reinforcement

Internal support B (Fig. 11.5–9(a)). From bending moment envelope (Fig. 11.5–6),

$$M = 520 \text{ kNm}$$

$$\beta_b = 0.83; \text{ hence } x/d \leq 0.43 \text{ (eqn 4.9–3)}.$$

% moment redistribution = $100(1 - \beta_b)$ = 17%

From Table 4.7–1 (or eqn 4.7–5),

$$K' = 0.139$$

$$K = \frac{M}{f_{cu}bd^2} = \frac{(520)(10^6)}{(40)(350)(475^2)} = 0.165 > 0.139$$

Compression reinforcement is required.
From Fig. 11.5–9(a),

$$d' = 50 \text{ mm}$$

From Table 4.7–2 (or eqns 4.7–7 and 4.6–3),

$$z/d = 0.81 \qquad x/d = 0.42$$

From eqn (4.7–10),

$$A_s' = \frac{M - M_u}{0.87f_y(d - d')} \quad \text{(where } M_u = K'f_{cu}bd^2\text{)}$$

$$= \frac{(520)(10^6) - (0.139)(40)(350)(475^2)}{(0.87)(460)(475-50)}$$

$$= 476 \text{ mm}^2$$

(From Section 4.10, when compression steel is required, the minimum area to be provided = 0.2% bh = 385 mm^2 < 476 mm^2 OK)
From eqn (4.7–11),

$$A_s = \frac{M_u}{0.87f_yz} + A_s' \quad \text{(where } M_u = K'f_{cu}bd^2\text{)}$$

$$= \frac{(0.139)(40)(350)(475^2)}{(0.87)(460)(0.81)(475)} + 476$$

$$= 3327 \text{ mm}^2$$

<u>Top 5T32 (4021 mm^2)</u>

<u>Bottom 2T32 (1608 mm^2)</u>—see Step 9

Span AB (Fig. 11.5–9(b)). From Section 4.8,

$$\text{effective flange width} = 350 + 0.2(0.7 \text{ of } 9000)$$

$$= 1610 \text{ mm}$$

From bending moment envelope (Fig. 11.5–6),

$$M = 522 \text{ kNm}$$

$$\frac{M}{f_{cu}bd^2} = \frac{(522)(10^6)}{(40)(1610)(500^2)}$$

$$\text{(where } d = 500 \text{ from Fig. 11.5–9(b))}$$

$$= 0.033$$

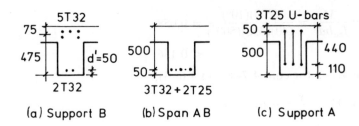

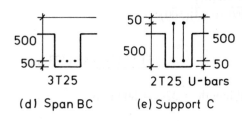

(a) Support B   (b) Span AB   (c) Support A

(d) Span BC   (e) Support C

Fig. 11.5–9

From Table 4.7–2,

$$\frac{z}{d} = 0.94 \qquad \frac{x}{d} = 0.13$$

Note that:

(a) As stated at the end of Section 4.5, BS 8110 does not allow z/d to be taken as greater than 0.95 in any case.

(b) $x/d = 0.13$. Therefore $0.9x = (0.9)(0.13)(500) = 59$ mm $< h_f$, i.e. stress block within flange thickness.

From eqn (4.8–3),

$$A_s = \frac{M}{0.87f_y z} = \frac{(522)(10^6)}{(0.87)(460)(0.94)(500)}$$

$$= 2775 \text{ mm}^2$$

<u>Bottom 3T32 + 2T25 (3394 mm²)</u>

End support A (Fig. 11.5–9(c)). Assume nominal fixing moment equal to 40% of the initial fixed end moment—see Comments on Step 5 at the end:

$$M = 40\% \text{ of } 507 \text{ kNm} \quad (\text{Fig. 11.5–5: Case 1})$$

$$= 203 \text{ kNm}$$

From Table 4.7–1 (or eqn 4.7–5),

$$K' = 0.156$$

$$K = \frac{M}{f_{cu}bd^2} = \frac{(203)(10^6)}{(40)(350)(500^2)} = 0.058$$

$$< 0.156$$

From Table 4.7–2 (or eqns 4.7–6 and 4.6–3),

$$\frac{z}{d} = 0.93 \qquad \frac{x}{d} = 0.16$$

From eqn (4.7–9),

$$A_s = \frac{M}{0.87f_y z} = \frac{(203)(10^6)}{(0.87)(460)(0.93)(500)}$$

$$= 1091 \text{ mm}^2$$

3T25 U-bars (1473 mm^2)

From Table 4.10–2,

ultimate anchorage bond length $= 32\phi = 800$ mm

Using eqns (6.6–3(a), (b)),

$$\text{required anchorage length} = \left[\frac{1091}{1473}\right](800)$$

$$= 593 \text{ mm}$$

Span BC (Fig. 11.5–9(d)). From Section 4.8,

$$\text{effective flange width} = 350 + (0.2)(0.7 \text{ of } 7000)$$

$$= 1330 \text{ mm}$$

From bending moment envelope (Fig. 11.5–6),

$$M = 236 \text{ kNm}$$

$$\frac{M}{f_{cu}bd^2} = \frac{(236)(10^6)}{(40)(1330)(500^2)} = 0.018$$

$$< K' \text{ of Table 4.7–1}$$

From Table 4.7–2,

$$\frac{z}{d} = 0.94 \qquad \frac{x}{d} = 0.13 \quad (\text{i.e. } 0.9x < h_f)$$

From eqn (4.8–3),

$$A_s = \frac{M}{0.87f_y z} = \frac{(236)(10^6)}{(0.87)(460)(0.94)(500)}$$

$$= 1255 \text{ mm}^2$$

Bottom 3T25 (1473 mm^2)

End support C (Fig. 11.5–9 (e)). Design for 40% of the initial fixed-end moment—see Comments on Step 5 at the end.

$$M = 40\% \text{ of } 307 \text{ kNm} \quad \text{(Fig. 11.5–5: Case 2)}$$
$$= 123 \text{ kNm}$$

$$\frac{M}{f_{cu}bd^2} = \frac{(123)(10^6)}{(40)(350)(500^2)} = 0.035$$

$$< K' \text{ of Table 4.7–1}$$

From Table 4.7–2,

$$\frac{z}{d} = 0.94 \qquad \frac{x}{d} = 0.13$$

From eqn (4.7–9),

$$A_s = \frac{M}{0.87f_y z} = \frac{(123)(10^6)}{(0.87)(460)(0.94)(500)}$$
$$= 654 \text{ mm}^2$$

<u>2T25 U-bars (982 mm^2)</u>

From Table 4.10–2

ultimate anchorage bond length $= 32\phi = 800$ mm

Using eqns (6.6–3(a) (b)),

$$\text{required anchorage length} = \frac{654}{982}(800) = 533 \text{ mm}^2$$

The **curtailment diagram** for the longitudinal reinforcement is shown in Fig. 11.5–10, which should be read in conjunction with the reinforcement details in Fig. 11.5–12 (see p. 444).

Step 6   Shear reinforcement

For further explanations of the calculations, see Comments on Step 6 at the end.

(a) From Fig. 11.5–12, the minimum tension reinforcement along the beam may reasonably be taken as 2T25, i.e. $A_s = 982$ mm^2.

$$\frac{100A_s}{b_v d} = \frac{(100)(982)}{(350)(500)} = 0.56$$

From Table 6.4–1,

$$v_c = 0.61 \text{ N/mm}^2$$

(b) Minimum links (see Section 6.4: 'Shear resistance in design calculations: Step 4') will be provided where $v \leq (v_c + 0.4)$, i.e. where $V \leq (v_c + 0.4)b_v d$.

$$V \leq (0.61 + 0.4)(350)(500) \text{ N}$$
$$= 177 \text{ kN}$$

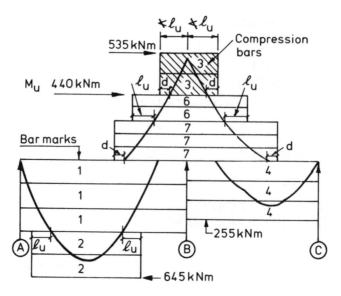

Notes :

1.   d = Effectve depth of beam

2. ℓ_u = Ultimate anchorage length (Eqn 6.6-3(b))

Fig. 11.5–10   Curtailment diagram

In Fig. 11.5–11, the 177 kN limits for minimum links are superimposed on the shear force envelope.

(c)   From eqn (6.4–2),

$$\frac{A_{sv}}{s_v} \text{ (min. links)} = \frac{0.4 b_v}{0.87 f_{yv}}$$

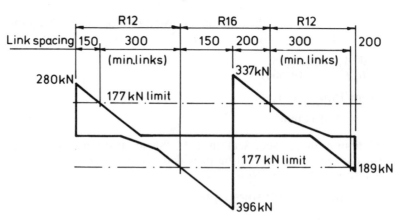

Fig. 11.5–11   Link diagram

$$= \frac{(0.4)(350)}{(0.87)(250)} = 0.64 \text{ mm}$$

Minimum links R12 at 300 (Table 6.4–2: $A_{sv}/s_v = 0.75$ mm)
(d) Where $V > 177$ kN, shear reinforcement will be provided in accordance with eqn (6.4–3). In Table 11.5–1, the shear force V has been taken at a distance from the support face equal to the effective depth d (see Comments on Step 6 at the end.)

Table 11.5–2 shows the link provision, and Fig. 11.5–11 shows the link diagram.

Step 7 Deflection
From Table 5.3–1,

basic span/depth ratio = 20.8 (for $b_w/b \leq 0.3$)

For span AB, $M = 522$ kNm and the effective flange width b is 1610 mm. Hence

$$\frac{M}{bd^2} = \frac{(522)(10^6)}{(1610)(500^2)} = 1.30 \text{ N/mm}^2$$

From Table 5.3–2,

modification factor = 1.28

allowable span/depth ratio = (20.8)(1.28)

$$= 26.6$$

Table 11.5–1 Link requirement

Location	V (kN)	v (N/mm²)	Effective reinf.	$A_s/b_v d$ (%)	v_c (N/mm²)	A_{sv}/s_v (eqn 6.4–3)
Support A	266	1.52	3T25	0.84	0.70	1.32 mm
Support BA	382	2.18	3T32	1.38	0.82	2.19 mm
Support BC	323	1.85	3T32	1.38	0.82	1.66 mm
Support C	175	1.00	2T25	0.56	0.61	0.63 mm

Table 11.5–2 Link provision

Location	Links provided (Fig. 11.5–11)	$\frac{A_{sv}}{s_v}$ (reqd) (Table 11.5–1)	$\frac{A_{sv}}{s_v}$ (provided) (Table 6.4–2)
Support A	R12 at 150	1.32 mm	1.51 mm
Support BA	R16 at 150	2.19 mm	2.68 mm
Support BC	R16 at 200	1.66 mm	2.01 mm
Support C	R12 at 300	0.63 mm	0.75 mm

actual span/depth ratio = 9000/500 = 18.0

$$< 26.6$$

span/depth ratio OK

Step 8   Cracking
Table 11.5–3   Bar spacing and corner distance

Tension bars	Moment redistr. (Fig. 11.5–6)	Spacing a_b (Fig. 5.4–1)		Spacing a_c (Fig. 5.4–1)	
		Actual (Fig 11.5–12)	Allowed (Table 5.4–1)	Actual (Fig. 11.5–12)	Allowed (Table 5.4–1)
Support B: top	−17%	34 mm	135 mm	See Comments at end	
Span AB: bottom	0	34 mm	160 mm	60 mm	80 mm
Span BC: bottom	−19%	92 mm	132 mm	60 mm	66 mm

Crack widths OK

Step 9   Robustness (BS 8110: Clauses 3.1.4.3, 3.12.3.4, and 3.12.3.6)
Internal longitudinal ties (BS 8110: Clause 3.12.3.4.1).

F_t = 20 + 4 times No. of storeys = 36 kN/m

From Step 2 of the solution to Example 11.5–1,

g_k = 32.1/5.5 = 5.84 kN/m^2 of floor

q_k = 16.5/5.5 = 3.00 kN/m^2 of floor

Check:

$$\left[\frac{g_k + q_k}{7.5}\right]\left(\frac{l_r}{5}\right) F_t$$

$$= \left[\frac{5.84 + 3.00}{7.5}\right]\left[\frac{9}{5}\right](36)$$

$$= 76.4 \text{ kN/m} > 36 \text{ kN/m}$$

Hence

tie force = (76.4 kN/m)(5.5 m) = 420 kN

minimum continuous internal tie = $\dfrac{(420)(10^3)}{*(0.87)(460)}$ = 1049 mm^2

Bottom 2T32 (1608 mm^2 continuous through support B)

---

*γ_m could have been taken as 1.0. See BS 8110: Clause 3.12.3.2.

Check lap length (see Section 4.10, under the heading 'Tension laps (b)'):

$$\text{cover to lapped bars} = 20 + \text{link } \phi = 36 \text{ mm}$$
$$< 2\phi \ (64 \text{ mm})$$
$$\text{spacing between adjacent laps} \doteq 100 \text{ mm}$$
$$< 6\phi \ (192 \text{ mm})$$

Hence apply a factor of 1.4 to Table 4.10–2, so that:

$$\text{full tension lap length} = (1.4)(32\phi) = 45\phi$$

$$\text{required lap length} = \left[\frac{A_s(\text{req})}{A_s(\text{prov})}\right](45\phi)$$

$$= \left[\frac{1049}{1608}\right](45)(32) = 939 \text{ mm}$$

Extend the tie bars 1500 mm from the column face at the support B.

External column tie (BS 8110: Clause 3.12.3.6.1).

(a) $2F_t = 2(36) = 72$ kN.

$$\left[\frac{\text{Floor height}}{2.5}\right](F_t) = \left[\frac{4 - 0.180}{2.5}\right](36) = 55 \text{ kN}$$

Hence 72 kN force need not be considered further.

(b) 3% of total design ultimate vertical load carried by column

$$\doteq 3\% \text{ of } 1500 \text{ kN} \quad \text{(calculations not shown)}$$
$$= 45 \text{ kN} < 55 \text{ kN in (a)}$$

Hence, tie force = 55 kN. Minimum continuity external tie

$$= \frac{(55)(10^3)}{(0.87)(460)} = 137 \text{ mm}^2 \quad \text{(See remark on } \gamma_m \text{ for internal tie)}$$

The 25 mm U-bars at the external supports provide ample area.

Step 10   Reinforcement details
Figure 11.5–12 shows the reinforcement details.

Comments on Step 1
Note BS 8110's definition of the term 'nominal cover', as explained under Table 2.5–7 in Section 2.5(e).

Comments on Step 2
The unit weight of reinforced concrete is taken as 24 kN/m^3. As explained in the beginning of Section 2.5, this value includes an allowance for the weight of the reinforcement.

Comments on Step 3
In practice, the entire building frame can be readily analysed as a three-dimensional structure, using a computer package (BS 8110: Clause 2.5–1).

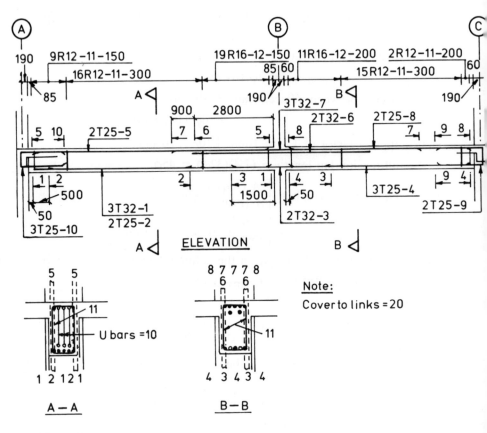

Fig. 11.5–12 **Typical floor main beam [3]**

BS 8110 : Clause 3.2.1.2.1 also permits the simplification into sub-frames, as explained earlier in Section 11.4(a). The method of sub-frames is used in Higgins and Rogers's book [3]; the sub-frame analysis is usually done by computer.

In Step 3 here, we have used the 'continuous beam simplification' of BS 8110 : Clause 3.2.1.2.3 as explained in Section 11.4(a). Of course, the moments and shears in the continuous beam can be obtained by any of the established methods of elastic analysis [19]. Calculations by the method of moment distribution [19] are shown in Fig. 11.5–5.

Moment redistribution and the construction of bending moment envelopes were explained in Section 4.9; study Example 4.9–1 again if necessary.

Comments on Step 4

To illustrate the method of calculation, consider Case 1 loading. The simple-span shear force is

$$(\tfrac{1}{2})(1.4g_k + 1.6q_k)(9) = (\tfrac{1}{2})(75.1)(9) = 338 \text{ kN}$$

The correction due to the continuity bending moment at support B is (520 kNm)/(9 m) ± 58 kN, where 520 kNm is the redistributed moment taken from Fig. 11.5–6. Therefore, the final end shears are 338 ± 58 = 396 kN (at B) and 280 kN (at A).

Comments on Step 5

Internal support B. The design procedure, using Tables 4.7–1 and 4.7–2, is that of the I.Struct.E. Manual [20]. All the design tables and design equations were derived in Section 4.7.

Span AB. The reinforcement for the flanged section is designed using the 'I.Struct.E. Manual's design procedure (Case II: 0–30% moment redistribution)', as explained in Section 4.8.

End supports A and C. The bending moments are in each case taken as 40% of the initial fixed-end moment, in accordance with a common practice in design—see the final paragraph in the solution to Example 11.4–1(a). (If the reader wishes, he might of course obtain these support moments by analysing the building as a complete structure or else analyse the appropriate subframes.) What is important is the understanding that the bending moments as shown in Fig. 11.5–6 are already in equilibrium with the design ultimate loads. The safety of the beam is not in question; the 40% fixed-end moments assumed for the end supports A and B are really to ensure serviceability.

Curtailment diagram (Fig. 11.5–10). See under the heading 'Curtailment and anchorage of bars' in Section 4.10.

The resistance moments as shown in Fig. 11.5–10 have been obtained as follows. Consider, for example, the internal support B:

$$M_u = K' f_{cu} b d^2$$
$$= (0.139)(40)(350)(475^2) \text{ Nmm}$$
$$= 440 \text{ kNm}$$
$$A_s = 4021 \text{ mm}^2 \text{ (5T32)}$$
$$A'_s = 1608 \text{ mm}^2 \text{ (2T32)}$$
$$\frac{A_s}{bd} = \frac{3327}{(350)(475)} = 2\%$$
$$\frac{A'_s}{bd} = \frac{1608}{(350)(475)} = 1\%$$

From Fig. 4.5–2 (BS 8110 design chart),

$$M = (6.8)(350)(475^2) \text{ Nmm} = 535 \text{ kNm}$$

Of course, in the calculations for the spans AB and BC, b should be taken as the effective flange width.

Comments on Step 6

Shear design to BS 8110 is explained in Section 6.4. With reference to Step 6(d), the shear forces have been taken, not at the respective support face, but at a distance from the support face equal to the effective depth d. This is permitted by BS 8110—for details, see the Comments on Step 5 of the solution to Example 6.4–2.

Comments on Step 7
See also Example 5.3–2 on possible remedial actions if the actual span/depth ratio had exceeded the allowable.

Comments on Step 8
The clear distance a_b between the bars, and the corner distance a_c, are calculated in the same manner as in Example 5.7–2. The a_c values in Table 11.5–3 include an allowance for the fact that the links can only be bent to an internal radius of 2ϕ (see Fig. 6.6–1).

In Table 11.5–3, corner distances a_c are not shown for the top bars at the support B; the reason is as explained in Comment (b) at the end of Example 5.7–2.

For the span AB, the moment redistribution is indicated as zero in Table 11.5–3. The reason can be found in Fig. 11.5–6, which shows that the redistributed curve IIIR in fact coincides with the elastic curve I.

Comments on Step 9
The internal tie requirement dictates the provision of the bottom bars 2T32 at the internal support B. This explains why, in Step 5, 2T32 (1608 mm^2) were provided even though the area required at the time was only 476 mm^2.

Comments on Step 10
The main bars in Fig. 11.5–12 have been curtailed in accordance with the curtailment diagram in Fig. 11.5–10. (See also the 'Simplified rules for curtailment of bars' in Section 4.10.)

Example 11.5–3
With reference to the braced building frame in Fig. 11.5–1, design and detail an internal column, B9.

SOLUTION
See Fig. 11.5–13 ($f_{cu} = 40$ N/mm^2), and Comments at the end.

Step 1 Durability and fire resistance
From Table 2.5–7,

nominal cover for mild exposure = 20 mm   (use 30 mm)

cover to main bars = 30 + link ϕ = 40 mm, say

From Table 3.5–1,

for 380 square column with 40 mm cover to main bars,

the fire resistance exceeds 1 hour

Nominal cover = 30 mm

Fire resistance OK

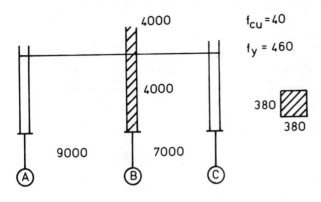

Fig. 11.5–13

Step 2   Column and beam stiffnesses
Columns: all floors

$$I = \frac{bh^3}{12} = \frac{380 \times 380^3}{12} = 1.74 \times 10^9 \text{ mm}^4$$

$$\frac{I}{l} = \frac{1.74 \times 10^9}{4000} = 435 \times 10^3 \text{ mm}^3$$

Floor beams

$$I = \frac{bh^3}{12} = \frac{350 \times 550^3}{12} = 4.85 \times 10^9 \text{ mm}^4$$

9 m span: $\dfrac{I}{l} = \dfrac{4.85 \times 10^9}{9000} = 539 \times 10^3 \text{ mm}^3$

7 m span: $\dfrac{I}{l} = \dfrac{4.85 \times 10^9}{7000} = 693 \times 10^3 \text{ mm}^3$

Roof beams

$$I = \frac{bh^3}{12} = \frac{350 \times 500^3}{12} \text{ (say)} = 3.65 \times 10^9 \text{ mm}^4$$

9 m span: $\dfrac{I}{l} = \dfrac{3.65 \times 10^9}{9000} = 406 \times 10^3 \text{ mm}^3$

7 m span: $\dfrac{I}{l} = \dfrac{3.65 \times 10^9}{7000} = 521 \times 10^3 \text{ mm}^3$

Step 3   Moments in column
Floor junctions (Fig. 11.5–14(a))

$$\sum \left(\frac{I}{l} \right) = (435 + 435 + 270 + 347) \times 10^3 = 1487 \times 10^3 \text{ mm}^3$$

From Fig. 11.5–5: Case 1,

out-of-balance moment = 507 − 142 = 365 kNm

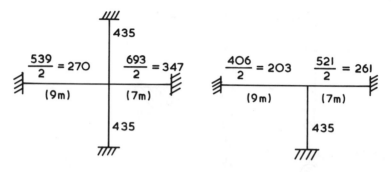

(a) Floor junction   (b) Roof junction

Fig. 11.5–14

$$M \text{ in column } = 365 \times \frac{435}{1487} = 107 \text{ kNm}$$

Roof junction (Fig. 11.5–14(b))

$$\sum \left(\frac{I}{l}\right) = (435 + 203 + 261) \times 10^3 = 899 \times 10^3 \text{ mm}^3$$

Out-of-balance moment = 226 kNm   (see Comments at end)

$$M \text{ at top of 3rd floor column } = 226 \times \frac{435}{899} = 109 \text{ kNm}$$

Step 4 Effective column height
Foundation to 1st floor
From eqn (7.2–2),

effective column height $l_e = \beta l_0$

From Table 7.2–1,
North–South direction:

$\beta = 0.9$ (end conditions: top = 1, bottom = 3)

Hence

$$l_{ex} = 0.9 l_0$$
$$= (0.9)(4000) = 3600 \text{ mm}$$
$$l_{ex}/h = 3600/380$$
$$= 9.5 < 15$$

East-West direction:

$\beta = 1.0$ (end condition: top = 3, bottom = 3)

Hence

$$l_{ey} = (1.0)(4000) = 4000 \text{ mm}$$

$$l_{ey}/b = 4000/380 = 10.5 < 15$$

<u>short column</u>

1st to 2nd floor; 2nd to 3rd floor; 3rd floor to roof
From Table 7.2–1:
North–South direction:

$$\beta = 0.75 \quad \text{(end conditions: top} = 1, \text{ bottom} = 1)$$

$$l_{ex} = \beta l_0 = (0.75)(4000)$$

$$= 3000 \text{ mm}$$

$$l_{ex}/h = 3000/380$$

$$= 7.90 < 15$$

East–West direction:

$$\beta = 1.0 \quad \text{(end conditions: top} = 3, \text{ bottom} = 3)$$

$$l_{ey} = (1.0)(4000) = 4000$$

$$l_{ey}/b = 4000/380 = 10.5 < 15$$

<u>short column</u>

Step 5 Axial loads
Table 11.5–4   Axial loads on column

		Column design loads (kN)			
		Imposed[a]		*Dead*[a]	
Floor supported	*Beam load V*(kN) (See Fig. 11.5–7: Case 3)	$V \times \dfrac{1.6q_k}{1.4g_k + 1.6q_k}$	Σ	$V \times \dfrac{1.4g_k}{1.4g_k + 1.6q_k}$	Σ
Roof	617[b]	168†	168	449[b]	
				(self-wt) 12	461
3rd	733	258	426	475	
				(self-wt) 12	948
2nd	733	258	684	475	
				(self-wt) 12	1435
1st	733	258	942	475	
				(self-wt) 12	1922

[a] For typical floors, g_k and q_k are taken from Step 2 of Example 11.5–2.
[b] For roof, see Comments at the end.

Reduced imposed loads (see Table 11.2–3)

3rd floor to roof	100% of 168 = 168 kN
2nd to 3rd floor	90% of 426 = 382 kN
1st to 2nd floor	80% of 684 = 547 kN
Foundation to 1st floor	70% of 942 = 659 kN

Total axial loads (see Table 11.5–4)

3rd floor to roof	$N_{3r} = 168 + 461 = 629$ kN
2nd to 3rd floor	$N_{23} = 382 + 948 = 1330$ kN
1st to 2nd floor	$N_{12} = 547 + 1435 = 1982$ kN
Foundation to 1st floor	$N_{f1} = 659 + 1922 = 2581$ kN

$N_{3r} = 629$ kN $N_{23} = 1330$ kN $N_{12} = 1982$ kN : $N_{f1} = 2581$ kN

Step 6 *Design bending moments*
BS 8110's design minimum eccentricity (see Comments at the end) = $0.05h$.

Minimum design moment = $0.05hN$

Foundation to 1st floor level:

$$0.05h \, N_{f1} = 0.05 \times 380 \times 10^{-3} \times 2581 \quad \text{(see Step 5)}$$

$$= 49 \text{ kNm}$$

Elsewhere, $N < N_{f1}$ and hence $0.05hN < 49$ kNm.

Therefore the column design is governed by the column moments in Step 3 as these are larger than $0.05hN$.

M (roof junction) = 109 kNm M (floor junction) = 107 kNm

Step 7 *Reinforcement*
$f_{cu} = 40$ N/mm²; $f_y = 460$ N/mm²

$$\frac{d}{h} = \frac{380 - 40 - (\text{bar } \phi)/2}{380} = 0.85 \text{ approx.}$$

Hence the design chart in Fig. 7.3–1 applies, and the results are as shown in Table 11.5–5.

Step 8 *Reinforcement details* (see Fig. 11.5–15)

Comments on Step 1
Note BS 8110's definition of the term 'nominal cover', as explained under Table 2.5–7 in Section 2.5(e).

Table 11.5–5 Column reinforcement

Location	$\dfrac{N}{bh}$ (N/mm²)	$\dfrac{M}{bh^2}$ (N/mm²)	A_{sc} required[a] (Fig. 7.3–1)	A_{sc} provided (Fig. 11.5–15)
3rd fl. to roof	4.4	2.0	0.4% bh = 578 mm²	4T25 (1960 mm²)
2nd to 3rd fl.	9.2	2.0	0.4% bh = 578 mm²	4T25 (1960 mm²)
1st to 2nd fl.	13.8	2.0	0.4% bh = 578 mm²	4T25 (1960 mm²)
Fndn to 1st fl.	17.9	2.0	1.4% bh = 2021 mm²	4T25 (3216 mm²)

[a] From Section 3.5, the minimum steel ratio is 0.4%.

Job No. : 1959/65/67

Trinity and Newnham Colleges

Date : 4 December 1987

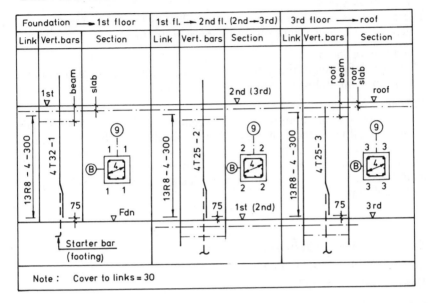

Fig. 11.5–15   Internal column—Trinity and Newnham Colleges

Comments on Step 2
The column and beam stiffnesses, as shown in Fig. 11.5–14, have been calculated using BS 8110 : Clause 3.2.1.2.5, as explained earlier in Section 11.4(a):

(a)   The column moments may be calculated by simple moment distribution procedure, on the assumption that the column and beam ends remote from the junction under consideration are fixed and that the beams possess *half* their actual stiffness (see Fig. 11.4–1(h)).

(b)   The arrangement of the design ultimate imposed load should be such as to cause the maximum moment in the column.

See also **second moment of area** in Section 11.4(a).

Comments on Step 3
The bending moments in the columns have been calculated from BS 8110 : Clause 3.2.1.2.5, referred to above. The out-of-balance moment of 226 kNm at the roof junction is taken from the design calculations for the roof main beam, which have not been presented in this book owing to space restriction. The calculations are similar to those in Fig. 11.5–5 except that g_k is 25.4 kN/m and q_k is 8.3 kN/m (imposed load = 1.5 kN/m^2).

Comments on Step 4
As explained (in the first sentence) in Section 7.3, BS 8110 defines a short column as one for which the ratios l_{ex}/h and l_{ey}/b are both less than 15.

Comments on Step 5
The roof loads are taken from design calculations for the roof main beam, which are not shown here owing to space restriction.

Comments on Step 6
BS 8110's definition of the design minimum eccentricity is explained in the definition of M following eqn (7.3–2).

Comments on Step 7
See 'Limits on main reinforcement' in Section 3.5. For the column lengths from the 1st floor to the roof, we could have used smaller bars than size 25. However, size 25 bars have the advantage of providing a robust cage with the links. Besides, it is good practice to limit the spacing of the column main bars to not exceeding 250 mm. With size 25 bars, the spacing is just about 250 mm.

Site Ref. : Job No. 1959/65/67

Trinity and Newnham Colleges

Date : 4 December 1987

Member	Bar mark	Type and size	No of mbrs	No in each	Total no.	Length each bar mm	Shape	Shape Code (BS 4466)	A mm	B mm	D mm
Column B9	1	T32	1	4	4	4800	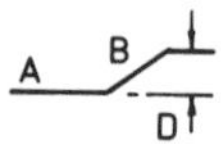	41	3075	700	70
	2	T25	2	4	8	4800	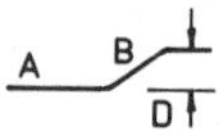	41	3450	550	55
	3	T25	1	4	4	3875	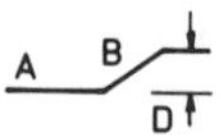	41	2535	550	55
	4	R8	4	13	52	1400	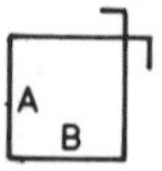	60	310	310	

Fig. 11.5–16   **Bar bending schedule—Trinity and Newnham Colleges** (see also Fig. A2–2)

Comments on Step 8
See 'Lateral ties or links' in Section 3.5. Note the restriction on link size and link spacings. Figure 11.5–15 conforms to the standard method of detailing [1] as explained in Example 3.6–3. If necessary, study again Figs 3.6–1 and 3.6–2.

In Fig. 11.5–15, the starter bars are indicated as broken lines, because they are to be detailed with the footing—see Example 3.6–3: Comment (d).

Example 11.5–4
Prepare a bar **bending schedule** for the reinforced concrete column in Example 11.5–3. Conform to BS 4466: *Bending Dimensions and Scheduling of Bars for the Reinforcement of Concrete*.

SOLUTION
The bending schedule is shown in Fig. 11.5–16.

11.6 Typical reinforcement details

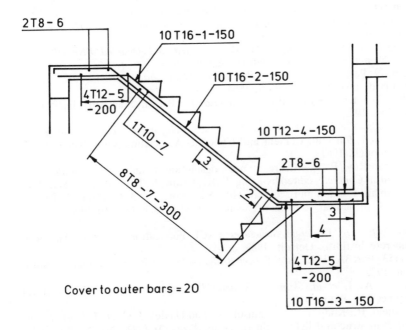

Fig. 11.6–1 Typical section through stairs [1]

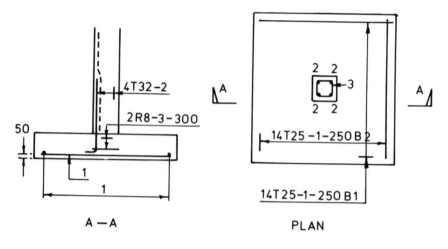

4T32–2

2R8–3–300

50

1
1

A — A

2 2
2 2
–3

14T25–1–250 B 2

14T25–1–250 B1

PLAN

Cover to outer bars = 40

Fig. 11.6–2   Typical foundation [1]

References

1 *Joint Committee Report on Standard Method of Detailing Structural Concrete.* Concrete Society and the Institution of Structural Engineers, London, 1987.
2 *Model Procedure for the Presentation of Calculations.* Concrete Society Technical Report No. 5, London, 1981.
3 Higgins, J. B. and Rogers, B. R. *Designed and Detailed BS 8110:1985.* Cement and Concrete Association, Slough, 1986.
4 BS 6399 : Part 1 : 1984. *Code of Practice for Dead and Imposed Loads.* British Standards Institution, London, 1984.
5 CP 3 : Chapter V : Part 2 : 1972. *Wind Loads.* British Standards Institution, London, 1972.
6 Reynolds, C. E. and Steedman, J. C. *Reinforced Concrete Designer's Handbook*, 9th edn. Cement and Concrete Association Viewpoint Publication, Slough, 1981.
7 Bardhan-Roy, B. K. Fire resistance—design and detailing. In *Handbook of Structural Concrete*, edited by Kong, F. K., Evans, R. H., Cohen, E. and Roll, F. Pitman, London and McGraw-Hill, New York, 1983, Chapter 14. Taylor, H. P. J. Structural performance as influenced by detailing. In ibid., Chapter 13.
8 *Fire Safety of Concrete Structures* (ACI Publication SP–80). American Concrete Institute, Detroit, 1983.
9 BS 153 : Part 3A : 1972. *Steel Girder Bridges—Loads.* British Standards Institution, 1972.
10 Harris, A. J. *et al. Aims of Structural Design.* Institution of Structural Engineers, London, 1975.
11 Mayfield, B., Kong, F. K., Bennison, A. and Davies, J. C. D. T. Corner joint details in structural lightweight concrete. *Proc. ACI*, **68**, No. 5, May 1971, pp. 366–72.

12   Mayfield, B., Kong, F. K. and Bennison, A. Strength and stiffness of lightweight concrete corners. *Proc. ACI*, **69**, No. 7, July 1972, pp. 420–7.

13   Somerville, G. and Taylor, H. P. J. The influence of reinforcement detailing on the strength of concrete structures. *The Structural Engineer*, **50**, No. 1, Jan. 1972, pp. 7–19. Discussion: **50**, No. 8, Aug. 1972, pp. 309–21.

14   ACI Committee 315. *ACI Detailing Manual*. American Concrete Institute, Detroit, 1980.

15   Noor, F. A. Ultimate strength and cracking of wall corners. *Concrete*, **11**, No. 7, July 1977, pp. 31–5.

16   Kong, F. K., Evans, R. H., Cohen, E. and Roll, F. In *Handbook of Structural Concrete*, edited by Tall buildings—1, Coull, A. and Stafford Smith, B. Pitman, London and McGraw-Hill, New York, 1983, Chapter 37. Cheung, Y. K. Tall buildings—2. In ibid., Chapter 38.

17   Kong, F. K. and Charlton, T. M. The fundamental theorems of the plastic theory of structures. *Proceedings*, Michael R. Horne Conference on Instability and Plastic Collapse of Steel Structures (Editor: L. J. Morris). Manchester, 1983. Granada Publishing, London, 1983, pp. 9–15.

18   Kong, F. K. Discussion of: 'Why not WL/8?' by A. W. Beeby. The Structural Engineer, **64A**, No. 7, July 1986, pp. 184–6.

19   Coates, R. C., Coutie, M. G. and Kong, F. K. *Structural Analysis*, 3rd edn. Van Nostrand Reinhold, London, 1987.

20   I.Struct.E./ICE Joint Committee. *Manual for the Design of Reinforced Concrete Building Structures*. Institution of Structural Engineers, London, 1985.

Chapter 12
Computer programs

In collaboration with **Dr H. H. A. Wong**, University of Newcastle upon Tyne

12.1   Notes on the computer programs

12.1(a)   Purchase of programs and disks

The complete FORTRAN listings (with commentaries) of all the computer programs in this chapter, together with the floppy disks incorporating these programs, can be purchased from the Publishers [1]. See Appendix 1 for details.

12.1(b)   Program language and operating systems

All the programs are written in PRO FORTRAN, for Microsoft's MS-DOS and IBM's PC-DOS microcomputers [2]. These include a fairly wide range of microcomputers, such as RM Nimbus, IBM-PC, Amstrad, Apricot, etc. The programming language, PRO FORTRAN, is an implementation based on the standard published by ANSI as X3.9–1966. This standard is in very wide use and forms the basis, too, for the various 'Fortran IV' implementations. PRO FORTRAN incorporates a number of extensions, notably in the area of file handling; these have generally been defined in the light of 'Fortran 77'. For full details, see Reference 2.

12.1(c)   Program layout

In writing each of the programs, we have borne in mind three points:

(a)   The program should be easy to read.
(b)   Its purpose should be clear to the reader.
(c)   Its structure should be clear to the reader.

We also realize that some readers might prefer to use other languages such as BASIC or PASCAL. Hence, in writing the programs, we have devoted the efforts to make them easier to translate into BASIC [3] or PASCAL if necessary. Efforts have also been devoted to make it easier for the programs to be amended by the reader to do slightly different jobs as required.

To make the programs easy to read and easy to understand, each one is written in modular form. The program layout has been designed to convey two points quickly to the reader:

(a)   The purpose of each program module;
(b)   How it achieves that purpose.

The layout of each program is illustrated in Fig. 12.1–1.

Each program consists essentially of three blocks: (i), (ii) and (iii) as shown in Fig. 12.1–1(a). Each program is followed by a table (Fig. 12.1–1(b)), which shows all the variables and their meanings. Block (i) in Fig. 12.1–1 is the Header of the Main Program and corresponds to Lines 1–27 in the program listing in Fig. 12.1–2. The Header gives the following information:

(a)   The program name and what it stands for (Line 3);
(b)   The purpose of the program (Lines 6–9);
(c)   The reference (Lines 11–14).

The rest of the Header (Lines 18–25) states the authorship, the program language and the operating systems of the microcomputers for which the program has been written.

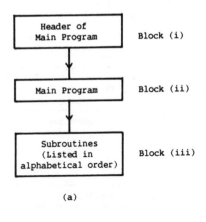

(a)

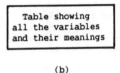

(b)

Fig. 12.1–1   Program layout

Fig. 12.1–2   Listing of program BMBRSR

```
  1    Com  **********************************************************************
  2    Com  *                                                                    *
  3    Com  *  Program Unit Name : BMBRSR (= BeaM; Bending Reinforcement;         *
  4    Com  *                              Simplified Rectangular)                *
  5    Com  *                                                                    *
  6    Com  *  Purpose : Design of bending reinforcement for a rectangular        *
  7    Com  *            or flanged beam section in accordance with               *
  8    Com  *            BS 8110 : 1985, using simplified rectangular             *
  9    Com  *            stress block.                                            *
 10    Com  *                                                                    *
 11    Com  *  Reference : This program refers to Sections 4.7, 4.8 and           *
 12    Com  *              4.12 of Chapter 4 of Kong and Evans : Reinforced       *
 13    Com  *              and Prestressed Concrete, Van Nostrand Reinhold,       *
 14    Com  *              3rd Edition, 1987.                                     *
 15    Com  *                                                                    *
 16    Com  *   *   *   *   *   *   *   *   *   *   *   *   *   *   *   *         *
 17    Com  *                                                                    *
 18    Com  *  Authors : Dr H. H. A. Wong in collaboration with                  *
 19    Com  *            Professors F. K. Kong and R. H. Evans                   *
 20    Com  *                                                                    *
 21    Com  *  Programming Language : PRO FORTRAN                                 *
 22    Com  *                                                                    *
 23    Com  *  Operating Systems : IBM's PC-DOS and Microsoft's MS-DOS.           *
 24    Com  *                                                                    *
 25    Com  *  Version : KE3-4.12-8S23P6                                          *
 26    Com  *                                                                    *
 27    Com  **********************************************************************
 28
 29          PROGRAM BMBRSR
 30
 31          INTEGER*1  BMTYPE, F, R
 32          REAL       AS, ASDASH, B, BEFF, BETAB, BW, D, DDASH, FCU
 33          REAL       FSDASH, FY, HF, K, KDASH, KF, M, MU, X, Z
 34          DATA       F, R /'F', 'R'/
 35
 36    Com  ------------------------------------------------------------------
 37    Com  | Read in the section and material properties, etc.              |
 38    Com  ------------------------------------------------------------------
 39          CALL INIT (BMTYPE,
 40         &           B, BEFF, BETAB, BW, D, DDASH, FCU, FY, HF, M)
 41
 42    Com  ------------------------------------------------------------------
 43    Com  | Calculate the constants K, K', z and x for subsequent use. |
 44    Com  ------------------------------------------------------------------
 45          CALL CONSTA (BMTYPE,
 46         &             B, BEFF, BETAB, D, FCU, M,
 47         &             K, KDASH, X, Z)
 48
 49    Com  ------------------------------------------------------------------
 50    Com  | Check the type of section (rectangular or flanged).            |
 51    Com  ------------------------------------------------------------------
 52          IF (BMTYPE .EQ. R) GO TO 100
 53          IF (BMTYPE .EQ. F) GO TO 200
 54
 55    Com  ------------------------------------------------------------------
 56    Com  | Calculate the areas of main reinforcement for rectangular      |
 57    Com  | section.                                                       |
 58    Com  ------------------------------------------------------------------
 59      100 CALL RECTAN (B, BETAB, D, DDASH, FCU, FY, KDASH, M, X, Z,
 60         &             AS, ASDASH, FSDASH, MU)
 61          GO TO 300
 62
 63    Com  ------------------------------------------------------------------
 64    Com  | Calculate the areas of main reinforcement for flanged          |
```

```
65    Com  | section.                                                      |
66    Com  ----------------------------------------------------------------
67      200 CALL FLANGE (BEFF, BETAB, BW, D, DDASH, FCU, FY, HF,
68          &                 KDASH, M, X, Z,
69          &                 AS, ASDASH, FSDASH, KF)
70
71    Com  ----------------------------------------------------------------
72    Com  | Print out the results.                                        |
73    Com  ----------------------------------------------------------------
74      300 CALL RESULT (BMTYPE,
75          &                 AS, ASDASH, D, DDASH, FSDASH, FY, HF,
76          &                 K, KDASH, KF, M, MU, X, Z)
77
78          STOP
79          END
80
81
82    Com  ****************************************************************
83    Com  *                                                              *
84    Com  *  Subroutine Name : CONSTA (= CONSTAnts)                      *
85    Com  *                                                              *
86    Com  *  Purpose : To calculate the constants K, K', z and x for     *
87    Com  *            subsequent use.  K', z and x are to be used by     *
88    Com  *            subroutine FLANGE or RECTAN.                       *
89    Com  *                                                              *
90    Com  *  References : Sections 4.5 to 4.7 of Chapter 4.              *
91    Com  *                                                              *
92    Com  ****************************************************************
93
94          SUBROUTINE CONSTA (BMTYPE,
95          &                 B, BEFF, BETAB, D, FCU, M,
96          &                 K, KDASH, X, Z)
97
98    Com  . . . . . . . . . . . Input parameters . . . . . . . . . . . .
99          INTEGER*1  BMTYPE
100         REAL       B, BEFF, BETAB, D, FCU, M
101   Com  . . . . . . . . . . Output parameters . . . . . . . . . . . .
102         REAL       K, KDASH, X, Z
103   Com  . . . . . . . . . . . Local  parameters . . . . . . . . . . .
104         INTEGER*1  F, R
105         DATA       F, R /'F', 'R'/
106
107   Com  ----------------------------------------------------------------
108   Com  | Calculate K from design ultimate moment.                      |
109   Com  ----------------------------------------------------------------
110         IF (BMTYPE .EQ. R) K = M * 1.0E6 / (FCU * B    * D ** 2)
111         IF (BMTYPE .EQ. F) K = M * 1.0E6 / (FCU * BEFF * D ** 2)
112
113   Com  ----------------------------------------------------------------
114   Com  | Calculate K' from Eqn 4.7-5 of chapter 4.                     |
115   Com  ----------------------------------------------------------------
116         IF (BETAB .LE. 0.9) GO TO 100
117         WRITE (6,5000)
118         BETAB = 0.9
119     100 KDASH = 0.40 * (BETAB - 0.4) - 0.18 * (BETAB - 0.4) ** 2
120
121   Com  ----------------------------------------------------------------
122   Com  | Calculate the lever-arm distance z from Eqn 4.7-6 or 4.7-7. |
123   Com  | Note that K > K' is equivalent to M > Mu.                     |
124   Com  ----------------------------------------------------------------
125         IF (K .LE. KDASH) Z = (0.5 + SQRT (0.25 - (K    / 0.9))) * D
126         IF (K .GT. KDASH) Z = (0.5 + SQRT (0.25 - (KDASH / 0.9))) * D
127   Com  ----------------------------------------------------------------
128   Com  | z should not > 0.95d, see penultimate paragraph of Sect 4.5.|
129   Com  ----------------------------------------------------------------
130         IF (Z .LE. 0.95 * D) GO TO 200
131         WRITE (6,5010) Z
```

```
132            Z = 0.95 * D
133
134    Com    ----------------------------------------------------------------
135    Com    | Calculate the neutral axis depth x from Eqns 4.6-3 and 4.7-3|
136    Com    ----------------------------------------------------------------
137      200 X = (D - Z) / 0.45
138            IF (X .GT. (BETAB - 0.4) * D) X = (BETAB - 0.4) * D
139
140            RETURN
141     5000 FORMAT (//, ' * * * *          The moment redistribution ratio',
142          &              ' exceeds 0.9.          * * * * *', //,
143          &              ' * * * * In subsequent calculations, a value of',
144          &              ' 0.9 is to be used. * * * * *', //)
145     5010 FORMAT (//, ' * * * * The calculated lever arm distance z = ',
146          &     3PE9.3, ' mm > 0.95d. * * * * *', //,
147          &              ' * * * * In subsequent calculations, a value of',
148          &              ' 0.95d is to be used. * * * * *', //)
149            END
150
151
152    Com ********************************************************************
153    Com *                                                                  *
154    Com *   Subroutine Name : FLANGE (= FLANGEd section)                   *
155    Com *                                                                  *
156    Com *   Purpose : To determine the amount of bending reinforcement     *
157    Com *             for a flanged beam section.                          *
158    Com *                                                                  *
159    Com *   Reference : Section 4.8 of Chapter 4.                          *
160    Com *                                                                  *
161    Com ********************************************************************
162
163            SUBROUTINE FLANGE (BEFF, BETAB, BW, D, DDASH, FCU, FY, HF,
164          &                    KDASH, M, X, Z,
165          &                    AS, ASDASH, FSDASH, KF)
166
167    Com    . . . . . . . . . . . Input parameters . . . . . . . . . . . .
168            REAL  BEFF, BETAB, BW, D, DDASH, FCU, FY, HF
169            REAL  KDASH, M, X, Z
170    Com    . . . . . . . . . . Output parameters . . . . . . . . . . . .
171            REAL  AS, ASDASH, FSDASH, KF
172    Com    . . . . . . . . . . Local  parameters . . . . . . . . . . .
173            REAL  ASF, ASW, MUF, MUW
174
175    Com    ----------------------------------------------------------------
176    Com    | Check whether the rectangular stress block is within the     |
177    Com    | flange thickness or not.                                      |
178    Com    ----------------------------------------------------------------
179            IF (0.9 * X .LE. HF) GO TO 100
180            IF (0.9 * X .GT. HF) GO TO 200
181
182    Com    ----------------------------------------------------------------
183    Com    | Case 1 : 0.9x <= hf - Singly reinforced section, since the    |
184    Com    | stress block lies wholly within the flange thickness.         |
185    Com    | Calculate the minimum amount of tension steel required from   |
186    Com    | Eqn 4.8-3, see Step 2 of Section 4.8.                         |
187    Com    ----------------------------------------------------------------
188      100 AS = M * 1.0E6 / (0.87 * FY * Z)
189            RETURN
190
191    Com    ----------------------------------------------------------------
192    Com    | Case 2 : 0.9x > hf                                            |
193    Com    | The stress block lies partly outside the flange thickness.    |
194    Com    | Then calculate the following :                                |
195    Com    | (a) Muf from Eqn 4.8-4 of Chapter 4, see Step 3 of Sect 4.8   |
196    Com    | (b) Kf  from Eqn 4.8-5 of Chapter 4, see Step 4 of Sect 4.8   |
197    Com    ----------------------------------------------------------------
```

```
198        200 MUF = 0.45 * FCU * (BEFF - BW) * HF * (D - 0.5 * HF)
199            KF  = (M * 1.0E6  - MUF) / (FCU * BW * D ** 2)
200
201    Com      ------------------------------------------------------------
202    Com      | Check the capacity of the web.                            |
203    Com      | Note : Kf > K' is equivalent to (M - Muf) > Mu, and so on. |
204    Com      ------------------------------------------------------------
205            IF (KF .LE. KDASH) GO TO 210
206            IF (KF .GT. KDASH) GO TO 250
207
208    Com      ------------------------------------------------------------
209    Com      | Case 2 : 0.9x > hf - Singly reinforced section, since the  |
210    Com      | web is adequate to resist the moment (M - Muf).            |
211    Com      | Determine the minimum amount of tension steel required from|
212    Com      | Eqn 4.8-6 of Chapter 4, see Step 4 of Section 4.8.         |
213    Com      ------------------------------------------------------------
214        210 ASF = MUF / (0.87 * FY * (D - 0.5 * HF))
215            ASW = (M * 1.0E6 - MUF) / (0.87 * FY * Z)
216            AS  = ASF + ASW
217            RETURN
218
219    Com      ------------------------------------------------------------
220    Com      | Case 2 : 0.9x > hf - Doubly reinforced section, since the  |
221    Com      | web is inadequate to resist the moment (M - Muf).          |
222    Com      | However, it needs to check whether As' reaches yield       |
223    Com      | strength at ULS or not, using Eqn 4.6-10 of Chapter 4.     |
224    Com      ------------------------------------------------------------
225        250 IF ((DDASH / X) .LE. (1.0 - FY / 800.0)) GO TO 260
226            IF ((DDASH / X) .GT. (1.0 - FY / 800.0)) GO TO 270
227
228    Com      ------------------------------------------------------------
229    Com      | The compression steel As' reaches yield strength at ULS.   |
230    Com      | Calculate the minimum amount of As' from Eqn 4.8-8 and As  |
231    Com      | from Eqn 4.8-9 of Chapter 4. See also Example 4.8-1.       |
232    Com      ------------------------------------------------------------
233        260 MUW    = KDASH * FCU * BW * D ** 2
234            ASDASH = (M * 1.0E6 - MUF - MUW) / (0.87 * FY * (D - DDASH))
235            AS     = (0.45 * FCU * (BEFF - BW) * HF  +
236           &            0.45 * FCU * BW * (0.9 * X)) / (0.87 * FY) +
237           &              ASDASH
238            RETURN
239
240    Com      ------------------------------------------------------------
241    Com      | Here the compression steel As' remains elastic at the      |
242    Com      | ultimate limit state. Calculate the minimum As' and As from|
243    Com      | Eqns 4.8-8 and 4.8-9 of Chapter 4, using a reduced stress  |
244    Com      | fs' for As'. fs' is calculated from Eqn 4.6-11 of Chapter 4.|
245    Com      ------------------------------------------------------------
246        270 MUW    = KDASH * FCU * BW * D ** 2
247            FSDASH = 700.0 * (1.0 - DDASH / X)
248            ASDASH = (M * 1.0E6 - MUF - MUW) / (FSDASH * (D - DDASH))
249            AS     = (0.45 * FCU * (BEFF - BW) * HF  +
250           &            0.45 * FCU * BW * (0.9 * X) +
251           &              FSDASH * ASDASH) / (0.87 * FY)
252
253            RETURN
254            END
255
256
257    Com **********************************************************************
258    Com *                                                                    *
259    Com *  Subroutine Name : INIT (= INITialization)                         *
260    Com *                                                                    *
261    Com *  Purposes : To initialize the program by reading in the            *
262    Com *                following input data :                              *
263    Com *                   1. Beam section title and type of beam section   *
```

```
264     Com *                    2. Section details                         *
265     Com *                    3. Characteristic strengths of the materials *
266     Com *                    4. Loading information                      *
267     Com *                                                                *
268     Com *   Remarks : In the following READ statements, the option       *
269     Com *                 "ERR=" is specified. If an error condition is   *
270     Com *                 detected during READ, an error message will be  *
271     Com *                 printed out, followed by termination of subsequent *
272     Com *                 program execution.                             *
273     Com *                                                                *
274     Com ***********************************************************************
275
276           SUBROUTINE INIT (BMTYPE,
277          &                  B, BEFF, BETAB, BW, D, DDASH, FCU, FY, HF, M)
278
279     Com  . . . . . . . . . . . Output parameters . . . . . . . . . . . .
280           INTEGER*1  BMTYPE
281           REAL       B, BEFF, BETAB, BW, D, DDASH, FCU, FY, HF, M
282     Com  . . . . . . . . . . Local parameters . . . . . . . . . . .
283           LOGICAL*1  BMTIT (40)
284           INTEGER*1  F, R
285           DATA       F, R /'F', 'R'/
286
287     Com  ------------------------------------------------------------------
288     Com  | Read in the beam section title and type of section.           |
289     Com  ------------------------------------------------------------------
290           PRINT 1000
291           READ (5,1,ERR=400) BMTIT
292           PRINT 1010
293        50 READ (5,2,ERR=400) BMTYPE
294           IF (BMTYPE .EQ. F) GO TO 100
295           IF (BMTYPE .EQ. R) GO TO 200
296           PRINT 1020
297           GO TO 50
298
299     Com  ------------------------------------------------------------------
300     Com  | Read in the section details for flanged beam.                 |
301     Com  | Note : By default, DDASH = 0.15 * D.                          |
302     Com  ------------------------------------------------------------------
303       100 PRINT 2010
304           READ (5,3,ERR=400) BEFF
305           PRINT 2020
306           READ (5,3,ERR=400) HF
307           PRINT 2030
308           READ (5,3,ERR=400) BW
309           PRINT 2040
310           READ (5,3,ERR=400) D
311           PRINT 2050
312           READ (5,3,ERR=400)  DDASH
313           IF (DDASH .EQ. 0.0) DDASH = 0.15 * D
314           GO TO 300
315     Com  ------------------------------------------------------------------
316     Com  | Read in the section details for rectangular beam.             |
317     Com  | Note : By default, DDASH = 0.15 * D.                          |
318     Com  ------------------------------------------------------------------
319       200 PRINT 2510.
320           READ (5,3,ERR=400) B
321           PRINT 2520
322           READ (5,3,ERR=400) D
323           PRINT 2530
324           READ (5,3,ERR=400)  DDASH
325           IF (DDASH .EQ. 0.0) DDASH = 0.15 * D
326
327     Com  ------------------------------------------------------------------
328     Com  | Read in the characteristic strengths of the materials.        |
329     Com  ------------------------------------------------------------------
330       300 PRINT 3010
```

```
331           READ (5,3,ERR=400) FCU
332           PRINT 3020
333           READ (5,3,ERR=400) FY
334
335    Com    ----------------------------------------------------------------
336    Com    | Read in the ultimate design moment in kNm and the moment    |
337    Com    | redistribution ratio.                                       |
338    Com    ----------------------------------------------------------------
339           PRINT 4010
340           READ (5,3,ERR=400) M
341           PRINT 4020
342           READ (5,3) BETAB
343           GO TO 500
344
345    400 PRINT 10
346           STOP
347
348    Com    ----------------------------------------------------------------
349    Com    | Print out the input data for checking.                      |
350    Com    ----------------------------------------------------------------
351    500 PAUSE
352           WRITE (6,5000)
353           WRITE (6,5010) BMTIT
354           IF (BMTYPE .EQ. F) WRITE (6,5020) BEFF, HF, BW, D, DDASH
355           IF (BMTYPE .EQ. R) WRITE (6,5025) B, D, DDASH
356           WRITE (6,5030) FCU, FY
357           WRITE (6,5040) M, BETAB
358
359           PAUSE
360
361           RETURN
362       1 FORMAT (40A1)
363       2 FORMAT (1A1)
364       3 FORMAT (F80.10)
365
366      10 FORMAT (' Input error detected. Please rerun the program',
367         &          ' again.')
368
369    1000 FORMAT (' Beam section title? (up to 40 characters) ')
370    1010 FORMAT (' Type of section: Enter R for rectangular section',
371         &          ' or F for flanged section? ')
372    1020 FORMAT (' Incorrect beam type. Please enter R or F. ')
373
374    2010 FORMAT (' Effective flange width beff in mm? (eg. 710.0) ')
375    2020 FORMAT (' Flange thickness hf in mm? (eg. 175.0) ')
376    2030 FORMAT (' Web width bw in mm? (eg. 325.0) ')
377    2040 FORMAT (' Effective depth of tension reinforcement d in mm?',
378         &          ' (eg. 325.0) ')
379    2050 FORMAT (' Depth of compression reinforcement d'' in mm, if',
380         &          ' necessary:'/,
381         &          ' Enter value (eg. 47.8) or press return for default',
382         &          ' value of 0.15d? ')
383    2510 FORMAT (' Beam width b in mm? (eg. 250.0) ')
384    2520 FORMAT (' Effective depth of tension reinforcement d in mm?',
385         &          ' (eg. 700.0) ')
386    2530 FORMAT (' Depth of compression reinforcement d'' in mm, if',
387         &          ' necessary:'/,
388         &          ' Enter value (eg. 60.0) or press return for default',
389         &          ' value of 0.15d? ')
390
391    3010 FORMAT (' Characteristic strength of concrete fcu in N/mm**2?',
392         &          ' (eg. 40.0) ')
393    3020 FORMAT (' Characteristic strength of steel fy in N/mm**2?',
394         &          ' (eg.460.0) ')
395
396    4010 FORMAT (' Design ultimate moment M in kNm? (eg. 900.0) ')
397    4020 FORMAT (' Moment redistribution ratio? (eg. 0.85) ')
```

```
398
399    5000 FORMAT (//, ' ***********************************************',/,
400       &                  ' * Summary of Input Data for Program BMBRSR *',/,
401       &                  ' ***********************************************',/)
402    5010 FORMAT (' Title for beam section : ', 40A1, /)
403    5020 FORMAT (' Details of the Flanged Section :' /,
404       &                  ' ------------------------------  ' /,
405       &                  ' Effective flange width, beff = ', 3PE10.3, ' mm', /,
406       &                  ' Flange thickness,        hf = ', 3PE10.3, ' mm', /,
407       &                  ' Web thickness,           bw = ', 3PE10.3, ' mm', /,
408       &                  ' Effective depth,          d = ', 3PE10.3, ' mm', /,
409       &                  ' Depth of comp. steel,    d'' = ', 3PE10.3, ' mm', /)
410    5025 FORMAT (' Details of the Rectangular Section :' /,
411       &                  ' ----------------------------------  ' /,
412       &                  ' Beam width,               b = ',    3PE10.3, ' mm', /,
413       &                  ' Effective depth,          d = ',    3PE10.3, ' mm', /,
414       &                  ' Depth of comp. steel, d'' = ',    3PE10.3, ' mm', /)
415    5030 FORMAT (' Characteristic Strengths :'/,
416       &                  ' -------------------------'/,
417       &                  ' Concrete,       fcu = ', 2PE10.2, ' N/mm**2', /,
418       &                  ' Reinforcement, fy = ', 3PE10.3, ' N/mm**2', /)
419    5040 FORMAT (' Loading Information :' /,
420       &                  ' -------------------'/,
421       &                  ' Design ultimate moment,   M = ', 3PE10.3, ' kNm', /,
422       &                  ' Moment redistribution ratio = ', 0PE10.2, ///)
423          END
424
425
426    Com ****************************************************************
427    Com *                                                              *
428    Com *  Subroutine Name : RECTAN (= RECTANgular section)            *
429    Com *                                                              *
430    Com *  Purpose : To determine the amount of bending reinforcement  *
431    Com *                  for a rectangular beam section.             *
432    Com *                                                              *
433    Com *  Reference : Section 4.7 of Chapter 4.                       *
434    Com *                                                              *
435    Com ****************************************************************
436
437          SUBROUTINE RECTAN (B, BETAB, D, DDASH, FCU, FY, KDASH, M, X, Z,
438       &                     AS, ASDASH, FSDASH, MU)
439
440    Com . . . . . . . . . . Input parameters . . . . . . . . . . .
441          REAL  B, BETAB, D, DDASH, FCU, FY, KDASH, M, X, Z
442    Com . . . . . . . . . . Output parameters . . . . . . . . . . .
443          REAL  AS, ASDASH, FSDASH, MU
444
445    Com  -------------------------------------------------------------
446    Com  | Calculate the moment capacity due to concrete Mu in Nmm.  |
447    Com  | from Eqn 4.7-4 of Chapter 4.                              |
448    Com  -------------------------------------------------------------
449          MU = KDASH * FCU * B * D ** 2
450
451    Com  -------------------------------------------------------------
452    Com  | Check the moment capacity of the concrete Mu.             |
453    Com  -------------------------------------------------------------
454          IF (M * 1.0E6 .LE. MU) GO TO 100
455          IF (M * 1.0E6 .GT. MU) GO TO 200
456
457    Com  -------------------------------------------------------------
458    Com  | Case 1 : Ultimate moment M is less than or equal to Mu.   |
459    Com  | The section is to be singly reinforced.                   |
460    Com  | Calculate the minimum amount of tension steel required from|
461    Com  | Eqn 4.7-9, see Step 2 of Section 4.7 in Chapter 4.        |
462    Com  -------------------------------------------------------------
463    100 AS = M * 1.0E6 / (0.87 * FY * Z)
464          RETURN
```

```
465
466    Com    ------------------------------------------------------------
467    Com    | Case 2 : Ultimate moment M exceeds Mu.                    |
468    Com    | The section is to be doubly reinforced.                   |
469    Com    | However, it needs to check whether As' reaches yield      |
470    Com    | strength at ULS or not, using Eqn 4.6-10 of Chapter 4.    |
471    Com    ------------------------------------------------------------
472    200 IF (DDASH / X .LE. (1.0 - FY / 800.0)) GO TO 210
473        IF (DDASH / X .GT. (1.0 - FY / 800.0)) GO TO 220
474
475    Com    ------------------------------------------------------------
476    Com    | The compression steel As' reaches the design strength at  |
477    Com    | the ultimate state. Calculate the minimum amount of As'   |
478    Com    | from Eqn 4.7-10 and As from Eqn 4.7-11 of Chapter 4.      |
479    Com    ------------------------------------------------------------
480    210 ASDASH = (M * 1.0E6 - MU) / (0.87 * FY * (D - DDASH))
481        AS     = MU / (0.87 * FY * Z) + ASDASH
482        RETURN
483
484    Com    ------------------------------------------------------------
485    Com    | Here As' does not reach the design strength at the ultimate|
486    Com    | limit state. Calculate the minimum amounts of As' & As from|
487    Com    | Eqns 4.7-10 and 4.7-14 of Chapter 4, using a reduced stress|
488    Com    | fs' for As'. fs' is calculated from Eqn 4.6-11 of Chapter 4.|
489    Com    ------------------------------------------------------------
490    220 FSDASH = 700.0 * (1.0 - DDASH / X)
491        ASDASH = (M * 1.0E6 - MU) / (FSDASH * (D - DDASH))
492        AS     = (MU * 1.0E6 / Z + FSDASH * ASDASH) / (0.87 * FY)
493
494        RETURN
495        END
496
497
498    Com    *****************************************************************
499    Com    *                                                               *
500    Com    *  Subroutine Name : RESULT (= RESULTs)                         *
501    Com    *                                                               *
502    Com    *  Purpose : To print out the results.                          *
503    Com    *                                                               *
504    Com    *****************************************************************
505
506
507        SUBROUTINE RESULT (BMTYPE,
508       &                   AS, ASDASH, D, DDASH, FSDASH, FY, HF,
509       &                   K, KDASH, KF, M, MU, X, Z)
510
511    Com    . . . . . . . . . . . Input parameters . . . . . . . . . . . .
512        INTEGER*1  BMTYPE
513        REAL       AS, ASDASH, D, DDASH, FSDASH, FY, HF
514        REAL       K, KDASH, KF, M, MU, X, Z
515    Com    . . . . . . . . . . . Local parameters . . . . . . . . . . . .
516        INTEGER*1  F, R
517        REAL       XOD, ZOD
518        DATA       F, R /'F', 'R'/
519
520        XOD = X / D
521        ZOD = Z / D
522
523        WRITE (6,5000)
524    Com    ------------------------------------------------------------
525    Com    | Print out K, K', x/d, z/d for rectangular or flanged section|
526    Com    ------------------------------------------------------------
527        WRITE (6,5010) K, KDASH
528        WRITE (6,5020) XOD, ZOD
529
530    Com    ------------------------------------------------------------
531    Com    | Check the type of section (rectangular or flanged).        |
```

```
532    Com    ----------------------------------------------------------------
533           IF (BMTYPE .EQ. R) GO TO 100
534           IF (BMTYPE .EQ. F) GO TO 200
535
536    Com    ----------------------------------------------------------------
537    Com    | Print out the results for rectangular section.              |
538    Com    ----------------------------------------------------------------
539    100 MU = MU * 1.0E-6
540           IF  (M .LE. MU) WRITE (6,6000) M, MU, AS
541           IF ((M .GT. MU) .AND. (DDASH / X .LE. (1.0 - FY / 800.0)))
542         &                WRITE (6,6010) M, MU, ASDASH, AS
543           IF ((M .GT. MU) .AND. (DDASH / X .GT. (1.0 - FY / 800.0)))
544         &                WRITE (6,6020) M, MU, FSDASH, ASDASH, AS
545           RETURN
546
547    Com    ----------------------------------------------------------------
548    Com    | Print out the results for flanged section.                  |
549    Com    ----------------------------------------------------------------
550    200 IF  (0.9 * X .LE. HF) WRITE (6,7000) AS
551           IF ((0.9 * X .GT. HF) .AND. (KF .LE. KDASH))
552         &                WRITE (6,7010) KF, KDASH, AS
553           IF ((0.9 * X .GT. HF) .AND. (KF .GT. KDASH) .AND.
554         &    (DDASH / X .LE. (1.0 - FY / 800.0)))
555         &                WRITE (6,7020) KF, KDASH, ASDASH, AS
556           IF ((0.9 * X .GT. HF) .AND. (KF .GT. KDASH) .AND.
557         &    (DDASH / X .GT. (1.0 - FY / 800.0)))
558         &                WRITE (6,7030) KF, KDASH, FSDASH, ASDASH, AS
559
560           RETURN
561    5000 FORMAT (///, ' ******************************', /,
562         &              ' * Output from Program BMBRSR *', /,
563         &              ' ******************************', /)
564    5010 FORMAT (' Ratio due to ultimate moment M, K  = ', 0PE10.3, /
565         &        ' Ratio due to concrete capacity, K'' = ', 0PE10.3, /)
566    5020 FORMAT (' At ultimate limit state :', /,
567         &        ' Neutral axis depth ratio, x/d = ', 0PE10.3, /,
568         &        ' Lever arm distance ratio, z/d = ', 0PE10.3, //)
569
570    6000 FORMAT (' Design ultimate moment M = ', 3PE10.3, ' kNm < Mu =',
571         &        3PE10.3, ' kNm', ///,
572         &        ' The rectangular section is to be singly',
573         &        ' reinforced :', /,
574         &        ' ---------------------------------------',
575         &        '-----------', /,
576         &        ' Tension steel area required, As = ', 4PE10.3.
577         &        ' mm**2', ///)
578    6010 FORMAT (' Design ultimate moment M = ', 3PE10.3, ' kNm > Mu =',
579         &        3PE10.3, ' kNm', ///,
580         &        ' The rectangular section is to be doubly',
581         &        ' reinforced :', /,
582         &        ' ---------------------------------------',
583         &        '-----------', /,
584         &        ' Compression steel area required, As'' = ', 4PE10.3,
585         &        ' mm**2', /,
586         &        ' Tension steel area required,     As = ', 4PE10.3,
587         &        ' mm**2', ///)
588    6020 FORMAT (' Design ultimate moment M = ', 3PE10.3, ' kNm > Mu =',
589         &        3PE10.3, ' kNm', ///,
590         &        ' The rectangular section is to be doubly',
591         &        ' reinforced :', /,
592         &        ' ---------------------------------------',
593         &        '-----------', /,
594         &        ' At ultimate limit state, the compression steel',
595         &        ' does not reach yield strength.', /,
596         &        ' Compression steel stress fs'' = ', 3PE10.3,
597         &        ' N/mm**2  < or = 0.87fy', /,
598         &        ' Compression steel area required, As'' = ', 4PE10.3,
```

```
599        &        ' mm**2', /,
600        &        ' Tension steel area required,      As  = ',  4PE10.3,
601        &        ' mm**2', ///)
602
603    7000 FORMAT (' The rectangular stress block lies wholly within',
604        &        ' flange thickness: 0.9x < or = hf', ///,
605        &        ' The flanged section is to be singly reinforced :', /,
606        &        ' ---------------------------------------------- ', /,
607        &        ' Tension steel area required, As  = ',  4PE10.3,
608        &        ' mm**2', ///)
609    7010 FORMAT (' The rectangular stress block lies partly outside',
610        &        ' flange thickness: 0.9x > hf', /,
611        &        ' And, Kf = ', 0PE10.3, ' < or = K'' = ', 0PE10.3, ///,
612        &        ' The flanged section is to be singly reinforced :', /,
613        &        ' ---------------------------------------------- ', /,
614        &        ' Tension steel area required, As = ',  4PE10.3,
615        &        ' mm**2', ///)
616    7020 FORMAT (' The rectangular stress block lies partly outside',
617        &        ' flange thickness: 0.9x > hf', /,
618        &        ' And, Kf = ', 0PE10.3, ' > K'' = ', 0PE10.3, ///,
619        &        ' The flanged section is to be doubly reinforced :', /,
620        &        ' ---------------------------------------------- ', /,
621        &        ' Compression steel area required, As'' = ', 4PE10.3,
622        &        ' mm**2', /,
623        &        ' Tension steel area required,    As  = ',  4PE10.3,
624        &        ' mm**2', ///)
625    7030 FORMAT (' The rectangular stress block lies partly outside',
626        &        ' flange thickness (0.9x > hf)', /,
627        &        ' And, Kf = ', 0PE10.3, ' > K'' = ', 0PE10.3, ///,
628        &        ' The flanged section is to be doubly reinforced :', /,
629        &        ' ---------------------------------------------- ', /,
630        &        ' At ultimate limit state, the compression steel does',
631        &        ' not reach yield strength.', /,
632        &        ' Since compression steel stress fs'' = ', 3PE10.3,
633        &        ' N/mm**2 is < or = 0.87fy', //,
634        &        ' Compression steel area required, As'' = ', 4PE10.3,
635        &        ' mm**2', /,
636        &        ' Tension steel area required,     As  = ',  4PE10.3,
637        &        ' mm**2', ///)
638        END
```

Table 12.1-1   Definition and type of variables used in program BMBRSR

Name of variables	Type of variables	*Equivalent symbol	Definitions
AS	REAL*4	A_s	Area of tension steel
ASDASH	REAL*4	A_s'	Area of compression steel
ASF	REAL*4	A_{sf}	See Eqn 4.8-6(a)
ASW	REAL*4	A_{sw}	See Eqn 4.8-6(b)
B	REAL*4	b	Width of rectangular section
BEFF	REAL*4	b_{eff}	Effective width of flanged section
BETAB	REAL*4	β_b	Moment redistribution ratio
BMTIT(I)	LOGICAL*1 array		BeaM section TITle (up to 40 characters)
BMTYPE	INTEGER*1		BeaM TYPE - rectangular (R) or flanged (F)
BW	REAL*4	b_w	Width of web of flanged section
D	REAL*4	d	Effective depth of tension reinforcement
DDASH	REAL*4	d'	Depth of compression reinforcement
F	INTEGER*1		Character to indicate Flanged section

Table 12.1–1   (continued)

FCU	REAL*4	f_{cu}	Characteristic strength of concrete
FSDASH	REAL*4	$f_s{}'$	See Eqn 4.6-11
FY	REAL*4	f_y	Characteristic strength of steel
HF	REAL*4	h_f	Flange thickness
K	REAL*4	K	$K = M/bd^2 f_{cu}$
KDASH	REAL*4	K'	See Eqn 4.7-5
KF	REAL*4	K_f	See Eqn 4.8-5
M	REAL*4	M	Design ultimate moment
MU	REAL*4	M_u	Moment capacity of concrete (Eqn 4.7-8)
MUF	REAL*4	M_{uf}	Moment capacity of flange (Eqn 4.8-4)
MUW	REAL*4	M_{uw}	Moment capacity of web (Eqn 4.8-7)
R	INTEGER*1		Character to indicate Rectangular section
X	REAL*4	x	Neutral axis depth
XOD	REAL*4	x/d	See Eqn 4.7-3
Z	REAL*4	z	Lever arm distance
ZOD	REAL*4	z/d	See Eqn 4.7-6 or Eqn 4.7-7

* See list of symbols in the "Notation" at the beginning of the book.

Block (ii) in Fig. 12.1–1 represents the Main Program itself, and corresponds to Lines 29–79 of the program listing in Fig. 12.1–2. The program begins with the program name BMBRSR on Line 29, followed by the declarations of the variables on Lines 31–34.

Lines 36–79 of Fig. 12.1–2 shows that the Main Program does the various jobs by calling on Subroutines. Thus Lines 36–40 refers to the Subroutine INIT (= INITialization) to initialize the program; similarly, Lines 42–47 refers to the Subroutine CONSTA (= CONSTAnts) and so on. All the Subroutines are listed alphabetically after the Main Program, i.e. in Block (iii) of Fig. 12.1–1. Block (iii) corresponds to lines 82–638 of the program listing in Fig. 12.1–2. The Subroutines are listed in alphabetical order so that each can be easily located by the program user. Thus, the Subroutine CONSTA is listed on Lines 82–149, followed by the Subroutine FLANGE on Lines 152–254 and so on.

The layout of each Subroutine is similar to that of the Main Program. Thus, Lines 82–92 are the Header for the Subroutine CONSTA, and Lines 94–149 are the Subroutine program statements.

The program listing in Fig. 12.1–2 also shows some other features common to all the computer programs in this chapter:

(a)   Each 'Call Subroutine' module in the Main Program includes a Header, which tells the user the reason for calling that Subroutine. See, for example, Lines 42–47.

(b)   Similarly, each coherent group of statements within a Subroutine also includes a Header which describes the action of that group of statements. See, for example, Lines 113–119.

It is appropriate here to remind the reader not to confuse Line numbers with **Label numbers**. For example, Line 116 says that if BETAB (which stands for β_b as explained in Table 12.1–1) is less than or equal to 0.9, go to Label 100. Label 100 occurs on Line 119. The Label number is for the computer's internal use; the Line number is purely for the convenience of the human user of the program.

(c) The Main Program is always initialized by a Subroutine INIT (= INITialization) which reads in the input data. See, for example, Lines 36–40.

(d) The Main Program always ends by calling the Subroutine RESULT to output the results. See, for example, Lines 71–76.

(e) All variables are explicitly declared. With reference to Line 32, for example, readers familiar with FORTRAN [4] will realize that the variables AS, ASDASH, etc. need not have been declared. However, on balance it is much better to declare all of them.

(f) All the variables in the Main Program and the Subroutines are listed alphabetically and explained in a table, which appears at the end of the program (through it does not form part of the program). See, for example, Table 12.1–1.

12.1(d)   How to run the programs

The programs are written to interact with the user. That is, they assume that the input comes from the keyboard and the output goes to the screen. Suppose the user has purchased (see Section 12.1(a) and Appendix 1) the set of two floppy disks:

(1) The first disk (Disk I) contains all the programs listed in this chapter in the form of 'source files'. A source-file name is the program name plus '.FOR'. Thus, the program BMBRSR (of Fig. 12.1–2 and Section 12.4(a)) will be stored under the source-file name BMBRSR.FOR. Similarly, any other program such as NMDDOE of Section 12.2(a) will be stored under the source-file name NMDDOE.FOR.

(2) The second disk (Disk II) contains the so-called executable files of all the source files of Disk I. Briefly, an executable file is the source file stored in machine code, ready for execution. An executable-file name is simply the program name plus '.EXE'. Thus the executable-file name for the program BMBRSR will be BMBRSR.EXE, that for the program NMDDOE will be NMDDOE.EXE and so on. That is:

Program name	Source-file name	Executable-file name
BMBRSR	BMBRSR.FOR	BMBRSR.EXE
NMDDOE	NMDDOE.FOR	NMDDOE.EXE
etc.		

Suppose the user wants to run the program BMBRSR of Fig. 12.1–2 (and Section 12.4(a)). The procedure is simple [2]:

Step 1

Place Disk II in, say, Drive A of the computer.

Step 2

Type the command BMBRSR (or the command A:BMBRSR, if the current disk drive is not Drive A).

Step 3

The computer will then prompt the user to input the data interactively.

Step 4

The output then appears on the monitor console screen (see below for printer or file output).

Figure 12.1–3 shows some typical console input and output.

Comments

(a) It is thus clear that of the set of two disks purchased from the Publishers, Disk II is the one required to run all the programs.

(b) Disk I is used only if the user wishes to output the original program listings.

Another way of running the programs

When the user becomes familiar with the sequence of data input, a more efficient way of running the program is to redirect the flow of the input and output. Suppose the file BMBRSR.DAT contains the input data for the program BMBRSR (see Fig. 12.1–4) and note that the sequence of data input is exactly as shown in Table 12.1–2. The command

BMBRSR < BMBRSR.DAT > BMBRSR.OUT

would cause the input data for the program BMBRSR to be extracted from the data file BMBRSR.DAT and the output from the program to be stored in the file named BMBRSR.OUT. Similarly, the following command would cause the output to be routed to a printer:

BMBRSR < BMBRSR.DAT > PRN

For more information on redirection of input and output on MS-DOS, see References 5–7.

Alternatively, the user may modify the program by inserting on OPEN statement before the Subroutine INIT and a CLOSE statement after the Subroutine RESULT; full details of this procedure are given in Reference 2.

12.1(e)   Program documentation

For each of the program listed in Sections 12.2–12.9, there are three main kinds of program documentation [1]:

(1) Internal documentation:
     COMMENT statements within the programs (e.g. Lines 36–38, 42–44 of Fig. 12.1–2)

(2) External documentation:
    (a) Computer flow charts (see Reference 1);
    (b) Summary of definition and type of variables used in the program (e.g. Table 12.1–1);
    (c) Summary of input data (e.g. Table 12.1–2).
(3) Captions and titles in the printed output (e.g. Fig. 12.1–3).

12.1(f)   Worked example

Example 12.1–1
Repeat Example 4.7–4 using the program BMBRSR.

SOLUTION
Figure 12.1–3 shows the console input and output.

Comment
It can be seen that the results obtained from the program BMBRSR are slightly different from those calculated in Example 4.7–4 due to rounding error.

Table 12.1–2 Summary of input data for program BMBRSR

Description of input data on each line	* Name of variable	Remarks
General data:		
1(a) Beam section title	BMTIT	Up to 40 characters
1(b) Beam type	BMTYPE	BMTYPE = R or F
Section details for flanged beam:		
2(a) Effective width	BEFF	1. Omit 2(a) to 2(e)
2(b) Flange thickness	HF	if BMTYPE = R
2(c) Web width	BW	2. All in mm
2(d) Effective depth	D	
2(e) Depth of compression steel	DDASH	
Section details for rectangular beam:		
2(f) Beam width	B	1. Omit 2(f) to 2(h)
2(g) Effective depth	D	if BMTYPE = F
2(h) Depth of compression steel	DDASH	2. All in mm
Characteristic strengths:		
3(a) Concrete	FCU	All in N/mm^2
3(b) Reinforcement	FY	
Loading information:		
4(a) Design ultimate moment	M	kNm
4(b) Moment redistribution ratio	BETAB	Dimensionless

Response to program PAUSEs:
  First PAUSE: Enter Y (or y) to obtain a summary of input data, or
                  enter N (or n) to terminate program execution.
  Second PAUSE: Enter Y (or y) to continue program execution, or
                  enter N (or n) to terminate program execution.

* See Table 12.1-1 for definition of the variables

Beam section title? (up to 40 characters) Rectangular beam section

Type of section: Enter R for rectangular section or F for flanged section? R

Beam width b in mm? (eg. 250.0) 250.0

Effective depth of tension reinforcement d in mm? (eg. 700.0) 700.0

Depth of compression reinforcement d' in mm, if necessary:
Enter value (eg. 60.0) or press return for default value of 0.15d? 60.0

Characteristic strength of concrete fcu in N/mm**2? (eg. 40.0) 40.0

Characteristic strength of steel fy in N/mm**2? (eg.460.0) 460.0

Design ultimate moment M in kNm? (eg. 900.0) 900.0

Moment redistribution ratio? (eg. 0.85) 0.85

PAUSE
Continue ? (Y/N) Y


```
************************************************
* Summary of Input Data for Program BMBRSR  *
************************************************
```

Title for beam section : Rectangular beam section

Details of the Rectangular Section :

Beam width,   b = 250.0E+00 mm
Effective depth,   d = 700.0E+00 mm
Depth of comp. steel, d' = 600.0E-01 mm

Characteristic Strengths :

Concrete,   fcu =   40.0E+00 N/mm**2
Reinforcement, fy = 460.0E+00 N/mm**2

Loading Information :

Design ultimate moment,   M = 900.0E+00 kNm
Moment redistribution ratio =   0.85E+00

 PAUSE
 Continue ? (Y/N) Y


```
******************************
* Output from Program BMBRSR *
******************************
```

Ratio due to ultimate moment M, K = 0.184E+00
Ratio due to concrete capacity, K' = 0.144E+00

At ultimate limit state :
Neutral axis depth ratio, x/d = 0.443E+00
Lever arm distance ratio, z/d = 0.801E+00

Design ultimate moment M = 900.0E+00 kNm > Mu = 703.4E+00 kNm

```
The rectangular section is to be doubly reinforced :
---------------------------------------------------
Compression steel area required, As' =  7676.E-01 mm**2
Tension steel area required,      As =  3903.E+00 mm**2


STOP
```

Fig. 12.1–3   Console input and output for Example 12.1–1

```
Rectangular beam section
R
250.0
700.0
60.0
40.0
460.0
900.0
0.85
YY
```

Fig. 12.1–4   Content of data file BMBRSR.DAT

12.2   Computer program for Chapter 2

12.2(a)   Program NMDDOE (= Normal Mix Design; Department Of the Environment)

Figure 12.2–1 shows the Header of the program NMDDOE, which designs normal concrete mixes in accordance with the method of the Department of the Environment, as described in Section 2.7(b). Each concrete mix is designed on the assumption that the density for uncrushed aggregate is 2600 kg/m^3 and that for crushed aggregate is 2700 kg/m^3 (see Step 4 of Section 2.7(b)).

The input data for the program NMDDOE are: the target mean strength at the specified age, the level of workability, the details about the cement, the coarse aggregate and the fine aggregate, and the allowable limits on the w/c ratio and the cement content. The output data from the program NMDDOE are: the intermediate results of the mix design (e.g. the required w/c ratio, etc) and the final mix properties in terms of weights of materials per cubic metre of fully compacted fresh concrete.

Comment
The complete listing of the above program (with commentary), together with the floppy disks, are obtainable from the Publishers. See Section 12.1(a) and Appendix 1.

```
 1    Com  *****************************************************************
 2    Com *                                                               *
 3    Com *   Program Unit Name : NMDDOE (= Normal Mix Design; the        *
 4    Com *                                Department Of the Environment)  *
 5    Com *                                                               *
 6    Com *   Purpose : Design of normal concrete mix using the method of *
 7    Com *             the Department of the Environment - the DoE mix    *
 8    Com *             design method.                                     *
 9    Com *                                                               *
10    Com *   Reference : This program refers to Sections 2.7(b) and 2.9   *
11    Com *               of Chapter 2 of Kong and Evans : Reinforced and  *
12    Com *               Prestressed Concrete, Van Nostrand Reinhold,     *
13    Com *               3rd Edition, 1987.                               *
14    Com *                                                               *
15    Com *    *   *   *   *   *   *   *   *   *   *   *   *   *   *   *    *
16    Com *                                                               *
17    Com *   Authors : Dr H. H. A. Wong in collaboration with            *
18    Com *             Professors F. K. Kong and R. H. Evans              *
19    Com *                                                               *
20    Com *   Programming Language : PRO FORTRAN                          *
21    Com *                                                               *
22    Com *   Operating Systems : IBM's PC-DOS and Microsoft's MS-DOS.    *
23    Com *                                                               *
24    Com *   Version : KE3-2.9-8S22P6                                    *
25    Com *                                                               *
26    Com  *****************************************************************
```

Fig. 12.2–1   Header of program NMDDOE

```
 1    Com  *****************************************************************
 2    Com *                                                               *
 3    Com *   Program Unit Name : SSCAXL (= Short Square Column;          *
 4    Com *                                AXially Loaded)                 *
 5    Com *                                                               *
 6    Com *   Purpose : Design of axially loaded short square columns.    *
 7    Com *                                                               *
 8    Com *   Reference : This program refers to Sections 3.4 and 3.7 of  *
 9    Com *               Chapter 3 of Kong and Evans : Reinforced and    *
10    Com *               Prestressed Concrete, Van Nostrand Reinhold,    *
11    Com *               3rd Edition, 1987.                              *
12    Com *                                                               *
13    Com *    *   *   *   *   *   *   *   *   *   *   *   *   *   *   *    *
14    Com *                                                               *
15    Com *   Authors : Dr H. H. A. Wong in collaboration with            *
16    Com *             Professors F. K. Kong and R. H. Evans             *
17    Com *                                                               *
18    Com *   Programming Language : PRO FORTRAN                          *
19    Com *                                                               *
20    Com *   Operating Systems : IBM's PC-DOS and Microsoft's MS-DOS.    *
21    Com *                                                               *
22    Com *   Version : KE3-3.7-8D10C6                                    *
23    Com *                                                               *
24    Com  *****************************************************************
```

Fig. 12.3–1   Header of program SSCAXL

12.3   Computer program for Chapter 3

12.3(a)   Program SSCAXL (= Short Square Column; AXially Loaded)

Figure 12.3–1 shows the Header of the program SSCAXL, which designs short axially loaded square columns, as described in Section 3.4. The input data for the program SSCAXL are: the design ultimate axial load, the characteristic strengths of the materials and the fire resistance requirement. The program first determines the minimum dimension of the column (see Table 3.5–1), before it outputs a list of possible dimensions and the corresponding amount of longitudinal reinforcement required for the column.

Comment
The complete listing of the above program (with commentary), together with the floppy disks, are obtainable from the Publishers. See Section 12.1(a) and Appendix 1.

12.4   Computer programs for Chapter 4

12.4(a)   Program BMBRSR (= BeaM; Bending Reinforcement; Simplified Rectangular)

Figure 12.4–1 shows the Header of the program BMBRSR, which determines the main steel required for a rectangular or flanged beam section using BS 8110's simplified rectangular stress block, as described in Sections 4.7 and 4.8. Figure 12.1–2 and 12.1–3 respectively show the listing and typical console input and output of the program BMBRSR. The definitions of the variable used in the program are summarized in Table 4.12–1. Table 4.12–2 also gives a summary of the required input data.

Comment
The complete listing of the above program (with commentary), together with the floppy disks, are obtainable from the Publishers. See Section 12.1(a) and Appendix 1.

12.4(b)   Program BMBRPR (= BeaM; Bending Reinforcement; Parabolic–Rectangular)

Figure 12.4–2 shows the Header of the program BMBRPR, which determines the main tension steel for a rectangular beam section using BS 8110's parabolic-rectangular stress block as shown in Fig. 4.4–3. The program uses an iterative procedure described in Reference 1.

The input data for the program BMBRPR are: the section details, the desired amount of compression steel, the characteristic strengths of the

```
1     Com ******************************************************************
2     Com *                                                                *
3     Com *  Program Unit Name : BMBRSR (= BeaM; Bending Reinforcement;    *
4     Com *                             Simplified Rectangular)            *
5     Com *                                                                *
6     Com *  Purpose : Design of bending reinforcement for a rectangular   *
7     Com *            or flanged beam section in accordance with          *
8     Com *            BS 8110 : 1985, using simplified rectangular         *
9     Com *            stress block.                                       *
10    Com *                                                                *
11    Com *  Reference : This program refers to Sections 4.7, 4.8 and      *
12    Com *              4.12 of Chapter 4 of Kong and Evans : Reinforced   *
13    Com *              and Prestressed Concrete, Van Nostrand Reinhold,   *
14    Com *              3rd Edition, 1987.                                 *
15    Com *                                                                *
16    Com *   *   *   *   *   *   *   *   *   *   *   *   *   *   *   *      *
17    Com *                                                                *
18    Com *  Authors : Dr H. H. A. Wong in collaboration with              *
19    Com *            Professors F. K. Kong and R. H. Evans               *
20    Com *                                                                *
21    Com *  Programming Language : PRO FORTRAN                            *
22    Com *                                                                *
23    Com *  Operating Systems : IBM's PC-DOS and Microsoft's MS-DOS.      *
24    Com *                                                                *
25    Com *  Version : KE3-4.12-8S23P6                                     *
26    Com *                                                                *
27    Com ******************************************************************
```

Fig. 12.4–1   Header of program BMBRSR

```
1     Com ******************************************************************
2     Com *                                                                *
3     Com *  Program Unit Name : BMBRPR (= BeaM; Bending Reinforcement;    *
4     Com *                             Parabolic-Rectangular)             *
5     Com *                                                                *
6     Com *  Purpose : Design of bending reinforcement for a rectangular   *
7     Com *            beam section in accordance with BS 8110 : 1985,     *
8     Com *            using parabolic-rectangular stress block.           *
9     Com *                                                                *
10    Com *  Reference : This program refers to Sections 4.5 and 4.12 of   *
11    Com *              Chapter 4 of Kong and Evans : Reinforced and      *
12    Com *              Prestressed Concrete, Van Nostrand Reinhold,      *
13    Com *              3rd Edition, 1987.                                *
14    Com *                                                                *
15    Com *   *   *   *   *   *   *   *   *   *   *   *   *   *   *   *      *
16    Com *                                                                *
17    Com *  Authors : Dr H. H. A. Wong in collaboration with              *
18    Com *            Professors F. K. Kong and R. H. Evans               *
19    Com *                                                                *
20    Com *  Programming Language : PRO FORTRAN                            *
21    Com *                                                                *
22    Com *  Operating Systems : IBM's PC-DOS and Microsoft's MS-DOS.      *
23    Com *                                                                *
24    Com *  Version : KE3-4.12-8N03V6                                     *
25    Com *                                                                *
26    Com ******************************************************************
```

Fig. 12.4–2   Header of program BMBRPR

materials and the loading information. The program first checks that the percentage of compression steel specified is within BS 8110's allowable limits for rectangular beams (see Section 4.10), before it outputs the areas of reinforcement, and the ratios of x/d and z/d at the ultimate limit state. If the percentage of tension steel required, or the x/d value or z/d value, does not satisfy BS 8110's requirements, warning messages will be printed out.

Comment
The complete listing of the above program (with commentary), together with the floppy disks, are obtainable from the Publishers. See Section 12.1(a) and Appendix 1.

12.5 Computer programs for Chapter 5

12.5(a) Program BDFLCK (= Beam DeFLection ChecK)

Figure 12.5–1 shows the Header of the program BDFLCK, which checks whether the span/depth ratio of a rectangular or flanged beam section is within BS 8110's allowable limit, as described in Section 5.3. Instead of using the data in Tables 5.3–2 and 5.3–3, the program calculates the modification factors for the tension reinforcement and the compression reinforcement from the equations given in BS 8110: Clauses 3.4.6.5 and 3.4.6.6 (see also eqns 5.3–1(a) and (b)).

```
 1   Com *********************************************************************
 2   Com *                                                                   *
 3   Com *  Program Unit Name : BDFLCK (= Beam DeFLection CHecK)             *
 4   Com *                                                                   *
 5   Com *  Purpose : Check whether the ratio of the span to the            *
 6   Com *            effective depth of a rectangular or flanged beam      *
 7   Com *            section is within BS 8110's allowable limit or not.*
 8   Com *                                                                   *
 9   Com *  Reference : This program refers to Sections 5.3 and 5.8 of       *
10   Com *              Chapter 5 of Kong and Evans : Reinforced and         *
11   Com *              Prestressed Concrete, Van Nostrand Reinhold,         *
12   Com *              3rd Edition, 1987.                                   *
13   Com *                                                                   *
14   Com *    *   *   *   *   *   *   *   *   *   *   *   *   *   *   *        *
15   Com *                                                                   *
16   Com *  Authors : Dr H. A. Wong in collaboration with                    *
17   Com *            Professors F. K. Kong and R. H. Evans                  *
18   Com *                                                                   *
19   Com *  Programming Language : PRO FORTRAN                               *
20   Com *                                                                   *
21   Com *  Operating Systems : IBM's PC-DOS and Microsoft's MS-DOS.         *
22   Com *                                                                   *
23   Com *  Version : KE3-5.8-8S22P6                                         *
24   Com *                                                                   *
25   Com *********************************************************************
```

Fig. 12.5–1 Header of program BDFLCK

The input data for the program BDFLCK are: the type of section, the support condition, the section details, the characteristic strength of the reinforcement and the loading information. The program output data are: the basic span/depth ratio, the modification factors, the actual and allowable span/depth ratios.

Comment
The complete listing of the above program (with commentary), together with the floppy disks, are obtainable from the Publishers. See Section 12.1(a) and Appendix 1.

12.5(b)   Program BCRKCO (= Beam CRacK COtrol)

Figure 12.5–2 shows the Header of the program BCRKCO, which calculates the maximum tension bar spacings a_b and a_c in Fig. 5.4–1, as described in Section 5.4. If the overall depth of the section exceeds 750 mm, the size of side bars required is also calculated. The program calculates the allowable tension bar spacings from eqn (5.4–2) instead of using the data in Table 5.4–1.

The input data for the program BCRKCO are: the section details, the

```
 1    Com  **********************************************************************
 2    Com  *                                                                    *
 3    Com  *  Program Unit Name : BCRKCO (= Beam CRacK COntrol)                  *
 4    Com  *                                                                    *
 5    Com  *  Purposes : To calculate the following bar spacing limits for      *
 6    Com  *             a rectangular or flanged beam section, below           *
 7    Com  *             which the BS 8110's 0.3 mm crack-width is              *
 8    Com  *             complied with.                                         *
 9    Com  *                1. Maximum permissible clear spacing between         *
10    Com  *                   tension bars;                                    *
11    Com  *                2. Maximum permissible clear spacing between         *
12    Com  *                   corner and the nearest bar.                       *
13    Com  *             The program also calculates the size of side           *
14    Com  *             bars when the beam depth h exceeds 750 mm.             *
15    Com  *                                                                    *
16    Com  *  Reference : This program refers to Sections 5.4 and 5.8 of        *
17    Com  *              Chapter 5 of Kong and Evans : Reinforced and          *
18    Com  *              Prestressed Concrete, Van Nostrand Reinhold,          *
19    Com  *              3rd Edition, 1987.                                    *
20    Com  *                                                                    *
21    Com  *   *   *   *   *   *   *   *   *   *   *   *   *   *   *   *   *      *
22    Com  *                                                                    *
23    Com  *  Authors : Dr H. H. A. Wong in collaboration with                  *
24    Com  *            Professors F. K. Kong and R. H. Evans                   *
25    Com  *                                                                    *
26    Com  *  Programming Language : PRO FORTRAN                                 *
27    Com  *                                                                    *
28    Com  *  Operating Systems : IBM's PC-DOS and Microsoft's MS-DOS.          *
29    Com  *                                                                    *
30    Com  *  Version : KE3-5.8-8S23P6                                          *
31    Com  *                                                                    *
32    Com  **********************************************************************
```

Fig. 12.5–2   Header of program BCRKCO

characteristic strength of the reinforcement and the moment redistribution ratio. The output data from the program BCRKCO are: the allowable bar spacings, and the size and spacing of the side bars, if any.

Comment
The complete listing of the above program (with commentary), together with the floppy disks, are obtainable from the Publishers. See Section 12.1(a) and Appendix 1.

12.6   Computer programs for Chapter 6

12.6(a)   Program BSHEAR (= Beam SHEAR)

Figure 12.6–1 shows the Header of the program BSHEAR, which designs the shear reinforcement for a rectangular or flanged beam section, using the procedures described in Section 6.4. The program BSHEAR allows the user to specify:

(1)   the type of shear reinforcement (i.e. either a combination of bent-up bars and links or links only);
(2)   the size or spacing of links.

With reference to (2), if the user specifies the link size, then the program calculates and outputs the link spacing; however, if the user specifies the

```
 1     Com  *********************************************************************
 2     Com  *                                                                   *
 3     Com  *  Program Unit Name : BSHEAR (= Beam SHEAR)                         *
 4     Com  *                                                                   *
 5     Com  *  Purposes : Design of shear reinforcement for a rectangular        *
 6     Com  *             or flanged beam section in accordance with             *
 7     Com  *             BS 8110 : 1985.  The program allows the user to         *
 8     Com  *             choose the size or spacing of the shear                 *
 9     Com  *             reinforcement required.                                *
10     Com  *                                                                   *
11     Com  *  Reference : This program refers to Sections 6.4 and 6.13 of        *
12     Com  *              Chapter 6 of Kong and Evans : Reinforced and           *
13     Com  *              Prestressed Concrete, Van Nostrand Reinhold,           *
14     Com  *              3rd Edition, 1987.                                    *
15     Com  *                                                                   *
16     Com  *    *   *   *   *   *   *   *   *   *   *   *   *   *   *   *        *
17     Com  *                                                                   *
18     Com  *  Authors : Dr H. H. A. Wong in collaboration with                  *
19     Com  *            Professors F. K. Kong and R. H. Evans                   *
20     Com  *                                                                   *
21     Com  *  Programming Language : PRO FORTRAN                                *
22     Com  *                                                                   *
23     Com  *  Operating Systems : IBM's PC-DOS and Microsoft's MS-DOS.          *
24     Com  *                                                                   *
25     Com  *  Version : KE3-6.13-8027T6                                         *
26     Com  *                                                                   *
27     Com  *********************************************************************
```

Fig. 12.6–1   Header of program BSHEAR

link spacing, then the program will output an appropriate link size up to size 16 mm.

The input data for the program BSHEAR are: the type of beam section, the section details, the characteristic strengths of the materials and the design ultimate shear force at the section considered. The output data from the program BSHEAR are: the design shear stress v, the concrete design shear stress v_c, the ratio of link area to link spacing A_{sv}/s_v, the size and spacing of the links required and the details of the bent-up bars, if they are used.

Comment

The complete listing of the above program (with commentary), together with the floppy disks, are obtainable from the Publishers. See Section 12.1(a) and Appendix 1.

12.6(b)   Program BSHTOR (= Beam SHear and TORsion)

Figure 12.6–2 shows the Header of the program BSHTOR, which designs the shear and torsion reinforcement for a solid rectangular beam section, as described in Sections 6.4, 6.10(a) and 6.11. The program is of course applicable to the web of a flanged beam. For simplicity, the program does not consider a combination of bent-up bars and links as shear reinforcement (compare with (1) of Section 12.6(a)), but allows the user to choose the size or spacing of links as in (2) of Section 12.6(a).

```
 1    Com  *******************************************************************
 2    Com  *                                                                 *
 3    Com  *  Program Unit Name : BSHTOR (= Beam SHear and TORsion)          *
 4    Com  *                                                                 *
 5    Com  *  Purpose : Design of shear and torsion reinforcement for a      *
 6    Com  *            solid rectangular beam section in accordance with    *
 7    Com  *            BS 8110 : 1985, Part 2, Clause 2.4. The section      *
 8    Com  *            may be the web of a flanged beam.                    *
 9    Com  *                                                                 *
10    Com  *  Reference : This program refers to Sections 6.10(a) and 6.13   *
11    Com  *              of Chapter 6 of Kong and Evans : Reinforced and    *
12    Com  *              Prestressed Concrete, Van Nostrand Reinhold,       *
13    Com  *              3rd Edition, 1987.                                 *
14    Com  *                                                                 *
15    Com  *    *    *    *    *    *    *    *    *    *    *    *    *    *  *
16    Com  *                                                                 *
17    Com  *  Authors : Dr H. H. A. Wong in collaboration with              *
18    Com  *            Professors F. K. Kong and R. H. Evans                *
19    Com  *                                                                 *
20    Com  *  Programming Language : PRO FORTRAN                            *
21    Com  *                                                                 *
22    Com  *  Operating Systems : IBM's PC-DOS and Microsoft's MS-DOS.       *
23    Com  *                                                                 *
24    Com  *  Version : KE3-6.13-8010T6                                     *
25    Com  *                                                                 *
26    Com  *******************************************************************
```

Fig. 12.6–2   Header of program BSHTOR

The input data for the program BSHTOR are: the type of beam section, the section details, the characteristic strengths of the materials, the dimensions of the rectangular links, and the design ultimate shear force and torsional moment. The program output data are: the design stresses due to shear and torsion, the ratios of link area to link spacing due to shear and torsion, the size and spacing of the links required, and the details of the longitudinal torsion reinforcement required.

Comment
The complete listing of the above program (with commentary), together with the floppy disks, are obtainable from the Publishers. See Section 12.1(a) and Appendix 1.

12.7 Computer programs for Chapter 7

12.7(a) Program RCIDSR (= Rectangular Column; Interaction Diagram; Simplified Rectangular)

Figure 12.7–1 shows the header of the program RCIDSR, which calculates the coordinates on the column interaction curve for a rectangular column section using BS 8110's simplified rectangular stress block, as described in Section 7.1. The program checks that the total percentage of the

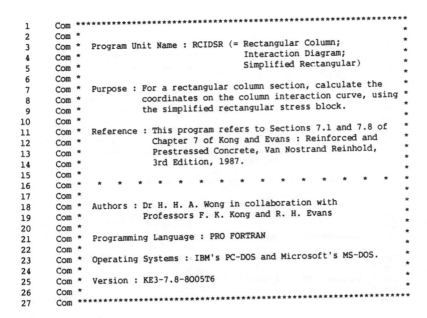

```
 1    Com  ********************************************************************
 2    Com  *                                                                  *
 3    Com  *  Program Unit Name : RCIDSR (= Rectangular Column;               *
 4    Com  *                                 Interaction Diagram;             *
 5    Com  *                                 Simplified Rectangular)          *
 6    Com  *                                                                  *
 7    Com  *  Purpose : For a rectangular column section, calculate the       *
 8    Com  *            coordinates on the column interaction curve, using    *
 9    Com  *            the simplified rectangular stress block.              *
10    Com  *                                                                  *
11    Com  *  Reference : This program refers to Sections 7.1 and 7.8 of      *
12    Com  *              Chapter 7 of Kong and Evans : Reinforced and        *
13    Com  *              Prestressed Concrete, Van Nostrand Reinhold,        *
14    Com  *              3rd Edition, 1987.                                  *
15    Com  *                                                                  *
16    Com  *    *   *   *   *   *   *   *   *   *   *   *   *   *   *   *      *
17    Com  *                                                                  *
18    Com  *  Authors : Dr H. H. A. Wong in collaboration with               *
19    Com  *            Professors F. K. Kong and R. H. Evans                 *
20    Com  *                                                                  *
21    Com  *  Programming Language : PRO FORTRAN                              *
22    Com  *                                                                  *
23    Com  *  Operating Systems : IBM's PC-DOS and Microsoft's MS-DOS.        *
24    Com  *                                                                  *
25    Com  *  Version : KE3-7.8-8005T6                                        *
26    Com  *                                                                  *
27    Com  ********************************************************************
```

Fig. 12.7–1 Header of program RCIDSR

longitudinal reinforcement of the section is not less than BS 8110's minimum limit of 0.4% of *bh* (see Section 3.5). When the total percentage of the reinforcement exceeds 6% of *bh* (see Section 3.5), a warning message will be printed out.

The input data for the program RCIDSR are: the section details, the material properties, the increment of x/h and the maximum x/h to be used. The program output data are: the steel stresses, α $(= N/f_{cu}bh)$ and β $(= M/f_{cu}bh^2)$ for various x/h values. The results are printed out in tabulated form.

Comment
The complete listing of the above program (with commentary), together with the floppy disks, are obtainable from the Publishers. See Section 12.1(a) and Appendix 1.

12.7(b) Program RCIDPR (= Rectangular Column; Interaction Diagram; Parabolic–Rectangular)

Figure 12.7–2 shows the Header of the program RCIDPR, which calculates the coordinates on the column interaction curve for a rectangular column section using BS 8110's parabolic–rectangular stress block, as described in Reference 1.

The input and output for the program RCIDPR are identical to those of the program RCIDSR (see Section 12.7(a)).

```
 1     Com ***********************************************************************
 2     Com *                                                                    *
 3     Com *   Program Unit Name : RCIDPR (= Rectangular Column;                *
 4     Com *                                Interaction Diagram;                *
 5     Com *                                Parabolic-Rectangular)              *
 6     Com *                                                                    *
 7     Com *   Purpose : For a rectangular column section, calculate the        *
 8     Com *             coordinates on the column interaction curve, using     *
 9     Com *             the parabolic-rectangular stress block.                *
10     Com *                                                                    *
11     Com *   Reference : This program refers to Sections 7.1 and 7.8 of       *
12     Com *               Chapter 7 of Kong and Evans : Reinforced and         *
13     Com *               Prestressed Concrete, Van Nostrand Reinhold,         *
14     Com *               3rd Edition, 1987.                                   *
15     Com *                                                                    *
16     Com *    *    *    *    *    *    *    *    *    *    *    *    *    *     *
17     Com *                                                                    *
18     Com *   Authors : Dr H. H. A. Wong in collaboration with                 *
19     Com *             Professors F. K. Kong and R. H. Evans                  *
20     Com *                                                                    *
21     Com *   Programming Language : PRO FORTRAN                               *
22     Com *                                                                    *
23     Com *   Operating Systems : IBM's PC-DOS and Microsoft's MS-DOS.         *
24     Com *                                                                    *
25     Com *   Version : KE3-7.8-8006T6                                         *
26     Com *                                                                    *
27     Com ***********************************************************************
```

Fig. 12.7–2   Header of program RCIDPR

Comment
The complete listing of the above program (with commentary), together with the floppy disks, are obtainable from the Publishers. See Section 12.1(a) and Appendix 1.

12.7(c) Program CTDMUB (= Column; Total Design Moment; Uniaxial bending; Biaxial bending)

Figure 12.7–3 shows the Header of the program CTDMUB, whch determines the total design moment(s) for slender columns under uniaxial bending or biaxial bending, as described in Section 7.5.

The input data for the program CTDMUB are: the type of loading (uniaxial or biaxial bending), the column details, the loading information and the reduction factor K. The output data from the program CTDMUB are: the respective height/depth ratio, the additional moment M_{add}, the initial moment M_i, the minimum design eccentricity e_{min}, the four possible total design moments defined by eqns (7.5–1) to (7.5–4) and the total design moment M_t.

Comment
The complete listing of the above program (with commentary), together with the floppy disks, are obtainable from the Publishers. See Section 12.1(a) and Appendix 1.

```
1    Com  ***********************************************************
2    Com  *                                                         *
3    Com  *  Program Unit Name : CTDMUB (= Column; Total Design Moment;  *
4    Com  *                              Uniaxial bending; Biaxial   *
5    Com  *                              bending)                    *
6    Com  *                                                          *
7    Com  *  Purpose : The program determines the total design moment(s)  *
8    Com  *            for slender columns under uniaxial bending or  *
9    Com  *            biaxial bending.                              *
10   Com  *                                                          *
11   Com  *  Reference : This program refers to Sections 7.5 and 7.8 of  *
12   Com  *              Chapter 7 of Kong and Evans : Reinforced and  *
13   Com  *              Prestressed Concrete, Van Nostrand Reinhold,  *
14   Com  *              3rd Edition, 1987.                          *
15   Com  *                                                          *
16   Com  *   *   *   *   *   *   *   *   *   *   *   *   *   *   *    *
17   Com  *                                                          *
18   Com  *  Authors : Dr H. H. A. Wong in collaboration with        *
19   Com  *            Professors F. K. Kong and R. H. Evans         *
20   Com  *                                                          *
21   Com  *  Programming Language : PRO FORTRAN                      *
22   Com  *                                                          *
23   Com  *  Operating Systems : IBM's PC-DOS and Microsoft's MS-DOS. *
24   Com  *                                                          *
25   Com  *  Version : KE3-7.8-8020T6                                *
26   Com  *                                                          *
27   Com  ***********************************************************
```

Fig. 12.7–3   Header of program CTDMUB

12.7(d)   Program SRCRSR (= Symmetrically reinforced **R**ectangular **C**olumn; **R**einforcement; Simplified **R**ectangular)

Figure 12.7–4 shows the Header of the program SRCRSR, which determines the amount of reinforcement for a symmetrically reinforced rectangular column section under uniaxial bending or biaxial bending using BS 8110's simplified rectangular stress block, as described in Reference 8. Full details about the program implementation are given in Reference 1.

The input data for the program SRCRSR are: the type of loading (uniaxial or biaxial bending), the section details, the characteristic strengths of the materials and the loading information. For slender columns, the total design moment(s) may be determined by using the program CTDMUB (see Section 12.7(c)). The program output data are: the respective height/depth ratios, the x/h value at the ultimate limit state, the reduction factor K (if the column considered is slender), and the area and percentage of the longitudinal reinforcement required. If the percentage of the reinforcement required is outside BS 8110's minimum and maximum limits (see Section 3.5), warning messages will be printed out.

Comment

The complete listing of the above program (with commentary), together with the floppy disks, are obtainable from the Publishers. See Section 12.1(a) and Appendix 1.

```
 1    Com  ******************************************************************
 2    Com  *                                                                *
 3    Com  *   Program Unit Name : SRCRSR (= Symmetrically reinforced        *
 4    Com  *                                Rectangular Column;              *
 5    Com  *                                Reinforcement; Simplified        *
 6    Com  *                                Rectangular)                     *
 7    Com  *                                                                *
 8    Com  *   Purpose : The program determines the amount of reinforcement *
 9    Com  *             required for a symmetrically reinforced            *
10    Com  *             rectangular column section under uniaxial bending  *
11    Com  *             or biaxial bending, using BS 8110's simplified     *
12    Com  *             rectangular stress block.                          *
13    Com  *                                                                *
14    Com  *   Reference : This program refers to Section 7.8 of Chapter 7  *
15    Com  *               of Kong and Evans : Reinforced and Prestressed   *
16    Com  *               Concrete, Van Nostrand Reinhold, 3rd Edition,    *
17    Com  *               1987.                                            *
18    Com  *                                                                *
19    Com  *     *    *    *    *    *    *    *    *    *    *    *    *    *
20    Com  *                                                                *
21    Com  *   Authors : Dr H. H. A. Wong in collaboration with            *
22    Com  *             Professors F. K. Kong and R. H. Evans             *
23    Com  *                                                                *
24    Com  *   Programming Language : PRO FORTRAN                           *
25    Com  *                                                                *
26    Com  *   Operating Systems : IBM's PC-DOS and Microsoft's MS-DOS.     *
27    Com  *                                                                *
28    Com  *   Version : KE3-7.8-8023T6                                     *
29    Com  *                                                                *
30    Com  ******************************************************************
```

Fig. 12.7–4   Header of program SRCRSR

12.7(e)   Program SRCRPR (= Symmetrically reinforced Rectangular Column; Reinforcement; Parabolic–Rectangular)

Figure 12.7–5 shows the Header of the program SRCRPR, which determines the amount of reinforcement for a symmetrically reinforced rectangular column section under uniaxial bending or biaxial bending using BS 8110's parabolic–rectangular stress block, as described in Reference 8. Full details about the program implementation are given in Reference 1.

The input and output of the program SRCRPR are identical to those of the program SRCRSR (see Section 12.7(d)).

Comment
The complete listing of the above program (with commentary), together with the floppy disks, are obtainable from the Publishers. See Section 12.1(a) and Appendix 1.

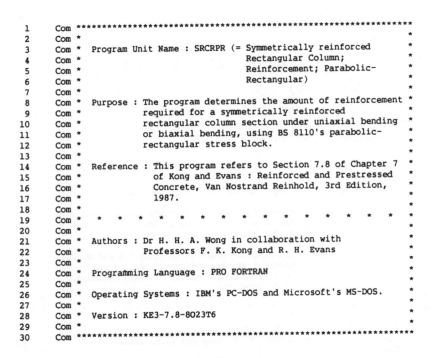

```
 1    Com  *************************************************************
 2    Com  *                                                           *
 3    Com  *  Program Unit Name : SRCRPR (= Symmetrically reinforced    *
 4    Com  *                              Rectangular Column;           *
 5    Com  *                              Reinforcement; Parabolic-     *
 6    Com  *                              Rectangular)                  *
 7    Com  *                                                           *
 8    Com  *  Purpose : The program determines the amount of reinforcement *
 9    Com  *            required for a symmetrically reinforced         *
10    Com  *            rectangular column section under uniaxial bending *
11    Com  *            or biaxial bending, using BS 8110's parabolic-  *
12    Com  *            rectangular stress block.                      *
13    Com  *                                                           *
14    Com  *  Reference : This program refers to Section 7.8 of Chapter 7 *
15    Com  *              of Kong and Evans : Reinforced and Prestressed *
16    Com  *              Concrete, Van Nostrand Reinhold, 3rd Edition, *
17    Com  *              1987.                                        *
18    Com  *                                                           *
19    Com  *    *   *   *   *   *   *   *   *   *   *   *   *   *        *
20    Com  *                                                           *
21    Com  *  Authors : Dr H. H. A. Wong in collaboration with          *
22    Com  *            Professors F. K. Kong and R. H. Evans           *
23    Com  *                                                           *
24    Com  *  Programming Language : PRO FORTRAN                        *
25    Com  *                                                           *
26    Com  *  Operating Systems : IBM's PC-DOS and Microsoft's MS-DOS.  *
27    Com  *                                                           *
28    Com  *  Version : KE3-7.8-8023T6                                  *
29    Com  *                                                           *
30    Com  *************************************************************
```

Fig. 12.7–5   Header of program SRCRPR

12.8   Computer programs for Chapter 8

12.8(a)   Program SDFLCK (= Slab DeFLection ChecK)

Figure 12.8–1 shows the Header of the program SDFLCK, which checks whether the span to depth ratio of a slab is within BS 8110's allowable limit or not, as described in Section 8.8. As for program BDFLCK in Section 12.5(a), the modification factor for the tension reinforcement is calculated from eqns (5.3–1(a), (b)).

The input data for the program SDFLCK are: the support condition, the section details, the characteristic strength of reinforcement and the loading information. The program output data are: the basic span/depth ratio, the modification factor for the tension reinforcement and the actual and allowable span/depth ratios.

Comment
The complete listing of the above program (with commentary), together with the floppy disks, are obtainable from the Publishers. See Section 12.1(a) and Appendix 1.

12.8(b)   Program SCRKCO (= Slab CRacK COntrol)

Figure 12.8–2 shows the Header of the program SCRKCO, which determines the maximum permissible spacing between tension bars of a slab, as described in Section 8.8.

```
 1    Com ********************************************************************
 2    Com *                                                                 *
 3    Com *  Program Unit Name : SDFLCK (= Slab DeFLection ChecK)           *
 4    Com *                                                                 *
 5    Com *  Purpose : This program checks whether the ratio of the span    *
 6    Com *            to the effective depth of a slab is within the       *
 7    Com *            BS 8110's allowable limit or not.                     *
 8    Com *                                                                 *
 9    Com *  Reference : This program refers to Sections 8.8 and 8.10 of    *
10    Com *             Chapter 8 of Kong and Evans : Reinforced and        *
11    Com *             Prestressed Concrete, Van Nostrand Reinhold,        *
12    Com *             3rd Edition, 1987.                                  *
13    Com *                                                                 *
14    Com *   *   *   *   *   *   *   *   *   *   *   *   *   *   *   *   *   *
15    Com *                                                                 *
16    Com *  Authors : Dr H. H. A. Wong in collaboration with              *
17    Com *            Professors F. K. Kong and R. H. Evans               *
18    Com *                                                                 *
19    Com *  Programming Language : PRO FORTRAN                            *
20    Com *                                                                 *
21    Com *  Operating Systems : IBM's PC-DOS and Microsoft's MS-DOS.      *
22    Com *                                                                 *
23    Com *  Version : KE3-8.10-8S29P6                                     *
24    Com *                                                                 *
25    Com ********************************************************************
```

Fig. 12.8–1   Header of program SDFLCK

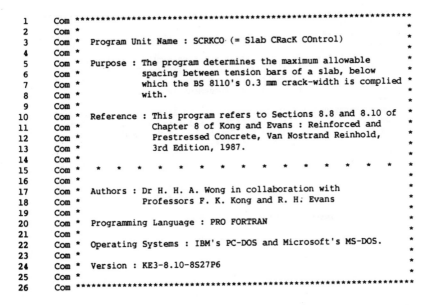

```
 1    Com  ***********************************************************
 2    Com  *                                                         *
 3    Com  *  Program Unit Name : SCRKCO (= Slab CRacK COntrol)       *
 4    Com  *                                                         *
 5    Com  *  Purpose : The program determines the maximum allowable  *
 6    Com  *            spacing between tension bars of a slab, below *
 7    Com  *            which the BS 8110's 0.3 mm crack-width is complied *
 8    Com  *            with.                                         *
 9    Com  *                                                         *
10    Com  *  Reference : This program refers to Sections 8.8 and 8.10 of *
11    Com  *              Chapter 8 of Kong and Evans : Reinforced and *
12    Com  *              Prestressed Concrete, Van Nostrand Reinhold, *
13    Com  *              3rd Edition, 1987.                          *
14    Com  *                                                         *
15    Com  *   *   *   *   *   *   *   *   *   *   *   *   *   *   *    *
16    Com  *                                                         *
17    Com  *  Authors : Dr H. H. A. Wong in collaboration with        *
18    Com  *            Professors F. K. Kong and R. H. Evans         *
19    Com  *                                                         *
20    Com  *  Programming Language : PRO FORTRAN                       *
21    Com  *                                                         *
22    Com  *  Operating Systems : IBM's PC-DOS and Microsoft's MS-DOS. *
23    Com  *                                                         *
24    Com  *  Version : KE3-8.10-8S27P6                                *
25    Com  *                                                         *
26    Com  ***********************************************************
```

Fig. 12.8–2   Header of program SCRKCO

The input data for the program SCRKCO are: the section details, the characteristic strength of the reinforcement and the moment redistribution ratio. The program outputs the allowable tension bar spacing.

Comment
The complete listing of the above program (with commentary), together with the floppy disks, are obtainable from the Publishers. See Section 12.1(a) and Appendix 1.

12.8(c)   Program SSHEAR (= Slab SHEAR)

Figure 12.8–3 shows the Header of the program SSHEAR, which designs the shear reinforcement for a slab, as described in Section 8.7. The program SSHEAR allows the user to specify:

(1)   the type of shear reinforcement (i.e. either links or bent-up bars);
(2)   the size or spacing of links.

With reference to (2), if the user specifies the link size, then the program calculates and outputs the link spacing; however, if the user specifies the link spacing, then the program will output an appropriate link size up to size 16 mm. The bent-up bars are designed using the equation given in Table 3.17 of BS 8110.

The input data for the program SSHEAR are: the section details, the characteristic strengths of the materials and the design ultimate shear force. The output data from the program SSHEAR are: the design shear

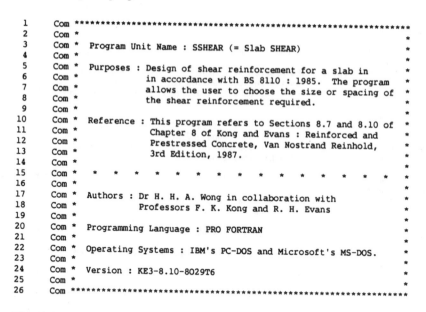

```
 1    Com  ******************************************************************
 2    Com  *                                                                *
 3    Com  *  Program Unit Name : SSHEAR (= Slab SHEAR)                     *
 4    Com  *                                                                *
 5    Com  *  Purposes : Design of shear reinforcement for a slab in        *
 6    Com  *             in accordance with BS 8110 : 1985. The program     *
 7    Com  *             allows the user to choose the size or spacing of   *
 8    Com  *             the shear reinforcement required.                  *
 9    Com  *                                                                *
10    Com  *  Reference : This program refers to Sections 8.7 and 8.10 of   *
11    Com  *              Chapter 8 of Kong and Evans : Reinforced and      *
12    Com  *              Prestressed Concrete, Van Nostrand Reinhold,      *
13    Com  *              3rd Edition, 1987.                                *
14    Com  *                                                                *
15    Com  *    *   *   *   *   *   *   *   *   *   *   *   *   *   *   *     *
16    Com  *                                                                *
17    Com  *  Authors : Dr H. H. A. Wong in collaboration with              *
18    Com  *            Professors F. K. Kong and R. H. Evans               *
19    Com  *                                                                *
20    Com  *  Programming Language : PRO FORTRAN                            *
21    Com  *                                                                *
22    Com  *  Operating Systems : IBM's PC-DOS and Microsoft's MS-DOS.      *
23    Com  *                                                                *
24    Com  *  Version : KE3-8.10-8029T6                                     *
25    Com  *                                                                *
26    Com  ******************************************************************
```

Fig. 12.8–3   Header of program SSHEAR

stress v, the concrete design shear stress v_c, the ratio of link area to link spacing A_{sv}/s_v, the size and spacing of the links required and the details of the bent-up bars, if they are used.

Comment
The complete listing of the above program (with commentary), together with the floppy disks, are obtainable from the Publishers. See Section 12.1(a) and Appendix 1.

12.9   Computer programs for Chapter 9

12.9(a)   Program PSBPTL (= Prestressed Simple Beam; Permissible Tendon Limits)

Figure 12.9–1 shows the Header of the program PSBPTL, which calculates the permissible tendon limits at a prestressed simple beam section, as described in Section 9.2. The permissible tendon limits are determined from eqns (9.2–18) to (9.2–21).

The input data for the program PSBPTL are: the section details, the allowable compressive and tensile stresses, the effective prestressing force and the moments due to the imposed and dead loads. The output data from the program PSBPTL are: the upper and lower permissible limits of the tendon.

Comment
The complete listing of the above program (with commentary), together

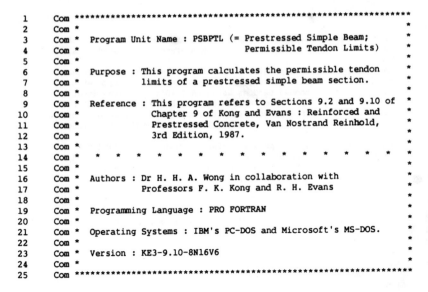

```
 1    Com  ****************************************************************
 2    Com  *                                                              *
 3    Com  *  Program Unit Name : PSBPTL (= Prestressed Simple Beam;      *
 4    Com  *                             Permissible Tendon Limits)       *
 5    Com  *                                                              *
 6    Com  *  Purpose : This program calculates the permissible tendon    *
 7    Com  *            limits of a prestressed simple beam section.      *
 8    Com  *                                                              *
 9    Com  *  Reference : This program refers to Sections 9.2 and 9.10 of *
10    Com  *              Chapter 9 of Kong and Evans : Reinforced and    *
11    Com  *              Prestressed Concrete, Van Nostrand Reinhold,    *
12    Com  *              3rd Edition, 1987.                              *
13    Com  *                                                              *
14    Com  *    *    *    *    *    *    *    *    *    *    *    *    *    *
15    Com  *                                                              *
16    Com  *  Authors : Dr H. H. A. Wong in collaboration with            *
17    Com  *            Professors F. K. Kong and R. H. Evans             *
18    Com  *                                                              *
19    Com  *  Programming Language : PRO FORTRAN                          *
20    Com  *                                                              *
21    Com  *  Operating Systems : IBM's PC-DOS and Microsoft's MS-DOS.    *
22    Com  *                                                              *
23    Com  *  Version : KE3-9.10-8N16V6                                   *
24    Com  *                                                              *
25    Com  ****************************************************************
```

Fig. 12.9–1   Header of program PSBPTL

with the floppy disks, are obtainable from the Publishers. See Section 12.1(a) and Appendix 1.

12.9(b)   Program PBMRTD (= Prestressed Beam; Moment of Resistance; Tabulated Data)

Figure 12.9–2 shows the Header of the program PBMRTD, which calculates the moment of resistance of a rectangular prestressed beam section with bonded tendons using tabulated data, as described in Section 9.5. The program is also applicable to flanged sections in which the neutral axis lies within the flange.

The input data for the program PBMRTD are: the section details, the characteristic strengths of the materials and the effective prestress. The program output data are: the ratios of $f_{pu}A_{ps}/f_{cu}bd$, $f_{pb}/0.87f_{pu}$, f_{pe}/f_{pu} and x/d, and the moment of resistance M_u.

Comment
The complete listing of the above program (with commentary), together with the floppy disks, are obtainable from the Publishers. See Section 12.1(a) and Appendix 1.

12.9(c)   Program PBSUSH (= Prestressed Beam; Symmetrical; Unreinforced; SHear)

Figure 12.9–3 shows the Header of the program PBSUSH, which calculates the required ratio of link area to link spacing for a symmetrical and unreinforced prestressed beam section, as described in Section 9.6.

```
 1    Com  *******************************************************************
 2    Com  *                                                                 *
 3    Com  *  Program Unit Name : PBMRTD (= Prestressed Beam; Moment of      *
 4    Com  *                              Resistance; Tabulated Data)        *
 5    Com  *                                                                 *
 6    Com  *  Purpose : This program calculates the moment of resistance     *
 7    Com  *            of a rectangular prestressed beam section with       *
 8    Com  *            bonded tendons using BS 8110's tabulated data,        *
 9    Com  *            which are reproduced as Table 9.5-1 of Chapter 9.     *
10    Com  *            This program is also applicable to flanged section   *
11    Com  *            in which the neutral axis lies within the flange.     *
12    Com  *                                                                 *
13    Com  *  Reference : This program refers to Sections 9.5 and 9.10 of     *
14    Com  *              Chapter 9 of Kong and Evans : Reinforced and        *
15    Com  *              Prestressed Concrete, Van Nostrand Reinhold,        *
16    Com  *              3rd Edition, 1987.                                  *
17    Com  *                                                                 *
18    Com  *    *    *    *    *    *    *    *    *    *    *    *    *    *   *
19    Com  *                                                                 *
20    Com  *  Authors : Dr H. H. A. Wong in collaboration with               *
21    Com  *            Professors F. K. Kong and R. H. Evans                *
22    Com  *                                                                 *
23    Com  *  Programming Language : PRO FORTRAN                             *
24    Com  *                                                                 *
25    Com  *  Operating Systems : IBM's PC-DOS and Microsoft's MS-DOS.       *
26    Com  *                                                                 *
27    Com  *  Version : KE3-9.10-8N17V6                                      *
28    Com  *                                                                 *
29    Com  *******************************************************************
```

Fig. 12.9–2   Header of program PBMRTD

```
 1    Com  *******************************************************************
 2    Com  *                                                                 *
 3    Com  *  Program Unit Name : PBSUSH  (= Prestressed Beam; Symmetrical;*
 4    Com  *                              Unreinforced; SHear)              *
 5    Com  *                                                                 *
 6    Com  *  Purpose : This program calculates the required ratio of        *
 7    Com  *            link area/link spacing for a symmetrical and         *
 8    Com  *            unreinforced (i.e. As = 0) prestressed beam          *
 9    Com  *            section in accordance with BS 8110 : 1985.           *
10    Com  *                                                                 *
11    Com  *  Reference : This program refers to Sections 9.6 and 9.10 of     *
12    Com  *              Chapter 9 of Kong and Evans : Reinforced and        *
13    Com  *              Prestressed Concrete, Van Nostrand Reinhold,        *
14    Com  *              3rd Edition, 1987.                                  *
15    Com  *                                                                 *
16    Com  *    *    *    *    *    *    *    *    *    *    *    *    *    *   *
17    Com  *                                                                 *
18    Com  *  Authors : Dr H. H. A. Wong in collaboration with               *
19    Com  *            Professors F. K. Kong and R. H. Evans                *
20    Com  *                                                                 *
21    Com  *  Programming Language : PRO FORTRAN                             *
22    Com  *                                                                 *
23    Com  *  Operating Systems : IBM's PC-DOS and Microsoft's MS-DOS.       *
24    Com  *                                                                 *
25    Com  *  Version : KE3-9.10-8N18V6                                      *
26    Com  *                                                                 *
27    Com  *******************************************************************
```

Fig. 12.9–3   Header of program PBSUSH

The input data for the program PBSUSH are: the section details, the characteristic strengths of the materials, the effective prestress and the design ultimate shear force and moment. The program output data are: the shear resistance of the uncracked section V_{c0}, the shear resistance of the crack section V_{cr}, the moment M_0, the adjusted design shear force V, the design ultimate shear resistance V_c and the ratio of link area to link spacing A_{sv}/s_v, if shear reinforcement is required.

Comment
The complete listing of the above program (with commentary), together with the floppy disks, are obtainable from the Publishers. See Section 12.1(a) and Appendix 1.

References

1 Wong, H. H. A. (in collaboration with Kong, F. K. and Evans, R. H.) *Complete listings of the computer programs (with Commentary and User Instructions) in 'Kong and Evans: Reinforced and Prestressed Concrete, Van Nostrand Reinhold, 3rd Edition, 1987'.* Van Nostrand Reinhold, Wokingham, 1987. This publication, together with the associated set of two floppy disks can be obtained from the Publishers—see Section 12.1(a) and Appendix 1 of Kong and Evans's book.

2 *Pro Fortran User Manual.* Prospero Software, London, March 1985. (Obtainable from Prospero Software Limited, 190 Castelnau, London SW13 9DH, England.)

3 Holloway, R. T. *Structural Design with the Microcomputer.* McGraw-Hill, Maidenhead, 1986.

4 Cope, R. J., Sawko, F. and Tickell, R. G. *Computer Methods for Civil Engineers.* McGraw-Hill, Maidenhead, 1982.

5 Hoffman, P. and Nicoloff, T. *MSDOS User's Guide.* Osborne, McGraw-Hill, New York, 1984.

6 Wolverton, V. *Running MSDOS.* Microsoft Press, 1984.

7 Norton, P. *Inside the IBM PC.* Prentice Hall, New Jersey, 1983.

8 Wong, H. H. A. and Kong, F. K. Concrete Codes—CP 110 and BS 8110. The *Structural Engineer*, **64A**, No. 12, Dec. 1986, pp. 391–3.

Appendix 1
How to order the program listings and the floppy disks

The complete listings of all the computer programs in Chapter 12, together with the associated floppy disks, may be ordered from the Publishers:

> Van Nostrand Reinhold (UK) Co Ltd
> Molly Millars Lane
> Wokingham
> Berkshire
> England

A1.1   Instructions

1. Study Section 12.1, before you place an order for any or all of the items below.
2. If you need only the printout of the program listings (with Commentary and User Instructions), order Item (a) below.
3. If you need only the floppy disk to generate the printout of program listings yourself, order Item (b1) or (b2) below.
      Items (b1) and (b2) are floppy disks containing all the programs stored as 'source files'—see explanation in Section 12.1(d).
4. If you need only the floppy disk to run the programs, order Item (c1) and (c2) below.
      Items (c1) and (c2) are floppy disks containing all the programs stored in machine code as 'executable files'—see explanation in Section 12.1(d).

ITEM (a):
Wong, H. H. A. (in collaboration with Kong, F. K. and Evans, R. H.), *Complete listings of the computer programs (with Commentary and User Instructions) in 'Kong and Evans: Reinforced and Prestressed Concrete, Van Nostrand Reinhold, 3rd Edition, 1987'*, Van Nostrand Reinhold, Wokingham, 1987.

ITEM (b1):
Floppy Disk I: Source files of the computer programs in 'Kong and Evans: Reinforced and Prestressed Concrete, Van Nostrand Reinhold, 3rd Edition, 1987', for RM Nimbus.

ITEM (b2):
Floppy Disk Ia and Ib: Source-files of the computer programs in 'Kong and Evans: Reinforced and Prestressed Concrete, Van Nostrand Reinhold, 3rd Edition, 1987', for IBM-PC/XT, IBM-PC/AT or IBM Compatibles.

ITEM (c1):
Floppy Disk II: Executable files of the computer programs in 'Kong and Evans: Reinforced and Prestressed Concrete, Van Nostrand Reinhold, 3rd Edition, 1987', for RM Nimbus.

ITEM (c2):
Floppy Disks IIa and IIb: Executable files of the computer programs in 'Kong and Evans: Reinforced and Prestressed Concrete, Van Nostrand Reinhold, 3rd Edition, 1987', for IBM-PC/XT, IBM-PC/AT or IBM Compatibles.

A1.2   System requirements

The system requirements necessary to run the Authors' computer programs are summarized as follows.

Personal computers:	RM Nimbus, IBM-PC/XT, IBM-PC/AT and IBM Compatibles (e.g. Amstrad-PC)	
Memory	:	512 K bytes are more than adequate
Operating system	:	MS-DOS
Disk drive	:	Single disk drive is adequate
Monitor	:	Monochrome is adequate
Printer	:	Optional

Appendix 2
Design tables and charts

Table A2–1   Areas of groups of reinforcement bars (mm²)

Bar size (mm)	Number of bars									
	1	2	3	4	5	6	7	8	9	10
8	50	101	151	201	251	302	352	402	452	503
10	79	157	236	314	393	471	550	628	707	785
12	113	226	339	452	565	679	792	905	1017	1131
16	201	402	603	804	1005	1206	1407	1608	1809	2011
20	314	628	942	1257	1571	1885	2199	2513	2827	3142
25	491	982	1473	1963	2454	2945	3436	3927	4418	4909
32	804	1608	2412	3216	4021	4825	5629	6433	7237	8042
40	1256	2513	3769	5026	6283	7539	8796	10050	11310	12570

Table A2–2   Reinforcement-bar areas (mm²) per metre width for various bar spacings

Bar size (mm)	Bar spacing (mm)									
	75	100	125	150	175	200	225	250	275	300
8	671	503	402	335	287	252	223	201	183	168
10	1047	785	628	523	449	393	349	314	286	262
12	1508	1131	905	754	646	566	503	452	411	377
16	2681	2011	1608	1340	1149	1005	894	804	731	670
20	4189	3142	2513	2094	1795	1571	1396	1257	1142	1047
25	6545	4909	3927	3272	2805	2454	2182	1963	1785	1636
32	—	8042	6434	5362	4596	4021	3574	3217	2925	2681
40	—	—	10050	8378	7181	6283	5585	5027	4570	4189

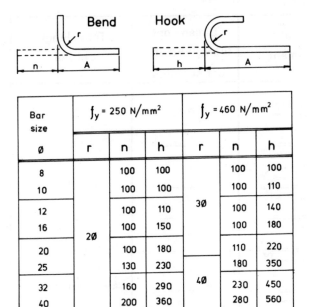

Bar size	f_y = 250 N/mm²			f_y = 460 N/mm²		
Ø	r	n	h	r	n	h
8		100	100		100	100
10		100	100		100	110
12		100	110	3Ø	100	140
16	2Ø	100	150		100	180
20		100	180		110	220
25		130	230		180	350
32		160	290	4Ø	230	450
40		200	360		280	560

Fig. A2–1 **Minimum bend and hook allowances (mm)—BS 4466**

Shape code	Method of measurement	Dimensions to be given in bar schedule	Total length of bar
20	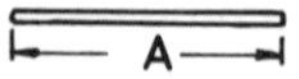	straight	A
32	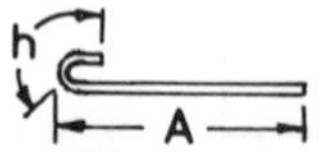	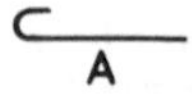	A + h
33	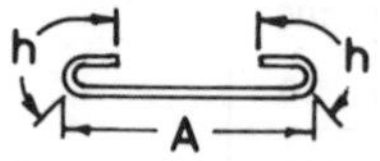	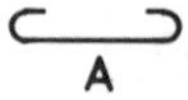	A + 2h
34	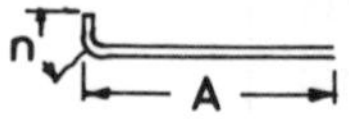	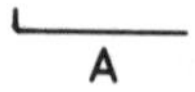	A + n
35	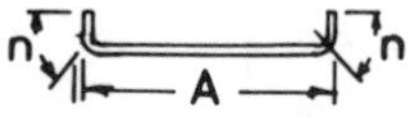	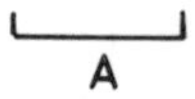	A + 2n
37	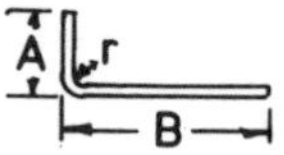		$A + B - \frac{1}{2}r - \emptyset$
38	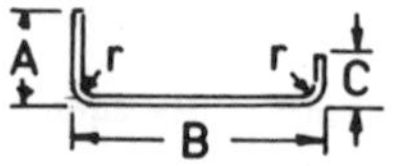	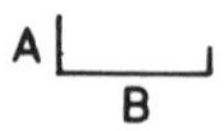	$\begin{cases} A + B + C \\ -r - 2\emptyset \end{cases}$
41	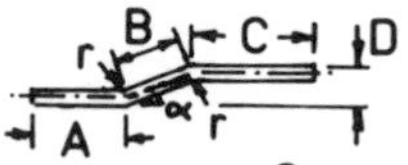	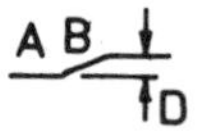	$\begin{cases} A + B + C \\ \text{if } \propto \ \leq 45° \end{cases}$
	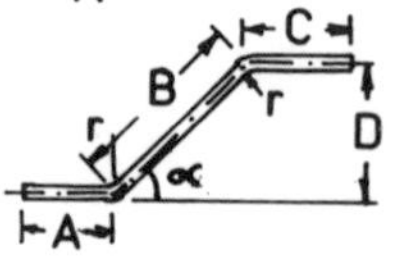	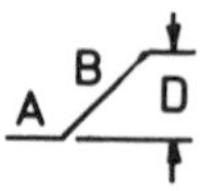	$\begin{cases} A + B + C - r - 2\emptyset \\ \text{if } \propto \ > 45° \end{cases}$

$\emptyset$ = bar size

h, n and r : see Fig. A2 –1

Fig. A2–2(a)   Bending dimensions—BS 4466

Shape code	Method of measurement	Dimensions to be given in bar schedule	Total length of bar
43	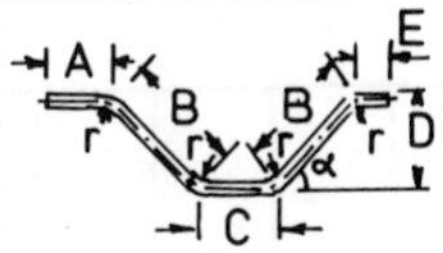	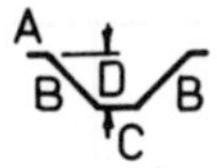	$\begin{cases} A+2B+C+E \\ (\text{if } \propto \le 45^0) \end{cases}$
51	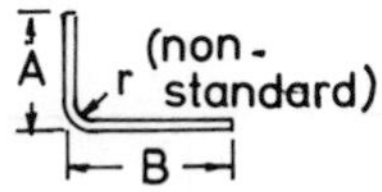 (non-standard)		$A+B-\frac{1}{2}r-\varnothing$
60			$2(A+B)+20\,\varnothing$
62	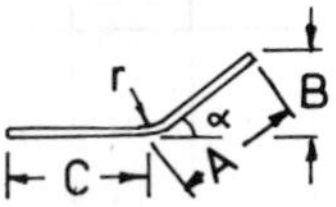	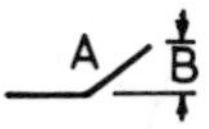	$\begin{cases} A+C \\ (\text{if } \propto \le 45^0) \end{cases}$
81	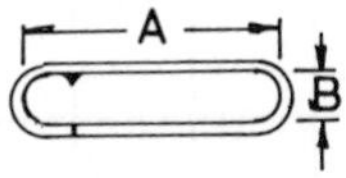	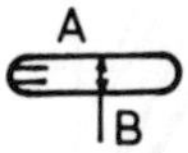	$2A+3B+22\varnothing$
83	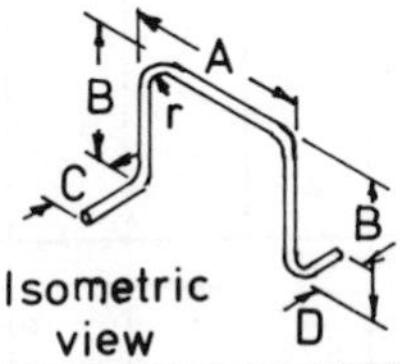 Isometric view	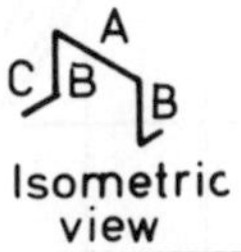 Isometric view	$A+2B+C+D$ $-2r-4\varnothing$

$\varnothing$ = bar size

h, n and r: see Fig A2-1

Fig. A2-2(b)   Bending dimensions—BS 4466

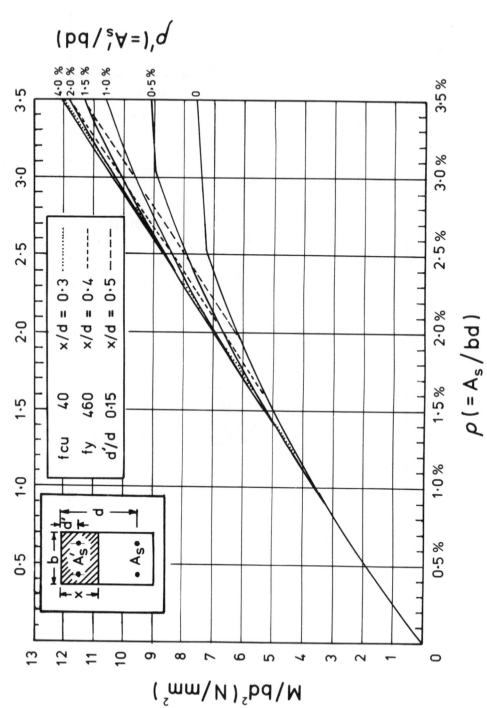

Fig. A2–3   Beam design chart—ultimate limit state (BS 8110)

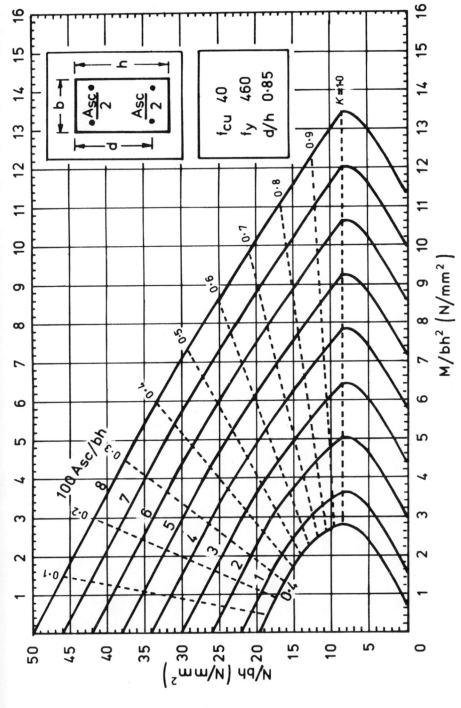

Fig. A2–4   Column design chart—BS 8110

Index

Note: **Bold numbers** indicate main references